AF517783

# GENETICS AND PLANT BREEDING

# GENETICS AND PLANT BREEDING

**E. B. Babcock**

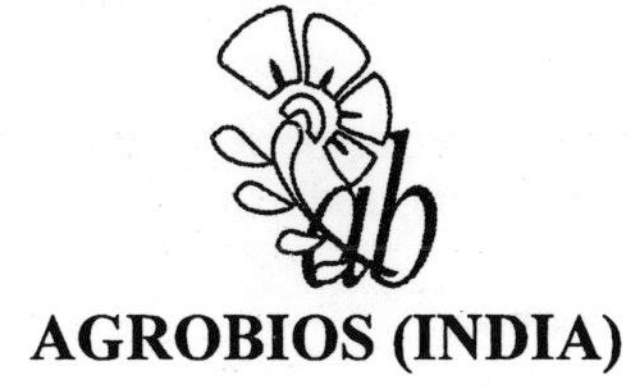


**AGROBIOS (INDIA)**

*Published by:*
**AGROBIOS (INDIA)**
Behind Nasrani Cinema
Chopsani Road, Jodhpur 342 003
Phone: +91 291 2643993, 2643994
Fax: +91 291 2642319
E. Mail: agrobiosindia@gmail.com

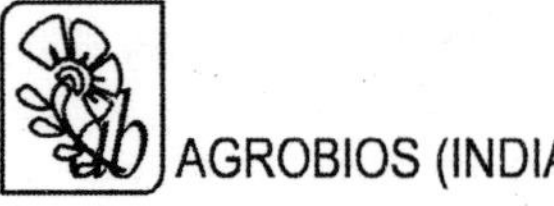

**ISBN: 978-81-7754-105-2**

***Published by:*** Dr. Updesh Purohit for Agrobios (India)
***Laser Typesetting:*** Yashee Computers, Jodhpur
***Cover Design:*** Reena
***Printed by:*** Bhawani Offset, Jodhpur

## PREFACE TO SECOND EDITION

Unexpected postponement of the revision of this text has made it necessary to rewrite it almost entirely. During the nine years that have elapsed since the publication of the first edition, tremendous advances have been made in the fields of both theoretical and applied genetics.

To illustrate the former, it is only necessary to remind the reader that, while in 1918 the majority of geneticists were perhaps fairly well satisfied as to the fundamental importance of the chromosomes in genetic theory, nevertheless some of the more conservative biologists saw fit to express surprise at the full commitment of this book to the chromosome theory of heredity. It would seem that no one who has kept abreast of recent progress in cytogenetic research would now question such commitment.

Progress in applied genetics has been equally rapid and important. For example, in discussing the future application of genetics to plant breeding, we ventured to say, "it is not too much to expect that eventually our more important crop plants at least will have been subjected to such thorough germinal analysis that the establishment of desired strains will become largely a matter of reference to breeding records and the repetition of certain crosses and selections." While this has scarcely yet been realized, already plans have been seriously proposed looking toward the synthesizing of "superlines" of maize and of wheat varieties resistant to all the physiologic races of stem rust. At the same time the genetic analysis of several important crop plants is well under way. Although progress in the application of genetics to animal production naturally cannot be as rapid as in plant production, yet here also considerable progress has been made even in the genetic analysis of the larger domestic animals.

It is the broad advance in fundamental research, however, which characterizes the close of the first quarter century of the science of genetics; and it is the hope of conserving somewhat the gains achieved and assisting if possible in still greater progress on the part of another generation of geneticists, that has been the primary motive in the preparation of this book.

To express our indebtedness to all those who have furnished the subject matter of the book is manifestly impossible. We have drawn freely from the contributions of others and have endeavored to give due credit in citing all recent and the more outstanding older works. Citations by number or date have been omitted from the text in order to save space and to gain in directness of presentation. But much care has been given to the inclusion of all cited works in the bibliography. Oversights in this respect will be gladly rectified.

It is a pleasure also to acknowledge the assistance of all who have furnished or permitted the reproduction of illustrations. Due acknowledgment has been made in each instance. When only the author's name appears at the end of a legend it will be understood that the source was not a copyrighted publication and reference should be made to the bibliography for place of publication. We are particularly indebted to Professor Thomas Hunt Morgan for furnishing drawings for the figures used in the frontispiece and in plate 4; to the journal, GENETICS, for the loan of the four engravings used in printing plate 2; and to the Cambridge University Press for the engravings used in plate 3.

The work of editing the book having fallen to his lot, the writer takes this opportunity to thank Professor J. L. Collins and Dr. J. W. Lesley for a critical reading of the manuscript, the former having taken Parts I and III and the latter, Part II, and both having made many helpful suggestions. Thanks are due also to Professor R. W. Hodgson for helpful advice in connection with Chapter XXXI.

E. B. B.

BERKELEY, CALIFORNIA,

# PREFACE TO FIRST EDITION

Of all the sciences that contribute to the great, tertiary composite which is known as agriculture none is more important economically than genetics. One may not overlook the fundamental relation borne by the primary sciences, mathematics, physics, and chemistry, and by the secondary sciences, botany, zoology, geology, meteorology, and economics, to the production and distribution of raw materials. But we confidently assert that the science which underlies the improvement of plants and animals for agricultural purposes is destined to receive increasing attention in agricultural education and in agricultural practice. Without doubt vast possibilities await realization through the more thorough and systematic development of our living economic resources. Such development is directly dependent upon the successful utilization of genetic principles in plant and animal breeding. The science of genetics is still very young, but it is firmly established and is developing rapidly. It claims the attention of the producer of today and invites the most serious study of the agriculturist of tomorrow. It lays claim also to the interest of the eugenist, the sociologist, the philanthropist, and all students of biology.

This text has been prepared in response to a real and widely recognized need. The experience of the authors in teaching the principles of breeding to undergraduate students has forced home the conviction that an adequate presentation in a single text of the facts and principles of genetics and their practical applications is a prime necessity. Those familiar with the literature of the subject will appreciate the magnitude of the task and, we trust, will be lenient in criticizing our choice of subject matter. It is impossible to include many things of mutual interest to genetics and agriculture if the work be limited to a single volume. We are keenly aware of many deficiencies and it is our desire to prepare a revised edition of the book in the near future. With this in view the suggestions of others are earnestly solicited.

We take this opportunity to express gratitude to all who have rendered assistance, especially to those who have read portions of the manuscript or assisted in proofreading and to all who donated or loaned photographs or who assisted otherwise with the illustrations. The onus of the work has been lessened in no small degree by the interest and encouragement of our colleagues.

THE AUTHORS.

BERKELEY, CALIFORNIA,

# CONTENTS

## PART II

### Plant Breeding

# GENETICS
IN
# RELATION TO AGRICULTURE

## PART I—FUNDAMENTALS

### CHAPTER I

### INTRODUCTION

Prior to 1900 the laws which govern heredity and variation were unknown. Heredity appeared to be lawless and fortuitous; variation seemed to be too complex for accurate analysis. Despite this situation, however, or probably because of it, and in recognition of the fundamental importance to all biology of the problems connected with these two subjects, they engaged the attention of master minds of biology who formulated many stimulating theories and sought to support them by appeal to facts of observation and experiment. Unfortunately, however, a satisfactory method of experimental approach had not been devised and properly applied; it was a period of speculation during which the way was cleared for application of stricter quantitative methods of analysis. During this period an Austrian monk, Gregor Mendel, working with garden peas in the cloister garden at Brünn in Austria, performed a set of now famous experiments which provided the method needed for scientific progress. He announced his discoveries in 1865 and they were published in 1866, but biological investigation had apparently not developed to a point where it could realize the full significance of his contribution. Consequently his work lay neglected until 1900, when it was rediscovered independently by three botanists, De Vries, Correns, and Tschermak, who had conducted experiments leading to the same general conclusions. Immediately after its rediscovery his results were heralded with enthusiasm in marked contrast to the silent reception which they had been accorded at their first announcement, and it was then clearly perceived that they provided the scientific method which was needed for exploitation of problems of heredity and variation. In the year 1900, then, a new era in the history of heredity and variation opened, an epoch which may be called the period of scientific development, the inauguration of which had to wait upon the discovery of a proper experimental method. The field thus opened to exploration was immediately stirred into a feverish activity which has lasted to the present time and which has resulted in remarkable and astounding progress. The laws of heredity and variation formerly were shrouded in mystery; now they are found to be capable of formulation and verification by means of properly planned and executed experiments.

It soon became evident as knowledge increased that a new term was needed to designate this particular portion of biological science, and accordingly Bateson proposed the word genetics at the Third International Congress on Hybridization at Paris in 1906. It is a coined work, derived from the Greek root, γεν, meaning "to be born," and it was designed to cover those interrelated matters comprised in the phrase, *physiology of heredity and variation.* It is difficult to define a science in terms comprehensible to those who are not already acquainted with its subject matter, but in order to satisfy the demand for a definition which implies no previous knowledge of the subject, the following statement is suggested:

*Genetics is the science which seeks to account for the resemblances and differences exhibited among organisms related by descent.*

**The Content of Genetics.**—If genetics be defined in the manner just stated, it may be said in a preliminary way that heredity is that portion dealing with resemblances and variation with differences. But this simple subdivision must be qualified, as may readily be perceived from a consideration of certain very obvious facts. Organisms universally tend to reproduce their kind, a tendency which results in the creation and maintenance of organic groups, the individuals of which are more or less closely identical in their characters. It is also a well-known fact that the members of a given group exhibit individual differences, and that when two individuals which are different are mated together, the offspring cannot be identical with both of them. The problem of heredity must, therefore, be considered from a more fundamental point of view than that of simple outward resemblance.

The basis for this more fundamental viewpoint arises from a knowledge of modes of reproduction. The individual begins its life as a single cell, which is usually the product of union of two sex cells, one contributed by the male and the other by the female parent. The material contributions which an individual receives by heredity obviously must be contained in these two sex cells, and in turn the materials which the sex cells contain must have been derived from materials present in the parents. The different materials which are contributed by the two cells interact in ways as yet little understood and govern the development of the individual. An individual may contain materials which are expressed and others which are not expressed in it, and it may likewise transmit materials which are expressed in it and others which are not. In defining heredity, therefore, it is necessary to distinguish between the expressed characters of an individual, its soma, and the materials which it contains in its cells and which it contributes to its sex cells, the germinal material. Heredity is concerned with germinal materials rather than with somatic characters; it is concerned particularly with germinal materials which are transmitted from parents to offspring, which is the central thought in the following definition:

*Heredity is genetic continuity of germinal material between parents and offspring.*

Heredity thus defined includes instances in which individuals differ from one or both parents when those variations depend upon the germinal materials which the individuals received from their parents. It is recognized in this statement that individuals may differ in their germinal substances even though they are identical in outward appearance. The first category of variation to be recognized, then, are differences caused by heredity, to which the term Mendelian variation or combination has been applied. The individual begins its life with certain material contributions from its parents. These materials govern the development of the organism within certain limits and represent a set of potentialities; but the particular manner in which these potentialities are realized depends upon conditions under which the organism develops. If these conditions differ for different organisms, as in fact they always do, then differences may arise in individuals despite identity of germinal materials. The conditions under which an individual develops are collectively termed the environment, and a second class of variation, called developmental variation or modification, may then depend upon differences in the environment. It will also be found as variation is studied more closely that the germinal substance itself is not immutable, but that occasionally elements in it may change, as a consequence of which an individual occasionally transmits to its offspring germinal materials which it did not itself receive from its parents, or some of the materials in the sex cells may become changed prior to development. These changes in germinal material are relatively rare, but they constitute a third class of variations to which the term "mutation" has been applied. Variation, therefore, must take into account these several distinctive sources of diversity, as in the following definition:

*Variation is difference, whether in the expression of somatic characters or in the elements of the germinal substance, exhibited among groups of organisms related by descent.*

It is impossible to explain more precisely the differences among the three categories of variation—combination, modification, and mutation—before the fundamental principles of heredity are understood. Consequently further consideration of them is deferred to Chapter XVI.

It is also necessary to state that the term variation has several connotations in biological literature. The definition which is here proposed seems to be most nearly in accord with present usage among geneticists. According to this usage heredity and variation are not mutually exclusive terms but represent overlapping categories. Another usage would make heredity practically synonymous with combination and variation with mutation, thus subdividing genetics into three exclusive categories—heredity, modification, and variation. This proposal has many advantages; perhaps it will prevail eventually.

**The Problems of Genetics.**—Obviously the problems of genetics are those which arise in a study of the resemblances and differences exhibited by organisms related by descent. Perhaps some further light may be thrown on their nature by considering certain statements which have received much favor among geneticists. One of the best is that of Wilson, who proposes the question, "How do the adult characteristics lie latent in the egg and how do they become patent as development proceeds?" Pearl has voiced very much the same thought in his statement that the critical problem of inheritance is the problem of the cause, the material basis, and the maintenance of the somatogenic specificity of germinal substance. The student will find it helpful as he advances in his study of genetics to analyze these statements carefully so that he may grasp their full implications.

**Modes of Research in Genetics.**—There are many methods of attacking problems of heredity, all of which yield necessary data for a well-rounded presentation of the subject. In general it may be stated that most of the information is obtained from five rather distinctive methods of research, namely, those of observation, biometry, experimental breeding, cytology, and experimental morphology. Brief statements of the characteristics and relative values of these methods are given below.

By the method of observation living organisms are studied as they occur in nature, and attempts are made to formulate and test hypotheses as an outcome of such study. It is the kind of research by which Darwin sought to demonstrate the occurrence of evolution by descent with modification and to determine the methods by which evolution had been brought about. It is a method which is subject to extreme difficulties of interpretation, because the material is not under experimental control but must be taken as it exists and subjected to a rigid scrutiny in order to detect the interacting variables which underlie it. Although rightly subject to criticism, it is, however, necessary to use it in certain cases. In human inheritance, in heredity in our domestic animals, and in other instances in which experiment is slow or out of the question entirely, there would be little knowledge, if information were not used from recorded observations.

The method of biometry is usually considered as an observational method, refined and improved by the application of statistical methods. In the method of biometry, as applied to genetic problems, observations are stated accurately in quantitative terms, and large numbers of individuals are dealt with before any attempt is made to deduce laws and principles concerning them. The data are subjected to mathematical treatment by which their various features are expressed in the form of statistical constants or formulae, which are much more precise than general impressions and which make it possible to compare distinct sets of data more accurately. The application of statistical methods to problems of genetics has had an important effect in directing attention to the mathematical requirements of accuracy in experimental studies; but as a method of stating the results of genetic investigations, it is open to the serious objection that it does not arrive at a determination of genetic behavior in individuals, but merely describes the average behavior in groups of individuals.

The method of experimental breeding is a method of study based on hybridization of races of plants and animals, and a careful analytic study of hybrid progenies in successive generations. It is a laboratory method of study, because the races which are employed are "purified" prior to hybridization, as a chemist purifies his reagents preparatory to mingling them. It is sometimes called the pedigree method because the pedigrees of all individuals are carefully kept. It has been the method par excellence of genetic research. The initiation of the period of scientific development awaited its introduction, and the remarkable advances which have since been made are largely attributable to its employment.

Cytology is concerned with the construction of cells, their mode of multiplication, and the manner in which the sex cells—the ova and spermatozoa of animals and the egg cells and pollen grains of plants—come into existence and unite to form new individuals. It is, therefore, obviously a subject intimately concerned with problems of heredity and variation, for it seeks to determine the nature of the material which bridges the gap between generations and the way in which that material behaves in the cell divisions following fertilization as the fertilized ovum, a single cell, develops into a multicellular organism of extremely great complexity. The conceptions arising from cytological studies have of late years illuminated the whole field of genetics; they have supplied a physical mechanism which perfectly accounts for many facts which had been discovered by experimental breeding.

In Wilson's statement of the central problem of heredity, he not only calls attention to the necessity for information on how the adult characteristics lie latent in the egg, but he also emphasizes the importance of determining how they become patent in the course of development.

This last feature of the problem is one which has been neglected by geneticists, but doubtless as the subject develops and new avenues of approach are discovered, it will be given greater attention. The method of experimental morphology is concerned with this problem. The contributions which the individual receives from its parents represent certain potentialities of development which are contained in the fertilized egg. The manner in which these potentialities are realized depends upon the particular conditions under which the individual develops. Its adult characters then are the result of a realization of its innate potentialities under a certain set of environmental conditions. In most investigations of heredity judgment is based upon adult characters, and inferences are drawn with respect to conditions in the sex cells from a study of the distribution of adult characters among the individuals of populations. There is a gap existing between the two; potentialities as they exist in the sex cells and the fertilized ovum, and characters as they are realized in the adult, which some day may be bridged. As yet little or nothing is known of the manner in which the germinal elements function in development. There is need for the elaboration of an experimental technique which will throw light upon this portion of the problem.

The student must not conclude from the discussion of methods of studying genetic problems that the results from the different methods correspondingly subdivide the province of genetics into distinct specialized branches. As a matter of fact all of these methods may be brought to bear upon a single problem. The experimenter often obtains valuable hints by observation as to the kind of problems which may be subjected to experimental study. After he has obtained his experimental data, he may find it necessary to employ statistical methods in analyzing his results. In certain problems he may find it necessary not only to carry out properly planned breeding experiments, but to subject the material to a thoroughgoing cytological study as well. Finally, if he can obtain, by means of experimental studies of development, some knowledge of the physiological processes which bring about the results he has secured, his studies will be much more satisfactory. It is only by a search for information from all possible angles that complete solution of a given problem may be expected.

**The Applications of Genetics.**—Genetics of course does not need to be justified by its applications, for it deals with matters which are full of human interest. Its problems, however, lie at the root of so many biological processes that the subject does not lack applications; in fact at times their importance has threatened to overshadow its scientific development.

Genetics has both scientific and practical applications. As an example of its scientific applications, its relation to evolution comes instantly to mind. Evolution rests fundamentally upon heredity, which accounts for the maintenance of types, and upon variation, which deals at least in part with the origin of new forms. In other branches of biology, particularly in physiology and in taxonomy, and in cytology and embryology, genetic conceptions are constantly receiving greater recognition. Its practical applications are found in agriculture and in human affairs. Its relation to animal and plant improvement is too obvious to require elaboration. In human affairs its conceptions are employed not only in an attempt to devise methods of improving the race, but also in a study of disease, of criminology, of sociology, and of history. As the subject of genetics develops, it cannot fail to affect our thoughts and relations to these broader problems of human welfare, although it must be recognized that its contributions to these matters are often very subordinate.

**Study of Genetics.**—From the foregoing discussion, it is easy to see that the thorough student of genetics or the prospective investigator must have a varied equipment in order to appreciate its problems. Since genetics is a biological science, intelligent study of it presupposes a thorough grounding in general biology such as may be obtained in general courses in botany and zoölogy. Inasmuch as the animal and plant forms upon which most work has been done belong to the higher orders, particular attention should be given to them, especially to a study of development and reproduction. Studies of physiology are of prime importance, because of the attitude towards life phenomena which they awaken in the student and because genetics itself is primarily a physiological science. A knowledge of cytology is of the utmost importance. Every prospective investigator should have mastered cytological technique sufficiently to enable him to make cell studies when they are necessary. A knowledge of higher mathematics, differential and integral calculus, the theory of probabilities, and the methods of least squares is necessary for those who intend to pursue biometrical studies, and it is helpful to those engaged in other kinds of investigation. A knowledge of algebra is, however, sufficient for an understanding of most of the fundamental principles. Finally, it may be mentioned that original articles on genetics are published in practically every important modern language. English and German lead in the number of articles published; but numerous articles also appear in other languages, especially French, Dutch, Japanese, the Scandinavian languages, Italian, Polish, Russian, etc.

**The Literature of Genetics.**—The technical studies of genetics are prepared in the form of original articles which are published in a variety of journals, some devoted exclusively to genetics, others covering the broader fields of biology, botany, physiology, zoölogy, and the like; and in independently published monographs, books, bulletins, etc. Original articles are not in general prepared from an elementary standpoint. but they are written for professional geneticists. They usually contain a description of experiments or studies which have been made on some particular problem, and a discussion of the significance of the results. Methods and results are described in such detail as is necessary for a complete understanding. There are no secrets in science, and results to be of value must be capable of verification. The life of genetics, as of other sciences, lies in prompt and open publication of results. A list of the more important journals devoted largely or exclusively to genetics is given at the end of this chapter.

The number of articles on genetics which is now published is so great that no one student can expect to read them all. In order that investigators may, however, be informed of the titles of all articles which are published and of the general nature of their contents, abstracting journals have been established which contain citations to original articles and brief abstracts setting forth the more important results contained in them. The most useful one in English is *Botanical Abstracts,* which attempts in the Section for Genetics to present citations to and brief abstracts of all articles on genetics. Its work is now being continued in *Biological Abstracts,* an abstracting journal which proposes to cover the whole field of biology. Many of the technical journals also contain sections devoted to reviews of articles published elsewhere, and the *Zeitschrift für induktive Abstammungs- und Vererbungslehre* publishes at intervals a classified list of original articles which is a convenient source of reference. By reference to abstracting journals, lists of literature, and the bibliographies contained in original articles, it is possible to trace down most, if not all, of the published information on any particular problem.

For the student beginning a study of genetics it is necessary to develop a background before the significance of the results contained in original articles will have any meaning to him. It is the purpose of textbooks and monographs to aid him in securing a general knowledge of the subject and an idea of the interrelations of its different parts. Textbooks, however, should be considered only an aid in securing preliminary orientation in the subject. As soon as possible the student should seek original sources in order to learn how experiments are planned and conducted, and how results are analyzed. It is impossible to present all these details in textbooks; but a full appreciation of them is necessary to a real understanding of the subject. A list of the more important general textbooks is given at the end of this chapter. Specific references to them and to others are contained at the end of each chapter.

Special attention should be called to the extensive reviewing project undertaken in *Bibliographia Genetica* under the editorship of Lotsy. It is planned to present in ten volumes a complete review of work in special fields of genetics up to 1925. Each review is written by a specialist and it is followed by a complete bibliography of the subject. Under the editorship of Baur a similar series is being issued with the general title, *Bibliotheca Genetica*. The student will find these volumes particularly helpful as an introduction to the voluminous literature of genetics.

**Recent General Textbooks**

CASTLE: Genetics and Eugenics, 3d ed.
CONKLIN: Heredity and Environment, 3d ed.
COULTER: Plant Genetics.
CREW: Animal Genetics.
JONES: Genetics in Plant and Animal Improvement.
MORGAN: The Physical Basis of Heredity.
MORGAN, MULLER, STURTEVANT, and BRIDGES: The Mechanism of Mendelian Heredity, 2d ed.
MORGAN: The Theory of the Gene.
PUNNETT: Mendelism, 6th ed.
SHULL: Heredity.
SINNOTT and DUNN: The Principles of Genetics.
THOMSON: Heredity, 3d ed.
WALTER: Genetics, 2d ed.

**Periodicals, Devoted Exclusively or Principally to Genetics**

*American Naturalist.*
*Genetica.*
*Genetics.*
*Hereditas.*
*Journal of Genetics.*
*Journal of Heredity.*
*Zeitschrift für induktive Abstammungs- und Vererbungslehre.*
*Zeitschrift für Pflanzenzüchtung.*

**Abstracting Journals, Reviews, and Bibliographies**

*Bibliographia Genetica.*
*Bibliotheca Genetica.*
*Biological Abstracts.*
*Botanical Abstracts* (particularly the Section for Genetics).
*Botanische Centralblatt.*
*Experiment Station Record.*
*Resumptio Genetica.*
*Zeitschrift für induktive Abstammungs- und Vererbungslehre* (lists of new literature).
*Zoologische Anzeiger.*

## CHAPTER II

## DEVELOPMENT

The study of genetics is intimately concerned with problems of development and reproduction. In this and the succeeding chapter an attempt will be made to develop the essential features of these two processes. The account which is presented deals more specifically with the higher forms of life, because agriculture is concerned particularly with them. For a more comprehensive treatment, the student must consult standard texts on botany, zoölogy, and cytology.

**The Cell Theory.**—The study of the finer structure of animals and plants reveals the fact that they are composed of very minute, more or less independent units, the cells, and of non-living materials which are the products of cellular activity. The diversified tissues and structures which make up the body of a higher animal may all be resolved either into cells or depositions of cells. The cells of the body are probably not absolutely independent entities bound together in a cooperative organization based purely on physiological interchanges. In certain cases it is possible to demonstrate the existence of protoplasmic bridges connecting neighboring cells in tissues. Nevertheless a study of individual development establishes the essential correctness of the idea that morphologically the individual is built up of cells and the products of cells. At the beginning of its life, it is a single cell comparatively undifferentiated, but certainly highly complex. The development of the individual from a single cell is a process of successive cell divisions, accompanied by differentiation and deposition of materials as a result of cellular activity. All the cells which the adult possesses have an unbroken cell lineage extending back to the original fertilized egg, and through it, as will be seen later, to similar cell lines in the parents.

**The Cell.**—The cell (figure 1), although usually very minute in size, is far from simple in structure and organization. There are certain organs which are common to almost all cells; there are other organs which appear to be characteristic of certain cells only; and finally by differentiation endless modifications arise in cells by which they are fitted for the performance of specialized functions. The two most characteristic portions of the cell are the nucleus and the cytoplasm; in fact the cell has been defined as a nucleated bit of protoplasm. There are, however, certain cells which perform specialized functions, like red blood corpuscles, which do not possess nuclei.

The cytoplasm is a clear viscous material resembling egg white. It is probably not homogeneous in structure or composition, but agreement on these details has not been reached among cytologists. It is limited externally by the plasma membrane. In animal cells the cytoplasm constitutes the greater portion of the cell; but in plants, particularly in differentiated cells, large vacuoles filled with water and dissolved materials may be present. In such cases there is a vacuole membrane bounding the cytoplasm where it comes in contact with the vacuole.

The nucleus is embedded in the cytoplasm. It possesses a definite membrane, the *nuclear membrane*, which bounds off the clear *nuclear sap*

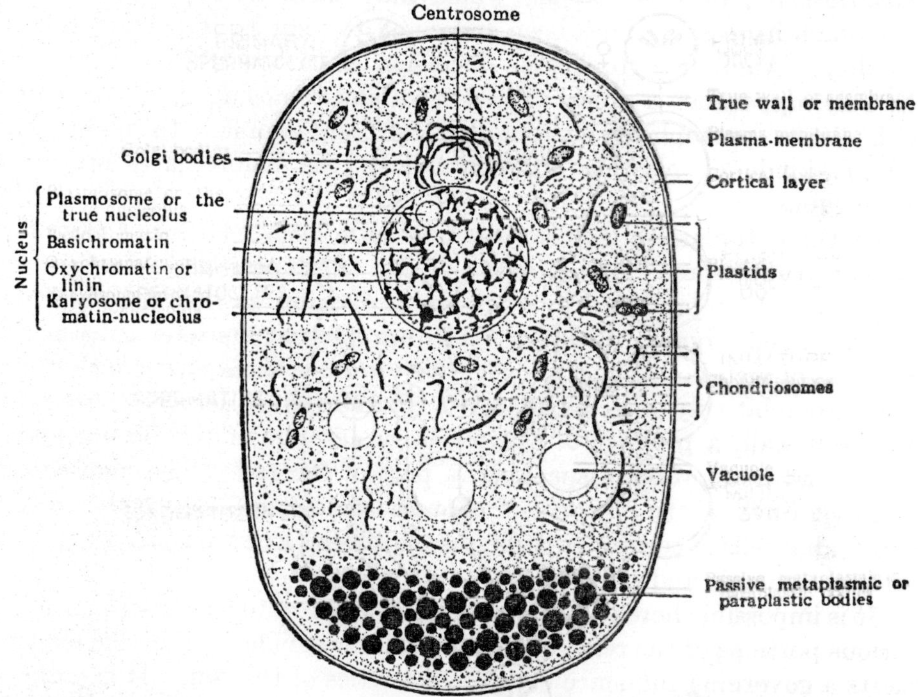


FIGURE 1.—General diagram of a cell. (*After Wilson, The Cell in Development and Inheritance, 3rd ed., copyright* 1925, *by The Macmillan Company. Reproduced by permission.*)

within from the cytoplasm outside. Suspended in the nuclear sap, or *karyolymph*, is a network of threads of *linin*, a substance which stains with acid dyes; and strung along this network are granules of more intensely staining substances, the *chromatin*. Linin network and chromatin granules together make up the *nuclear reticulum*. Besides these substances, the nucleus usually contains one or more relatively large *nucleoli* (singular, *nucleolus*) or *plasmosomes*, bodies which, like the linin, have an affinity for acid stains and which probably perform a nutritive function in the cell. Some nuclei may also contain a *karyosome*, a body which closely resembles the nucleolus in appearance but which is composed of chromatin material instead of plastin. The nucleus is, therefore, a complexly organized system, in which the relation of the various parts to each other is only beginning to be understood.

Besides the nucleus there are a number of other cytoplasmic inclusions which are found in most cells. In most animal cells there is a *centrosome* lying somewhere near the nucleus, which may be the center of a visibly differentiated portion of cytoplasm, the centrosphere, and from which during cell division a conspicuous system of *astral rays*, collectively termed the *aster*, radiates. Also lying near the nucleus in certain cells are clumps of granules, rods, or vesicles, variously termed *mitochondria*, *chondriosomes*, or *condrioconts*, which may have metabolic functions, and which have sometimes been suspected to play an important role in heredity. Another cytoplasmic inclusion of a speculative nature is the *Golgi apparatus*, a group or system of anastomosing rods, associated with dense protoplasm surrounding the centrosome. In plant cells various kinds of plastids occur of which the chloroplasts are most conspicuous.

Of the cytoplasmic inclusions which have been mentioned thus far, nucleus, centrosome, mitochondria, Golgi elements, and plastids are supposed to be reproduced by a process of division, so that they are true self-continued organs. Besides there are present various substances, products of metabolic activity collectively termed metaplastic inclusions, which are relatively passive in nature. Plant cells are usually surrounded by a cell wall, a product of the secretory activity of the plasma membrane, and not, properly speaking, a part of the cell. The demonstration by Bowen of the presence of chondriosomes and Golgi bodies in the antheridial cells of certain Bryophytes establishes a new basis of homology between animals and plants.

It is impossible here to enter into details relative to the functions of the various portions of the cell. There is abundant evidence that the nucleus exerts a governing influence over the activities of the cell. It is perhaps better, however, not to overemphasize the importance of any one cell organ but to regard the entire cell as a unit of function in which all the components are necessary parts of a unified system.

**Cell Division.**—Multiplication in number of cells takes place entirely by a process of division of preexisting cells. Cell division is a complex process which includes an organization of the chromatin material into a specific number of *chromosomes*, the disappearance of the nuclear membrane, the formation of a spindle upon which the chromosomes are arranged, the division of the chromosomes, the reconstruction of two daughter nuclei from the divided chromosomes, and finally the partition of the cytoplasmic mass. As a result two daughter cells, each of which usually contains all the structures and organs which were present in the mother cell, are produced.

The striking feature of cell division is the complex way in which the nuclear materials are apportioned to the daughter nuclei. It is customary to divide this process of nuclear division, or *mitosis*, into four phases,

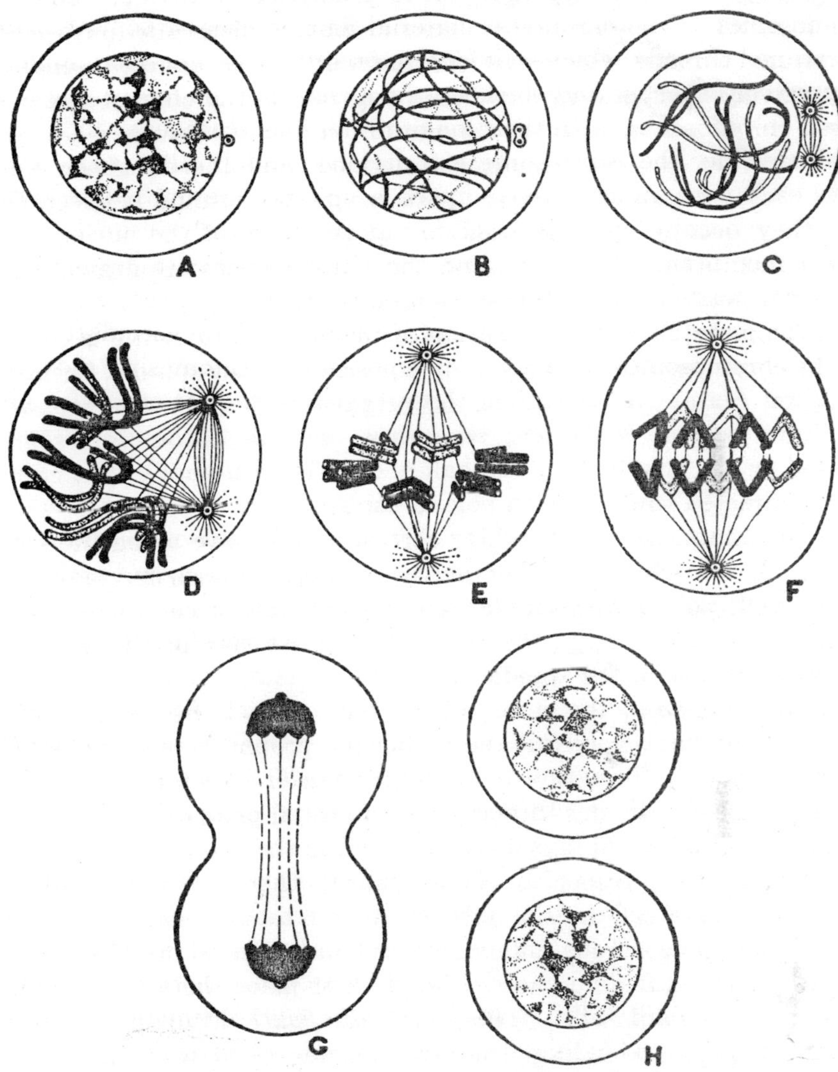


FIGURE 2.—Diagram of mitosis in an animal cell containing six chromosomes. *A*, resting nucleus; *B*, early prophase, individual chromosomes not yet distinguishable, centrosome dividing; *C*, middle prophase, appearance of spindle figure; *D*, late prophase, the nuclear membrane has disappeared and the chromosomes are becoming attached to the spindle fibers. *E*, metaphase; *F*, anaphase; *G*, telophase; *H*, nuclear and cell division complete, and daughter nuclei reconstituted. (*Adapted from Agar's Cytology, copyright 1920 by Macmillan and Company, Ltd. By permission.*)

known in order of sequence as prophase, metaphase, anaphase, and telophase (figure 2). These divisions are convenient helps in description, as long as they do not conflict with the conception of mitosis as a dynamic process, which ordinarily passes without interruption, from one stage to another.

In the prophase the chromatin, which is distributed diffusely through the resting nucleus, is aggregated into a number of distinct bodies, the chromosomes. The chromatin material first condenses to form slender intertwined threads, which split longitudinally. The double threads then progressively shorten and thicken, giving rise to the characteristic condensed chromosomes with their longitudinal line of cleavage.

Meanwhile, the centrosome has divided, and the daughter centrosomes each with an aster move apart, a spindle forming between them, until they occupy opposite positions at the poles of the nucleus. The nuclear membrane disappears, and the chromosomes are drawn into an equatorial position in the system of spindle fibers.

In the metaphase, the formation of the achromatic figure is complete and the chromosomes have taken up a position on the spindle, forming an equatorial plate. At this time the preexisting line of cleavage, which may have become indistinct, reappears, and the chromosomes are so oriented on the spindle that one-half of each longitudinally split chromosome is directed towards each pole of the spindle.

In the anaphase, the daughter chromosomes move along the spindle towards the centrosomes. The halves of each chromosome move apart so that each pole of the spindle receives one-half of each chromosome. At the close of this stage the chromosomes are clumped together in a mass at each pole of the spindle.

In the telophase, the chromosomes enlarge and pass into a diffused condition; the nuclear membrane reappears, probably as a precipitation phenomenon; the formation of daughter nuclei is completed. At the same time the cell divides through the equatorial plate of the spindle, the spindle fibers and astral rays disappear, and in the end two daughter cells equivalent in every respect to the original mother cell are formed.

The process of cell division just described is that typical of animal cells. It is subject to numerous differences in detail. In plant cells there are two rather constant differences. In the first place there are no centrosomes in most plant cells. The *achromatic figure* develops by the production of fibers extending from two masses of more or less hyaline material located in the cytoplasm at opposite sides of the nucleus. In the second place, the method of cytoplasmic division is usually different. A new cell wall is laid down in the equatorial plate of the spindle, apparently by a process of flowing of the spindle fiber material to this region during the telophase.

Mitosis is ordinarily studied in carefully prepared material, stained to differentiate the various structures and materials in the cell. A dynamic process is studied, therefore, by a static method; and as a consequence there is a marked divergence of opinion as to the significance of the various structures observed in it. In certain favorable material, however, it is possible to watch division in the living cell, and there is no doubt that the figures observed in stained preparations accurately represent conditions in the living, dividing cell. There is no question that the central feature of the whole process is the accurate division of chromatin material, so that each daughter cell contains not only the same number of chromosomes as the mother cell but also, by reason of the exact longitudinal division of each chromosome, the same chromatin elements as the mother cell. Inasmuch as all cells in the body arise by division according to this process and trace back to a single fertilized egg, it also follows that every cell in the body typically contains the same number of chromosomes and a similar, if not identical, complement of chromatin elements.

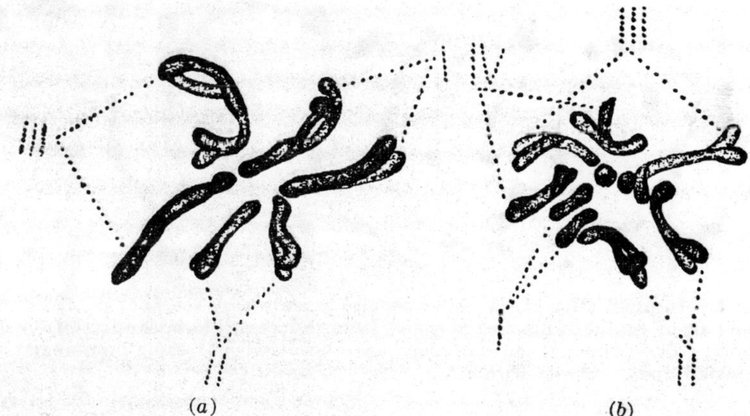


FIGURE 3.—Chromosome individuality as shown in somatic metaphase in *a*, *Crepis capillaris*, and in *b*, *C. setosa*. The pairs of chromosomes are indicated by Roman numerals. (*From Collins and Mann.*)

**Individuality of the Chromosomes.**—The process of cell division emphasizes the importance of the chromatin material, and certain other details of cellular organization and behavior, not stressed in this account of it, also argue to this end.

The number of chromosomes into which the chromatin is aggregated during mitosis is constant for the species. It differs widely, however, in different species; *Ascaris megalocephala*, a thread worm, has only two chromosomes; *Crepis capillaris*, a plant species, has six; *Drosophila melanogaster*, the common vinegar fly, has eight. These species have unusually low numbers of chromosomes. In most plants and animals the number is greater than this. In tobacco and man, it is apparently forty-eight; in corn it is twenty; in the mouse thirty-six. Usually there is an even number of chromosomes in the cells, a consequence, as will be seen, of the fact that each parent contributes an equal number of chromosomes in the production of a new individual. In many animal and plant species the exact number of chromosomes has not been determined, or it is still in dispute; for it is no easy task to make an exact count when the number is great, or when the chromosomes are large and intertwined.

The chromosomes within a species often differ in size and shape, and when such differences occur, they appear, with striking regularity, in every division figure. When the differences are marked, it is possible to determine that there are two chromosomes of every kind in every cell (figure 3). The differences are particularly striking in certain insect species, in which they are sometimes so extensive that every kind of chromosome is distinct. The constancy of number, of size, and of shape has led to the doctrine of individuality of the chromosomes, the conception that the chromosomes are self-perpetuating entities, qualitatively different from one another in the kind of chromatin which they contain.

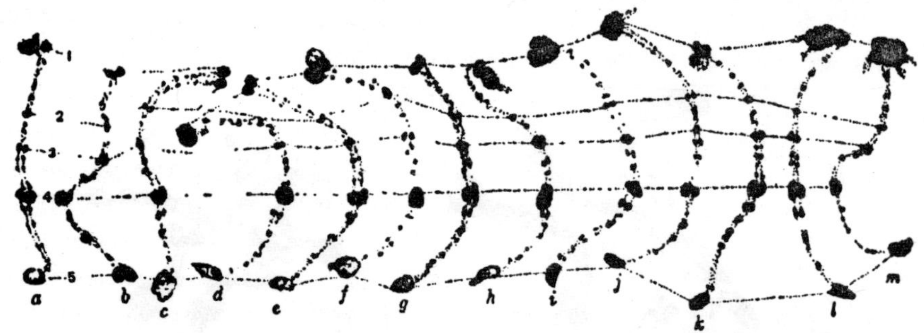


FIGURE 4.—Individuality of the chromosomes as shown by the finer details of structure of the same pair of conjugating chromosomes from thirteen different individuals of *Phrynotettix magnus*. The principal chromomeres are numbered and their homologies are indicated by dotted lines. (*From Wenrich.*)

The doctrine of individuality of chromosomes receives further support from studies of the finer structure of the chromatin material. These studies seem to show clearly that a chromosome is not merely a mass of chromatin, but that it consists of a definitely organized series of chromatin granules, the *chromomeres*, arranged in linear order, and that the number and arrangement of these elements is a persistent feature of individual chromosomes. In the grasshopper, *Phrynotettix magnus* (figure 4), Wenrich has shown that homologous pairs of chromosomes in comparable stages of conjugation present a remarkable constancy in number, size, and order of chromatin granules in different individuals; and within an individual many investigators have presented similar figures. In the organization of chromatin, therefore, the chromosomes are not the ultimate units, but rather the chromatin elements which they contain, and in mitosis each of these elements divides so that every cell in the body has a comparable set of them.

Those who oppose the doctrine of chromosome individuality point to the diffuse condition of the chromatin in the resting stage as an insuperable difficulty. If the chromosomes actually break down completely and coalesce in the intervals between cell divisions, it would be difficult to understand how individuality could be maintained, but more accurate observation in particularly favorable forms indicates that the chromosomes probably maintain a material continuity through the resting nucleus. Thus Wenrich has observed in certain cells of the grasshopper, *Phrynotettix magnus*, that the chromosomes in the telophase enlarge by a process of vesicle formation, and that these vesicles may be shown to be distinct even in the most diffused condition of the chromatin (figure 5). The formation of chromosomes for the succeeding division begins

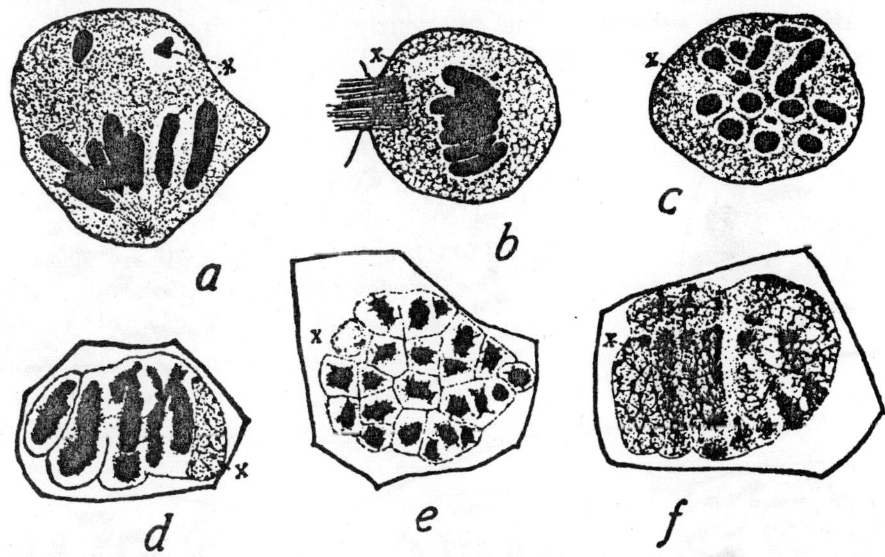


FIGURE 5.—Maintenance of chromosome individuality through the resting stage of the nucleus in *Phrynotettix magnus*. *a*, spermatogonial cell in late anaphase; *b*, similar cell in early telophase; *c*, the same stage in cross-section (in *a*, *b*, and *c* the cytoplasm was drawn in order to show the hyaline region around the chromosomes); *d*, vesicular expansion of the chromosomes in the telophase; *e*, the same stage in cross-section; *f*, completion of the telophase, the vesicles are still visible at the time of greatest diffusion of the nuclear material. (*From Wenrich.*)

within the vesicles by the formation of a thin coiled thread of chromatin material which divides longitudinally and condenses to form the double chromosomes of the division figure (figure 6). It is not difficult to account for chromosome individuality, if the nucleus in the resting stage is merely packed full of vesicles, each of which represents a diffused condition of a chromosome.

The doctrine of chromosome individuality is also supported by observations on cells with abnormal chromatin content. If a chromosome or a portion of a chromosome is lost, the cell is unable to regenerate the missing chromatin elements, and all the daughter cells which arise from it preserve the abnormal constitution. If an additional chromosome or portion of a chromosome becomes included in a cell by some irregular process, the cell does not cast out the excess material, but the daughter cells all exhibit a like deviation in chromatin content. The cell is not self-regulating with respect to its chromatin content; there is a specificity in the chromatin material which argues for the importance of this substance in the cell economy.

**Differentiation.**—The characteristic feature of development in multicellular animals and plants is the differentiation of cells or groups of cells into tissues for the performance of specialized functions. In the present state of knowledge, however, it is impossible to evaluate accurately the part which the various portions of the cell play in differentiation. The subject is one frequently avoided by biologists as a consequence of the difficulty of devising critical experiments bearing on it.

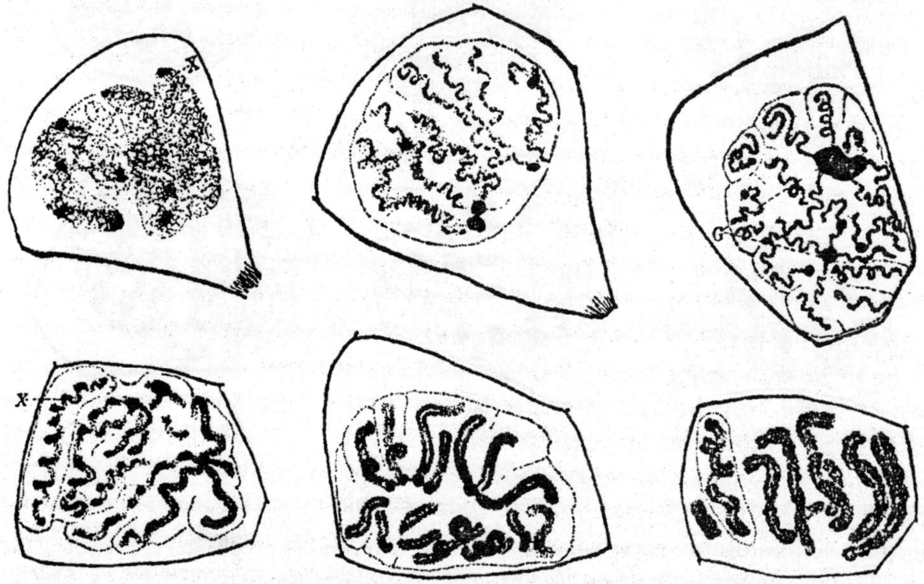


FIGURE 6.—Condensation of chromatin material within the vesicles of the resting nucleus in *Phrynotettix magnus*, stages following figure 5. (*From Wenrich.*)

In plants the visible results of differentiation are largely seen in the cell walls rather than in the protoplast itself. In meristematic tissue the cells are simple, thin-walled boxes of cellulose, and the protoplasts are relatively undifferentiated. Successive divisions follow each other rapidly in these cells. Such regions of meristematic cells persist during the life of the individual in the growing points of root and stem and in the cambium, and cell division is largely confined to them. Differentiated cells, in contrast, are usually larger and vacuolate, and the cell wall in certain tissues is highly thickened and modified in curious ways. In some cases the protoplast even disappears entirely, leaving lifeless cell walls which act as conducting channels. Differentiation of function occurs in various tissues whereby certain cells devote their energies largely to activities of a definite kind, such for example, as that of starch formation; but the visible changes connected with this differentiation are for the most part localized in cell inclusions rather than in the protoplast itself. Cells which perform obviously different functions may have protoplasts strikingly alike in appearance. In animals, however, differentiation gives rise to cells which are very diverse in structure and in function. Nerve cells, muscle cells, secretory cells, and the like are highly modified structurally so that they may be easily recognized in properly prepared material.

While it is impossible to state what the motivating factors in differentiation are, the main features of it may be traced with some assurance. The process of somatic mitosis would seem to indicate that differentiation is not primarily a matter of distribution of chromatin elements of different potentialities, for this process is so ordered that it distributes to the daughter cells all the chromatin elements which the mother cells possess. There is no sorting out of chromatin elements of different potentialities to different portions of the body as a result of mitosis. On the other hand the constituents of the cytoplasm, if there is regional distribution in it, may be unequally divided in cell division; and it is a well-known fact that a complicated regional distribution of cytoplasmic materials may be exhibited by the unsegmented egg. The nucleus resides in the cytoplasm; its functions are performed in connection with the cytoplasm in which it is embedded; and it is conceivable that if the cytoplasm of two cells is different, even though the nuclear elements remain the same, the activities of the cells will differ correspondingly. Differentiation is, therefore, possible without a corresponding nuclear modification; and it is at least significant that the visible results of differentiation are largely resident in the cytoplasm and the materials which it produces.

It is impossible to deny absolutely that nuclear differentiation occurs, for nuclear modifications are possible in the intervals between cell divisions without involving the assumption of distribution of chromatin elements unequally in cell division. It is, nevertheless, an attractive hypothesis to consider differentiation as localized in the cytoplasm, the nuclear elements remaining the same in every cell in the body, however far modification has progressed. In line with this assumption, the nucleus constitutes the governing body of the cell, intimately interacting with the particular cytoplasm in which it is embedded. The particular activities of a cell are, therefore, a function of the kind of cytoplasm which surrounds the nucleus and supplies it with the materials for its activities, and these activities are in turn modified by the internal environment in which the cell is situated, in which it obtains its materials, discharges the products of its activities, and performs its specialized functions. Should this conception be correct, then the individuality of the organism may well be considered merely a manifestation of the fact that all its cells are governed by nuclei containing identical sets of chromatin elements.

**The Germ Track.**—One of the features of differentiation which has exerted a great influence on theories of heredity is the early setting aside of those cells which eventually give rise to the reproductive cells. In both animals and plants the germ cells are ordinarily produced from undifferentiated cell lineages, that is, from cells which have never gone through, in any generation, the complex modification involved in the production of somatic tissues. In animals these cells are set aside in early development and localized in the sex glands, or gonads. In plants the corresponding undifferentiated lineage is preserved in the meristematic cells of the growing point or vegetative cone; but it is not marked off in the striking fashion characteristic of some animal species.

The classical example of germ-track differentiation is provided by the thread worm, *Ascaris megalocephala,* as studied by Boveri. In the variety *univalens*, which possesses two chromosomes, the steps involved are essentially as follows:

In the first division of the fertilized egg, the two chromosomes divide in the customary manner giving rise to two daughter cells each with two large chromosomes. One of these cells, however, undergoes a striking nuclear change during which the ends of the two large chromosomes are broken off and cast out into the cytoplasm, and the remaining portions segment into a large number of smaller bodies, which thenceforth act like chromosomes in subsequent mitoses (figure 7).

In the divisions of the two cells to form the four-cell stage, one of the cells then possesses two large chromosomes like the original cell, the other possesses a large number of small chromosomes. The cell which has the numerous smaller chromosomes gives rise to two cells of identical nuclear organization; the cell with two large chromosomes gives rise to two cells, one of which retains the two-chromosome organization, but the other is subjected to the same changes which took place in one of the cells of the two-celled stage; the ends of the chromosomes are cast off and the middle portions segment into smaller chromosomes. This process is repeated four successive times, so that in the sixteen-cell stage, one cell contains the original chromatin organization, one cell is undergoing chromatin diminution and chromosome fragmentation, and the remaining fourteen cells have numerous small chromosomes. Thereafter the daughter nuclei have the same chromatin organization as their mother nuclei. The cells which have the two intact chromosomes are localized in the gonads and eventually give rise to the reproductive cells; the cells with numerous smaller chromosomes develop into the soma. The line of cells with intact chromosomes leading from the undivided egg cell to the gametes is called the germ track. That portion of the

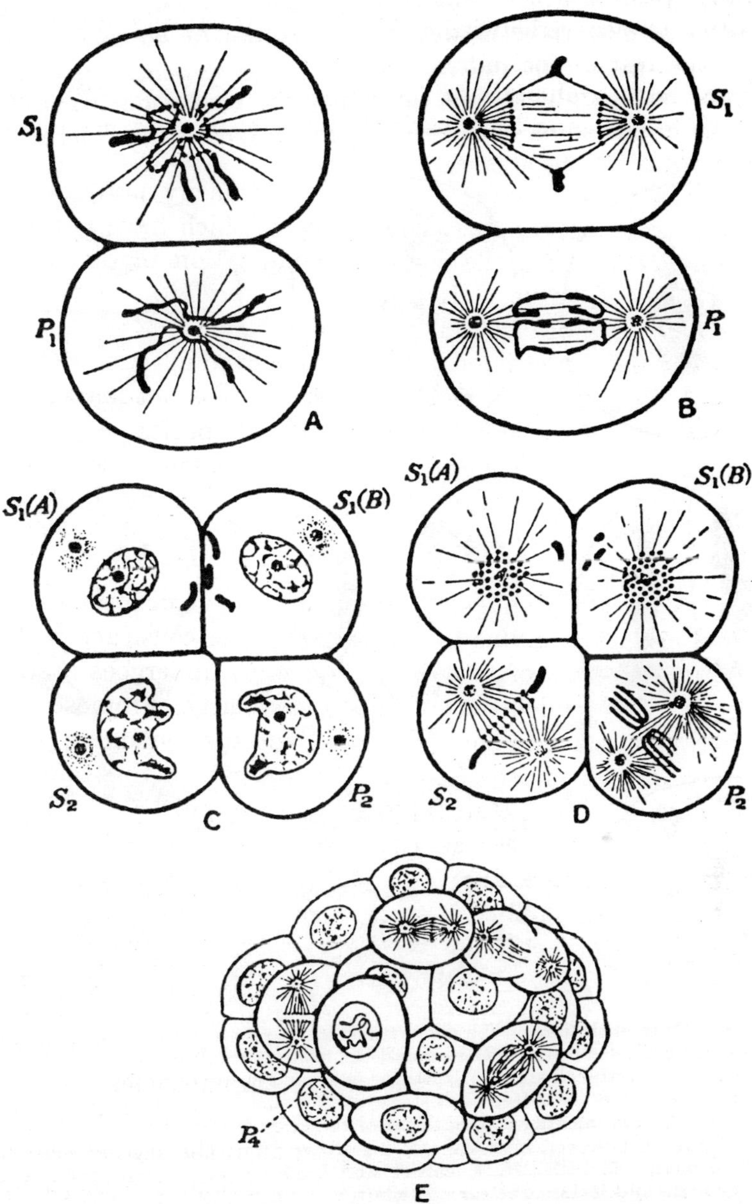


FIGURE 7.—Differentiation of the germ track in *Ascaris megalocephala univalens.* *A*, two-cell stage. Chromosome diminution and fragmentation under way in $S_1$; chromosomes intact in $P_1$. *B*, division to form the four-cell stage. Numerous small chromosomes in $S_1$, the broken-off ends outside the equator of the spindle. *C*, four-cell stage. The broken-off ends of the chromosomes lying in the cytoplasm of $S_1(A)$ and $S_1(B)$. *D*, division to form the eight-cell stage. Chromosome diminution and fragmentation repeated in $S_2$. *E*, later embryo. Chromosome diminution and fragmentation complete. The original chromosome organization maintained intact in $P_4$ only. (*After Boveri, from Agar, Cytology, copyright* 1920 *by Macmillan and Company, Ltd. Reprinted by permission.*)

body which possesses the cells with intact chromosomes is called the *germ plasm,* in contradistinction to the soma, or somatoplasm which makes up the rest of the individual.

It might be thought at first sight that the visible nuclear differentiation of cells in the soma and germ plasm of Ascaris is evidence that the

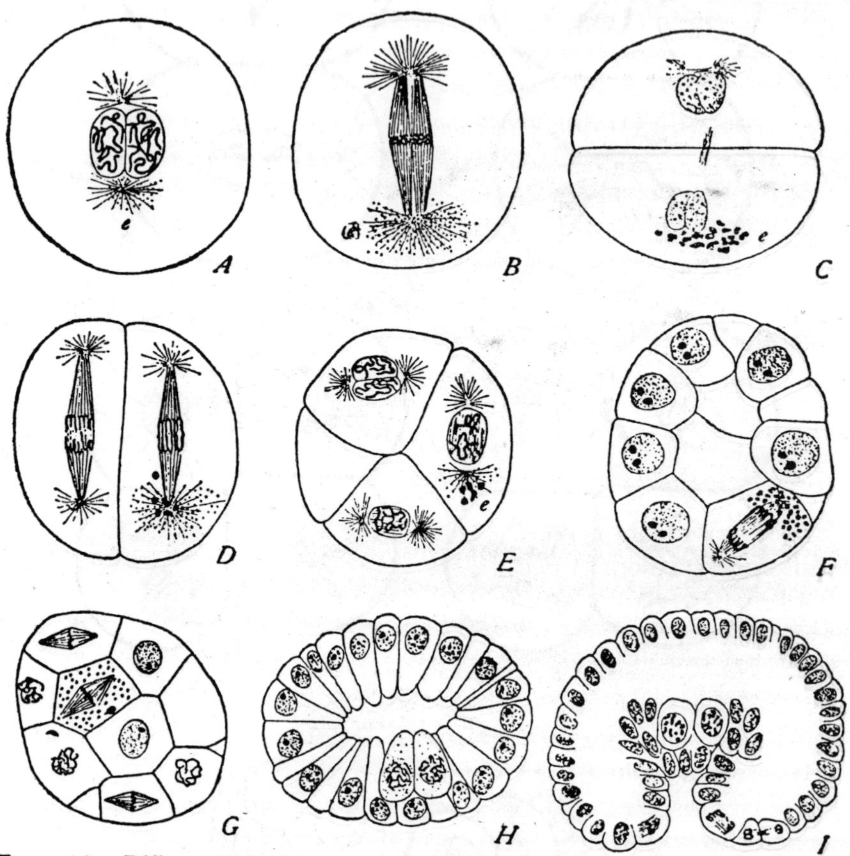


FIGURE 8.—Differentiation of the germ track in *Cyclops fuscus.* *A*, prophase of first cleavage mitosis, granules congregated around the lower attraction sphere. *B*, metaphase of same division. *C*, two-cell stage. *D*, stages in second cleavage mitoses. *E*, twelve-cell stage, note granule cell, *e*. *F*, fifteen-cell stage; all the nuclei have completed their division and entered into the resting stage, except that of the granule cell which is still in anaphase. *G*, division of the granule cell into the two primitive germ cells, the granules uniformly distributed through the cell. *H*, *I*, earlier and later gastrulas, primordial germ cells. Note that the male and female nuclear contributions remain distinct in *A*, *B*, *C*, and one cell in *E*. (*After Amma from Wilson, The Cell in Development and Inheritance, 3rd. ed., copyright* 1925 *by The Macmillan Company. Reprinted by permission.*)

nuclear elements are actually changed in the process of differentiation. But Boveri has shown by various ingenious experiments and observations, particularly on dispermic eggs, that the cytoplasm of the egg possesses a polarity such that cell division divides the cytoplasmic materials unequally and that the resulting cytoplasmic differences are responsible for the changes which take place in the chromatin material. The significance of the difference in chromatin content and organization in somatic cells has not been determined.

In line with this idea, it is found that the visible features of germ-track differentiation are usually cytoplasmic rather then nuclear. In *Cyclops fuscus*, for example, one of the poles of the spindle is distinguished from the other by a group of granules surrounding the attraction sphere (figure 8). For four successive cell divisions one attraction-sphere only contains the group of granules (*c*), so that at the sixteen-cell stage they are confined to one cell, while the other fifteen cells are entirely free from them. Thereafter the granules are distributed to both cells in division, the granule cells constitute the germ track, and the remaining cells free from granules constitute the soma.

There are endless details in the process of germ-track differentiation; but the problem of most interest is the situation in vertebrates. Apparently in these forms the primordial germ cells which are set aside early in development do not give rise to the definitive germ cells, but degenerate. In fact, in mammals these cells apparently are replaced by a second group of cells, which in turn degenerate and are replaced by a third group arising by proliferation from the germinal epithelium. These last then give rise to the definitive germ cells, but it has not been determined definitely whether they arise from primordial germ cells which fail to degenerate or from indifferent epithelial cells.

In his important speculations on heredity, variation, and evolution, Weismann laid great emphasis upon the distinction between germ track and soma. He regarded the early setting apart of the germ track and its subsequent existence as an isolated mass of undifferentiated tissue as evidence for its immunity from environmental conditions and from the effects of external agents acting upon the body during development. The germ plasm resident in the germ track thus represented an immortal cell lineage aptly expressed in the phrase, continuity of the germ plasm, and the soma, merely an evanescent product having no permanent influence upon the former. Even if, as in vertebrates, the early origin of the germ track is not so clearly demonstrable as in lower forms, the identification of the germ plasm with the chromatin material of the nucleus and the accurate partition of this substance at each successive mitosis provide a satisfactory physical basis for this conception.

**References**

AGAR: Cytology.
CONKLIN: Heredity and Environment, 3d ed., Chapter III.
COWDRY (Editor): General Cytology.
HEGNER: The Germ Cell Cycle in Animals.
SHARP: Introduction to Cytology, 2d ed.
WALTER: Genetics, Chapter XII.
WILSON: The Cell in Development and Heredity.

# CHAPTER III

## REPRODUCTION

The discussion of the development of the individual started with a single cell, the fertilized egg. The manner in which this cell comes into existence constitutes the fundamental problem of reproduction. In the discussion of development, the phenomena characteristic of Metazoa have been described; in the discussion of reproduction, likewise, the phenomena will be described as they occur in animals, and we shall be content merely to point out the analogies which occur in plants.

It has been shown that early in the life of the individual certain cells are set aside which constitute the germ track. These cells later give rise to the reproductive cells, or *gametes*, which are single cells produced by a complex process of cell division differing from ordinary somatic mitosis. The special processes concerned in the production of gametes are collectively termed *gametogenesis*. The processes of gametogenesis differ in detail in the two sexes; consequently it is convenient to consider them separately. In the female the gametes are known as *ova* (singular, *ovum*) and gametogenesis is given the special term, *oögenesis;* in the male the gametes are *spermatozoa* (singular, *spermatozoön*) and the process of their production is termed *spermatogenesis*.

**Spermatogenesis.**—The cells of the germ track divide by a process of ordinary mitosis by which the number of chromosomes and their characteristic chromatin content are preserved unchanged. When the individual reaches sexual maturity, certain of the cells, called *spermatogonia*, can be distinguished as those which will eventually give rise to spermatozoa. They divide a number of times by ordinary mitosis, and then they cease to divide and enter upon a period of growth. They are now called *primary spermatocytes*. At the close of the growth cycle, these primary spermatocytes prepare for division, but the ensuing divisions, although essentially concerned with the partition of chromatin elements, differ strikingly from somatic mitosis. The differences lie essentially in the fact that the prophase is complicated by a series of events connected with a conjugation of the chromosomes in pairs and in the ensuing metaphase whole chromosomes are separated to opposite poles so that the daughter nuclei each receive only half the number of chromosomes contained in the nucleus of the primary spermatocyte. This type of nuclear division is termed *meiosis*.

Starting with the resting condition of the primary spermatocyte, the sequence of events as they occur in the polychaete worm, *Tomopteris onisciformis* (figure 9), may be described as follows:

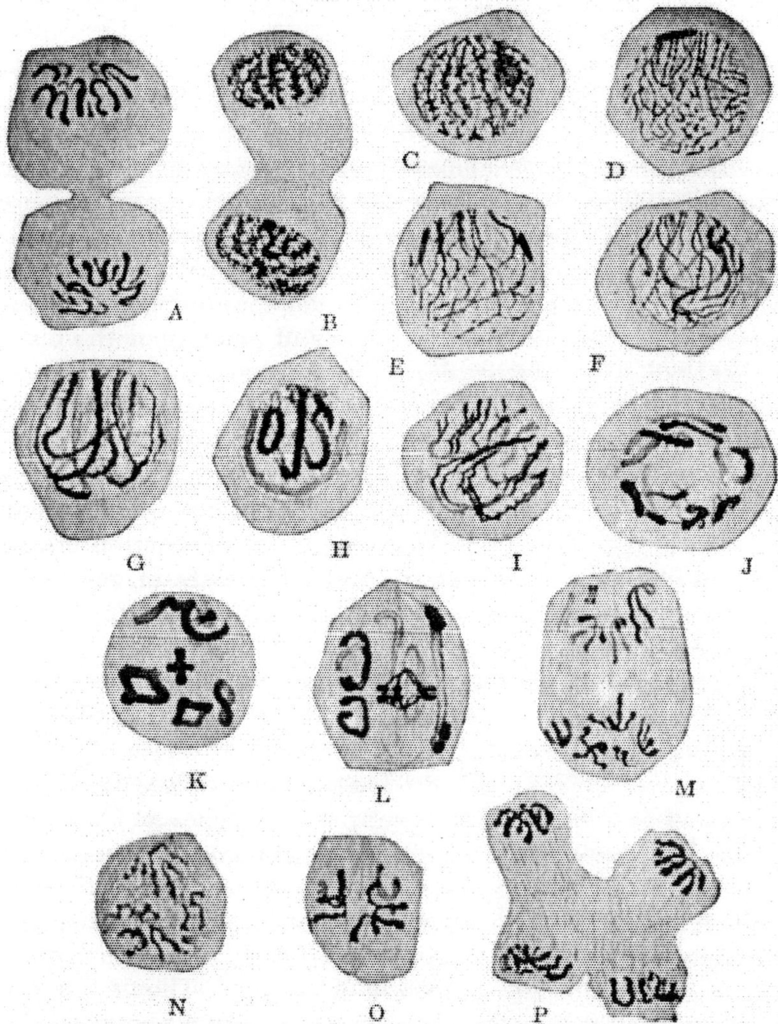


FIGURE 9.—Meiotic phase in the male *Tomopteris onisciformis*. *A*, *B*, telophases of last premeiotic (spermatogonial) mitoses; *C*, late telophase of same, passing into the primary spermatocyte; *D*, leptotene stage; *E*, *F*, *G*, zygotene stage; *H*, pachytene stage; *I*, diplotene stage; *J*, diakinesis; *K*, disappearance of nuclear membrane, immediate prophase of meiotic I; *L*, metaphase I; *M*, anaphase I; *N*, prophase II; *O*, metaphase II; *P*, telophase II. (*From Agar, Cytology, copyright* 1920 *by Macmillan and Company, Ltd. Reprinted by permission.*)

The chromatin at the end of the last premeiotic division is distributed through the nucleus in the form of a fine network in which, however, it is sometimes possible to distinguish areas corresponding to the diffuse chromosomes. As division approaches, the chromatin is first condensed into fine threads, the condensation beginning at the pole of the nucleus at which the centrosome is situated and extending gradually throughout the nucleus until all the chromatin material is included. This stage is called the *leptotene* stage. It passes into the *zygotene* stage in which the threads take up a parallel arrangement in pairs, followed by a fusion of the two members of each pair of threads. The pairing of the threads and their fusion extends gradually throughout the chromatin mass, accompanied by thickening of the threads, until the entire chromatin mass, in the *pachytene* stage, is condensed into long double loops converging at one pole of the nucleus. The process of fusion of the chromatin threads, which is interpreted as a conjugation of homologous chromosomes, is termed *synapsis;* and the chromatin loops are termed *bivalents*, because they consist of two conjugating chromosomes. Following the pachytene stage, the chromatin material threads out again, and the bivalents separate into their two components, giving rise to the *diplotene* stage, or, as some prefer to call it on account of the twisted condition of the threads as they emerge, the *strepsitene* stage. The separation of the bivalents into their two components, is not often complete, but they usually remain united at one or both ends to form a variety of characteristic figures. The chromatin material then condenses again, the chromosomes still remaining paired. This stage is known as *diakinesis*, and it is possible to observe that the members of each pair of chromosomes in diakinesis are of the same kind.

During the course of the nuclear changes which have been described above, the centrosome has divided, and the two centrosomes have moved apart until they have reached opposite poles of the nucleus. The nuclear membrane then disappears, and the bivalents of diakinesis are drawn into the equatorial plate of the spindle, completing the prophase and giving rise to the metaphase. In the ensuing anaphase whole chromosomes are separated from each other, so that each pole of the spindle receives half the number of chromosomes characteristic of somatic cells. The completion of the first meiotic division, which gives rise to the secondary spermatocytes, is followed almost immediately by the initiation of a second meiotic division. The chromosomes of the telophase, which may exhibit a longitudinal cleavage as early as the prophase of the first meiotic division, are maneuvered into place on the spindle of the second division, and one-half of each chromosome passes to each pole as in mitosis. The completion of the second division results in the production of four cells, the *spermatids*, which are ultimately transformed into the functional male sex cells, the spermatozoa, without further divisions.

In other forms the process of meiosis is subject to minor modifications from that described for Tomopteris. Perhaps the most important of these is the insertion of a contraction of the chromatin material between the diplotene stage and diakinesis, as is well shown in the American lungfish, *Lepidosiren paradoxa* (figure 10). During this stage, the double threads of the diplotene stage contract into a compact mass, to which process the term *synizesis* has been applied. After the contraction the chromatin knot loosens up and passes into diakinesis.

By most authors the term synapsis has been applied to the conjugation of the chromosomes during meiosis; but some confusion has arisen in the use of this term by reason of its application by some cytologists

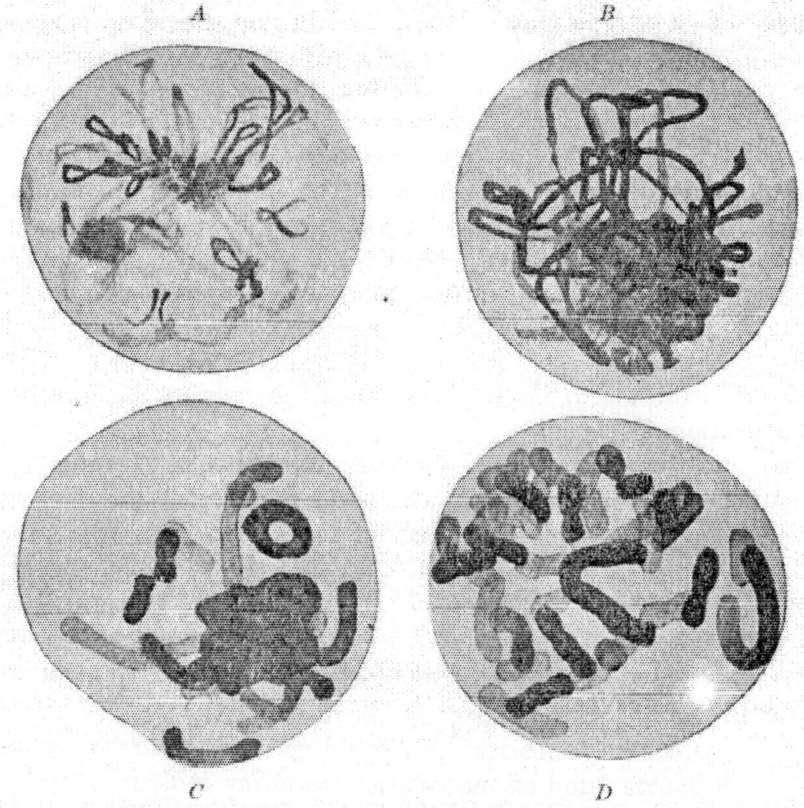


FIGURE 10.—Synizesis in the meiotic phase of *Lepidosiren paradoxa*. *A*, diplotene stage and beginning of synizesis; *B*, synizesis further advanced; *C*, synizesis breaking up; *D*, diakinesis. The bivalents of the zygotene stage have been completely resolved into their univalent constituents; but they later pair again before moving onto the spindle. *(From Agar, Cytology, copyright* 1920 *by Macmillan and Company, Ltd. Reprinted by permission.)*

to the contraction stage of the chromatin described above. On this account the term *syndesis* has been suggested for the pairing of the chromosomes and synizesis for the contraction above described. When the term synapsis is used, however, it should be understood to be synonymous with syndesis, for the conjugation of the chromosomes is the important feature from the standpoint of genetics.

The first meiotic division, which gives rise to two cells each of which possesses half the number of chromosomes of the mother cell, is termed a reductional or heterotype division. The second, in which the chromosomes split longitudinally as they do in mitosis, is called an *equational* or *homotype* division. The entire process described above is often termed *maturation of the germ cells*.

**Oögenesis.**—In oögenesis the sequence of events is closely comparable to that of spermatogenesis. The *oögonia*, which correspond to the spermatogonia, have the somatic number of chromosomes and divide in the usual fashion of somatic mitosis. They eventually cease to divide and pass into a growth phase, comparable to that described in spermatogenesis, by which the primary oöcytes are formed. In the division of the primary oöcytes, the prophase is marked by the same series of complex maneuvers of the chromatin material as that described for the heterotype mitosis in spermatogenesis. But when the primary oöcyte completes the strepsitene stage of emergence of the double chromosomes from synizesis, it enters upon a second growth period, which may be of long duration in some forms, during which the deposition of yolk takes place in the egg and its size may enlarge to many thousand times that of the original oögonium. During this stage the nucleus is very large and diffuse, doubtless as a consequence of its metabolic activities. When the egg becomes full sized, the nucleus shrinks in volume and prepares for the maturation divisions.

There are two maturation divisions in oögenesis, corresponding to the two in spermatogenesis. In the first, the heterotypic division, the number of chromosomes is reduced; the second is a homotypic division, in which the chromosomes split longitudinally. While the nuclear divisions are closely comparable to those of spermatogenesis, the cell divisions are conspicuously different. In the first maturation division, the cell division is extremely unequal; the chromosomes at one pole of the spindle are pinched off with a very small portion of cytoplasm to form a minute polar body. In the second division, likewise, this process is repeated, so that the two divisions result in one large cell which is the functional ovum, and two or three minute, non-functional polar bodies, the number depending on whether or not the first polar body divides. As a consequence of the reduction division, the ovum, like the spermatozoön, finally contains only half the number of chromosomes characteristic of somatic cells. The essential homology between spermatogenesis and oögenesis is illustrated diagrammatically in figure 11.

**The Significance of Meiosis.**—As a result of the processes which have just been described, the gametes which are formed contain half the number of chromosomes characteristic of somatic cells. By means of this provision the constancy of the number of chromosomes is preserved, for when the two gametes unite in fertilization to form the zygote, the somatic number of chromosomes is restored. It is customary to call the double number of chromosomes which is found in somatic cells the *diploid* or $2n$ number, and the reduced number which is characteristic of gametes, the *haploid* or $n$ number. Certain important details connected with reduction in the chromosome number remain to be considered.

When the chromosomes differ strikingly in size and shape, as they do in many forms, it can be observed that two chromosomes of each kind occur in somatic cells (figure 3). In certain insect forms these relations are particularly obvious because the members of the chromosome pairs lie side by side even in somatic division figures. It is possible to demonstrate further that the chromosomes which pair in synapsis are of the

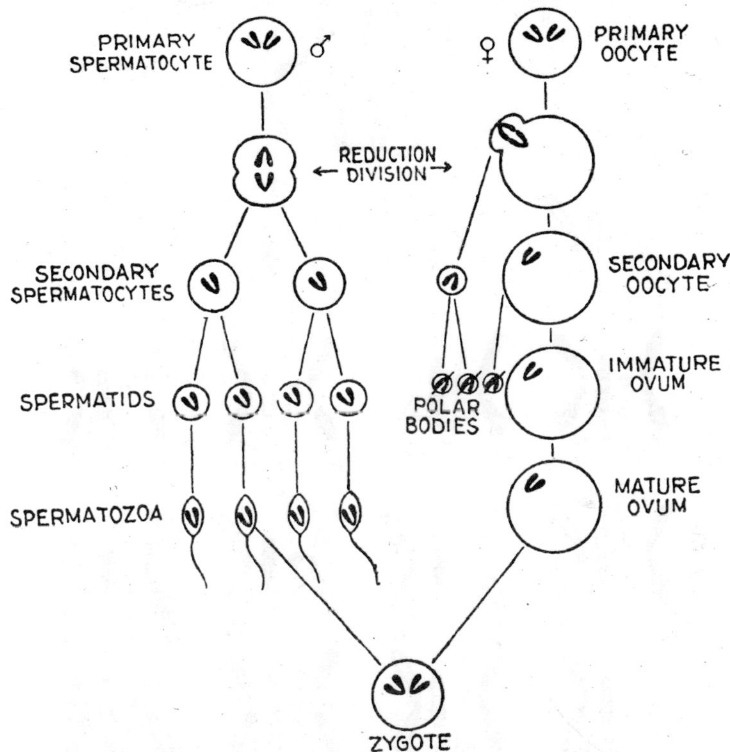


FIGURE 11.—Comparative diagram of spermatogenesis and oögenesis. (*From Mohr.*)

same kind; *i.e.*, they are homologous chromosomes. It follows that the ensuing reduction division separates the different kinds of chromosomes, and that the resulting germ cells receive not merely half the number of chromosomes characteristic of the somatic cell, but one of each of the kinds of chromosomes present in these cells. This conception is vital to the doctrine of individuality of the chromosomes. Arising from this conception of homology of the chromosomes, there is the further conception essential to a proper comprehension of phenomena of heredity that the cells of the soma contain two complete sets of chromatin elements, one derived from the male and one from the female. Gametes, on the contrary, contain only one complete set of chromatin elements.

Although homologous chromosomes separate in reduction and pass to opposite poles of the spindle, it does not follow that the gametes contain sets of chromosomes either entirely maternal or entirely paternal in

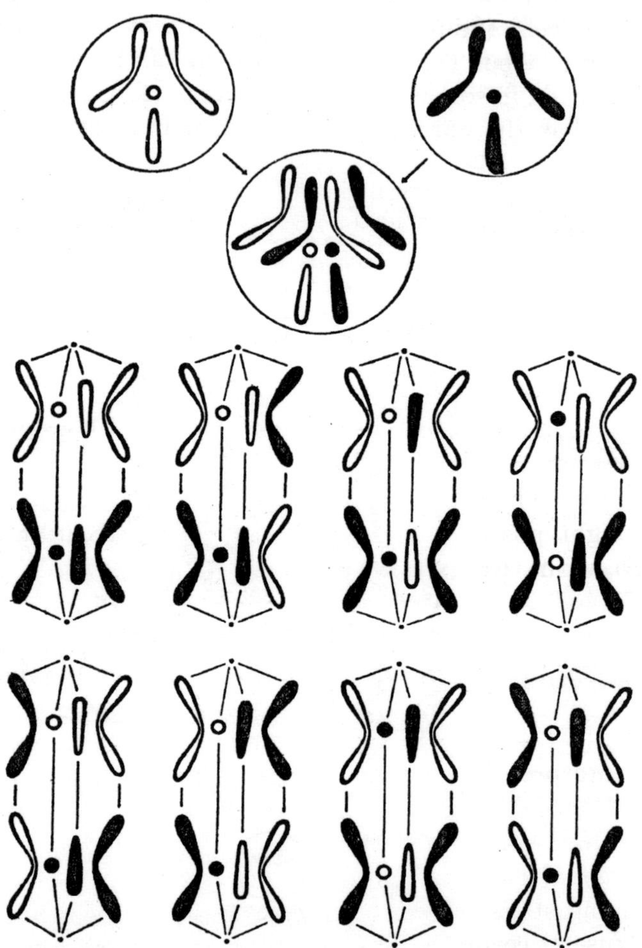

FIGURE 12.—Diagram illustrating random distribution of chromosomes in oögenesis in *Drosophila melanogaster*. Upper left, chromosomes contributed by the female parent; upper right, those contributed by the male parent to the zygote containing the diploid number of chromosomes. Below, the eight different types of distribution which occur in formation of gametes by such an individual.

origin. When the paired chromosomes are oriented on the spindle preparatory to reduction, it is a matter of chance whether a maternal or a paternal chromosome is directed toward a particular pole. As a consequence a given pole of the spindle will receive maternal members from some pairs and paternal members from others, so that the gametes which are produced usually contain mixtures of maternal and paternal chromosomes, but cnly one of each kind.

The result of random distribution of chromosomes is illustrated in figure 12 for the female of *Drosophila melanogaster*, which has four pairs of chromosomes. Two germ cells, one from the female with the chromosomes in outline and the other from the male with them in full black, unite to form the female zygote shown in the middle of the figure. The lower portion of the figure shows the eight different types of distribution of maternal and paternal chromosomes which occur when this individual in turn forms gametes. The eight different types of distribution are equally likely to occur. Since each one gives rise to two different kinds of gametes, there are sixteen different kinds of gametes with reference to origin of their chromosomes, formed in equal numbers in gametogenesis in *Drosophila melanogaster*. Only two of these sixteen kinds are identical in constitution with those which united to form the individual.

Obviously the number of different kinds of chromosome combinations possible in the gametes of a given species depends upon the number of pairs of chromosomes present in it. The number of combinations possible in a specific case is given by the equation, $x = 2^n$, in which $x =$ the number of combinations and $n =$ the number of pairs of chromosomes in the species. Thus in *Crepis capillaris* with three pairs of chromosomes there are eight possible combinations of chromosomes in the gametes; in tobacco with twenty-four pairs, there is the enormous number of 16,777,216, only two of which are identical with those which united to form the given individual.

Some further consideration of the significance of the peculiar events occurring in the prophase of the first meiotic division is also necessary. The pairing of the threads which occurs in the zygotene and following stages is interpreted as a conjugation of homologous chromosomes. Moreover, Wenrich's figures (figure 4), which show the close correspondence in number, size, and arrangement of the chromomeres, argue for the homology of the two members of a conjugating chromosome pair. In the ensuing stages the two chromatin threads which conjugate may be seen to fuse at certain points and often are twisted around each other. It is, therefore, believed that the two chromosomes which conjugate in synapsis do not necessarily emerge from that stage with their original set of chromatin elements, but that they may have exchanged equivalent portions, a phenomenon known as *crossing-over*, so that after synapsis the two chromosomes may consist of chromatin material derived from both of the original chromosomes. Important theories of heredity are based on crossing-over between homologues during conjugation.

There is not, however, a unanimity of opinion as to the method of conjugation of the chromosomes in synapsis. The majority of investigators favor the view that the chromosomes pair side by side in approximately the manner described above. This method of conjugation is known as *parasynapsis* or *parasyndesis*. There are others, however, who believe that the doubling of the thread in the zygotene stage is due, not to conjugation of two separate threads, but to a longitudinal cleavage of a single thread, and that the homologous chromosomes are united end to end during these stages. They explain the emergence of thick chromosomes side by side in the strepsitene stage as due to a bending of the long double chromosome thread back against itself. This method of conjugation is known as *telosynapsis* or *telosyndesis*.

If the chromosomes are united end to end instead of side by side, it would be difficult to account for exchange of chromatin material between homologous chromosomes. Certain observations have, however, been made which seem to show that these divergent views may be brought into agreement. The American lungfish, *Lepidosiren paradoxa*, appears to offer exceptionally favorable material for observation of the phenomena of the meiotic prophase. According to Agar's description the zygotene stage is clearly a side-by-side pairing of chromosomes and the number of loops formed is one-half that of the number of chromosomes. The line of union disappears during the pachytene stage, but when the chromosomes emerge from synizesis, the two united threads separate again everywhere except at the end to form loops and rings which consist of two chromosomes united end to end. In this form then there is first a side-by-side conjugation followed by an emergence of chromosomes united end to end. A further discussion of the significance of synapsis is deferred until the genetic evidence shall have been presented.

It should be noted, therefore, that the genetic continuity of chromosomes, which is so accurately preserved in each succeeding division, may possibly be disturbed during the meiotic prophase, but only to the extent of permitting exchange of corresponding portions of homologous chromosomes. When such exchange occurs, the chromosomes which are distributed in gametogenesis are not really maternal and paternal chromosomes at all, as was assumed in the foregoing account of chromosome distribution, and the mathematical relations there described are no longer applicable. As a further consequence of this conception, it is also obvious that the preservation of genetic continuity lies in the ultimate chromatin elements and not in the chromosomes themselves.

**The Two Meiotic Divisions.**—In the foregoing account the first division has been regarded as reductional and the second as equational. As a matter of fact, however, the problem as to the significance of these two divisions cannot even yet be considered as solved, and it may be possible that the behavior is not the same for all species. If, however, as seems probable, the chromosomes really complete the equational split in the prophase, each giving rise to two chromatids, and if as a result of crossing-over these chromatids consist partly of maternal and partly of paternal elements, then it is probable that both divisions are partly reductional and partly equational. Cytological evidence is difficult to secure, because the two conjugating chromosomes rarely exhibit any morphological differences. But Wenrich has studied some interesting cases in Phrynotettix (figure 13) in which the homologues of certain pairs of chromosomes differed in size. In one such pair of chromosomes it was shown, as judged by the sizes of the separating elements that in half the cases the first division was reductional; in the other half, equational. Hence in this instance chance determines the mode of orientation of the conjugating elements and their type of distribution in the ensuing

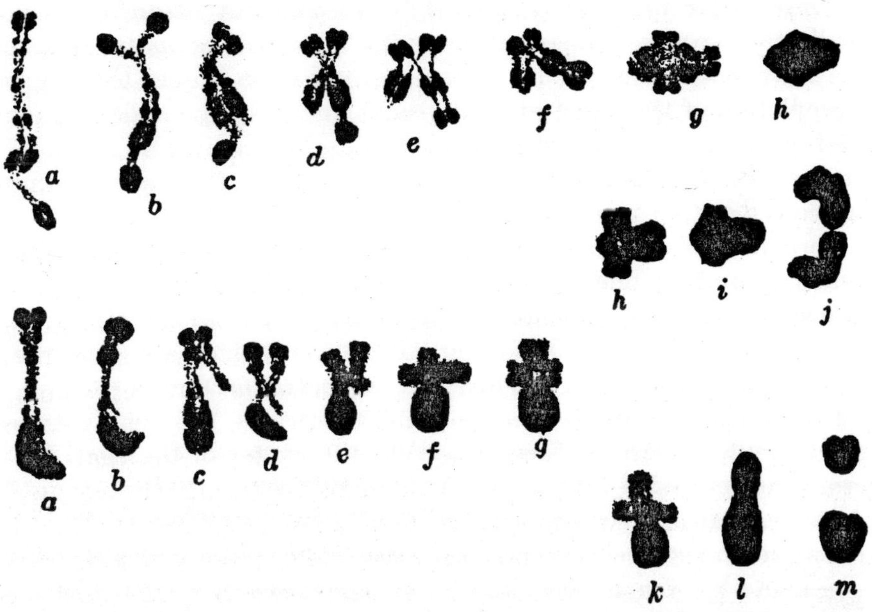


FIGURE 13.—Conjugation of unequal pairs of chromosomes in Phrynotettix. Row, 1, *a–h*, successive stages of conjugation in Wenrich's chromosome *B*, followed by equational division in meiosis I. Row 2, *a–g*, similar stages for chromosome *C* followed by equational, *h–j*, or reductional, *k–m*, division. (*From Wenrich.*)

divisions. In another instance in the same species an unequal pair of chromosomes appeared regularly to divide equationally in the first and reductionally in the second division. Interesting as these problems are, the essential feature from a genetic point of view is the separation of maternal and paternal elements, and whether it occurs in the first or second or partly in each division is relatively immaterial.

**Fertilization.**—In an account of fertilization it is necessary to distinguish between accessory and essential phenomena. All the various organs which are concerned in the act of mating and the great variety of methods by which the gametes are brought into contact are purely accessory and need not be considered here. The following account is devoted entirely to the essential phenomena of fertilization, the processes involved in fusion of the gametes.

The ultimate gametes of the two sexes usually differ strikingly in form and behavior as has already been noted. The ovum is commonly large and stationary and contains a great deal of food material. The spermatozoön on the contrary is very minute and motile; it consists generally of a head of condensed chromatin material which may have an anterior apical body or acrosome, a middle piece which contains or at least later gives rise to a centrosome; and a long actively motile tapering tail.

The moment at which the spermatozoön enters the egg differs with different species. In some it enters the primary oöcyte; in others it may enter after the first polar body is formed but before the second; in still other species it does not enter the egg until the maturation divisions have been completed. If the spermatozoön enters the egg prior to completion of the maturation divisions, fusion of the male and female nuclei is deferred until the meiotic phase is completed. Since the maturation divisions have already been described, they will be omitted from consideration in this account.

Normally only one spermatozoön enters an egg, and in those species in which more than one enters, only one fuses with the egg nucleus. The entire spermatozoön may enter the egg or only the head and middle piece; but in any event it is the head and middle piece which are concerned in the subsequent processes. When the spermatozoön enters the egg, the head precedes the middle piece, but after entrance the head rotates so that the middle piece is directed toward the center of the egg. As the sperm head, preceded by the middle piece, moves toward the egg nucleus, it becomes vacuolated and gradually assumes the form of a nucleus. Meanwhile an aster is formed around the middle piece, and a centrosome appears in it. When the two nuclei come together, the centrosome divides, a spindle is formed, and an ordinary somatic division, the first cleavage division, ensues.

Sometimes the two nuclei fuse before the sperm nucleus has attained the size of the egg nucleus, but studies of particularly favorable material reveal the fact, which follows naturally from the method of spermatogenesis and oögenesis, that the two uniting nuclei contain the same number and kinds of chromosomes. In particularly favorable cases (figure 8) the two chromatin contributions remain distinct in the first and sometimes in subsequent division figures so that it is possible to demonstrate that the chromatin contributions of the two nuclei, despite their original differences in volume, are exactly the same.

Although details of fertilization differ strikingly throughout the plant and animal kingdoms, the essential features are remarkably uniform. In the process two nuclei derived from the two parents unite and mingle their substance in such a way that the chromosomes from each parent are equally represented in the two daughter nuclei produced by the first segmentation division, and through them in all the cells of the developing individual.

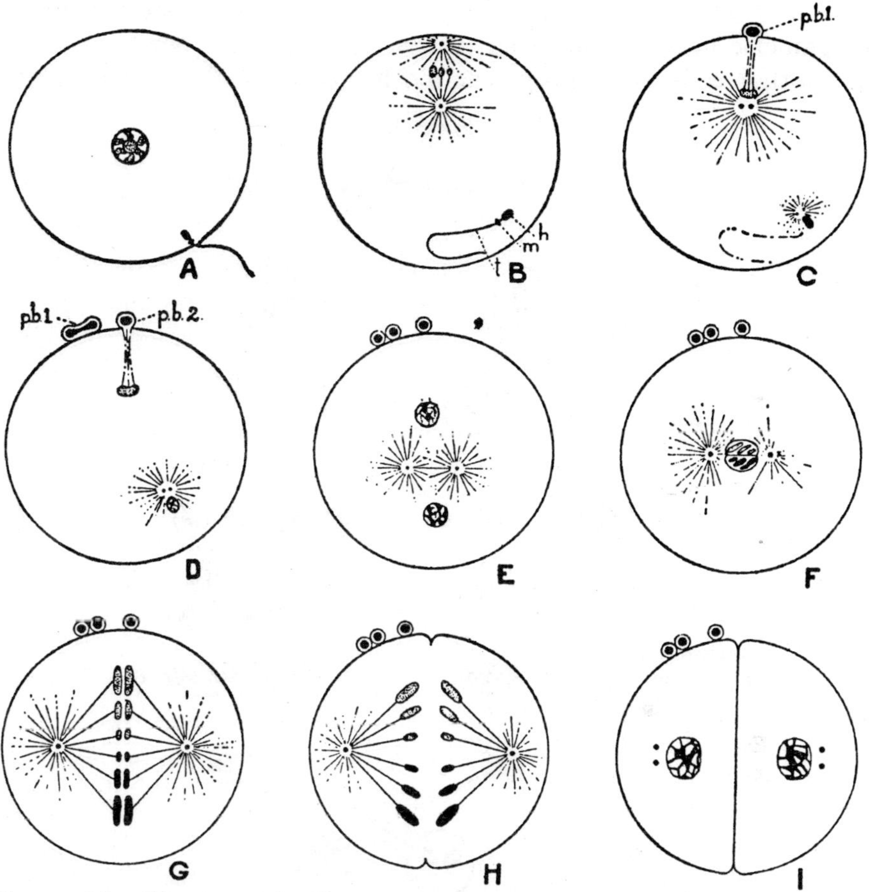


FIGURE 14.—Diagram of fertilization. *A*, entry of spermatozoön into the primary oöcyte. *B*, the female nucleus has moved to the surface and is in metaphase I. *C*, the female nucleus in telophase I. The tail of the spermatozoön has broken off and is degenerating. The head of the spermatozoön has rotated so that the two centrosomes which have disengaged from the middle piece now precede the male nucleus as it travels toward the center of the egg. *D*, the female nucleus in telophase II, first polar body dividing. *E*, approach of the male and female nuclei. *F*, the two nuclei, now of approximately equal size, and in late prophase, are in contact. *G*, the two nuclei have fused into a zygote nucleus which has reached the metaphase of the first cleavage accompanied by formation of spindle figure out of centrosomes and asters introduced by the spermatozoön. *H*, anaphase of first cleavage division of the zygote nucleus. *I*, two-celled stage.

*h*, head of spermatozoön (male nucleus); *m*, middle piece; *p.b.*1 and *p.b.*2, first and second polar bodies; *t*, tail of spermatozoön. (*Adapted from Agar's Cytology by permission.*)

**Reproduction in Plants.**—While gametogenesis in plants results in the production of gametes with the haploid number of chromosomes just as it does in animals, there are differences in detail dependent upon the fact that the nuclei which are formed as a result of the two meiotic divisions are not immediately concerned in fertilization but divide a number of times, giving rise to derivative nuclei which unite in the subsequent process of fertilization.

The differences may be appreciated by reference to figure 15 which contains a diagram of the processes involved in the ultimate production of the gametic nuclei in the seed plants. The process begins with the pollen mother cells and embryo-sac mother cells which are comparable

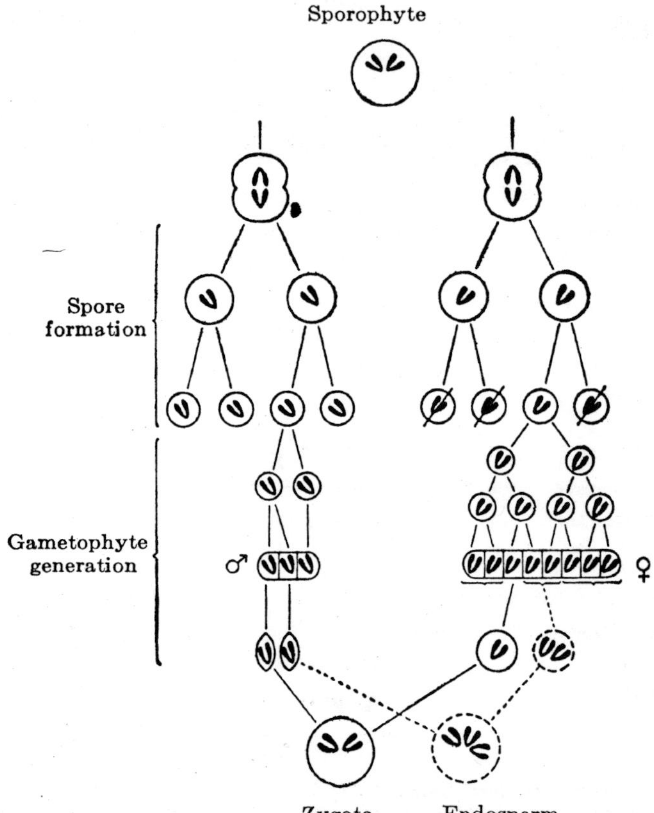


FIGURE 15.—Diagram illustrating the essential features of gametogenesis and fertilization in plants. (*From Mohr.*)

respectively to the primary spermatocytes and oöcytes. These pass through two meiotic divisions as in animals, giving rise in the case of the pollen mother cell to four functional microspores, each with the reduced number of chromosomes, and in the case of the embryo-sac mother cell to one functional megaspore and three cells, comparable to the polar bodies in animals, which degenerate.

The haploid nuclei thus formed subsequently divide a number of times prior to fertilization. In the microspores, the nucleus divides giving rise to two nuclei, a tube nucleus and a generative nucleus, which are both present in the mature pollen grain. When the pollen grain germinates, the generative nucleus passes down the pollen tube and again divides, giving rise to two nuclei which are the ultimate male gametes. In the megaspore there are three successive divisions giving rise to eight nuclei which are all contained in the enlarged embryo-sac (figure

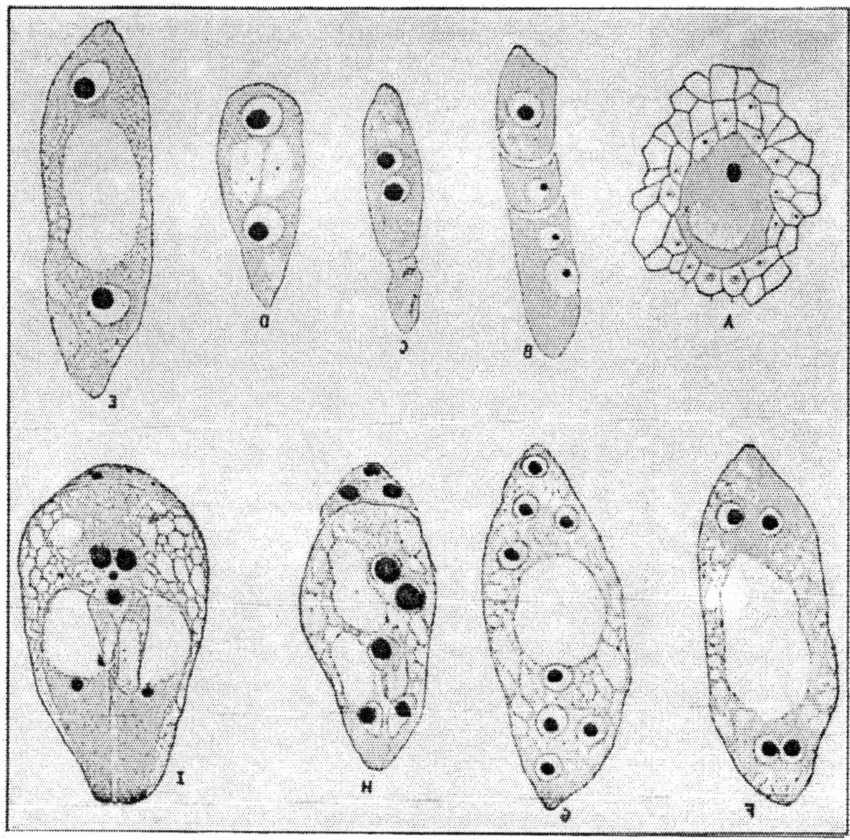

FIGURE 16.—Development of the embryo-sac, and fertilization in the squash. *A*, megaspore mother cell, which is situated in the middle of a very young ovule. *B*, the four spores which develop from the mother cell. Reduction in chromosome number takes place during the production of these cells, so that each possesses the haploid number of chromosomes. Three of these cells degenerate but the fourth, larger than the others, is the functional megaspore. *C*, the enlarged megaspore, in which the nucleus has now divided into two. *D*, a stage slightly later than *C*. *E*, a still later stage with the embryo-sac much increased in size. *F*, each of the two nuclei has divided into two. *G*, each of these four nuclei has again divided into two, producing eight nuclei in all. *H*, the embryo-sac, nearly mature. At the upper end (away from the micropyle of the ovule) are three antipodal cells. At the lower end (toward the micropyle) are the egg nucleus and the two synergids. The two endosperm nuclei in the middle have not yet fused. *I*, fertilization. One small male nucleus is about to unite with the endosperm nucleus (still double) and the other is about to fertilize the egg. (*After A. I. Weinstein, from Sinnott, Botany, Principles and Problems, copyright* 1923 *by the McGraw-Hill Book Company, Inc. Reproduced by permission.*)

16). Two of these nuclei, the *synergids*, take up a position at the end of the embryo-sac which the pollen tube enters; a third, the true female gamete, takes up a position near them; two others, the *polar nuclei*, move toward the center of the embryo-sac, and by fusion give rise to the *fusion nucleus;* and the remaining three, the *antipodals*, occupy the other pole of the embryo-sac. In fertilization one of the male gametes unites with the female gamete to form the new zygote; and the other unites with the double fusion nucleus to form the endosperm nucleus. In seed plants, therefore, there is a double fertilization; one fertilization giving rise to a diploid nucleus from which the new individual arises; the second gives rise usually to a triploid nucleus from which the endosperm develops. In spite of these differences in detail the new individual in plants, therefore, is comparable to the new individual in animals, for it arises from a single cell which is the product of union of a male and a female gamete each of which contains a haploid set of chromosomes.

The intervention of a number of nuclear divisions between the reduction divisions and the processes of fertilization is an interesting and distinctive feature in the life history of plants. The phase in the life history in which the cells have the diploid number of chromosomes is called the

FIGURE 17.—Diagram showing the characteristic pairing, size relations, and shapes of the chromosomes of *Drosophila melanogaster*. In the male an X- and a Y-chromosome correspond to the X-pair of the female. On the basis of X = 100 the length of each long autosome is 159, of each small autosome 12, of the whole Y 112, of the long arm of the Y 71, and of the short arm of the Y 41. (*From Bridges.*)

sporophyte, the intervening phase with the haploid number, the gametophyte generation. In the seed plants the gametophyte generation is reduced in the manner described above to a few cell divisions, but in lower forms the gametophyte generation becomes increasingly prolonged down the scale until in the mosses and liverworts it is the dominant phase of the life history. The regular recurrence of these two phases in plants is known as the *alternation of generations.*

**Inheritance of Sex in Drosophila melanogaster.**—In certain forms, particularly in animal species, there is a constant difference in the chromosome complex of the two sexes, which provides a splendid illustration of a somatic difference related to a cytological one. The chromosome differences in sex in *Drosophila melanogaster* are illustrated in figure 17. In the female there are four pairs of chromosomes, each of which consists of two exactly equivalent members. In the male, there are also four pairs of chromosomes, three of which are identical with those in the female.

but the fourth pair consists of two unlike members, one of which is like the two members of the corresponding pair in the female, but the other is morphologically different; it is longer than its mate and is J-shaped instead of straight. The members of the three pairs of chromosomes which are alike in the two sexes are called *autosomes;* those of the unlike pair are

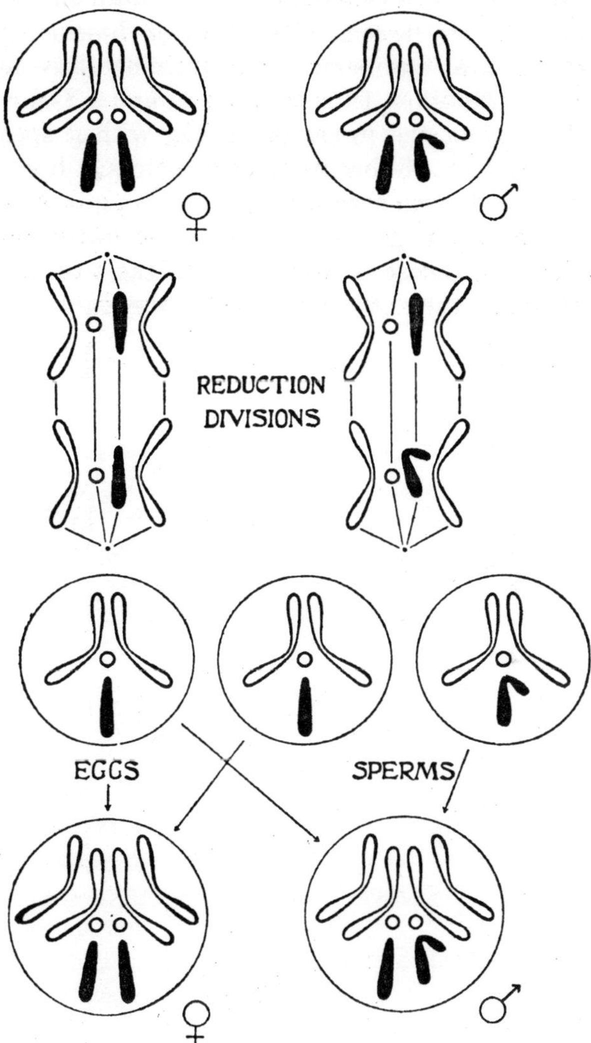


FIGURE 18.—Diagram to show chromosome relations in the inheritance of sex in *Drosophila melanogaster*. The sex-chromosomes are represented in full black in order to focus attention on them.

called *sex chromosomes.* The sex chromosomes of the female and the similar one in the male are called X-chromosomes, and the unlike member in the male is called the Y-chromosome.

The way in which the chromosome difference in the two sexes is preserved from generation to generation is shown diagramatically in figure

18. Beginning with the parents the diploid number is shown in the circles. In the female the three pairs of autosomes are outlined and the X-chromosomes are drawn in full black to focus attention upon them. Correspondingly in the male only the X- and Y-chromosomes are drawn in black. The reduction divisions in the female result in the production of eggs each of which contains two large curved and a small autosome and one X-chromosome. In the male the reduction divisions give rise to two kinds of sperm in equal numbers; one containing three autosomes and an X-chromosome; the other, three autosomes and a Y-chromosome. In the ensuing fertilization the union of an egg with a sperm containing three autosomes and an X-chromosome produces a female zygote; and the union of an egg cell with a sperm cell containing three autosomes and a Y-chromosome, a male zygote. Obviously the operation of this mechanism preserves the chromosome inequality of the two sexes, and at the same time accounts for their observed numerical equality.

**References**

See Chapter II.

## CHAPTER IV

## MENDELIAN METHOD

Before taking up an account of the results of genetic investigations, it is necessary to understand something about the methods of conducting breeding experiments. There are certain general principles involved in planning and conducting such experiments with plants and animals, a proper regard for which is essential to a satisfactory outcome of the work. These principles in turn depend upon certain features of variation which may properly be discussed in an elementary way at this point.

**Variation.**—It is a matter of common knowledge that individuals within a species are not all alike, but that they exhibit a variety of differences. If these differences are examined closely, it will be found possible in the simpler cases to resolve them into a limited number of character contrasts. A specific example may help to make this conception clearer. If, for example, grain characters in maize are considered, a great variety is found in differences in color, composition, and the like which at first sight may impress the student with a sense of utter confusion. But if a systematic study is made of them, it will be found possible to reduce them to a semblance of order by considering the character differences separately as alternatives. It will be found that the aleurone layer is either *colored* or *colorless;* if colored it is either *purple* or *red.* Likewise the endosperm color is either *yellow* or *white;* its texture is either *starchy* or *sugary.* This conception may be extended with proper modification to all the character differences existing in a given species. In genetics contrasts of this kind are studied. If all guinea pigs had black coat color, it would be impossible to study the transmission of coat color in them; but since, in addition to black, there are animals with yellow, agouti, chocolate, and other coat colors, it is possible to devise experiments which will show how these differences are preserved.

How these characters arise is a question which will be treated in subsequent chapters as fully as present knowledge permits. In many instances their origin has not been observed, and genetic investigation has, therefore, merely made use of existing variations. When large numbers of animals or plants are grown, however, occasional individuals are found which exhibit new characters; and from them it is usually possible to establish races which uniformly exhibit these characters. The spontaneous origin of new characters is known as mutation, and there is reason to believe that mutation primarily is responsible for the diversity existing within species. The remarkable series of characters of the fruit fly, *Drosophila melanogaster*, have practically all arisen as mutants in experimental cultures. Genetic investigation, then, employs both new mutant characters and existing variations of unknown origin but suspected to have arisen by mutation.

It is interesting to note at this point how the resolution of the diversities within a species into pairs of character contrasts simplifies the treatment of variation. An individual can possess only one member of each pair of characters, but it may possess either member of a given pair in combination with either member of any other pair. The problem of the number of possible combinations of characters which may be obtained in a species with a given number of pairs of characters is, therefore, mathematically the same problem which is met in the distribution of maternal and paternal chromosomes in the germ cells. The equation $x = 2^n$ gives the solution when $x =$ the number of possible combinations and $n =$ the number of pairs of characters. With ten pairs of characters, $x =$ 1,024; with twenty, $x =$ 1,048,576. In the snapdragon Baur recognizes the existence of over 150 such fundamental pairs of characters; in *Drosophila melanogaster*, Morgan and his associates have found about 400 of them. Such a treatment of variation obviously introduces at once a marked simplification of the elaborate diversity which occurs in some species. The important problem of genetics is that of resolving the chaos of diversity exhibited by species into order based on the operation of well-defined principles.

**Experimental Breeding.**—As a first conception it is necessary to realize that there are reasons behind all this diversity. Things do not just happen in genetics any more than they do in other branches of science. There is always a train of antecedent circumstances preceding any event. So in genetics in order to determine the reasons for a particular sort of behavior, it is necessary to study the events leading up to it. Now these antecedent circumstances are capable of simplification, and the successful outcome of breeding experiments depends upon careful control over them.

The first step in the simplification of a genetic problem is that of securing races constant for the characters which it is proposed to study. It is not difficult to obtain pure races; in principle their production depends merely upon selection of parents conforming to the desired type and rejection of all off-type individuals. This process continued through a number of generations will usually establish races which uniformly reproduce the desired characters. This preliminary step in genetic investigations is comparable to the purification of substances necessary before performing chemical experiments. The methods of purifying races may be refined in certain ways, so as to reduce the amount of time necessary to secure them, but the principle involved nevertheless remains the same. In plants which may be self-fertilized, the production of constant races is more readily effected than in animals where the germ cells must come from different individuals; but introduction of refinements of technique, which will become clear later, makes it possible to overcome this handicap in animal experimentation.

Domesticated animals and plants have been favorite subjects in breeding experiments for two reasons. In the first place they usually exhibit a high degree of diversity, probably because the necessities of economic employment have led to the preservation of a large number of characters to meet the variable conditions and requirements to which they are subjected. The second reason is even more important, namely, that they already exist in reasonably constant races. A measure of purity is necessary in practical operations. Breeders of animals and growers of plants object to mixtures, for very often they result in economic loss. The seedsman whose seeds do not produce plants conforming to the advertised standards is in danger of losing his reputation and his trade. It is often possible, therefore, to use cultivated varieties of plants and domestic breeds of animals directly in genetic investigations, because they have already been subjected to the process of purification.

When the geneticist is finally in possession of constant races, he is ready to begin his experiments. Here again in planning the experiments the utmost simplification compatible with a proper solution of the problem is necessary. If it is desired, for example, to study the transmission of two contrasted characters, two races are chosen differing only in the two selected characters. As a first step these races are hybridized and observations are made on the characters of the hybrids, particularly as to uniformity and relation to the characters of the parents. As a second step the progeny of the hybrids is studied. The individuals are divided into groups or classes depending on the characters which they exhibit; the individuals in each class are counted and the number recorded; then their characters are compared with those of the parents and of the hybrid. The members of this second generation are further tested, if possible, individually, and their progenies are separately classified and enumerated; and so on until a sufficient body of data has been accumulated. Here again the greater ease of manipulation of some plant material should be noted because, if self-fertilization is possible, individuals may be tested separately whereas in animals it is necessary to mate two individuals together in order to secure progeny; but the progress of genetics has resulted in the development of modifications of the above described technique which now make it possible to conduct investigations with animals quite as expeditiously as with plants.

A shorthand set of symbols has been generally adopted by geneticists to distinguish the various generations in the above described method of experimentation. The symbols ♀ = female and ♂ = male are used to designate the sex of individuals. The original parents are denoted by the symbol $P_1$; the hybrid between them, $F_1$, which is read, first filial generation; the progeny of $F_1$ is $F_2$, which is read, second filial generation, and so on.

The data secured from such an experiment are finally analyzed in order to determine the principles in operation in it and these principles are stated in the form of laws. If different interpretations are still possible, it is the task of the investigator to devise further experiments on which to base a further choice; if the data are not sufficient to establish the conclusion absolutely, further investigations should also be designed to supply the missing evidence.

Solution of the simplest cases serves as a basis for attack on the more complex ones. When the difference between two parents differing in one pair of contrasted characters has been studied and solved, it is possible to go on to the solution of problems in which the parents differ in two pairs of characters, or three, or four, and it will then be found possible to determine the various interrelations of different sets of characters. In this way as a final goal it may sometime be possible to fit every character difference within a species into the scheme and to present an orderly account of its entire genetic variability.

The above description of the method of conducting breeding investigations applies to the simplest conditions. The geneticist is constantly called upon to exercise his ingenuity in meeting unusual situations which call for appropriate modifications of these methods. Sometimes breeding experiments will not of themselves provide sufficient information to solve the problem; then recourse must be had to other sources of evidence, particularly to cytological examination. In genetics, as in other sciences, it is the exception, the instance of failure to conform to known principles the investigation of which throws a flood of new light on the biological processes at work in heredity. Just as such instances are exceptional, so the investigator must be prepared to devise original methods of meeting them. Behind all the discussion in this text lie experiments, a full account of which is impossible, but it will be exceedingly helpful for the student to try to picture in his own mind the details connected with the actual prosecution of them.

**Mendel's Experiments.**—The method which has just been described is due primarily to an Augustinian monk, Gregor Johann Mendel (1822–1884), of the House of St. Thomas, or Königskloster, at Brünn, Austria, of which he eventually became Abbot or Prälat. Mendel spent two years (1851–1853) at the University of Vienna studying mathematics, physics, and the natural sciences. After his return to Brünn he became a teacher of natural science in the Staatsrealschule in Brünn; and it was during this period that he conducted his experiments on hybridization of garden peas. In connection with the monastery there was a garden and greenhouse which Mendel used in his experiments. The results of the work with peas were read in February and March, 1865, before the Natural History Society of Brünn and were published in the *Memoirs* of that society in 1866. They apparently attracted no attention whatever at the time of publication, and even a subsequent correspondence between Mendel and the renowned botanist, Carl Nägeli, himself an enthusiastic student of heredity, failed to bring them a proper recognition. In 1900 three other botanists, de Vries, Correns and Tschermak, who had been independently performing experiments of the same kind, called attention to the paper, thus revealing it to a scientific world ready to appraise it properly and to exploit the field which it opened. Mendel's papers, originally published in German, have been since republished and translated into several other languages. An English translation is available in the appendix to Bateson's "Mendel's Principles of Heredity."

The student should read Mendel's paper in order to realize how fully he comprehended the experimental requirements of a study of heredity and how logically he arranged his investigations in conformance with them. He himself states in his introduction:

> Those who survey the work done in this department will arrive at the conviction that among all the numerous experiments made, not one has been carried out to such an extent and in such a way as to make it possible to determine the number of different forms under which the offspring of hybrids appear, or to arrange these forms with certainty according to their separate generations, or definitely to ascertain their statistical relations.

The three requirements which Mendel enumerates are fundamental to a successful prosecution of breeding experiments, and he kept them constantly in mind in choosing the particular plant with which to carry on the work and in performing the experiments with it.

After a survey of the plant kingdom he was attracted to the garden pea as a suitable object for his work, particularly because it existed in a number of distinct varieties and because its floral structure and mode of pollination ensured the production of seeds by self-fertilization only. He obtained seeds of thirty-four different varieties and grew them for a period of two years preceding the beginning of the experiments. He found they remained perfectly constant during this period and during the whole period of the experiments as well. Twenty-two varieties were selected upon which to base the experiments, because they possessed desirable combinations of seven pairs of differentiating characters which Mendel chose as most suitable for study by reason of their definiteness and ease of determination.

The characters selected were as follows:

1. Seed shape: either *smooth* or *wrinkled.*
2. Cotyledon color: either *yellow* or *green.*
3. Seed coat color: either *colored* with *purple* flowers or *colorless* with *white* flowers.

4. Pod type: either *hard,* as in most garden varieties or *soft* as in the kinds with edible pods.

5. Pod color: either *green* or *yellow,* with a like coloring in stalk, leaf, vines, and calyx.

6. Flower position: either *axial,* distributed along the main stem in the axils of the leaves, or *terminal,* borne in a false umbel at the top of the stem.

7. Stature: either *tall,* 6 to 7 feet, or *dwarf,* ¾ to 1½ feet.

Mendel first determined how these pairs of characters behaved when taken singly. As an example, he took two parents which differed only in stature and hybridized them, using the tall parent as female and the dwarf parent as male in some crosses and the reverse, or reciprocal, in others. The $F_1$ plants were uniformly tall, in fact slightly taller than the tall parent.

Similarly, taking all the other pairs of characters separately, Mendel determined that in every case one member of each pair of characters was expressed in the hybrid in the same form as it appeared in the parent; the other member of the pair disappeared completely from view, so that no effect of it could be detected at all. The character expressed in the hybrid, Mendel termed *dominant;* the one which disappeared from view, he called *recessive.* The results of the seven experiments were as follows:

| | $P_1$ | $F_1$ |
|---|---|---|
| 1. Seed form: | smooth × wrinkled | = smooth |
| 2. Cotyledon color: | yellow × green | = yellow |
| 3. Seed coat: | colored × white | = colored |
| 4. Pod type: | hard × soft | = hard |
| 5. Pod color: | green × yellow | = green |
| 6. Flower position: | axial × terminal | = axial |
| 7. Stature: | tall × dwarf | = tall |

Mendel next grew $F_2$ progenies from the self-fertilized $F_1$ and found that in every case $F_2$ consisted of two classes of individuals, one of which exhibited the dominant character and the other the recessive, both in the same form as they were shown in the parent races, without any transitional forms whatsoever. The number of individuals in each class was enumerated with results contained in table 1. It may be seen from these results that in every experiment the ratio of dominant to recessive was nearly 3:1, the normal fluctuations characteristic of statistical experiments giving in some cases a ratio of dominants slightly in excess of 3:1 and in others a ratio slightly lower. The results taken all together are almost exactly 3:1.

From these results Mendel decided that all these character contrasts were transmitted according to one fundamental law. It has already been noted how great a simplification in the treatment of variation was effected by resolving the variables into a number of pairs of character contrasts. Here a further simplification is effected by determining that all these character contrasts follow one fundamental law in their transmission.

But the experiments were not yet complete. From the results thus far obtained Mendel decided that the dominant character had a double signification; it might be either a parental character, in which case as in the parents it transmitted only the parental condition, or a hybrid character, in which case as in the hybrid it produced a progeny consisting of 3 dominant:1 recessive. The only way to determine whether an $F_2$ dominant was of the parental or hybrid type was to test it by growing its progeny separately in the next following generation. He, therefore, tested out a number of the dominant $F_2$ individuals in order to determine the ratio of pure dominants to hybrid dominants. The results are given in table 2.

TABLE 1.—$F_2$ RESULTS OBTAINED BY MENDEL IN HIS FIRST SEVEN EXPERIMENTS

| Character pair | Dominant | Recessive | Ratio |
|---|---|---|---|
| 1. Seed form | 5,474 smooth | 1,850 wrinkled | 2.99:1.01 |
| 2. Cotyledon color | 6,022 yellow | 2,001 green | 3.00:1.00 |
| 3. Seed coat color | 705 colored | 224 white | 3.04:0.96 |
| 4. Pod type | 882 hard | 299 soft | 2.99:1.01 |
| 5. Pod color | 428 green | 152 yellow | 2.95:1.05 |
| 6. Flower position | 651 axial | 207 terminal | 3.03:0.97 |
| 7. Stature | 787 tall | 277 dwarf | 2.96:1.04 |
| Totals | 14,949 dominant | 5,010 recessive | 3.00:1.00 |

TABLE 2.—$F_3$ RESULTS OBTAINED BY MENDEL BY TESTING $F_2$ INDIVIDUALS OF THE DOMINANT CLASS IN EACH EXPERIMENT SEPARATELY

| Character pair | Pure dominants | Hybrid dominants | Ratio |
|---|---|---|---|
| 1. Seed form | 193 | 372 | 1.02:1.98 |
| 2. Cotyledon color | 166 | 353 | 0.96:2.04 |
| 3. Seed coat color | 36 | 64 | 1.08:1.92 |
| 4. Pod type | 29 | 71 | 0.87:2.13 |
| 5. Pod color | 40 | 60 | 1.20:1.80 |
| 6. Flower position | 33 | 67 | 0.99:2.01 |
| 7. Stature | 28 | 72 | 0.84:2.16 |
| Totals | 525 | 1,059 | 0.99:2.01 |

From this further extension of the experiments it may be seen that the dominants of $F_2$ can be separated into two classes according to the kinds of progeny which they produce in $F_3$. One class breeds true for the dominant character and its descendants also breed true; the other class produces both dominants and recessives in the ratio found in $F_2$, namely 3 dominant: 1 recessive. The ratio of these two classes of dominants in $F_2$ is further shown to be that of 1 pure dominant:2 hybrid dominant in all the experiments. The ratio fluctuates more widely in the individual experiments because of the smaller numbers of individuals in these investigations; but the totals of all the experiments together are in quite as good agreement with a 1:2 ratio as those of $F_2$ are with a 3:1 ratio. It is thus finally shown that the $F_2$ ratio of 3 dominant:1 recessive must be resolved into the ratio of 1 pure dominant:2 hybrid dominant:1 recessive.

The recessives can be pure only for the recessive character, as Mendel determined by growing progeny from the $F_2$ recessive individuals. All the progeny in such cases exhibited the recessive character. All of the different pairs of characters with which Mendel worked conformed to this one given formula, so that the operation of one law in all seven cases was demonstrated.

The explanation of these results was very simple. Mendel simply assumed that the germ cells of the various kinds of plants differed in elements corresponding to their character differences. If in a given experiment, the dominant character is represented by $A$ and the recessive by $a$, then the germ cells—ovules and pollen grains—of the plants can be represented by the same symbols. Hybridization between dominant and recessive can then be symbolized as follows:

$$A \times a = Aa$$

and $F_1$ individuals develop the character which is dominant. Although they develop the dominant character, they form germ cells of two kinds in equal numbers, namely $A$ and $a$, which are present in both pollen grains and ovules. When these pollen grains and ovules form new individuals there are four possible ways in which they may unite, as follows:

1. Ovule $A$ × pollen grain $A$ = $AA$ – pure dominant.
2. Ovule $A$ × pollen grain $a$ = $Aa$ – hybrid dominant.
3. Ovule $a$ × pollen grain $A$ = $aA$ – hybrid dominant.
4. Ovule $a$ × pollen grain $a$ = $aa$ – recessive.

The four types of union are equally frequent; consequently the progeny are distributed as required by the experimental evidence in the ratio 1 pure dominant: 2 hybrid dominant:1 recessive.

From this analysis it is apparent that the paired characters of a group of individuals are represented by paired elements in the germ cells, that there are as many paired elements as there are pairs of characters, and that these elements are not affected by hybridization, for when they are brought together in a hybrid, they separate out in its germ cells in their original pure condition.

Mendel's first seven experiments have been reviewed in order to show how the Mendelian method of analysis, as it is called, was applied in the first instance. Mendel performed many other experiments in order to determine how transmission occurred when more than one character difference was present in the original parents and found that in such cases each pair of characters still behaved as above described, and that the increasing complexity of the results depended purely upon the statistical laws which are operative when several pairs of alternatives are present simultaneously in the experiment instead of one. He also subjected the conception that the number of possible combinations of the members of these seven pairs of characters is equal to $2^7 = 128$ to experimental verification by actually producing the required 128 different kinds of garden peas.

### References

Bateson: Mendel's Principles of Heredity, Part II.
Castle: Genetics and Eugenics, Appendix.
Punnett: Mendelism, Chapters I–III.

# CHAPTER V

## SEGREGATION

The preceding chapter has shown how Mendel applied an analytical method to problems of heredity. Essentially his contribution to methodology depended upon a preliminary simplification of the problem followed by hybridization between races exhibiting a limited number of character differences and a systematic study of successive generations of progeny.

When the original parents are constant and differ in one pair of characters, for example tall *vs.* dwarf stature in peas, the results obtained by crossing them may be diagrammed as shown in figure 19. $F_1$ consists entirely of tall plants. $F_2$ consists of tall and dwarf plants in the ratio 3 tall: 1 dwarf. When the dwarf $F_2$ plants are tested by growing their progeny, they are found to be pure for that character. The progeny is all dwarf, and all their descendants, in spite of their hybrid derivation, are dwarfs, On the other hand, tests of the tall $F_2$ plants individually disclose the fact that they are of two kinds: pure talls, which produce only tall offspring, and hybrid talls, which produce both tall and dwarf plants in the ratio 3 tall: 1 dwarf. The subsequent behavior of these two classes is also shown in the diagram.

| | | | | | |
|---|---|---|---|---|---|
| $P_1$ | | tall | × | dwarf | |
| $F_1$ | | | tall hybrids | | |
| $F_2$ | 1 tall | | 2 tall hybrid | | 1 dwarf |
| | ↓ | | | | ↓ |
| $F_3$ | tall | 1 tall | 2 tall hybrid | 1 dwarf | dwarf |
| | ↓ | ↓ | | ↓ | ↓ |
| $F_4$ | tall | tall | 1 tall: 2 tall hybrid: 1 dwarf | dwarf | dwarf |

FIGURE 19.—Diagram illustrating the results of hybridization of tall with dwarf peas. It is understood that $F_2$ is secured by self-fertilization of $F_1$ plants; $F_3$, by self-fertilization of $F_2$ plants, the progenies of which are grown separately, etc.

The explanation for this behavior depends upon the composition of the tall hybrids and upon the kinds of germ cells which they produce. The visible differences between races of plants and animals are assumed to depend upon differences in the elements contained in their cells. These elements of the germinal material are called *factors.* When two races differ in a single pair of characters, such as tall and dwarf, the germ cells which they produce likewise differ in a single pair of factors. The factors are usually represented by symbols, like chemical elements. In this instance the factor for tall is represented by $T$, and that for dwarf by $t$, in order to indicate the paired relation existing between them.

When tall and dwarf peas are hybridized, the germ cells from the tall race contain *T* and those from the dwarf race, *t* (figure 20). The cells of the resulting tall hybrid, therefore, are of the composition *Tt*. They contain both the factor for tall and that for dwarf, but they develop into tall plants, because *T* is dominant over *t*. When such plants in turn form germ cells, the two factors separate into different germ cells, so that some contain *T* and some *t*. The number of germ cells of each kind is

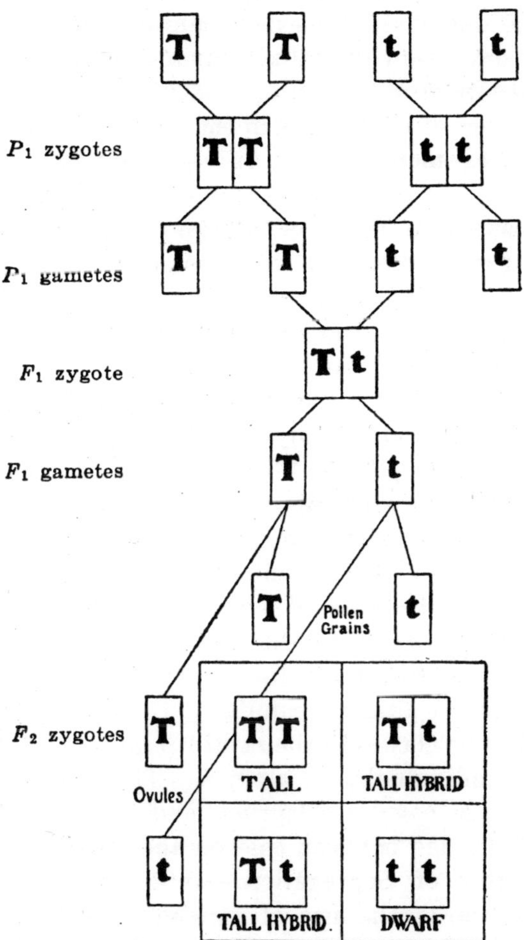


FIGURE 20.—Diagram of the factor analysis of a cross between tall and dwarf peas.

equal, and the two kinds are represented in both the pollen grains and ovules, so that their union to give the $F_2$ progeny results, as has been seen, in the $F_2$ ratio 1*TT* (pure tall): 2*Tt* (hybrid tall) : 1 *tt* (dwarf). The pure talls breed true because their germ cells all contain *T*; the dwarf plants likewise breed true because their germ cells all contain *t*; and the tall hybrids reproduce the $F_2$ ratio because they produce germ cells in the ratio 1*T*: 1*t*. The separation of the members of a pair of factors in germ-cell formation is called *segregation*. This phenomenon is the basis of Mendel's first law, the **Law of Segregation,** which may be stated thus:

*The members of a pair of factors separate into sister gametes in germ cell formation.*

It will be noted that the law is stated in general terms, so that it applies not only to hybrids in which the members of the pair of factors are different, but also to constant races in which the members are alike. Thus segregation occurs in a pure tall plant $TT$, separating $T$ from $T$, but it cannot be demonstrated experimentally because both germ cells are alike. It is important to realize that segregation is a normal process which takes place in germ-cell formation, and that it is not set in operation by the existence of differences in the pairs of factors. The presence of such differences merely gives an opportunity to demonstrate its occurrence.

**The Backcross.**—The fact that segregation occurs in a 1: 1 ratio may be demonstrated experimentally by crossing a hybrid with its recessive parent. If a tall hybrid pea is crossed with dwarf, the progeny consists of equal numbers of tall and dwarf plants, a result which obviously depends upon the fact that the tall hybrid produces gametes containing $T$ and $t$ in equal numbers, for the dwarf plant produces gametes bearing $t$ exclusively. This method of analysis is known as backcrossing. It is an important addition to experimental methods, for by its use it is possible to work as expeditiously with animals as with those plants which may be self-fertilized. Thus in guinea pigs, albinism is a simple recessive to colored. If a member of a constant colored race is mated to an albino, $F_1$ is colored, and $F_2$ consists of 3 colored: 1 albino. The $F_1$ colored individuals can be shown to produce two kinds of germ cells by backcrossing them to the recessive albino parent, whereupon the progeny consists of 1 colored: 1 albino. Similarly, the $F_2$ colored individuals can be shown to be of two classes in the ratio 1 pure colored: 2 hybrid colored, by mating them individually to albinos. The pure-colored individuals will then give only colored offspring, and the hybrid ones will give progeny equally divided between colored and albino classes.

**The Mechanism of Segregation.**—Inquiry may be made at this point as to what the mechanism of segregation is, what cell organs are involved, and how they operate. The connection of segregation with known cellular phenomena is not a difficult matter. The mere description of a simple Mendelian experiment shows at once a close analogy between cell history and factor transmission. Thus, as has been pointed out in the account of development and reproduction, the cells of the individual have a double or diploid number of chromosomes. In germ-cell formation the members of the pairs of chromosomes conjugate; thereafter a reduction division ensues, and eventually germ cells are produced which contain a single or haploid set of chromosomes. The process of fertilization restores the diploid number of chromosomes. So also in a Mendelian sense, the individual contains two members of each pair of factors; the members of the pairs of factors separate in germ-cell formation, so that germ cells contain only a single member of each pair of factors, and finally fertilization reconstitutes the double condition of the individual with respect to its factors. It should be clear from this comparison of the history of chromosomes in the life cycle of the individual and of factors in a Mendelian experiment that the chromatin system is the mechanism of segregation. If the factors are borne in the chromosomes; then the phenomena which have been found typical of a Mendelian experiment follow as a matter of course. In a tall hybrid pea, when *T* separates from *t* in germ-cell formation, what happens in terms of the chromatin mechanism is that the chromosome bearing *T* separates from the one bearing *t* in the reduction division.

It should be understood what the implications of this conception are. In order that it may work out consistently, a given factor must not only be contained in a definite chromosome, but it must also occupy a specific position or *locus* in that chromosome. It has been seen that prior to reduction homologous chromosomes conjugate and that during conjugation there may be an exchange of materials between homologous chromosomes. Now if the members of a pair of factors are borne by homologous chromosomes, then they must be so located that exchanges of material between homologues will never bring them into the same chromosome. Since the exchanges of material between homologues must obviously involve exactly corresponding portions of the conjugating chromosomes, this condition is satisfied if the members of a pair of factors occupy corresponding positions in homologous chromosomes.

The theory that the chromosomes bear the factors and that the maneuvers of the chromatin mechanism are responsible for the distribution of the factors is called the *chromosome theory of heredity*. As additional developments of Mendelian studies are presented, an attempt will be made to show how they can be interpreted in terms of the known behavior of the chromatin material. It is interesting at this point to realize that Mendel advanced his interpretation long before biologists had any knowledge of the existence of the elaborate cellular mechanism which cytological studies have disclosed.

**References**

DARBISHIRE: Breeding and the Mendelian Discovery, Chapters II, III, V.

MORGAN: The Physical Basis of Heredity, Chapters I–III.

MORGAN: The Theory of the Gene, Chapters I–IV.

MORGAN, MULLER, STURTEVANT, and BRIDGES: The Mechanism of Mendelian Heredity, Chapter I.

SINNOTT and DUNN: Principles of Genetics, Chapter III.

## CHAPTER VI

## MENDELIAN CONCEPTIONS

The law of segregation may appear to be a simple matter, but it leads to some very fundamental conceptions which the student should understand thoroughly before he moves on to more complex things. It has been pointed out that the diversity which occurs within a species may be reduced to a limited number of character contrasts or pairs of differences. Aside from a few differences in plants which depend upon the color of plastids or other contents of the cytoplasm (*cf.* Chapter XXX), all these variations are distributed according to the law of segregation, which is operative whether there are any differences or not.

**The Factors.**—Since the distribution of heritable differences is governed by the law of segregation, it is concluded that the germinal material is composed of more or less independent elements which are transmitted as units in heredity. For brevity this conception is called the particulate or, more appropriately, the Mendelian theory of heredity.

These elements of the germinal material are variously termed genes, determiners, Mendelian factors, or simply factors. The term gene is often preferred because it is a technical term invented expressly for the purpose. It was introduced by Johannsen and defined as an internal condition or element of the hereditary material upon which some morphological or physiological condition of the organism depends. In this text the term factor is most frequently employed, since this word has a meaning in English which is descriptive of the part these elements play in development. Nothing is known as to the material nature of factors; their existence is merely inferred from the effects which they exert upon the individual. Aside from their mode of transmission, it is known only that they are autocatalytic; *i.e.*, they have the power of self-propagation. It is not known whether or not they are living elements; they might conceivably be rather simple chemical substances.

Factors are highly constant, but not absolutely so, a matter which comes up for discussion later. Here it is sufficient to note that selection of grades of expression of a character for numerous successive generations has been shown to have no effect upon the factors. Moreover, the intimate association of pairs of contrasted factors in the same cell in the hybrid condition, as in peas of the constitution *Tt*, has been shown to result in no contamination of the factors. They emerge by segregation unchanged. Hybrid races have been kept in this condition for as many as seventy-five consecutive generations without showing any effect of the factors upon each other. Rarely, however, factors do change for reasons not as yet understood. Such changes are called factor mutations. They appear to involve a qualitative alteration, which is discontinuous like the change in composition of a chemical compound. Further information may be found in Chapter XXII.

**Factor Symbols.**—Factors, like chemical elements, are represented by symbols, usually one or two letter abbreviations of one member of the pair of characters with which they are connected. In order to simplify representation and to facilitate solution of problems, the same symbol is used for both members of a pair of characters; the dominant member being capitalized. For example, tall *vs.* dwarf stature in peas are contrasted characters; the dominant member, tall, is represented by $T$; the recessive member, dwarf, by $t$. Usage favors the consistent employment of the same symbols to represent a given pair of factors; but a few investigators prefer to use arbitrary symbols, which are employed simply for the purpose of solving problems and which may be changed from problem to problem. There are many other deviations in practice in this respect; but they should not confuse the student because they are mere details. What he should endeavor to do is to determine the significance of the symbols in a given account and then to operate with them in conformance with Mendelian principles.

**Allelomorphism.**—Gametes normally contain only one member of a pair of factors; consequently the members of a pair of factors are called *allelomorphs.* Allelomorphism is defined as a relation between two factors such that they are separated into sister gametes in germ-cell formation. This term by analogy is extended to the contrasted characters for which the allelomorphic factors are differentials. The members of a pair of characters are also said to be alternative, because every individual exhibits either one or the other of them. They are sometimes called unit characters, because they behave as units in heredity. This term, however, is falling into disrepute, because it leads to difficulties in complex cases. It is better to focus attention upon factors, not upon characters, in problems of heredity.

**The Classification of Individuals.**—It has also been seen that individuals may look alike and yet differ in genetic constitution, as for example, pure tall and hybrid tall peas. Individuals may, therefore, be classified on the basis of the characters which they exhibit or on the basis of genetic constitution. The classes into which individuals are separated on the basis of their visible characters are termed *phenotypes.* A phenotype is defined as the externally obvious characters of an individual or group of individuals. The classes into which individuals are separated on the basis of genetic constitution are termed *genotypes.* A genotype is defined as the constitution of an individual or group of individuals in terms of factors. A distinction must be made between genotype and phenotype. Thus all the tall peas in $F_2$ in the example described above belong to one phenotype, tall; but they are of two genotypes, *TT* and *Tt*. A genotype is said to be *homozygous*, if the two members of each pair of genes are alike; if they are different, it is *heterozygous*. *TT* and *tt* are homozygous genotypes, but *Tt* is heterozygous. Individuals which are homozygous breed true; those which are heterozygous normally do not. An individual may be homozygous for some pairs of factors and heterozygous for others, so that it is often necessary to qualify the description of its genetic constitution by a specific statement as to the factors under consideration.

The distinction between genotype and phenotype is one of far-reaching practical significance. In practical operations it is one of the chief desires of animal and plant breeders to establish races which will transmit their characters, if possible, uniformly to all their progeny. The ability to transmit depends upon the genotype; the only method of determining the genotype is through observation of the type of progeny which the individual gives. Consequently, the emphasis which is placed upon the progeny test as a measure of an individual's value from a breeding standpoint.

The demonstration of the existence of a pair of factors in a species, such for example as *T-t* in peas, carries with it the implication that every individual of the species normally must contain any two members of the pair or series of factors. Thus any garden pea plant must be *TT*, *Tt*, or *tt* in genetic constitution, and so for any other pair of factors. Usually, however, in genetic formulae no attempt is made to include all demonstrated factors; only those in which the parents exhibit allelomorphic differences are represented.

The student should not permit the unfamiliar terms which are used in genetics to confuse him. They are not numerous, and their use is necessary for accurate and expeditious treatment of genetic problems. Their meanings should be clearly understood and their employment should be restricted to conform to their several definitions.

**References**

General textbooks, and see Chapters I and V.

# CHAPTER VII

## SEX LINKAGE OR HETEROSOMAL INHERITANCE

In the instances which have been described in Chapter V sex need not be considered, because it is inherited independently of the characters in question. In hermaphroditic plants like peas, of course, all the individuals are alike in this respect; but in animals, where sexual differentiation occurs, the sex may or may not show some relation to the characters under discussion. In colored *vs.* white in guinea pigs, the two $F_2$ classes are approximately equally divided between males and females, and the same statement applies to the genotypes; consequently no account is taken of sex in the discussion. In certain instances, however, a consistent analysis can only be obtained by considering the sex as well as the character distribution.

**Red vs. White Eye Color in Drosophila.**—In Drosophila red and white eye colors constitute a pair of Mendelian characters with red dominant, but the transmission of the characters bears a consistent relation to sex as may be seen from the following account.

When crosses are made between red- and white-eyed flies, the results differ according to the direction of the cross as shown in the following scheme:

| | | |
|---|---|---|
| $P_1$ | red ♀ × white ♂ | white ♀ × red ♂ |
| $F_1$ | red ♀♀ + red ♂♂ | red ♀♀ + white ♂♂ |
| $F_2$ | 2 red ♀♀ :1 red ♂:1 white ♂ | 1 red ♀:1 white ♀:1 red ♂:1 white ♂ |

The second type of cross, white ♀ × red ♂, which produces an $F_1$ with the characters reversed in relation to sex is known as crisscross inheritance.

In Chapter III it was shown that the inheritance of sex in Drosophila depended upon a difference in one pair of chromosomes in the two sexes. The female possesses two X-chromosomes, constitution XX, and the male one X paired with an unequal chromosome called the Y-chromosome, constitution XY. In Chapter V it was shown that transmission may be accounted for on a mechanistic basis by assuming that a particular pair of factors occupies corresponding positions in a particular pair of chromosomes. In the present instance, the results may be explained consistently, if two assumptions are made: (1) that the factors for red and white are borne by the X-chromosomes; and (2) that the Y-chromosome is neutral in development; *i.e.*, it carries no factors which influence the development of these characters. In the male, therefore, the eye color will be determined by the factor contained in its single X-chromosome.

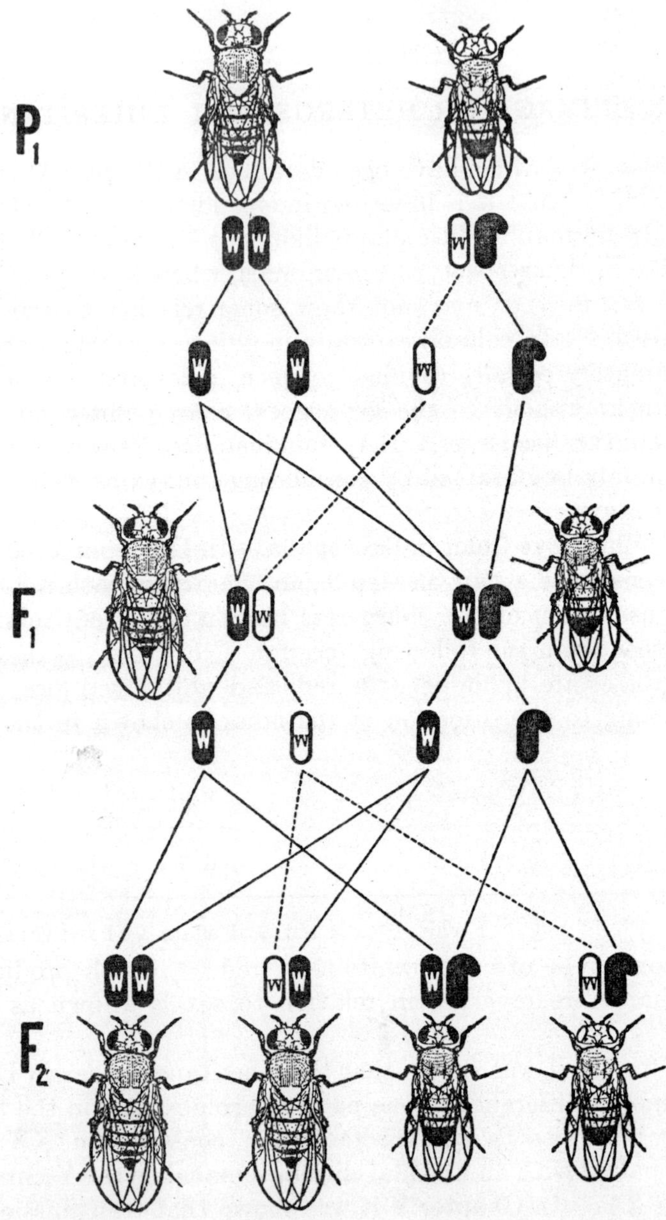


FIGURE 21.—Inheritance of white eye color in Drosophila. Red-eyed female mated to white-eyed male. Solid lines indicate history of chromosomes of $P_1$ female and the Y-chromosome; dotted lines, history of X-chromosome of the $P_1$ male. (*Adapted from Morgan.*)

The operation of these assumptions is shown by the diagrams contained in figures 21 and 22. The factor for white eye color is represented by $w$, that for red by $W$, and these factors are assumed to occupy corresponding positions in the X-chromosomes.

In figure 21, the two X-chromosomes of the red-eyed female each contain the factor *W* for red eye color. The constitution of such a fly

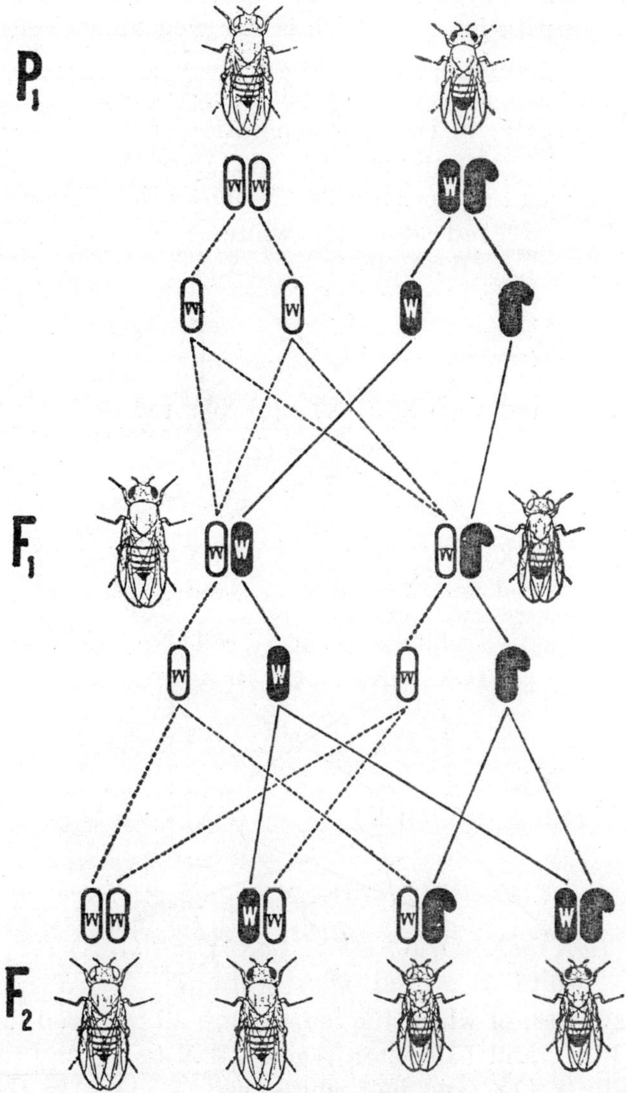


FIGURE 22.—Inheritance of white eye color in Drosophila. White-eyed female mated with red-eyed male. Dotted lines indicate history of chromosomes of $P_1$ female; solid lines, of the $P_1$ male. (*Adapted from Morgan.*)

may be represented symbolically as (*W*X)(*W*X), which simply means that the X-chromosomes bear *W* factors. Each egg from such a female contains one X-chromosome bearing a factor for red eye color, in short-hand notation (*W*X). On the other hand, the white-eyed male (*w*X)Y produces two kinds of sperm cells in equal numbers (*w*X) and Y. An egg cell (*W*X) from the red-eyed female fertilized by an X-bearing sperm cell (*w*X) gives an $F_1$ zygote (*W*X)(*w*X), which is a female because it has two X-chromosomes and is red eyed because $W$ is dominant over $w$. An egg cell ($W$X) from the red-eyed female fertilized by a Y-bearing sperm, Y, gives an $F_1$ zygote ($W$X)Y, which is a male because of its XY-chromosomal constitution and which is red eyed because of the $W$ factor present in its X-chromosome. In the $F_1$ female the two X-chromosomes separate in reduction, so that two kinds of egg cells are produced, ($WX$) and ($w$X). Similarly, in the male separation of the X- and Y-chromosomes gives two kinds of sperm cells, ($W$X) and Y. The four possible combinations of egg and sperm cells are shown in the figure to give four

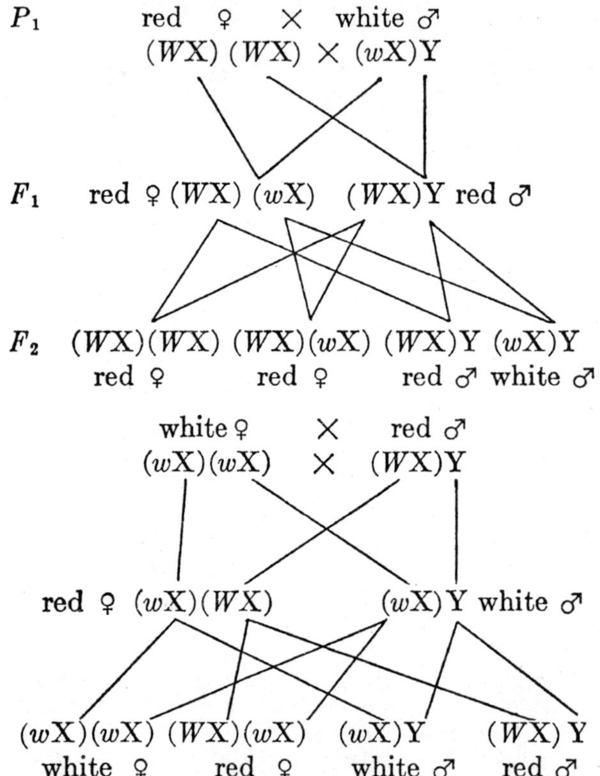


kinds of $F_2$ zygotes, of which the females are all red eyed, half homozygous ($W$X)($W$X), and half heterozygous ($W$X)($w$X); and the males are half red eyed ($W$X)Y, and half white eyed ($w$X)Y. In the reciprocal cross (figure 22) the white-eyed female is ($w$X)($w$X). All her eggs, therefore, are ($w$X). The red-eyed male is ($W$X)Y; consequently he produces two kinds of sperm, ($W$X) and Y. An egg ($w$X) fertilized by an X-bearing sperm ($W$X) gives a red-eyed $F_1$ female ($w$X)($W$X); and the same kind of egg fertilized by a Y-bearing sperm gives an $F_1$ white-eyed male ($w$X)Y. The $F_2$ distribution 1 red ♀ : 1 white ♀ : 1 red ♂ : 1 white ♂ follows logically as shown in the figure.

By way of recapitulation this analysis may be set forth in shorthand notation as in the above diagram. This method, if followed consistently, is a convenient way to work out a problem involving sex linkage.

Characters which exhibit the foregoing type of distribution, dependent upon location of the factors in the sex chromosomes, are termed sex-linked, and the type of distribution which the factors undergo is called sex linkage. Sex linkage has been introduced at this point, because it strengthens the conclusion reached in Chapter V that the factors are borne by the chromosomes and that the facts of transmission may be accounted for in terms of maneuvers of the chromosomes. The distribution of the sex

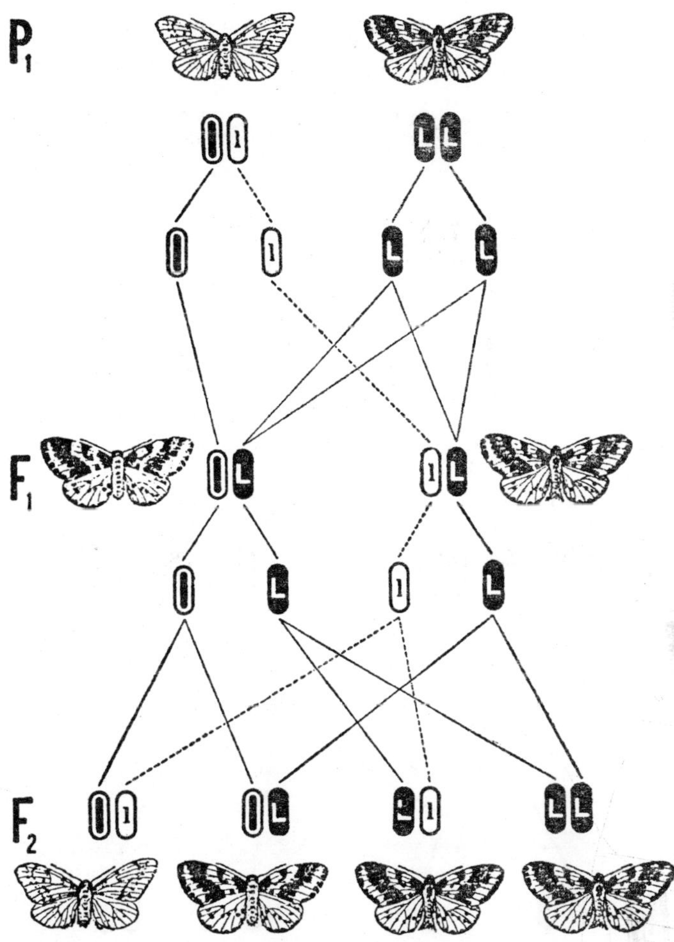


FIGURE 23.—Diagram illustrating the results of *lacticolor* ♀ × normal ♂ in Abraxas. *L* = normal, *l* = *lacticolor*, W(*l*Z) = *lacticolor* ♀, (*L*Z)(*L*Z) = normal ♂.

chromosomes differs from that of the autosomes, which are the same in both sexes; and in the same way the distribution of factors borne by the sex chromosomes differ from that of those borne by the autosomes.

**Barred vs. Black Plumage in Poultry.**—In poultry barred plumage such as is found in Barred Plymouth Rocks and similar breeds and the black of Black Plymouth Rocks, Black Langshans, etc. make up a pair of Mendelian characters with barred dominant. Here again the transmission of the characters is consistently related to sex, as may be seen from the following scheme:

| | barred ♀ × black ♂ | black ♀ × barred ♂ |
|---|---|---|
| $F_1$ | black ♀ ♀ + barred ♂ ♂ | barred ♀ ♀ + barred ♂ ♂ |
| $F_2$ | 1 barred ♀ : 1 black ♀ : 1 barred ♂ : 1 black ♂ | 1 barred ♀ : 1 black ♀ : 2 barred ♂ ♂ |

These results depend upon the fact that the mode of sex determination in poultry is different from that in Drosophila. In Drosophila the female

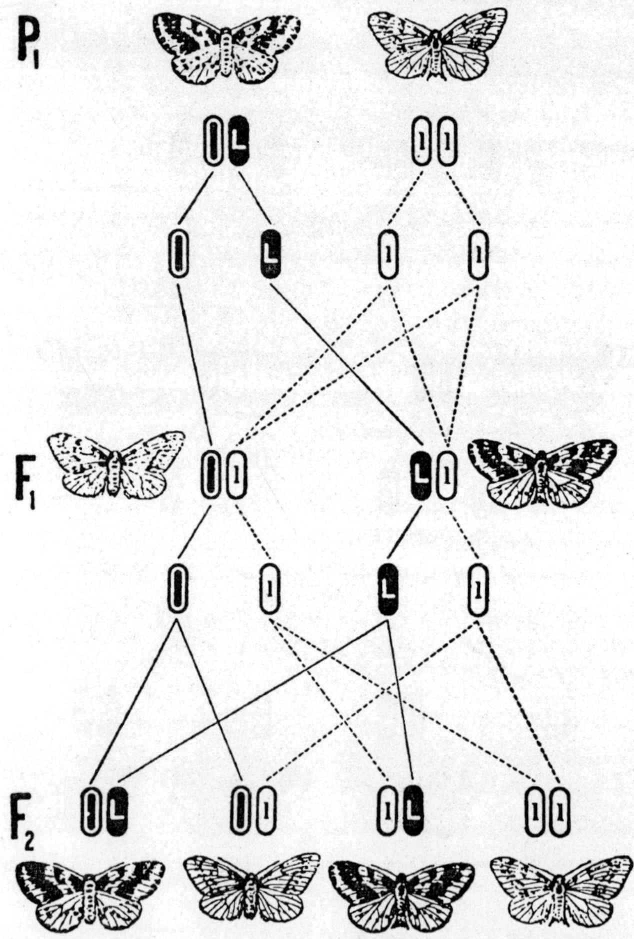


FIGURE 24.—Diagram illustrating the reciprocal of the cross represented in figure 23. W(*LZ*) = normal ♀, (*l*Z)(*l*Z) = *lacticolor* ♂.

is homozygous for sex and the male is heterozygous. In poultry the conditions are reversed; the female is heterozygous for sex and the male is homozygous, according to the following scheme:

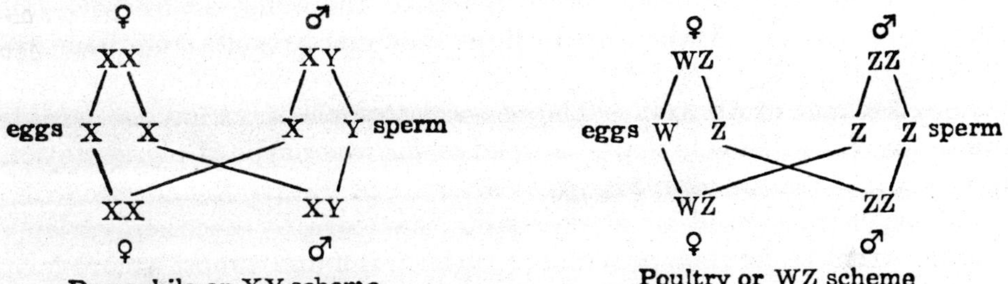


Drosophila or XY scheme

Poultry or WZ scheme

In poultry the sex chromosomes in the male are both alike; in the female, one member of the pair of sex chromosomos is identical with those found in the male but is paired with an unequal member. On account of this difference in the mode of sex determination, the sex chromosomes of the male are called Z-chromosomes, and the unequal member in the female is called the W-chromosome. Z-chromosomes of the poultry type of sex determination correspond to X-chromosomes of the Drosophila type, and W corresponds to Y. The two types of sex determination are briefly referred to as the XY and WZ types of sex determination.

The results which have been presented for barred *vs.* black in poultry are consistently accounted for on the assumptions that the genes for these characters are borne in the Z-chromosomes and that the W-chromosome is merely a neutral mate of the Z-chromosome in the female. If $B$ = barred and $b$ = black, the following scheme will show how these assumptions work out:

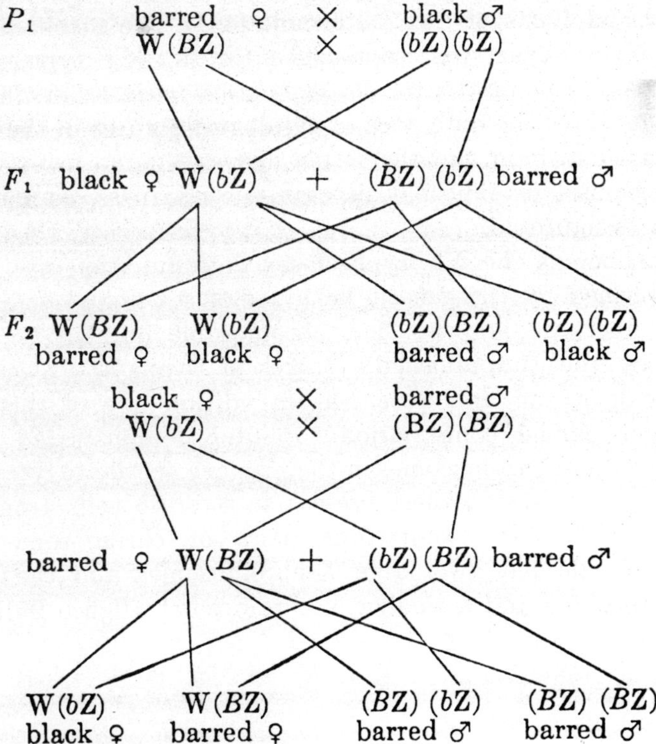

It should be obvious that barred males of the constitution (*BZ*)(*bZ*) are heterozygous and when bred will give different results from homozygous barred males (*BZ*)(*BZ*).

**Sex Linkage in Abraxas.**—The classical example of sex linkage according to the WZ scheme is that of normal *vs. lacticolor* type of pigmentation in the currant moth, *Abraxas grossulariata.* In nature the normal type, which is characterized by darker wing markings of a rather variable but nevertheless distinctive type and pattern, is found almost exclusively. Occasionally, however, a lighter variant is found which exhibits a marked reduction in number and size of the dark markings. This is the form styled *lacticolor.* It is interesting to note that according to Doncaster, save in one doubtful case, the *lacticolor* individuals which have been picked up in nature have all been females.

Breeding experiments have been conducted with these two forms. By appropriate methods it has been possible to establish races which breed true for normal and *lacticolor* type of pigmentation. Crosses between them demonstrate them to be a pair of Mendelian characters, with normal dominant to *lacticolor,* the factors for which are sex linked according to the WZ scheme. The details are represented diagramatically in figures 23 and 24, so that no further explanation is necessary. Obviously the reason why the *lacticolor* moths found in nature are females is due to the fact that a single dose of the *lacticolor* gene is sufficient to produce the character in females; while a double dose is necessary for the production of *lacticolor* males.

**The Physical Basis of Sex Determination.**—The mechanism of sex determination has been well established for the XY type. We have already discussed this matter for *Drosophila melanogaster* in Chapter III. Figure 17 illustrates the difference in cytological figures in the two sexes in a diagramatic fashion, but the actual figures secured from cytological studies are themselves almost of diagramatic clearness, as may be seen by reference to figure 25.

In human beings the XY type of sex determination has also been clearly established by the genetic behavior of sex-linked characters, of which a number are known. The chromosome number is now well established by the (unpublished) studies of Evans and Sweezy which definitely confirm the announcement in edition one of this text and Painter's later partial confirmation. Painter demonstrated the existence of two unequal chromosomes in the primary spermatocytes. Evans and Sweezy have found these present in somatic cells from various body parts. There are twenty-four pairs of chromosomes in both sexes. The female has two large X-chromosomes and the male one large X-chromosome and a smaller Y-chromosome (figure 26).

FIGURE 25.—Drawings of actual cytological preparations showing the differences in chromosome content in the two sexes of *Drosophila melanogaster*. In (1) preparation from a female showing two rod-shaped X-chromosomes; in (2) preparation from a male showing a somewhat bent rod-shaped X-chromosome lying above the characteristically hooked and longer Y-chromosome. Compare with figure 17. (*From Bridges.*)

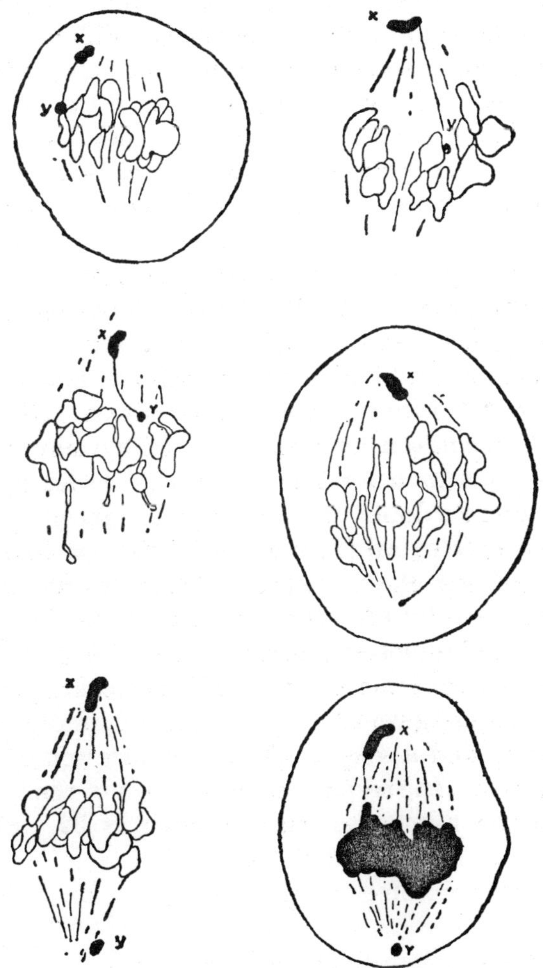

FIGURE 26.—Human spermatogenesis, the X- and Y-chromosomes are shown passing to opposite poles of the spindle in the division of primary spermatocytes. (*From Painter.*)

In the insects a great variety of differences has been determined for the relative sizes of the X- and Y-chromosomes. In some forms the Y-chromosome is absent, so that the chromosome formulae are XX = female and XO = male. From this condition there is a complete series leading up to those in which X- and Y-chromosomes are of the same size and indistinguishable morphologically. In addition to these there are species like *Drosophila melanogaster* in which the Y-chromosome is larger than the X-chromosome. There are also species in which the X-chromosome is represented by a group of small, more or less distinct elements. In spermatogenesis these smaller chromosomes, which may be called the X group, are loosely associated and are separated together to one pole of the spindle, the Y going to the other. In some well-established instances the X-chromosomes are attached to autosomes.

In some cases, even though a morphological difference between X- and Y-chromosomes is not demonstrable, they behave differently, as compared with autosomes, in spermatogenesis. Sometimes they pass to the poles before the autosomes; in other species they may lag characteristically on the spindle. In many species of Hemiptera, they have been found to divide equationally in the first spermatocyte division and to pair and undergo reduction in the second, whereas the reverse behavior is exhibited by the autosomes.

Despite these differences in detail, the mechanism of sex determination is essentially the same in all XY forms. The females produce but one kind of egg as respects chromatin content, and the males two kinds of sperm, one of which, containing an X-chromosome, is often called female determining; the other, containing a Y-chromosome, male determining.

As to the WZ type of sex determination, the evidence is far less satisfactory and abundant than for the XY type. So far apparently only birds and Lepidoptera exhibit this type of sex determination. In birds there are very few cytological studies which offer conclusive evidence as to the existence of sex chromosomes. Recent work of Shiwago and certain figures prepared by Miss Stevens seem to demonstrate the existence in poultry of a chromosome mechanism of the kind to be expected from the mode of distribution of the sex-linked characters. Shiwago's figures are presented in figure 27. The female chromosome complex has a large chromosome and a very small one, which apparently represent the sex chromosomes. In the male there are two of the large Z-chromosomes. The somatic number of chromosomes is 32. Shiwago states that other Passerine birds exhibit a similar chromosome complex, so that in birds the chromosome situation is in harmony with the genetic evidence.

In a moth, *Phragmatobia fuliginosa,* Seiler has shown that there are 56 chromosomes in both sexes, 54 of which, the autosomes, are small, and two, the sex chromosomes, much larger. There is no clear morophological distinction between the two sexes, but the sex chromosomes behave differently in the two sexes. In the male the sex chromosomes conjugate in spermatogenesis and separate like autosomes, so that each spermatid has 28 chromosomes. In the female, however, one of the sex chromosomes, interpreted to be the W-chromosome, splits up in the anaphase of the first division into two unequal parts, whereas the other passes undivided to the other pole. The final result is that there are two kinds of eggs, a male-determining type with 28 chromosomes, including one large Z-chromosome; and a female-determining type with 29 chromosomes, two of which represent a divided W-chromosome. In another moth, Talaeoporia, the same investigator has found that the males have 60 chromosomes and females 59. Sperm cells all have 30 chromosomes; and eggs either 29, female-determining; or 30, male-determining. Here the sex-chromosome formulae are evidently male = ZZ, female = OZ.

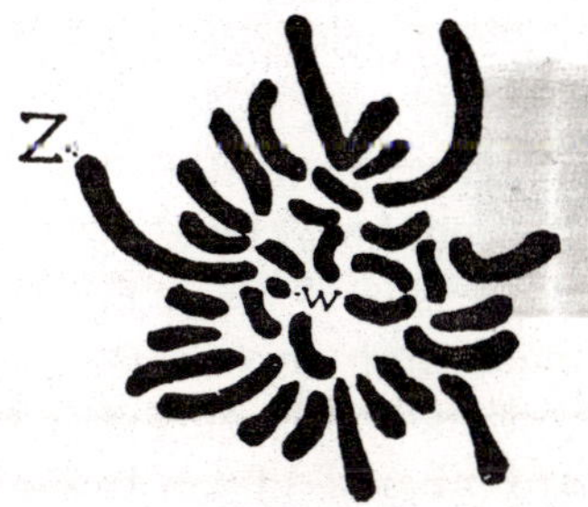


Male Female

FIGURE 27.—Male and female chromosome groups in the domestic fowl. (*After Shiwago from Morgan, in The Theory of the Gene, copyright* 1926 *by the Yale University Press. Reprinted by permission.*)

Despite the difficulty which has been encountered in securing cytological evidence for the WZ type of sex determination, there can be no question as to the adequacy of the genetic evidence. The cytological studies which have been made are in agreement with the genetic evidence, and they show further that similar differences in details of the cytological situation occur in the WZ type as in the XY.

**Sex Determination in Hymenoptera.**—The conclusion should not be drawn that all types of sex determination are included in one or the other of the foregoing schemes. As a matter of fact, some very intricate relations of the chromosomes to different portions of the life cycle have been demonstrated, especially in those insects in which a parthenogenetic cycle occurs in the life history. An interesting difference is found in the honey bee, and perhaps in Hymenoptera generally. The female produces one kind of egg only, with a full haploid complement of chromosomes; and the male also produces only one kind of sperm, containing a full haploid set of chromosomes. Eggs may, however, develop without fertilization, in which case males (drones) are produced; or with fertilization in which case females (queens or workers) are produced. The female, therefore, possesses the diploid number of chromosomes; the male, the haploid. In the male no reduction in chromosome number occurs in spermatogenesis, so that all the sperm cells contain a full haploid set of chromosomes.

Obviously if the chromosome theory of heredity is correct, the male should receive his total complement of factors from his mother. It has not been possible to perform adequate experiments on heredity in bees, because no method has been discovered for controlling matings; but the observations which have been made are, in the main, in agreement with theory. In certain other Hymenoptera it has been possible to conduct breeding experiments. In *Hadrobracon brevicornis*, a hymenopteran parasitic on caterpillars of the Mediterranean flour moth, the normal eye color is black, but a mutant race with orange eye color has been established. Black ♀ × orange ♂ gave black $F_1$ offspring of both sexes. The $F_1$ black wasps mated *inter se* produced black females and equal numbers of black and orange males. Orange females were obtained by mating heterozygous black females to orange males. Orange females mated to black males gave black females and orange males, for the most part; but an occasional black male was also produced. The black males are exceptional and apparently depend upon some disturbance in the normal operation of the chromosome mechanism. Evidently the male characters are inherited from the mother, as might be predicted from the known fact that they develop from unfertilized eggs; but further genetic evidence is necessary before the case can be considered established on a firm basis.

It has been proposed to call the type of inheritance found in hymenopterans sex-linkoid inheritance. Sex-linkoid inheritance is simply a special case of sex-linked inheritance in which all the characters are sex-linked, and not only certain ones which are borne in one pair of chromosomes.

**Sex Determination in Plants.**—While there are many dioecious plants, among them such important species commercially as the date palm, hemp, asparagus, spinach, etc., the problem of the mode of sex determination has not been so thoroughly studied as in animals. The conditions as to sex distinctions are obviously not so clear cut as in animals. Dioecious species are distributed more or less haphazardly throughout the plant kingdom, and in some instances male, female, and hermaphroditic forms occur in the same species. In some species as in strawberries, female (imperfect) and hermaphroditic (perfect) varieties occur. These features have led some geneticists to conclude that sex determination in plants is probably merely a matter of Mendelian factor differences as with other character contrasts, and that under these circumstances no cytological differences between the sexes are to be expected.

A number of cases have, however, been reported of well-defined chromosomal differences between the two sexes. The first demonstration was made by Allen in the liverwort, *Sphaerocarpos Donnellii*, which has an unequal pair of chromosomes in the sporophyte, and presumably the gametophytes, in which the sex distinction occurs, are 7 + X = ♀♀ and 7 + Y = ♂♂. In the Angiosperms the first demonstration was made by Santos in *Elodea canadensis* in which males have an unequal pair of chromosomes. A number of species have since been added to the list. In *Rumex acetosa* Kihara and Ono found that male plants have 15 and female plants 14 chromosomes. In the female there are two large chromosomes, interpreted as X-chromosomes; while in the male there is a single X which in reduction is V-shaped and to each arm of the V a somewhat smaller chromosome is attached. These two chromosomes are assumed to constitute a double Y element. Pollen mother cells in second metaphase show 7 chromosomes in one plate and 8 in the other. The relations here seem to be ♀♀ = 12 + XX, ♂♂ = 12 + XYY′. In *Lychnis alba* (= *Melandrium album*) Winge demonstrated the existence of an unequal pair of chromosomes in the male, so that here the formulae are ♀♀ = 22 + XX, ♂♂ = 22 + XY. Hitherto all cases examined in the higher plants cytologically appear to belong to the XY type.

Correns has published some results on controlled pollination in Lychnis (Melandrium) and other genera which point to male digamety. Under normal conditions of pollination many dioecious plants exhibit an excess, sometimes very great, of female plants. By using small quantities of pollen and by other ingenious methods, Correns obtained a marked increase in the proportion of males. These results indicate that there are two classes of pollen grains, female-determining and male-determining, and that under competitive conditions the former are favored.

Lychnis also appears to exhibit the only known instance of sex linkage in plants. The experiments deal with a mutant type having narrow rosette leaves, discovered in a single male individual (figure 28) crossed with the normal broad-leaved type, and the following results were secured:

| | |
|---|---|
| $P_1$ | broad ♀ × narrow ♂ |
| $F_1$ | broad ♀♀ + broad ♂♂ |
| $F_2$ | 2 broad ♀♀ : 1 broad ♂ : 1 narrow ♂ |

Evidently narrow is a typical sex-linked recessive character.

**Sex Linkage in General.**—It should be clearly understood that sex-linked characters differ in no essential feature from other characters. In *Drosophila melanogaster* over 100 sex-linked characters have been identified. They affect every part of the body: eye color, body color wing size, wing venation, eye shape, spine size and distribution, abdominal markings, etc.; in short they are exactly like autosomal characters. In particular they show no dependence on sex development; they usually occur in both sexes; their only relation to sex is exhibited in transmission and depends solely upon the fact that the factors are borne in the sex chromosomes. While sex linkage is limited to factors borne in the sex chromosomes, which usually constitute only a small portion of the germinal material. nevertheless it is the normal type of heredity for such factors.

It is a matter of debate whether or not the term sex chromosome should be used. On the one hand, studies of sex-linked inheritance show that the sex chromosomes bear many factors which have nothing to do with sex determination, and on the other hand, certainly some of the

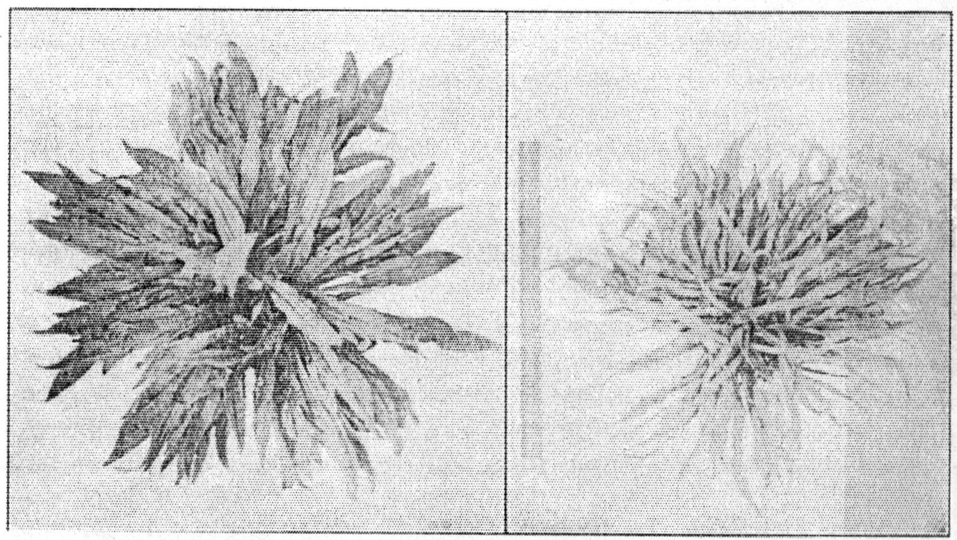

FIGURE 28.—Adult rosettes of *Lychnis dioica;* on the left a plant of the normal form, *typica;* on the right a plant of the narrow-leaved form, *angustifolia*. (*From Shull.*)

elements which participate in the development of sex are located in the autosomes. This last point, however, is not a very weighty one, for the question at issue is one of differentials and not of the totality of elements contributing to the development of sex; and certainly these are located in the sex chromosomes. Some people prefer, however, to call them heterosomes as a contrast to autosomes, the term applied to chromosomes which are equally paired in the two sexes. Correspondingly factors borne in the X- or Z-chromosomes may be termed heterosomal factors as those borne in autosomes are called autosomal. The same notation may be extended to characters, and sex-linked inheritance may appropriately be called heterosomal inheritance.

The distribution of the two types of sex determination has not been completely determined; but present evidence indicates that they are characteristic of greater groups of organisms. Birds and lepidopteran insects appear to be the only forms which follow the WZ scheme. Mammals, fishes, insects of the hemipteran, orthopteran, dipteran, coleopteran, and neuropteran orders, myriapods, spiders, and nematodes appear to follow the XY scheme. In some cases the evidence is purely cytological; in others purely genetic; and in some cases both types of evidence are available. In mammals and dipteran insects both cytological and genetic evidence has been presented; in fishes the evidence is hitherto wholly genetic. Hymenopteran insects may be considered as conforming to a modified XY scheme, and the same scheme of sex determination is found in rotifers.

**The Allosome.**—The Y-chromosome in the XY type of sex determination and the W-chromosome in the WZ scheme are found only in one sex, and they seem to differ in their genetic content from the other chromosomes. The question arises whether or not they have some determinable function. This matter will be taken up later, but here it may be stated that in Drosophila males which lack a Y-chromosome, although indistinguishable in appearance from normal males, are sterile.

The question also arises whether or not the Y-chromosome ever bears any factors. In Drosophila the normal allelomorph of bobbed is the only factor hitherto referred to it, but in fishes there is more evidence. Obviously, if it does, a peculiar type of transmission solely from father to son may be expected, because the Y-chromosome descends from father to son. Such inheritance was first demonstrated by Schmidt in the millions fish, *Lebistes reticulatus,* and more extended experiments with this subject have been made by Winge.

Inheritance through the Y-chromosome is well illustrated by the contrast between red and white pigmentation in ornamental varieties of a small Japanese fish, *Aplocheilus latipes.* The results secured by Aida should be compared and contrasted with those which have been described on the inheritance of red and white eye color in Drosophila, where the Y-chromosome is neutral. Pure races of both red and white have been established. Crosses between them yielded the following results:

| | | |
|---|---|---|
| $P_1$ | red ♀ × white ♂ | white ♀ × red ♂ |
| $F_1$ | red ♀♀ + red ♂♂ | red ♀♀ + red ♂♂ |
| $F_2$ | 2 red ♀♀ : 1 red ♂ : 1 white ♂ | 1 red ♀ : 1 white ♀ : 2 red ♂♂ |

Let $R$ = red and $r$ = white, since red is dominant, and assume that the factors are borne in both X- and Y-chromosomes. The analysis then works out as follows:

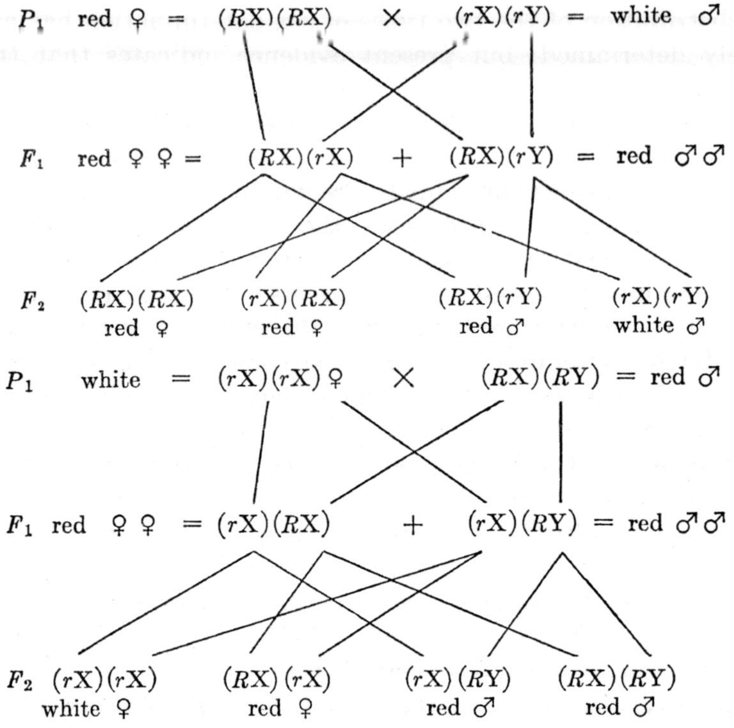


The W- and Y-chromosome mates of the Z- and X-chromosomes may be termed allosomes. Correspondingly when factors are borne in the allosome, they may be called allosomal factors, and the type of distribution which they exhibit may be styled allosomal inheritance. It has not, however, been demonstrated that any factors are borne exclusively by allosomes; for in the fishes, the only forms which have yielded acceptable experimental evidence, the factors are located both in the X- and Y-chromosomes, and the results are complicated by occasional exchange of factors between these two chromosomes, a matter to be considered again under linkage.

### References

Crew: Animal Genetics, Chapter VI.
Morgan: Heredity and Sex, Chapter II.
Morgan: The Mechanism of Mendelian Heredity, Chapter IV.
Morgan: The Physical Basis of Heredity, Chapter XIV.
Morgan: The Theory of the Gene, Chapter XIV.
Sharp: Introduction to Cytology, Chapter XIX.
Sinnott and Dunn: Principles of Genetics, Chapter IX.
Walter: Genetics, 2d ed., Chapter XIII.

# CHAPTER VIII

## HETEROZYGOUS EXPRESSIONS

The typical 3:1 $F_2$ ratio depends upon the dominance of one member of a pair of factors in development. In Mendel's seven pairs of characters, one member of each pair of characters was found to control the expression of the heterozygote, so that the principle seemed to be of general application; but with the growth of the knowledge of Mendelian character contrasts, many exceptions to this rule have been disclosed.

**Complete Dominance.**—Complete dominance is a condition in which the heterozygous form is fully equivalent in character expression to the homozygous dominant. According to Mendel's report such a condition was found to be particularly well illustrated by crosses between tall and dwarf peas, for the hybrid tall plants were, if anything, slightly taller than the parental type. The slight increase in the height of the hybrids as contrasted with the parents, however, probably is an evidence of hybrid vigor which in turn indicates that the two parental races differed not only in the *T-t* pair of factors but also in others which had a comparatively slight effect on height. In order to make an accurate determination of the completeness of dominance in a case like this, it is first necessary to be sure that the two parental races differ only in the factors under consideration, a stipulation difficult to satisfy under actual experimental conditions. Nevertheless in Drosophila, in which the conditions stipulated above are perhaps more nearly realized than in any other species upon which investigations have been performed, it has been found that many character pairs exhibit, in so far as available evidence is concerned, complete dominance of one member.

**Incomplete Dominance.**—An example of a slight degree of intermediacy in a hybrid is afforded by the character contrast of wild type *vs.* black body color in Drosophila. When wild type is crossed with black, $F_1$ is practically wild type, but it is distinctly darker than normal. Representing wild type by $B$ and black by $b$, the $F_2$ conforms to the ratio 1*BB* (wild type):2*Bb* (darker wild type):1*bb* (black). The difference between the homozygous and heterozygous wild-type flies is not sufficient to make it possible to classify the two types satisfactorily, so that in experimental work they are usually lumped together as one class and the populations are treated as conforming to the ratio 3 wild type:1 black.

In some other cases the degree of intermediacy is so slight that refined methods of technique must be resorted to in order to detect it. Thus in Mendel's experiments on round *vs.* wrinkled in peas, round was found to be the dominant character. Darbishire studied the starch content of the two types in order to determine the nature of the difference between them. He found (figure 29) that in the development of round seeds, the sugar is almost wholly converted into starch, so that when the seed is ripe, it retains water rather firmly, and on drying it shrinks uniformly to form a smooth round seed. The seeds of wrinkled varieties are also smooth before drying, but in consequence of an incomplete transformation of sugar into starch, they give up more water on drying, and they shrink unequally so that the matured seed becomes angular and wrinkled. He also found a difference in the character of the starch grains of the two kinds. In round peas the starch grains are numerous and entire; in wrinkled peas, they are few in number and exhibit irregular fissures and fragmentation, suggesting a partial disintegration after their formation. In the hybrid peas of round crossed with wrinkled, which to outward

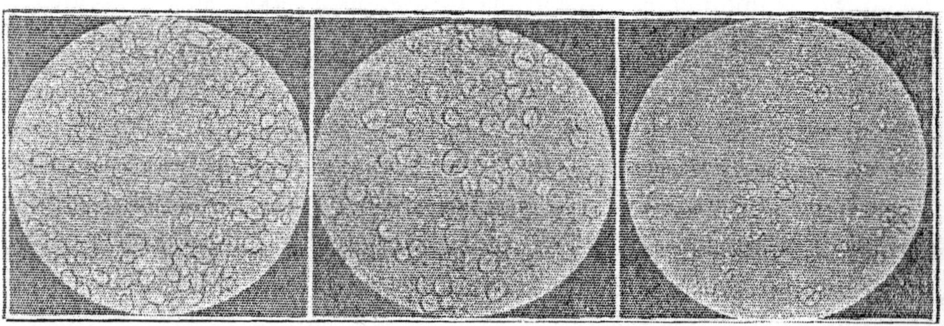

FIGURE 29.—At the left, starch grains of round pea; at the right, of the wrinkled pea; and in the middle, of a hybrid between round and wrinkled peas. (*From Darbishire, Breeding and the Mendelian Discovery, Cassell and Company, Ltd.*, 1911. *Reprinted by permission.*)

appearance are identical with round peas, Darbishire found an intermediate condition of the starch grains, both in number and in degree of disintegration, as shown by the existence of fissures and fragmentation. The heterozygote, therefore, in this instance, is really intermediate, if the condition of its starch grains is considered, although superficially smoothness is completely dominant to the wrinkled condition. These studies indicate that even in cases where dominance appears to be complete, more careful examination may disclose a difference between the homozygous and heterozygous classes.

In Drosophila it has been found that miniature wings, a short-winged condition, is a sex-linked character recessive to the normal long-winged condition of the wild type. In this case comparative measurements of length of wings with leg length in homozygous and heterozygous flies disclosed a slight but significant difference in length in favor of the homozygous flies. In this instance employment of precise biometrical methods is necessary to demonstrate a slight degree of intermediacy in the heterozygote.

In the cases which have just been described it is most convenient to treat homozygous and heterozygous genotypes as members of one phenotype to be contrasted with the homozygous recessive, because of the impracticability of making an accurate classification into three phenotypes. Such behavior is called incomplete dominance in contrast with complete dominance on the one hand and a readily recognizable intermediate condition on the other. These cases all illustrate the point that apparent instances of complete dominance may be found upon more careful examination to exhibit slight differences between the homozygote and the heterozygote. Some investigators go so far as to contend that sufficient refinement of technique will always disclose some difference.

**Intermediate Heterozygotes.**—In the common garden four o'clock, *Mirabilis jalapa*, when red- and white-flowered varieties are crossed, the $F_1$ is pink. When $F_1$ is selfed, the $F_2$ progeny conforms to the ratio 1 red:2 pink:1 white. When the $F_2$ red plants are selfed, they produce only red progeny; and the $F_2$ white plants likewise breed true for white. The $F_2$ pink plants, on the contrary, reproduce the $F_2$ ratio in $F_3$; *i.e.*, their progeny conforms to the ratio 1 red:2 pink:1 white.

The same scheme of factor notation may be used in this instance as was employed previously. If $R$ = red and $r$ = white, then the $F_1$ is $Rr$, which was found by experiment to be pink; and $F_2$ conforms to the ratio $1RR$ (red):$2Rr$ (pink):$1rr$ (white). It is then perfectly obvious that this modification of the 3:1 ratio depends upon the fact that neither member of the $R$-$r$ pair of factors is dominant, but the heterozygote is intermediate in color and distinguishable from both parental types.

Many other instances of this type of modification are known. Moreover, all degrees of intermediacy may be illustrated ranging from that of Mirabilis, which lies midway between the two parental forms, to others in which the degree of intermediacy is so slight as to be undetectable save by the most careful scrutiny.

**Mosaic Heterozygotes.**—In certain heterozygotes the character expression is a mosaic, rather than a true intermediate. In Shorthorn cattle, red and white breed true when mated *inter se*, but when crossed together they give roan individuals, a type of coloration which depends upon a fine intermixture of red and white hairs in the coat. Roan Shorthorns are always heterozygous, and when mated together they produce offspring in the ratio 1 red:2 roan:1 white.

A similar instance is afforded by the Andalusian fowl, a recognized breed of poultry. The Andalusian recognized as the type of the breed is the blue form, but when the blue ones are bred together they produce birds in the ratio of 1 black:2 blue:1 white splashed with blue. The black form is a self-black and it breeds true when mated *inter se*. The white splashed with blue approaches white but has colored feathers irregularly distributed through the plumage, which contain a blue pigment like that of the blue type. This form also breeds true when mated *inter se*. The blue form is a self-colored type like black, but the pigment is blue. Feathers of this type are uniformly distributed through the body except in males in which the long hackle feather and those of the saddle are black. The pigment in all cases is black, but in black feathers it occurs in the form of rod-shaped granules evenly distributed through all the cells of the feathers, whereas in blue feathers the granules are round and clumped in the cells which possess them, and the cells of the barbules are practically free from pigment. The blue form resembles the black in that all feathers are pigmented, and it is like the blue-splashed

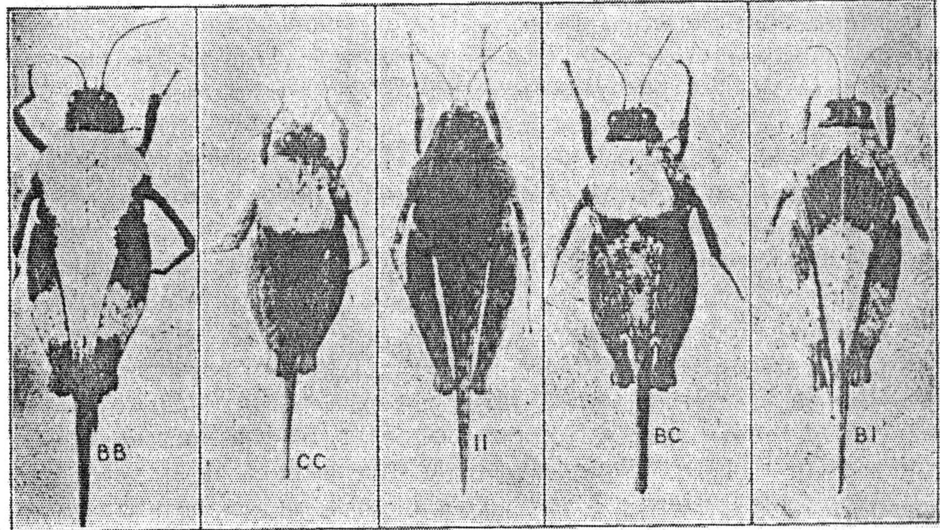


FIGURE 30.—Three types of Paratettix, *BB*, *CC*, *II*, and two of the hybrids between them. (*From Nabours.*)

white in the character of the pigment granules and in their distribution in the feathers.

The blue type does not breed true but continually segregates in the ratio stated above; nevertheless it is the recognized type of the Andalusian breed. It is interesting to note that it just comes within the requirements of the Standard of Perfection for poultry breeds, which refuses recognition to forms which reproduce themselves in less than 50 per cent of their progeny. The enterprising breeder who wishes to produce eggs which will give nothing but Andalusian chicks can do so very readily by mating black with blue-splashed white; but all three forms will appear in the second generation.

In the grouse locust, Paratettix, a large number of distinct patterns occur which differ not only in color but also in distribution of pigment.

Nabours has studied a number of distinct forms which breed true for particular color patterns. He found that whenever any two of them are bred together, the hybrids exhibit the entire color pattern of both parents superimposed one upon another. If, for example, two races of Paratettix $B$ and $C$ (figure 30) are crossed, the hybrid $BC$ will exhibit the color pattern of both parents superimposed, and it is possible by simple inspection to determine that it is a hybrid between $B$ and $C$. Such hybrids mated *inter se* give populations conforming to the ratio $1B$: $2BC$:$1C$, a proof of a unifactorial contrast between the two parental types; and if the factors are designated $A$ and $a$, in this instance, the heterozygous form $Aa$ exhibits the developmental effect of both parents.

In certain crosses which did not appear to conform to this scheme, a microscopic examination disclosed the difficulty. Thus in figure 30 the superficial characters of the hybrid $BI$ between the form called *leuconotus* $BB$ and the form called *nigronotatus* $II$ are for the most part those of *leuconotus* except for the broad black band across the pronotum, which is clearly derived from *nigronotatus.* In the posterior part of the pronotum particularly the characters of *leuconotus* appear to be dominant, but the microscopic examination revealed the existence of differences of distribution of the pigment in the parental races of such a nature as to account for the masking of *nigronotatus* characters by the development of those of *leuconotus.*

**Variable Heterozygotes.**—In some cases the variability of the heterozygote is so great as to obscure separation of $F_2$ into definitive classes. An example is afforded by crosses between ebony, a very dark body color, and sooty, a grayish body color, in Drosophila. Crosses between ebony and sooty give ebony-sooty compounds in $F_1$, which are on the whole intermediate in color, but the variability is so great that some of the $F_1$ flies are as dark as ebony and some are as light as sooty. In $F_2$ as a result of this variability in the heterozygotes, the flies cannot be classified accurately into the three expected classes in the ratio 1 ebony: 2 ebony-sooty compound: 1 sooty; but if such flies are subjected to proper tests, which can be done with the males by double mating them to pure-ebony and pure-sooty females, it can be demonstrated that the three forms occur in the proper proportions. The intermediate character here is variable, but this fact does not imply a corresponding variability in the factors. They are stable and segregate as cleanly as in other cases.

**Endosperm Characters in Maize.**—In maize Hayes and East have reported an interesting case which exhibits a modified Mendelian ratio. It has to do with the flinty-floury endosperm characters. In flinty varieties (figure 31) the endosperm is composed mostly of corneous tissue with only a small amount of soft starch in the center of the grain. In floury varieties most of the endosperm is of the soft starch type and there is only a thin shell of corneous material.

When a floury corn is pollinated with a flinty one, the $F_1$ kernels are floury. In the reciprocal cross the $F_1$ kernels are flinty. In both crosses, therefore, the $F_1$ kernels are purely of the maternal type. When $F_1$ floury kernels from floury ♀ × flinty ♂ are planted, they produce an $F_2$ progeny which exhibits distinct segregation according to the ratio 1 floury:1 flinty. $F_1$ flinty kernels from flinty ♀ × floury ♂ produce $F_2$ populations which conform to the same ratio. Evidently the $F_1$ embrycs are of the same genetic constitution despite the differences in the character of the endosperms.

As was explained in Chapter III, the endosperm nucleus arises from a separate union in fertilization of two female nuclei with a single male nucleus. The embryo on the contrary arises from a fertilization of the ordinary type in which a single female nucleus unites with one male nucleus. The occurrence of double fertilization makes it possible to offer a simple explanation for the peculiar results of crossing floury with

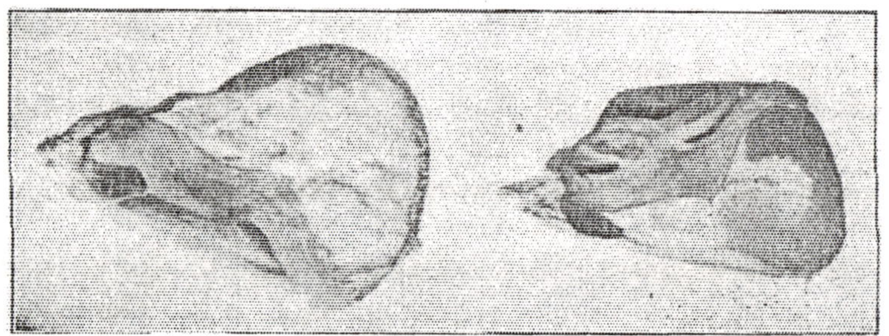

FIGURE 31.—Longitudinal section of maize grains showing differences in character of starch; left, floury; right, flinty.

flinty. Let $F$ be the symbol for flinty and $f$ that for floury. Then a pure flinty zygote is $FF$ and its endosperm is $FFF$; and floury zygote and endosperm are $ff$ and $fff$, respectively. In the cross of flinty ♀ × floury ♂, the $F_1$ zygote is $Ff$ and the endosperm $FFf$. In the reciprocal cross, floury ♀ × flinty ♂, the embryo is again $Ff$, but the endosperm is $ffF$. The two $F_1$ embryos are of the same constitution but the endosperms are different, and this difference provides a clue to the explanation of the case. The explanation evidently is that $F$ is dominant over $f$ in endosperm development, if present in a double dose, and $f$ is dominant if it is present in a double dose. If this conception is developed logically for $F_2$, the ratio of 1 floury:1 flinty is obtained from either $F_1$ and half the members of each class are heterozygous and half homozygous. Here an appeal to the nuclear phenomena in fertilization has led to a proper solution of the problem.

Another somewhat similar case is afforded by the contrast between colored and white aleurone in maize which depends upon the $R$-$r$ pair of factors. Colored ♀ × white ♂ gives $F_1$ solid colored; but white ♀ ×

colored ♂ gives $F_1$ seeds which are mottled, with colored and white areas irregularly mixed on individual grains. Evidently *RRr* endosperms are solid colored, and *rrR* endosperms are mottled. A single dose of *R* is insufficient to produce a full-colored aleurone.

The above relations do not hold for most endosperm characters in maize. Thus sugary endosperm, the wrinkled, translucent type characteristic of sweet corn, is recessive to starchy. Starchy ♀ × sugary ♂ gives starchy $F_1$, and the reciprocal cross, sugary ♀ × starchy ♂ gives the same type of $F_1$. In both cases $F_2$ is in the ratio 3 starchy: 1 sugary.

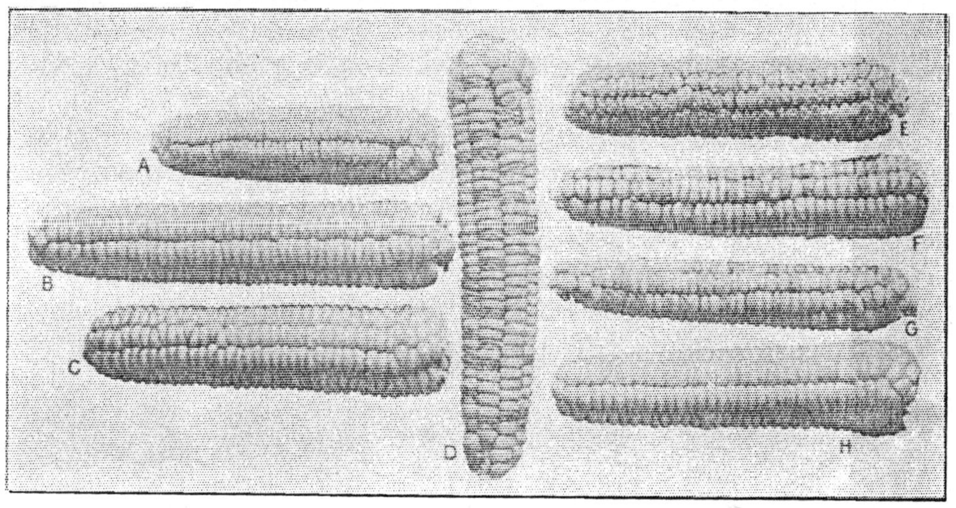


FIGURE 32.—Results of crossing starchy and sugary maize; *a*, sugary parent; *c*, starchy parent; *b*, the $F_1$ showing complete dominance of starchiness; *d*, the $F_2$ showing monohybrid segregation; *e*, *f*, *g*, and *h*, $F_3$ populations, the last three obtained by planting $F_2$ starchy grains, the sugary ear, *e*, by planting an $F_2$ sugary grain. (*From East and Hayes.*)

Segregation is clear and distinct (figure 32); and evidently a single dose of *Su*, the gene for starchy, is as effective in determining the development of starchy grains in endosperms of the genotype *Su su su*, as the double dose in those of the constitution *Su Su su*. The great majority of endosperm characters in maize apparently follow the starchy-sugary scheme.

**Heterozygous Characters.**—Since distinct characters such as roan in cattle and blue in Andalusian fowls may depend upon a heterozygous genotype, it is often convenient to distinguish them as *heterozygous characters* in contrast to those characters which may be obtained in a homozygous condition. Heterozygous characters are obviously incapable of fixation by ordinary breeding methods; consequently when the choice of the breeder is laid upon them, he must be prepared to discard a certain proportion of "wasters," usually 50 per cent, in his breeding operations.

**Heterozygous Expressions.** From the foregoing discussion it should be perfectly clear that the heterozygous genotype may exhibit a great variety of differences in character developement. The particular way in which a heterozygote will develop can be determined only by experiment. Perhaps the majority of instances fall in two classes, complete dominance or incomplete dominance. In Mendel's experiments one member of each of the seven pairs of characters with which he worked was dominant. For this reason dominance was once considered a general feature in Mendelian character pairs, and a law of dominance was formulated to cover it. With the extension of Mendelian investigations, however, it has become clear that dominance is only a special feature exhibited by certain pairs of Mendelian characters; consequently the inclusion of such a law among the general laws of heredity is no longer justified.

Whether a given character is dominant or not has no effect upon its mode of transmission. The law of segregation operates in exactly the same way for cases of intermediacy as for perfect dominance. Its operation is in fact more clearly seen in the former case, because the heterozygote may be distinguished by visual inspection, instead of solely by the breeding test, as is necessary when complete dominance is exhibited.

### References

General texts listed in Chapter I.
Darbishire: Breeding and the Mendelian Discovery, Chapter IX.

## CHAPTER IX

## LETHAL FACTORS

It has been found that certain factors may destroy the life of the individual, a conception first developed by Cuenot from his studies of the inheritance of coat color in mice. Yellow mice are always heterozygous. When outcrossed to non-yellow, the progeny are in the ratio of 1 yellow:1 non-yellow; when bred *inter se*, yellows give progeny conforming to the ratio 2 yellow:1 non-yellow. If yellow is represented by $Yy$, they obviously should produce, when bred *inter se*, zygotes in the ratio $1YY:2\,Yy:1yy$. The failure of homozygous yellow individuals to appear and the 2:1 ratio are both accounted for, if it is assumed that $YY$ mice die. There is some positive evidence that this is actually the case. Yellow females bred to non-yellows produce larger litters than when bred to yellow males, and the death of a certain percentage of young embryos in matings of the latter type has been demonstrated. The gene $Y$, therefore, has a dominant character effect which it exhibits in the heterozygote and a recessive lethal effect which is exhibited only in the homozygous condition.

A similar instance of some economic importance is afforded by Dexter cattle. The Dexter, the smallest breed of cattle in Great Britain, apparently originated in Ireland as an offshoot of the Kerry breed. As compared with Kerry cattle, Dexters are smaller and have shorter legs and broader heads, together with other very characteristic differences. The type is heterozygous as is shown by matings of Dexter × Dexter which give 1 "bulldog":2 Dexter type:1 Kerry type (normal); and of Dexter × Kerry which give half Dexter type and half Kerry type offspring. "Bulldog" calves (figure 33) are always stillborn, and exhibit very characteristic features, thus described by Mohr:

> The cranium is bulging, the nose markedly depressed, the lower jaw protruding, the upper lip split, baring the teeth, while the swollen tongue, thrust far out, curls up over the nose. Owing to the disproportionate development of the buttocks, the tail seems to have its origin far up on the back; usually there is a gaping deficiency of the abdominal wall through which the intestines pass to form a large umbilical hernia. The skin hangs loosely in folds; there is abundant subcutaneous fat. The limbs are ridiculously short and the digits unusually separated.

The simplest explanation of this case is that the Dexter is heterozygous for a factor which produces the "bulldog" calf in the homozygous condition. A more complex explanation has been offered, but as yet the correctness of it has not been ascertained. It is interesting to note that Dexter breeders in Ireland finally gave up the attempt to fix this type and have since devoted their attention to the Kerry type. In order, however, to satisfy a demand for Dexters in England the scheme of mating Dexters to Kerries was evolved as a means of avoiding the production of "bulldog" calves.

The probable existence of a similar lethal factor in man is indicated by extensive studies of Mohr and Wriedt on brachyphalangy in a Norwegian family. The trait is characterized by shortening of the middle digit of both hands and feet and is transmitted as a simple dominant character. In most matings, of course, one parent is normal; but a single marriage of two affected individuals yielded three children one of whom lacked all fingers and toes and died at the age of a year. Possibly this individual was homozygous for the factor, in which case a recessive lethal effect is associated with it.

FIGURE 33.—The "bull-dog" calf, representing the lethal homozygous dominant produced by segregation in matings of Dexter × Dexter type in cattle. (*From Mohr.*)

A large proportion of the dominant mutants in Drosophila are of the yellow-mouse type. Star, dichaete, beaded, notch, streak, etc. are all dominant characters which occur only in the heterozygous condition. In plants a similar condition is afforded by Baur's golden Antirrhinum. It is a form with leaves finely variegated with yellow and green stripes. When selfed, it produces approximately 2 golden : 1 green, but Baur found, when observations were made on very young seedlings, a third class which was entirely yellow, so that the actual ratio is 1 yellow : 2 golden : 1 green, and the death of the yellow seedlings accounts for the 2 : 1 ratio in adult populations.

There are also lethals which have no dominant character effect in the heterozygous condition. Various albino conditions in plants afford excellent illustrations. Thus the factor *w* in maize in the homozygous condition gives seedlings devoid of chlorophyll, which die as soon as they have exhausted the food material in the endosperm. Heterozygous plants *Ww* have a full green color and cannot be distinguished from homozygous green plants *WW*, except by application of the breeding test. The death of *ww* plants in the seedling stage necessitates the continuation of such strains through heterozygous individuals, which give seedlings in the ratio 3 green: 1 albino, when selfed. There are many such chlorophyll-deficient strains in maize, wheat, barley, and other plants; and in some cases differences may be observed in the color of the chlorophyll-deficient individuals of the different types. A few types

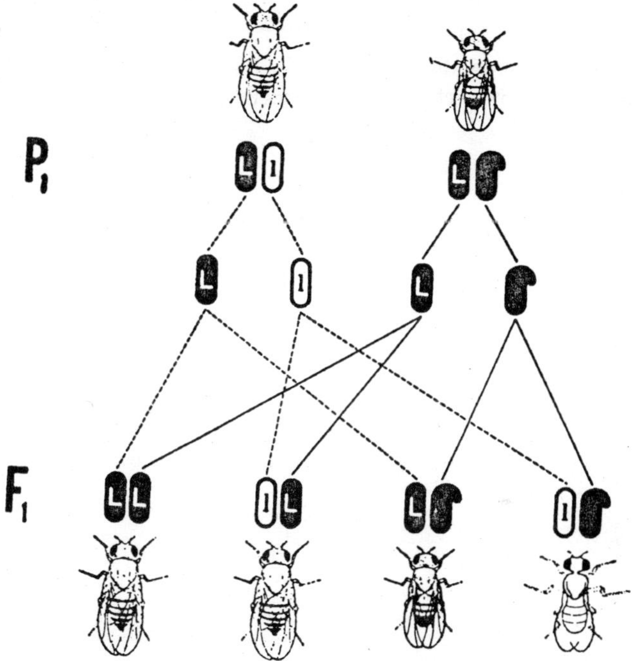


FIGURE 34.—Production of a 2 ♀♀ : 1 ♂ sex ratio in Drosophila by segregation in a female heterozygous for a sex-linked recessive lethal factor.

which exhibit chlorophyll deficiency in seedlings may gradually attain a full color and go on to maturity. Such plants when selfed give nothing but chlorophyll-deficient seedlings.

A large number of completely recessive lethal factors have been discovered and studied in Drosophila. Muller estimates that they occur about six times as frequently as non-lethal mutations. In general they can be detected only by studying their linkage relations with other non-lethal genes. If they are sex-linked, however, they give rise to a 2 ♀♀ : 1 ♂ sex-ratio; and they can be continued by breeding from females which give such a ratio, for half of the females will then be heterozygous for the lethal (figure 34).

The period at which death occurs has not been determined for most lethals but it apparently may vary from the early zygote (before hatching) to the late pupal stages. In some instances a few individuals of the lethal type survive to maturity (semilethals).

Dominant lethals are obviously incapable of continued existence because they would kill all individuals which receive them, leaving none to transmit the gene. Recessive lethals in most species tend to be eliminated in time in competition with their normal allelomorphs. Nevertheless in certain cases they persist for a long time and afford a convincing refutation of the contention that Mendelism deals only with superficial characters.

### References

Crew: Animal Genetics, pp. 127–132.

Morgan, Bridges, and Sturtevant: Genetics of Drosophila, Bibliographia Genetica 2, Chapter VII.

## CHAPTER X

## MULTIPLE ALLELOMORPHISM

Hitherto the discussion has dealt with pairs of characters and pairs of factors, but certain experiments which have been performed indicate that this conception is too narrow to fit all cases. There is a series of sex-linked eye colors in Drosophila which, when arranged in order of intensity of coloration from darkest to lightest are as follows: (1) wild type, (2) coral, (3) blood, (4) cherry, (5) eosin, (6) apricot, (7) ivory, (8) tinge, (9) buff, (10) ecru, and (11) white. These characters are so related genetically that any two of them taken together give results conforming to a unifactorial scheme. For example, apricot ♀ × white ♂ gives in $F_1$ apricot-white compound ♀ ♀ and apricot ♂ ♂, and $F_2$ consists of flies in the ratio 1 apricot ♀ : 1 apricot-white compound ♀ : 1 apricot ♂ : 1 white ♂. The reciprocal cross, white ♀ × apricot ♂, gives in $F_1$ apricot-white compound ♀ ♀ and white ♂ ♂, and in $F_2$, 1 apricot-white compound ♀ : 1 white ♀ : 1 apricot ♂ : 1 white ♂. Any other two will give similar results, except that when red is one member of the pair, it exhibits complete dominance and the results conform to the red *vs.* white scheme described in Chapter VII.

These results are explicable, if it is assumed that all these types differ in a single factor from the same wild-type factor. This wild-type factor is assigned the symbol $W$ in contrast to $w$ = white, the first discovered member of the series. The other members of the series are given symbols with definitive superscripts; *viz.*, coral = $w^{co}$, blood = $w^{bl}$, cherry = $w^{c}$, eosin = $w^{e}$, apricot = $w^{a}$, ivory = $w^{i}$, tinge = $w^{t}$, buff = $w^{b}$, and ecru = $w^{ec}$. The factorial analysis of the reciprocal crosses of apricot with white may then be portrayed according to the following scheme:

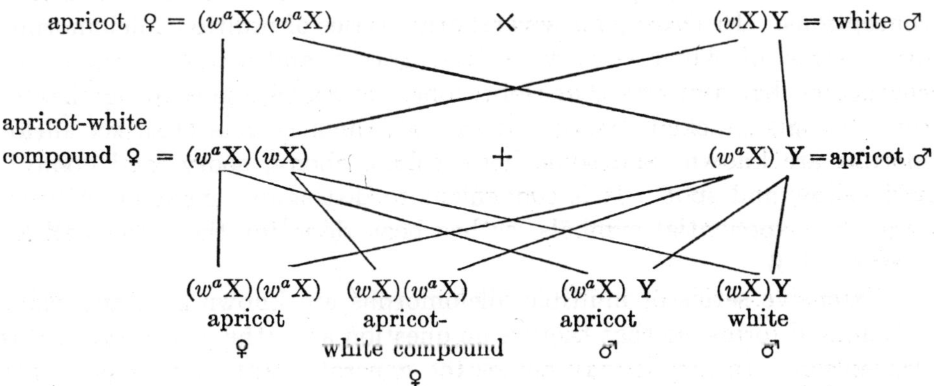

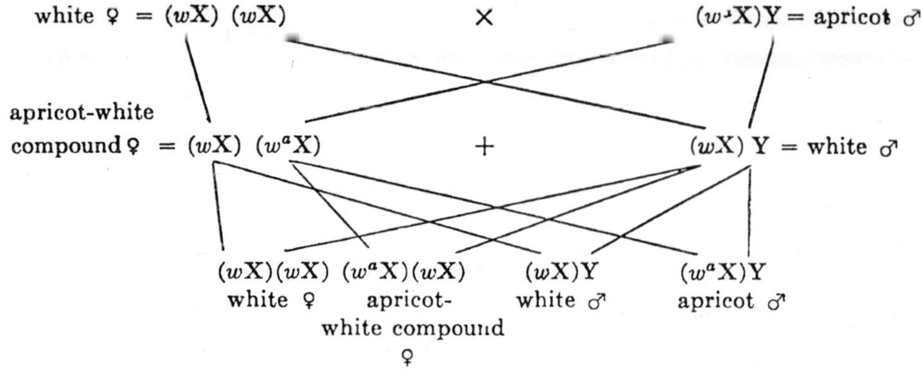


These results indicate that allelomorphism is not necessarily restricted to pairs of factors, but that any number of factors may make up an allelomorphic series. The conception is, perhaps, best grasped, if the manner in which these different allelomorphs arose in the first place is considered. The starting point was the factor *W* of the wild-type fly, which has a red eye color. In the white-eyed race, this gene *W* became changed to *w*. What caused the change, or mutation, or what is the nature of the difference between *W* and *w* is not known; but it has the effect in the homozygous condition of producing a white-eyed fly instead of a red-eyed one. Now other changes have also apparently occurred in this same factor *W*, so that it became changed to $w^c$, giving rise to cherry eye color; $w^a$, apricot, etc. Apparently all the members in the series arose directly from wild type by a change in the factor *W*, except eosin, which arose from white by conversion of *w* into $w^e$. These factors are allelomorphic to each other then, because they all arose by changes in the same wild-type factor. Similarly, secondary changes in them, as for example the change from *w* to $w^e$, producing eosin, will be allelomorphic to the entire series of primary changes from *W*. The term multiple allelomorphism is applied to such series of allelomorphs as consist of more than two members.

Multiple allelomorphism is an important extension of Mendelian conceptions. In Drosophila several other series of multiple allelomorphs exist, some of which are eye colors, garnet and garnet²; some eye characters, bar and infra-bar; lobe, lobe², and lobe³; some wing characters, vestigial, nicked, strap; and cut, of which no less than six allelomorphs are known; and some body colors, ebony, sooty, and ebony²; and yellow and spot. It is convenient to distinguish members of such series by exponential symbols, as has been done for the various allelomorphs of *W*.

Extensive series of multiple allelomorphs are known in many plant and animal forms, so that there is no question as to the generality of this conception. In fact it may not be too general a statement to say that all allelomorphic series may eventually be found to possess several members, or at least to be capable of producing this condition. In maize, for example, Emerson has demonstrated the existence of an allelomorphic series controlling the development of red color in the husks, silks, pericarp, and cob. Red color may occur in all these parts of the

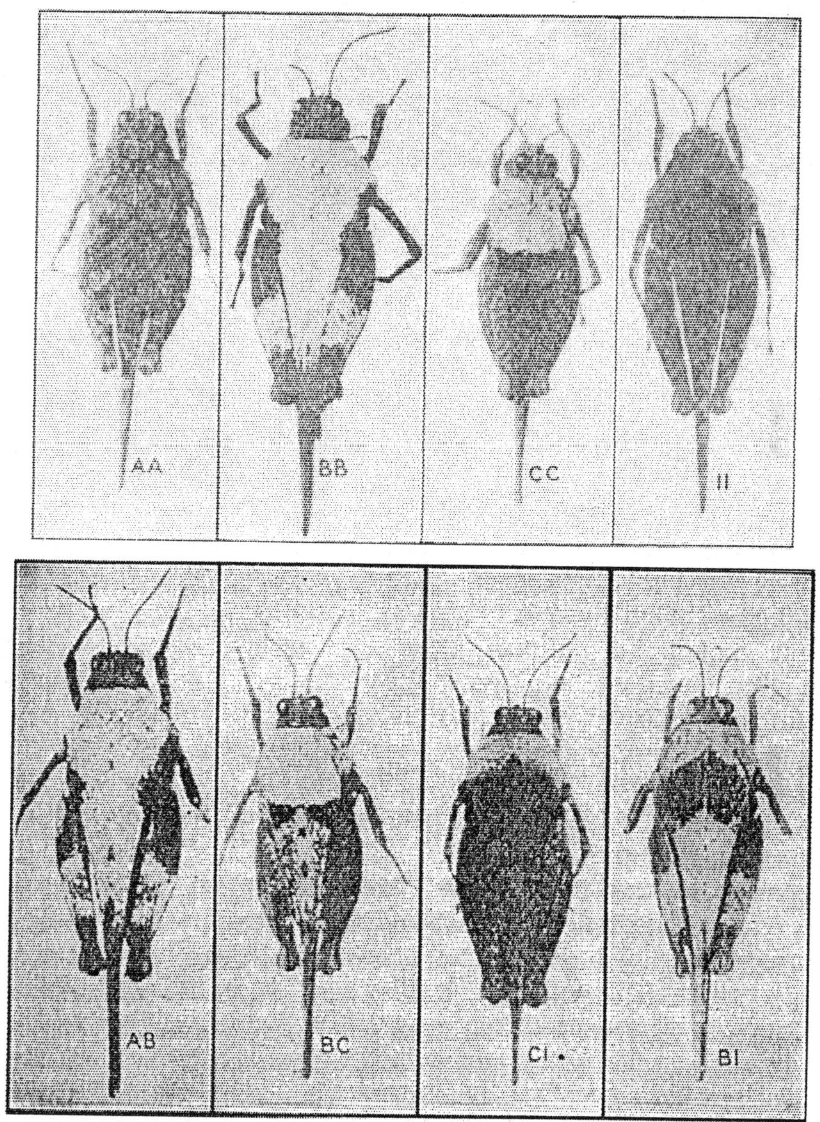


FIGURE 35.—Above, forms belonging to a series of multiple allelomorphs in Paratettix. Below, hybrids between them. *(From Nabours.)*

plant, or they may all be colorless, or any part or parts may be red and the rest colorless. Moreover, the red color may be uniformly distributed or variegated. Considering only red and white, it is possible to have sixteen different types. Any two of these races are allelomorphic and will give ordinary unifactorial segregation in crosses; *i.e.*, a single factor appears to control the development of the entire color pattern, no matter what combinations of colored and colorless portions it may exhibit. Red color is dominant and the $F_1$ exhibits the characters of the two parents superimposed. In $F_2$ the ratio is 1:2:1, the phenotypes being the two parental types and the $F_1$, if it is different from them. If it is like one of them, the ratio becomes 3:1. This result indicates that the $F_1$ hybrids produce gametes representing only the parental conditions. This fact has been tested by backcrossing $F_1$ hybrids to varieties colorless in all parts. Thus in one case an $F_1$ plant with red cobs and striped grains gave a progeny approximately equally distributed between two phenotypes: white cobs and striped grains, and red cobs and white grains. None were of the $F_1$ type, red cob and striped grains, and none had white cobs and white grains. These results are very curious, but thus far they have withstood all experimental tests which have been applied to them.

A similar set of multiple allelomorphs have been found in Paratettix, the grouse locust (figure 35). In this insect Nabours has demonstrated the existence of fourteen pattern types, any two of which give simple monohybrid results when crossed together. The case presents an extremely close parallel to that in maize, for the $F_1$ is merely a composite formed by superposition of the two parental types, and the color patterns, although they affect the entire body, are inherited as units.

Other examples of multiple allelomorphism have been demonstrated in the silkworm, a series of larval patterns; in the guinea pig, intense pigmentation, dilute pigmentation, red-eyed dilution, and Himalayan albinism; in the bean, green leaves with green pods, green leaves with yellow pods, to mention only a few. There can be no doubt that multiple allelomorphism is of frequent occurrence and may be expected in any animal or plant form.

**General Features of Multiple Allelomorphs.**—In most cases the members of a series of multiple allelomorphs have a similar effect upon some one portion of the individual or some one feature. The members of the *W* series in Drosophila are all eye colors of different intensities, which make up a fairly complete series of diminishing intensity from red to white in the order named above. The albino series in guinea pigs all affect coat and eye color; and the other series mentioned agree with them in this respect. There is, however, some evidence that members of a series of multiple allelomorphs need not necessarily bear this relation, for Muller has discovered members of the truncate series in Drosophila which have quite different effects.

**The Presence-absence Hypothesis.**—One of the first attempts to establish a logical basis for allelomorphism was the presence-absence hypothesis, first introduced by Correns and developed especially by

Bateson and his associates. According to this conception, dominance represents presence of a factor in the germinal system, and recessiveness, merely the loss of that factor. According to this view, new characters arise by loss of factors, hence the term loss mutation, to designate the mode of occurrence of some new Mendelian character.

It is impossible to enter into an extended presentation of the arguments for and against the presence-absence hypothesis at this point. Here it is sufficient to point out that multiple allelomorphism definitely disposes of it, at least in certain cases. According to the presence-absence hypothesis allelomorphism is restricted to a pair of conditions; either a given factor is present or it is absent. When a third member is added to an allelomorphic series, it becomes no longer tenable and, as has been seen, some series of allelomorphs contain as many as eleven members.

Since the presence-absence hypothesis is rejected, it is necessary to provide some substitute for it. Allelomorphism may now be said to depend upon the position of factors in the germinal system. A factor occupies a definite position in a certain chromosome. If two different factors occupy identical positions in homologous chromosomes, then they are allelomorphic, for the normal exchanges of material between homologous chromosomes in meiosis will never throw them into the same chromosomes. The mutation process still bears a relation to allelomorphism, but a different one from that postulated by the presence-absence hypothesis. A factor is regarded as a material entity, possibly a chemical compound, which is subject on rare occasions to changes in composition. A change in composition does not, however, give rise to a change in the position of the factor in the germinal system; consequently mutated factors are allelomorphic to the factors from which they arose. And since, *a priori*, there is no reason to believe that factors are subject to only one kind of change, multiple allelomorphism may simply be considered evidence that a given factor has mutated in a variety of different ways. Whether when two allelomorphic factors are present in an individual one will prove dominant and the other recessive in development is purely a matter of relationship between the two factors. Thus in the eye-color series in Drosophila, red is dominant to white and also to apricot; but apricot with white gives an intermediate eye color, a fact which can be determined only by experiment.

### References

BATESON: Mendel's Principles of Heredity, Chapter IV.

MORGAN, STURTEVANT, MULLER, and BRIDGES: Mechanism of Mendelian Heredity, Chapter VII.

MORGAN, BRIDGES, and STURTEVANT: Genetics of Drosophila, Bibliographia Genetica, 2, Chapter V.

PUNNETT: Mendelism, 6th ed., Chapter IV.

## CHAPTER XI

## SEX-LIMITED CHARACTERS

Chapter VII discussed at some length the type of transmission which appears when factors are borne in the heterosomes. Heterosomal and allosomal characters do not differ from autosomal characters in the kind of differences which they condition in the development of the individual, but they do differ in mode of transmission according as the genes which condition them are borne by the heterosomes or allosomes. In certain characters there is, however, a developmental relation to sex which is described below.

**"Bobbed" in Drosophila.**—In Drosophila there is a character called bobbed which appears only in females. Bobbed females have shortened, slender bristles on the thorax and scutellum, or some bristles may be entirely missing, and the abdomen usually has a scaly appearance. Males of a bobbed stock do not differ from wild-type males. Bobbed (symbol *bb*) therefore, exhibits its character effect only when associated with the female sex; but males of a bobbed stock, although they do not exhibit the character, do transmit it. In addition the case is somewhat complicated by the fact that bobbed is sex-linked. Accordingly, (*bb*X) (*bb*X) is a bobbed female exhibiting the differential characters described above, and (*bb*X)Y is genotypically a bobbed male, but phenotypically it is indistinguishable from an ordinary wild-type male.[1] Crosses between bobbed and wild type work out as follows:

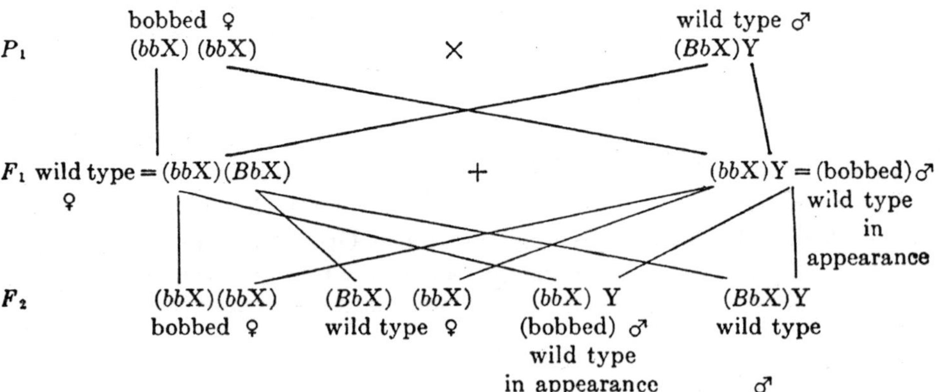


[1]Stern (*Biol. Zentr.*, 46: 344–348, 1926) has shown that the normal allelomorph of bobbed is also carried in the Y-chromosome, which explains the non-appeana of bobbed in males of a bobbed stock.

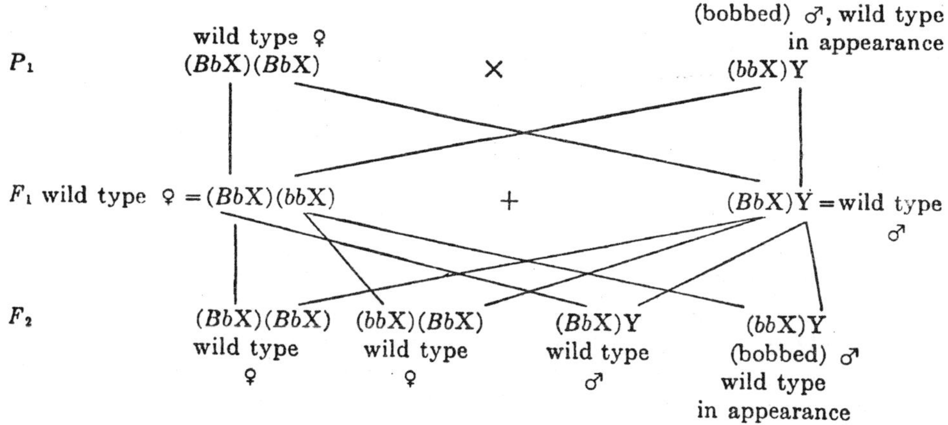


It should be remembered that bobbed males are always wild type in appearance. In the cross between wild-type female and bobbed male, therefore, the character is not seen at all in $P_1$, $F_1$, and $F_2$, but the student should be able to suggest matings which would bring it to expression in $F_3$.

**Eosin Eye Color in Drosophila.**—Eosin eye color, a member of the white series of allelomorphs, is a sex-linked character; but, like bobbed, it is also sex-limited. The sex limitation in this instance is shown only by the comparative degree of development of the character in the two sexes; the eye color in females is distinctly darker than that in males. A suggestion has been made that this difference in the development of the color in the two sexes depends upon the dosage of the factor, for females $(w^eX)$ $(w^eX)$ have two doses of it and males $(w^eX)Y$, only one. This may be the correct explanation, but since other members of the series do not show the same sexual bicolorism, it would seem to be more likely that the development of the character is associated in some way with the development of the sex differences.

Many other characters in Drosophila differ in degree of development in the two sexes, some only slightly, others to such an extent that the character develops only in one sex. The two examples which we have discussed happen to be sex-linked as well as sex-limited; but sex-limited characters may depend upon autosomal genes, as illustrated by the examples described below.

**Mahogany vs. Red Coat Color in Ayrshire Cattle.**—In Ayrshire cattle the favorite color is brown and white, the term including a number of different shades of red and almost any relative proportions of colored and white areas, although fashion seems to favor a large proportion of white. In the Ayrshire district of Scotland and occasionally also in America, a certain proportion of black and white animals occur. This black is not a true black, such as is found in Holstein-Friesian cattle, but a very dark mahogany. It is called mahogany, therefore, although this term is also used by Ayrshire breeders for a dark shade of brown. It shows the following relations to red in crosses:

| | | |
|---|---|---|
| $P_1$ | red ♀ × mahogany ♂ | mahogany ♀ × red ♂ |
| $F_1$ | red ♀♀ + mahogany ♂♂ | red ♀♀ + mahogany ♂♂ |
| $F_2$ | ♀♀ = 3 red: 1 mahogany<br>♂♂ = 1 red: 3 mahogany | ♀♀ = 3 red: 1 mahogany<br>♂♂ = 1 red: 3 mahogany |

These results simply show that this is a pair of Mendelian characters in which dominance is determined by sex. If $M$ = mahogany and $m$ = red, then $MM$ = mahogany and $mm$ = red in both sexes, but $Mm$ males are mahogany and $Mm$ females are red. This conception applied to the above results clears up all difficulties in connection with them.

**Horns in Sheep.**—In sheep different breeds exhibit three principal differences in development of horns. Breeds like Shropshire, Cotswold, and Suffolk are hornless in both sexes. In the American Merino and some other Merino breeds, the ewes are hornless and the rams have horns. In Dorset sheep (figure 36) both sexes are horned, but rams have heavier and larger horns than ewes. In some breeds the horned condition is not fixed. When Suffolk sheep, hornless in both sexes, are crossed with Dorsets, horned in both sexes, the results secured duplicate those of mahogany *vs.* red in Ayrshire cattle as shown below:

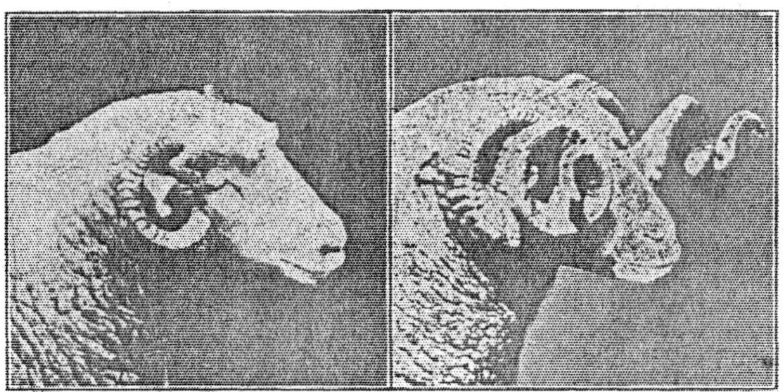

FIGURE 36.—Dorset sheep showing differences in development of horns in the two sexes. (*From Shaw and Heller.*)

| | | |
|---|---|---|
| $P_1$ | hornless ♀ × horned ♂ | horned ♀ × hornless ♂ |
| $F_1$ | hornless ♀♀ + horned ♂♂ | hornless ♀♀ + horned ♂♂ |
| $F_2$ | ♀♀ = 3 hornless: 1 horned<br>♂♂ = 1 hornless: 3 horned | ♀♀ = 3 hornless: 1 horned<br>♂♂ = 1 hornless: 3 horned |

Here again the results are readily accounted for, if it is assumed that the horned condition is dominant in males and recessive in females. Then, if $H$ = horned and $h$ = hornless, $HH$ individuals of both sexes are horned and $hh$ hornless; but $Hh$ males are horned and females hornless. The relation of those breeds which have hornless females and horned males to the above type has not been determined.

**Unisexual Polymorphism in Lepidoptera.**—In many species of butterflies in nature there are two or more distinct female types but only a single male type. More rarely the reverse condition is shown in which two or more distinct male types are opposed to a single female form. To this phenomenon the term unisexual polymorphism, has been given.

A simple illustration is afforded by the common yellow clover butterfly, *Colias philodice.* In nature the males are always yellow and most of the females are also yellow, but about 10 per cent are white. Breeding experiments conducted by Gerould show that the white females possess a dominant factor $W$, which exhibits its appropriate character effect only in females; males which possess it are yellow. Accordingly in a $WW$ stock, females will be white and males yellow, and in a $ww$ stock both sexes will be yellow. Although the actual experimental results were somewhat complicated by the existence of other factors, they show that the following results are to be expected from crosses between two such stocks:

| | | |
|---|---|---|
| $P_1$ | white ♀ × yellow ♂ | yellow ♀ × yellow ♂ |
| $F_1$ | white ♀♀ + yellow ♂♂ | white ♀♀ + yellow ♂♂ |
| $F_2$ | ♀♀ = 3 white: 1 yellow<br>♂♂ = all yellow | ♀♀ = 3 white: 1 yellow<br>♂♂ = all yellow |

Evidently yellow males may be $WW$, $Ww$, or $ww$; but $WW$ and $Ww$ females are white and only $ww$ females are yellow. Obviously since yellow females are most common in nature most of the males must be $ww$. These experiments demonstrate that unisexual polymorphism in the clover butterfly depends upon a sex-limited Mendelian character. The inheritance is comparable to that of bobbed in Drosophila, except that the character is dominant and autosomal, instead of being recessive and sex-linked.

The same white form appears in other species of Colias and the breeding experiments give identical results. It is also interesting to note that establishment of a homozygous $WW$ stock would give rise to a permanent dimorphism in color as respects the two sexes.

More complicated analyses are necessary where three or more female forms are found opposed to a single male type, but these also have given consistent results in breeding experiments. Thus in the butterfly, *Papilio polytes*, a native of the island of Ceylon, there are three types of females and only one type of male. The male type is called *polytes*, and one of the female types called by the same name is in all essential respects identical with it. The other two female types called *cyrus* and *romulus* are distinctly different from it and from each other, and the differences are in all cases discontinuous. Genetic studies indicate that males may occur in genotypes which correspond to all three of the female types, or in any heterozygous combination, to such an extent that a single *polytes* male mated to certain *cyrus* females will give all three types of females in the progeny, but only *polytes* males. The genetic analysis indicates the following relationship among the different types:

| romulus ♀ ♀ | polytes ♀ ♀ | cyrus ♀ ♀ |
|---|---|---|
| *AABB* | *AAbb* | *aaBB* |
| *AABb* | *Aabb* | *aaBb* |
| *AaBB* | | *aabb* |
| *AaBb* | | |

Here again it is perfectly clear from the genetic evidence that the remarkable feature of a number of different female types and a single male type within a given species depends upon sex-limited characters. Since this type of variation is not uncommon in butterflies and even in other species of animals, it is not without interest to have such a simple solution offered.

**Secondary Sexual Characters.**—The discussion of sex-limited characters leads naturally to a consideration of those characters, aside from sex differences proper, which normally distinguish the two sexes. In man the presence of a beard at maturity is a typical difference of this kind. In horses the stallion is commonly larger and more rugged than the mare; and a corresponding difference in degree of development is well known to occur in cattle, sheep, and swine. In some animals the differences are very striking, as in the Brown Leghorn fowl, the Rouen duck, and the ostrich. The male has a very highly developed, showy plumage and the female has a much more modest plumage.

In some breeds of poultry, such as Sebright bantams, the male has the same general type of plumage as the female; in other breeds, such as Hamburghs and Campines, two types of males appear, one with plumage like the female, called the hen-feathered type; the other with the more showy plumage normal for males, called the cock-feathered type. In many breeds only cock-feathered males are found. These differences in type of plumage make it possible to devise experiments to determine their genetic basis.

While the genetic basis of hen-feathering and cock-feathering has not been completely analyzed, certain experiments indicate that in some cases the relation is a simple Mendelian one with cock-feathering recessive to hen-feathering, and plumage type so related to sex in development that only males of the proper genetic constitution are cock-feathered. Let $H$ = hen-feathering and $h$ = cock-feathering. Then

$HH$ $Hh$, and $hh$ females all are hen-feathered; $HH$ and $Hh$ males are hen-feathered, and $hh$ males only are cock-feathered. With these relations established, the results of crosses work out as follows:

| | | | |
|---|---|---|---|
| $P_1$ | hen-feathered ($HH$) ♀ | × | cock-feathered ($hh$) ♂ |
| $F_1$ | hen-feathered ($Hh$) ♀♀ | + | hen-feathered ($Hh$) ♂♂ |
| $F_2$ | ♀♀ = all hen-feathered ($1HH$: $2Hh$: $1hh$)<br>♂♂ = 3 hen-feathered ($1HH$: $2\ Hh$): 1cock-feathered ($hh$) | | |
| $P_1$ | hen-feathered ($hh$) ♀ | + | hen-feathered ($HH$) ♂ |
| $F_1$ | hen-feathered ($Hh$) ♀♀ | + | hen-feathered ($Hh$) ♂♂ |
| $F_2$ | ♀♀ = all hen-feathered ($1HH$: $2Hh$: $1hh$)<br>♂♂ = 3 hen-feathered ($1HH$: $2Hh$): 1 cock-feathered ($hh$) | | |

Obviously the different genotypes of females which occur in $F_2$ may be demonstrated by mating them individually to cock-feathered males, and males may be tested by mating them individually to females from a stock which produces only cock-feathered males. The characters of main interest here are obviously sex-limited, as in the butterflies, except that the male sex is here the polymorphic one.

**The Nature of Secondary Sexual Characters.**—From the foregoing account secondary sexual differences between males and females of the same breed obviously may depend upon sex-limited characters. Males and females of such breeds may be of the same genetic constitution, aside from the sex determiners, but sex-limited characters are so related to sex that they develop differently in the two sexes. The factors which are responsible for these differences may or may not be sex-linked; in short they exhibit no differences from other genes in their mode of transmission.

That the differences are wholly developmental has been shown in certain cases by the removal of the sex glands. In birds removal of ovaries from females results in development of the cock-feathered type of plumage even when the operation is performed late in life. Castrated cock-feathered males do not change their plumage type significantly; but castrated hen-feathered cocks develop the cock-feathered type of plumage. Removal of ovaries from hens of breeds which produce hen-feathered cocks likewise results in development of cock-feathered plumage. These results presumably show that the ovaries of females produce a substance, possibly an internal secretion, which suppresses development of the cock-feathered type of plumage; and in hen-feathered males the testes have a similar effect. The developmental relations are, therefore, reduced to an effect which the sex glands themselves exert on plumage production. Further evidence as to the effect of secretions of the sex glands upon development is contained in Chapter XLIII.

**References**

MORGAN: Heredity and Sex, Chapters IV and V.

MORGAN, STURTEVANT, MULLER, and BRIDGES: Mechanism of Mendelian Heredity, pp. 106–114.

## CHAPTER XII

## INDEPENDENT ASSORTMENT

Mendel investigated not only crosses in which the parents differed in a single pair of characters but also others in which the parents differed in two or three pairs. As an example of the results which are secured in such cases, the data from his cross of smooth yellow with wrinkled green garden peas may be considered. The $F_1$ seeds were smooth yellow, exhibiting only the dominant characters, as might be expected from the results of monohybrid crosses. Fifteen plants were raised from hybrid seeds, and the $F_2$ seeds on these plants were distributed into four classes as follows:

| | Observed | Expected |
|---|---|---|
| Smooth yellow | 315 | 313 |
| Smooth green | 108 | 104 |
| Wrinkled yellow | 101 | 104 |
| Wrinkled green | 32 | 35 |

If the character pairs which enter into this hybrid are considered separately, it will be seen that they give 3:1 here as they do in unifactorial crosses. There were 423 smooth:133 wrinkled seeds which is in close agreement with an expectation of 417 smooth:139 wrinkled. For yellow *vs.* green, the numbers are 416 yellow:140 green, to be compared with an expectation of 417 yellow:139 green. Evidently the law of segregation, which accounts for the $F_2$ results in unifactorial crosses, is operative without modification when the original parents differ in more than one pair of characters.

Moreover, the inheritance in the two pairs of characters is independent, as may be seen by subjecting the data to analysis from this point of view. If the two pairs of characters are independent, the distribution into ultimate classes is expected to take the following form:

$$F_2 = \begin{cases} 3 \text{ smooth} \begin{cases} 3 \text{ yellow} = 9 \text{ smooth yellow} \\ 1 \text{ green} = 3 \text{ smooth green} \end{cases} \\ 1 \text{ wrinkled} \begin{cases} 3 \text{ yellow} = 3 \text{ wrinkled yellow} \\ 1 \text{ green} = 1 \text{ wrinkled green} \end{cases} \end{cases}$$

If Mendel's data are tested for conformance with this ratio, a very satisfactory agreement is obtained as shown in the second column above.

The 9:3:3:1 ratio then is typical of dihybrid results, when the two pairs of characters each taken separately give a 3:1 ratio.

In this instance, as in other cases, attention should be focused upon the factors rather than upon the characters which are finally secured. If the following symbols are used, $W$ = smooth, $w$ = wrinkled and $G$ = yellow, $g$ = green, then smooth yellow = $WWGG$ and wrinkled green = $wwgg$. The gametes of the smooth yellow plant are all $WG$ and of wrinkled green, $wg$. The $F_1$ zygote, therefore, is $WwGg$, a smooth yellow seed by reason of the dominance of smooth and yellow. The formation of gametes is governed by the law of segregation; *i.e.*, one member of each pair of factors passes into each gamete. But a gamete which received $W$ from the $Ww$ pair may obtain either $G$ or $g$ from the other pair, and likewise for gametes which receive $w$. As a consequence four kinds of gametes are formed in equal numbers, *viz.*,

$$WG + Wg + wG + wg.$$

| | $WG$ | $Wg$ | $wG$ | $wg$ |
|---|---|---|---|---|
| $WG$ | $WWGG$ smooth yellow | $WWGg$ smooth yellow | $WwGG$ smooth yellow | $WwGg$ smooth yellow |
| $Wg$ | $WWGg$ smooth yellow | $WWgg$ smooth green | $WwGg$ smooth yellow | $Wwgg$ smooth green |
| $wG$ | $WwGG$ smooth yellow | $WwGg$ smooth yellow | $wwGG$ wrinkled yellow | $wwGg$ wrinkled yellow |
| $wg$ | $WwGg$ smooth yellow | $Wwgg$ smooth green | $wwGg$ wrinkled yellow | $wwgg$ wrinkled green |

FIGURE 37.—$F_2$ checkerboard for Mendel's cross of smooth yellow with wrinkled green peas.

The four classes of gametes occur both in the ovules and in the pollen grains, so that the $F_2$ individuals arise as a result of random combination of the members of two such series of gametes. There are sixteen possible ways in which these four kinds of gametes may unite, as is most easily shown by the checkerboard method illustrated in figure 37. The female gametes are written along the left side of the checkerboard and the male gametes at the top. Each square is then filled in with the symbols of the factors which it receives from both the ovule and pollen grain; and the phenotype determined by these factors is then finally entered.

If like phenotypes are collected together from this checkerboard, writing the different genotypes under each one, the distribution shown in the third column below is secured:

| Observed | Expected | $F_2$ | $F_3$ expectation |
|---|---|---|---|
| | | 9 smooth yellow: | |
| 38 | 33 | 1 *WWGG* | all smooth yellow |
| 65 | 67 | 2 *WWGg* | 3 smooth yellow: 1 smooth green |
| 60 | 67 | 2 *WwGG* | 3 smooth yellow: 1 wrinkled yellow |
| 138 | 134 | 4 *WwGg* | 9 smooth yellow: 3 smooth green: 3 wrinkled yellow: 1 wrinkled green |
| | | 3 smooth green: | |
| 35 | 34 | 1 *WWgg* | all smooth green |
| 67 | 68 | 2 *Wwgg* | 3 smooth green: 1 wrinkled green |
| | | 3 wrinkled yellow: | |
| 28 | 32 | 1 *wwGG* | all wrinkled yellow |
| 67 | 63 | 2 *wwGg* | 3 wrinkled yellow: 1 wrinkled green |
| | | 1 wrinkled green: | |
| 30 | 30 | 1 *wwgg* | all wrinkled green |

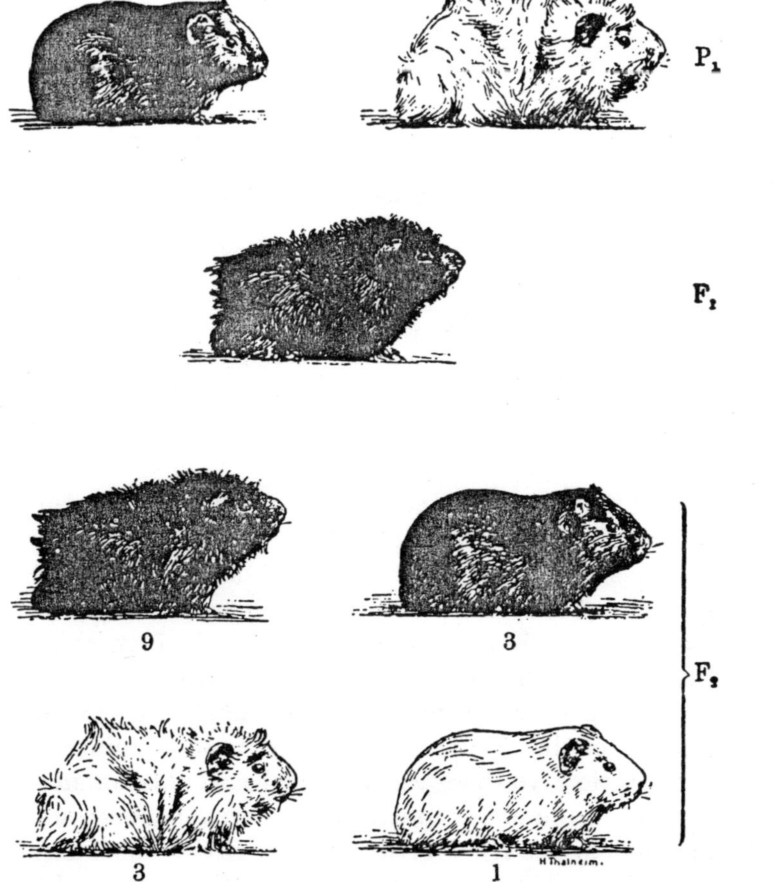


FIGURE 38.—Results of crossing smooth-black and rough-white guinea pigs. $F_1$ is rough black. $F_2$ is in the ratio 9 rough black : 3 rough white : 3 smooth black : 1 smooth white. (*After Castle from Baur, Vererbungslehre, 2nd. ed., copyrighted* 1914 *by Gebrüder Borntraeger. Reprinted by permission.*)

The distribution of individuals in the 9:3:3:1 ratio in $F_2$, therefore, has certain consequences in $F_3$ on account of the different genotypes included in the phenotypes which must be realized if the analysis is to be considered correct. Mendel planted all of his $F_2$ seeds in order to determine what types of $F_3$ progenies they would produce, and he found them distributed in the manner shown above. The data are in very satisfactory agreement with expectation on all points.

The distribution of the different members of two or more pairs of factors or characters with respect to one another is called *assortment.* From the foregoing case it appears that assortment is independent, although important exceptions to this behavior will be found in the further development of the subject. Mendel's second law, then, deals with independent assortment. It may be stated in general terms thus:

*The members of different pairs of factors segregate independently in the formation of gametes.*

| | *RC* | *Rc* | *rC* | *rc* |
|---|---|---|---|---|
| *RC* | *RRCC* rough black | *RRCc* rough black | *RrCC* rough black | *RrCc* rough black |
| *Rc* | *RRCc* rough black | *RRcc* rough white | *RrCc* rough black | *Rrcc* rough white |
| *rC* | *RrCC* rough black | *RrCc* rough black | *rrCC* smooth black | *rrCc* smooth black |
| *rc* | *RrCc* rough black | *Rrcc* rough white | *rrCc* smooth black | *rrcc* smooth white |

FIGURE 39.—Checkerboard showing $F_2$ results of the cross, smooth black × rough white in guinea pigs.

It is called the law of **Independent Assortment.** Independent assortment has been observed throughout the animal and plant kingdoms. Among animals important work has been done in the establishment of Mendelian principles by the use of rodents, in particular, mice, rats, guinea pigs, and rabbits. These animals are unusually favorable subjects for genetic investigations because they may be reared in large numbers and the time elapsing between generations is short. An excellent illustration (figure 38) of the results of a dihybrid cross in guinea pigs has been reported by Castle. In these animals rough coat $R$ is dominant to smooth $r$, and colored $C$ is dominant to white $c$. A cross between smooth colored $rrCC$ and rough white $RRcc$ gives an $F_1$ of $RrCc$, which is rough colored on account of the dominance of $R$ and $C$. The $F_1$ individuals of both sexes produce four kinds of gametes in equal numbers, *viz.*,

$$RC + Rc + rC + rc$$

The checkerboard for this cross (figure 39) shows how the $F_2$ ratio of 9 rough colored: 3 rough white: 3 smooth colored: 1 smooth white is accounted for. The results of this experiment taken together with those secured with the garden peas demonstrate that the type of combinations of characters exhibited by the parents has no effect on the final results. In the cross of smooth yellow with wrinkled green peas, both dominant characters entered from one parent, whereas in the cross of smooth black with rough white guinea pigs, one dominant and one recessive character entered through each parent. The $F_2$ progenies in both cases conform to the same ratio, *viz.*, 9:3:3:1.

If the $F_2$ genotypes are collected in their respective phenotypes, the distribution shown below occurs:

| | Backcross Expectation |
|---|---|
| 9 rough colored: | |
| 1 *RRCC* × *rrcc* | all rough colored |
| 2 *RRCc* × *rrcc* | 1 rough colored: 1 rough white |
| 2 *RrCC* × *rrcc* | 1 rough colored: 1 smooth colored |
| 4 *RrCc* × *rrcc* | 1 rough colored: 1 rough white<br>1 smooth colored: 1 smooth white |
| 3 rough white: | |
| 1 *RRcc* × *rrcc* | all rough white |
| 2 *Rrcc* × *rrcc* | 1 rough white: 1 smooth white |
| 3 smooth colored: | |
| 1 *rrCC* × *rrcc* | all smooth colored |
| 2 *rrCc* × *rrcc* | 1 smooth colored: 1 smooth white |
| 1 smooth white: | |
| 1 *rrcc* × *rrcc* | all smooth white |

The question now arises as to how the existence of the different genotypes may be demonstrated. The method used for garden peas is obviously impracticable, because it is necessary to mate two individuals together in order to get progeny, and it is impossible to be sure that the two individuals are of the same genotype. $F_2$ individuals can, however, be tested expeditiously by mating them singly to smooth white guinea pigs. The results of such matings differ in a diagnostic way for each genotype, and the number of individuals which it is necessary to secure when the test is made this way is small enough so that the method is practicable. This is the backcross method applied to analysis of an $F_2$ population. The same method can be used in order to demonstrate that the $F_1$ produces gametes of four kinds in equal numbers, for if $F_1$ is backcrossed to smooth white, the progeny fall into four classes of equal frequency; *viz.*, rough colored, rough white, smooth colored, and smooth white, corresponding to the four classes of gametes produced by $F_1$ hybrids. In bisexual animals and in dioecious plants, this test has been applied with $F_1$ individuals as both male and female parents. The results which are identical in the two cases indicate that sex has no effect upon assortment.

**Multifactor Hybrids.**—When the number of pairs of factors in which the original parents differ is greater than two, the same principles apply in segregation and assortment as for two pairs of factors, but the practical difficulties attending operations with them are so great that they are rarely used. In actual experimentation it is generally found advisable to reduce them to a simpler basis and to investigate them in sections, as it were. It very rarely happens that all the possible classes of individuals in multifactor hybrids are distinguishable in $F_2$, or that each pair of characters exhibits the complete dominance necessary to illustrate the typical conditions.

The method of deriving the $F_2$ terms of a more complex hybrid may be illustrated with an example from guinea pigs. The results of crossing smooth colored with rough white have been discussed. To these two pairs of characters a third may be added for length of hair; short-haired *S* being dominant over long-haired *s*. Castle describes the results of crossing a short-haired smooth colored race with a long-haired rough white one. $F_1$ was short-haired and rough colored, exhibiting the dominant character of each pair of factors. $F_2$ consisted of eight phenotypes, which may be determined by the following method:

| | | | |
|---|---|---|---|
| 3 short-haired | 3 rough | 3 colored | = 27 short-haired rough colored |
| | | 1 white | = 9 short-haired rough white |
| | 1 smooth | 3 colored | = 9 short-haired smooth colored |
| | | 1 white | = 3 short-haired smooth white |
| 1 long-haired | 3 rough | 3 colored | = 9 long-haired rough colored |
| | | 1 white | = 3 long-haired rough white |
| | 1 smooth | 3 colored | = 3 long-haired smooth colored |
| | | 1 white | = 1 long-haired smooth white |

The student should be able to secure the same results from application of the checkerboard method. The latter method is of value more particularly in instructional work. After the student has become acquainted with Mendelian ratios, he should, by appeal to the symmetry exhibited by them, be able to write them offhand. This statement applies to genotypes as well as phenotypes, as may be demonstrated by considering one of the above phenotypes, say short-haired rough colored, which is *SRC*, expressed in symbols. The genotypes which occur in this phenotype and their proportions are as follows:

27 *SRC* = short-haired rough colored

| | |
|---|---|
| 1 *SSRRCC* | 1 homozygous for all its factors |
| 2 *SSRRCc*<br>2 *SSRrCC*<br>2 *SsRRCC* | 2 each of every possible kind heterozygous for one pair of factors |
| 4 *SSRrCc*<br>4 *SsRrCC*<br>4 *SsRRCc* | 4 each of every kind heterozygous for two pairs of factors |
| 8 *SsRrCc* | 8 each of every kind heterozygous for three pairs of factors |

This method of writing the genotypes corresponding to given phenotypes is sometimes called the method of progression. It applies to any phenotype of any Mendelian hybrid, however complex, if the pairs of factors all assort independently. In the application of this method, it is advisable to start with one individual which is homozygous for all its factors, double this coefficient, and take all kinds heterozygous for one pair of factors; then double it and again take all possible combinations heterozygous for two pairs of factors, and so on until all combinations are exhausted. The student will find that the scheme works out automatically in such a way as to give the number of individuals called for in the ratio. By its use it is possible to write out any desired part of a Mendelian ratio without working out the whole of it.

There are many ways of dealing with more complex ratios, which are sometimes discussed at great length in textbooks. They are more interesting as algebraic problems, however, than they are useful in actual genetic experimentation. The principle involved in all of them is, of course, the same. They all depend upon a gametic ratio in which the number of classes is determined by the number of combinations of the different members of each heterozygous pair of genes, and the frequency of the different kinds of gametes is the same. The number of different classes of gametes, obviously, then is equal to $2^n$, if $n$ = the number of pairs of heterozygous factors, and the particular classes may be written out systematically by following a scheme of dichotomy, such as is illustrated herewith for a general three-factor hybrid:

$$AaBbCc\begin{cases} A\begin{cases} B\begin{cases} C = ABC \\ c = ABc \end{cases} \\ b\begin{cases} C = AbC \\ c = Abc \end{cases} \end{cases} \\ a\begin{cases} B\begin{cases} C = aBC \\ c = aBc \end{cases} \\ b\begin{cases} C = abC \\ c = abc \end{cases} \end{cases} \end{cases}$$

The phenotypes may be written out in the same manner by simply placing the coefficient 3 before each dominant factor and multiplying coefficients together in order to obtain the proper coefficients for the final terms. There are obviously three distinct series involved in every Mendelian analysis; (1) the gametic series, the classes and ratios in which the gametes are produced; (2) the phenotypic series, the classes and ratios in which the phenotypes are produced; and (3) the genotypic series, the classes and ratios in which the genotypes are produced. Stated in algebraic terms, these series may be expressed as follows:

Gametic series $= (A + a)(B + b)(C + c) \ldots$
Phenotypic series $= (3A + a)(3B + b)(3C + c) \ldots$
Genotypic series $= (AA + 2Aa + aa)(BB + 2Bb + bb)(CC + 2Cc + cc) \ldots$

the expressions to be continued for as many terms as there are pairs of heterozygous factors in the hybrid. The student will be able to appreciate the difficulties involved in dealing with complex hybrids by simply expanding these expressions for four or five terms. Table 3 contains a summary of some of the features of $F_2$ populations arising from hybrids heterozygous for from one to six pairs of factors and in the last column general terms which may be used for computation with any number of pairs of factors.

TABLE 3.—PROPORTIONS EXISTING IN MENDELIAN EXPERIMENTS INVOLVING VARIOUS NUMBERS OF FACTOR DIFFERENCES

| Number of pairs of factors | 1 | 2 | 3 | 4 | 5 | 6 | $n$ |
|---|---|---|---|---|---|---|---|
| Number of different kinds of gametes | 2 | 4 | 8 | 16 | 32 | 64 | $2^n$ |
| Number of combinations of gametes.. | 4 | 16 | 64 | 256 | 1,024 | 4,096 | $4^n$ |
| Number of homozygotes in $F_2$....... | 2 | 4 | 8 | 16 | 32 | 64 | $2^n$ |
| Number of heterozygotes in $F_2$....... | 2 | 12 | 56 | 240 | 992 | 4,032 | $4^n - 2^n$ |
| Number of kinds of genotypes in $F_2$... | 3 | 9 | 27 | 81 | 243 | 729 | $3^n$ |
| Number of kinds of homozygous genotypes....................... | 2 | 4 | 8 | 16 | 32 | 64 | $2^n$ |
| Number of kinds of heterozygous genotypes....................... | 1 | 5 | 19 | 65 | 211 | 665 | $3^n - 2^n$ |

**Modifications of the Typical Ratio.**—The 9:3:3:1 ratio for dihybrids and the corresponding higher ratios for multifactor hybrids are, of course, to be expected only when the individual pairs of Mendelian characters conform to the 3:1 ratio of perfect dominance. If they conform to some one of the other ratios described in Chapters VIII to X, a corresponding modification is to be expected in the multifactor ratio,

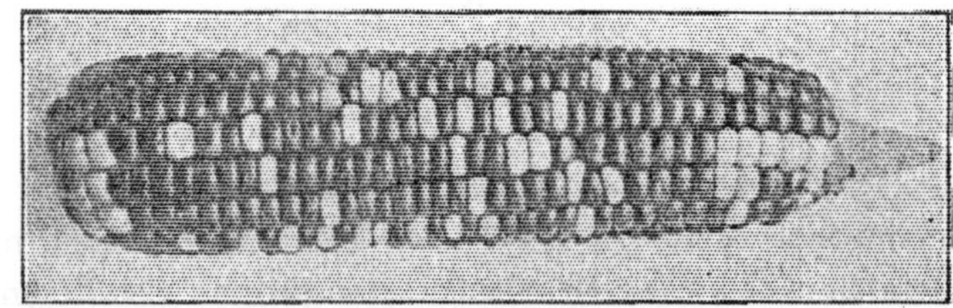

FIGURE 40.—Maize ear showing $F_2$ segregation of grains in the ratio of 3 purple : 1 white.

when they are studied in association with other factors. The simple principle to be observed is to carry over into the multifactor problem those idiosyncrasies of behavior which are characteristic of the component pairs of Mendelian characters. A few examples should be sufficient to clarify this situation.

There are white and purple races of corn which differ in a single pair of genes; $C$ = purple, $c$ = white. Purple is dominant to white in $F_1$, no matter which way the cross is made, and the $F_2$ ratio is 3 purple:1 white (figure 40). The relation of floury to flinty has been described in Chapter VIII. Here the maternal parent is dominant in $F_1$, and the $F_2$ ratio is 1 flinty:1 floury, which, as has been seen, depends upon the fact that two doses of either factor are dominant over one dose of the other in endosperm development. Now if these facts regarding the two pairs of characters are kept in mind, it can readily be seen that:

purple flinty ♀ × white floury ♂ = $F_1$, purple flinty
purple floury ♀ × white flinty ♂ = $F_1$, purple floury
white flinty ♀ × purple floury ♂ = $F_1$, purple flinty
white floury ♀ × purple flinty ♂ = $F_1$, purple floury

Each $F_1$ will give the same $F_2$ results according to the following scheme:

$$F_2 = \begin{cases} 3 \text{ purple} \begin{cases} 1 \text{ flinty} = 3 \text{ purple flinty} \\ 1 \text{ floury} = 3 \text{ purple floury} \end{cases} \\ 1 \text{ white} \begin{cases} 1 \text{ flinty} = 1 \text{ white flinty} \\ 1 \text{ floury} = 1 \text{ white floury} \end{cases} \end{cases}$$

The $F_2$ ratio to be expected in these crosses then is 3 purple flinty:3 purple floury:1 white flinty:1 white floury, not 9:3:3:1. The student should be able to arrive at the same results by assigning the proper factor symbols to the parental races and following the analysis out to $F_2$.

The same method may be used when one of the characters is sex-linked. In Drosophila long (wild type) is dominant to vestigal wing type, and inheritance is autosomal. Red (wild type) *vs.* white eye color is sex-linked, and follows the scheme described in detail in Chapter VII. Consequently, when the two pairs of characters are followed together, the results shown below are to be expected:

$P_1$, red long ♀ × white vestigial ♂ = $F_1$, red long ♀♀ + red long ♂♂
$P_1$, red vestigial ♀ × white long ♂ = $F_1$, red long ♀♀ + red long ♂♂
$P_1$, white long ♀ × red vestigial ♂ = $F_1$, red long ♀♀ + white long ♂♂
$P_1$, white vestigial ♀ × red long ♂ = $F_1$, red long ♀♀ + white long ♂♂

In deriving $F_2$ the first two crosses may be treated together, for they give the same results, as follows:

$$F_2 = \begin{cases} 3 \text{ long} \begin{cases} 2 \text{ red } ♀♀ = 6 \text{ red long } ♀♀ \\ 1 \text{ red } ♂ = 3 \text{ red long } ♂♂ \\ 1 \text{ white } ♂ = 3 \text{ white long } ♂♂ \end{cases} \\ 1 \text{ vestigial} \begin{cases} 2 \text{ red } ♀♀ = 2 \text{ red vestigial } ♀♀ \\ 1 \text{ red } ♂ = 1 \text{ red vestigial } ♂ \\ 1 \text{ white } ♂ = 1 \text{ white vestigial } ♂ \end{cases} \end{cases}$$

Similarly, $F_2$ from the second two crosses, each of which gives $F_1$ consisting of red long females and white long males, is determined in the following way:

$F_2$ =

- 3 long
  - 1 red ♀ = 3 red long ♀ ♀
  - 1 white ♀ = 3 white long ♀ ♀
  - 1 red ♂ = 3 red long ♂ ♂
  - 1 white ♂ = 3 white long ♂ ♂
- 1 vestigial
  - 1 red ♀ = 1 red vestigial ♀
  - 1 white ♀ = 1 white vestigial ♀
  - 1 red ♂ = 1 red vestigial ♂
  - 1 white ♂ = 1 white vestigial ♂

The factorial method of arriving at the $F_2$ in these two cases is shown in figures 41 and 42, respectively.

| | *V*(*w*X) | *v*(*w*X) | *V*Y | *v*Y |
|---|---|---|---|---|
| *V*(*W*X) | *VV*(*W*X)(*w*X)<br>long red<br>♀ | *Vv*(*W*X)(*w*X)<br>long red<br>♀ | *VV*(*W*X)Y<br>long red<br>♂ | *Vv*(*W*X)Y<br>long red<br>♂ |
| *V*(*w*X) | *VV*(*w*X)(*w*X)<br>long white<br>♀ | *Vv*(*w*X)(*w*X)<br>long white<br>♀ | *VV*(*w*X)Y<br>long white<br>♂ | *Vv*(*w*X)Y<br>long white<br>♂ |
| *v*(*W*X) | *Vv*(*W*X)(*w*X)<br>long red<br>♀ | *vv*(*W*X)(*w*X)<br>vestigial<br>red ♀ | *Vv*(*W*X)Y<br>long red<br>♂ | *vv*(*W*X)Y<br>vestigial<br>red ♂ |
| *v*(*w*X) | *Vv*(*w*X)(*w*X)<br>Long white<br>♀ | *vv*(*w*X)(*w*X)<br>vestigial<br>white ♀ | *Vv*(*w*X)Y<br>long white<br>♂ | *vv*(*w*X)Y<br>vestigial<br>white ♂ |

FIGURE 41.—Checkerboard analysis of $F_2$ obtained from crossing vestigial white ♀ with long red ♂ or long white ♀ with vestigial red ♂ and mating $F_1$ *inter se* in Drosophila.

| | *V*(*W*X) | *v*(*W*X) | *V*Y | *v*Y |
|---|---|---|---|---|
| *V*(*W*X) | *VV*(*W*X)(*W*X)<br>long red<br>♀ | *Vv*(*W*X)(*W*X)<br>long red<br>♀ | *VV*(*W*X)Y<br>long red<br>♂ | *Vv*(*W*X)Y<br>long red<br>♂ |
| *V*(*w*X) | *VV*(*W*X)(*w*X)<br>long red<br>♀ | *Vv*(*W*X)(*w*X)<br>long red<br>♀ | *VV*(*w*X)Y<br>long white<br>♂ | *Vv*(*w*X)Y<br>long white<br>♂ |
| *v*(*W*X) | *Vv*(*W*X)(*W*X)<br>long red<br>♀ | *vv*(*W*X)(*W*X)<br>vestigial<br>red ♀ | *Vv*(*W*X)Y<br>long red<br>♂ | *vv*(*W*X)Y<br>vestigial<br>red ♂ |
| *v*(*w*X) | *Vv*(*W*X)(*w*X)<br>long red<br>♀ | *vv*(*W*X)(*w*X)<br>vestigial<br>red ♀ | *Vv*(*w*X)Y<br>long white<br>♂ | *vv*(*w*X)Y<br>vestigial<br>white ♂ |

FIGURE 42.—Checkerboard analysis of $F_2$ obtained by crossing vestigial red ♀ with long white ♂ or long red ♀ with vestigial white ♂ and mating $F_1$ *inter se* in Drosophila.

From these results the principle is established that prediction in complex cases depends upon a knowledge of the behavior of the individual character pairs.

**The Mechanism of Assortment.**—The appeal to a cytological basis for segregation was made to reduction, which provides an obvious mechanism for the separation of allelomorphs in gamete production. For assortment the obvious appeal is to random orientation of bivalents in gametogenesis. The implication of this conception, in addition to those

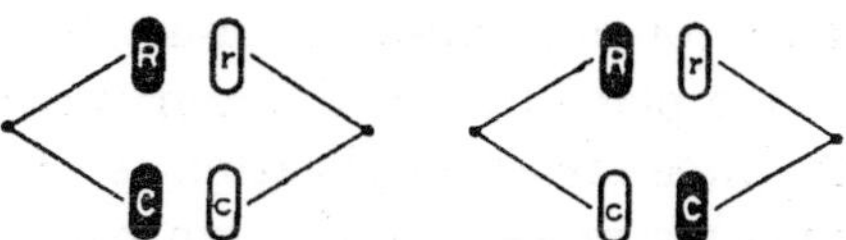


FIGURE 43.—Orientation of chromosomes as a basis for independent assortment in a two-factor guinea-pig cross.

which are operative in segregation, is that the different pairs of allelomorphs are borne in different pairs of chromosomes. If then the different pairs of chromosomes are oriented at random on the reduction spindle, different combinations of members of the different pairs of genes will result in accordance with the requirements of Mendelian experimentation.

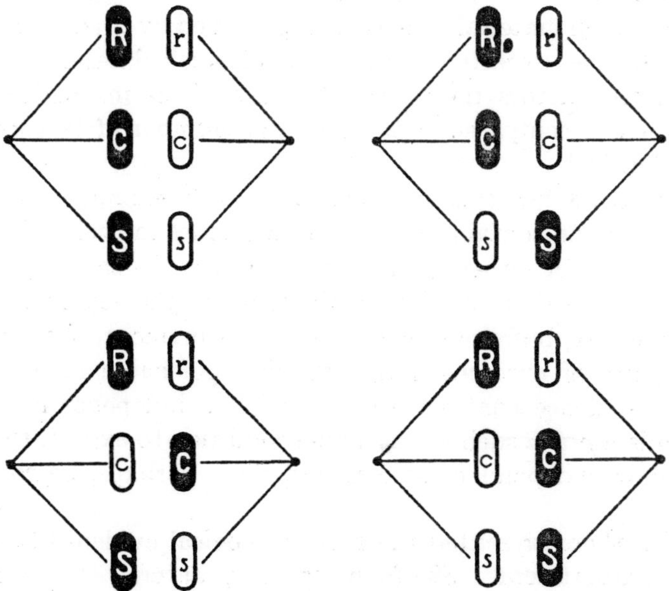


FIGURE 44.—Orientation of chromosomes as a basis for independent assortment in a three-factor guinea-pig hybrid.

An illustration of the application of random orientation as a physical explanation for independent assortment is shown in figure 43 for a two-factor hybrid. In the guinea-pig cross described on page 100, the two pairs of factors *R-r* and *C-c* are assumed to be borne in different pairs of chromosomes. Obviously the orientation of the two pairs of chromosomes may take place in two different ways, giving rise to the four kinds of gametes which have been shown to be produced by such two-factor hybrids. Segregation occurs in each pair of factors in the same fashion as was depicted for one-factor hybrids.

This conception may be applied to more complex hybrids and fulfils all the requirements of independent assortment. In a three-factor hybrid, for example, three pairs of chromosomes must be taken into account, the orientation may take place in four different ways, and eight kinds of gametes result. The three-factor application is illustrated in figure 44 for the trihybrid guinea-pig cross described on page 102. In these illustrations one member of each pair of chromosomes is in full black and the other in outline. This mode of representation has been adopted simply to make the diagrams more effective. It should not mislead the student into thinking that the entire chromosomes are different, for ideally it is assumed in these cases that the two members of a pair differ solely in the pair of factors represented by the symbols.

**Cytological Observations on Assortment.**—In most cases it is too much to expect to be able to observe the process of assortment in cytological preparations. Homologous chromosomes are usually identical in appearance, and in material which is being studied by Mendelian methods, it can hardly be expected that morphological differences would be observed in the factors they carry, because of the insignificance of the differences when compared with the large number of factors in which they are identical.

Nevertheless a few species which have been examined cytologically have been found to contain pairs of chromosomes in their cells the members of which differ either in size or more often in shape. In bisexual animals of course, there is often a difference in the sex chromosomes of the digametic sex, but in order to observe assortment, there must be at least two pairs of chromosomes with distinguishable members. Since there is no evidence contrary to the belief in independent assortment, and since it is *a priori* such a reasonable condition to expect, the few cases which do show independent assortment, may, perhaps, be considered as crucial.

Miss Carothers presented the first cytological evidence in support of independent assortment. She found in some material of grasshoppers of the genus Trimerotropis that in addition to the sex chromosome which is unpaired in males, one pair of chromosomes had unequal members which may be distinguished from each other. She counted 300 first spermatocytes in this material and found that in 51.3 per cent of the figures the large member was going to the same pole as the sex chromosome and in 48.7 per cent the smaller member was accompanying the sex chromosome. Confirmation of this evidence has been provided by Voinov who studied assortment in *Gryllatalpa vulgaris;* by Wenrich who counted 472 primary

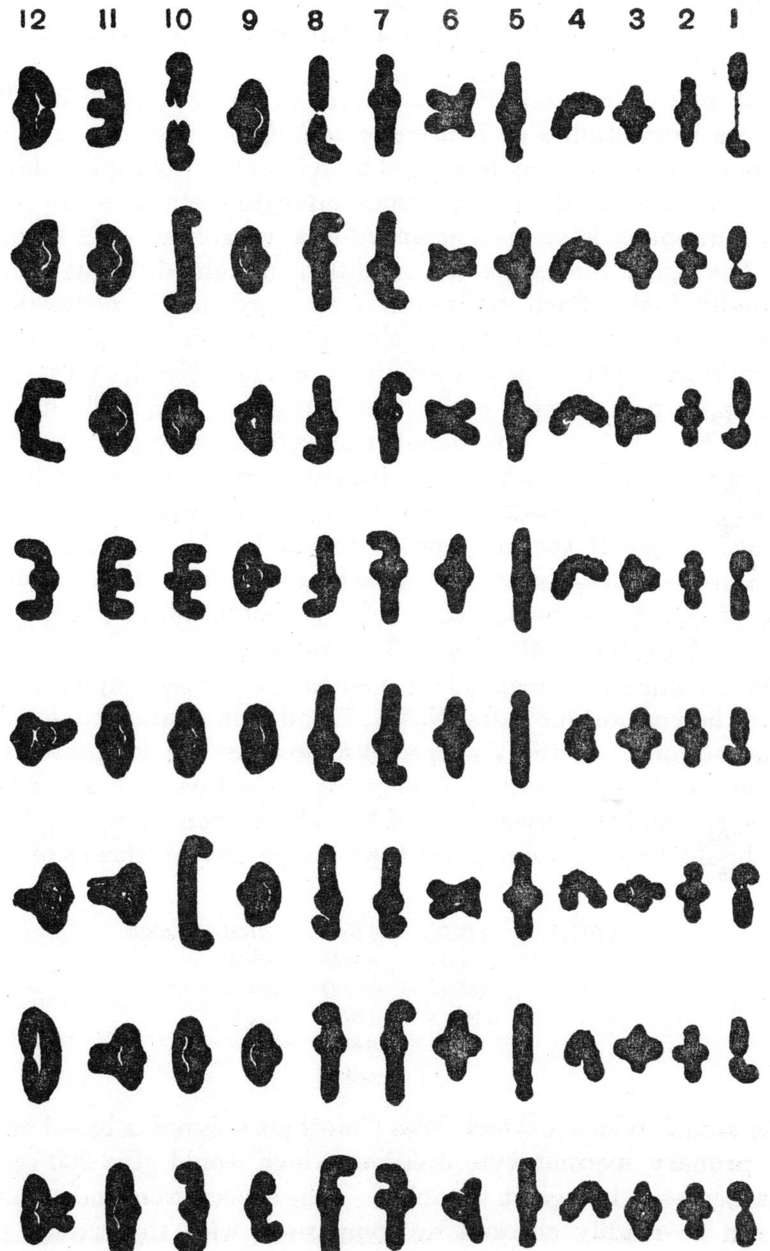


**Figure 45.—Lateral views of chromosomes in the first spermatocyte of eight cells from a single individual. The chromosomes of each cell are arranged horizontally in a series according to size. The vertical rows represent the same pair of chromosomes in different cells. Note the difference in shape of separating members in pairs Nos. 1, 7, and 8, and the unpaired sex-chromosome, No. 4.** (*From Miss Carothers.*)

spermatocytes in *Phrynotettix magnus* in which a distinguishable pair was present in addition to the unpaired sex chromosome; and by Robertson for studies in the related species *Tettigidae parvipennis* and *Acridium granulatum.*

Confirmation in more complex material has been published by Miss Carothers from studies of *Trimerotropis fallax.* The nature of the evidence in this example may be judged by reference to figure 45, which shows the chromosomes of the primary spermatocyte metaphases in side view. The chromosomes have been arranged in a horizontal series according to size. The figure represents the condition in eight different cells in the same individual. Each horizontal row represents a complete set of chromosomes of a cell, and the columns represent the same tetrads in different cells. The unpaired sex chromosome is No. 4, and it is passing to the upper pole in each case. The tetrads in which the homologues differ are Nos. 1, 7, 8. One member of each of these pairs has terminal spindle attachment, and hence is straight in form; the other has a subterminal attachment, which pulls it out into a V shape. There is a slight difference in size of the members of tetrad 1. The members of tetrad 7 and 8 are indistinguishable from each other. The other chromosomes consist of equal members; either both are subterminal (Nos. 9, 10, 11, and 12) or both terminal (Nos. 2, 3, 5, and 6).

The evidence for independent assortment is based upon a study of the three heteromorphic pairs (Nos. 1, 7, and 8) in relation to the unpaired sex chromosome. If the V-shaped chromosomes are designated by capital letters *A, B, C, D,* and their homologues in Nos. 1, 7, and 8 by small letters *a, b, c,* and the absent mate of No. 4 by *d,* then obviously the assortment should be of a four-factor type giving sixteen classes of gametes, *viz.,*

| *ABCD* | *ABCd* | *ABcd* | *Abcd* | *abcd* |
|---|---|---|---|---|
| | *ABcD* | *AbcD* | *aBcd* | |
| | *AbCD* | *abCD* | *abCd* | |
| | *aBCD* | *aBCd* | *abcD* | |
| | | *AbCd* | | |
| | | *aBcD* | | |

The actual evidence which Miss Carothers secured is based on counts of 100 primary spermatocyte divisions which would give 200 secondary spermatocytes. It is given in table 4. The expectations indicated in this table can be readily checked by comparison with the sixteen possible types of gametes listed above. The agreement between theory and expectation is on the whole excellent. In the absence of any contradictory evidence, it may be concluded that cytological results favor the assumption of independent assortment in reduction.

**Practical Applications.**—The foregoing experiments illustrate under the simplest conditions the methods employed in obtaining new forms in animal and plant breeding. If two existing varieties contain characters which it is desired to unite in a single variety, the direct method of procedure is (1) to cross them, (2) to grow a comparatively small $F_1$, (3) to grow as large an $F_2$ as possible in order to be sure to obtain the proper combinations, and (4) to establish the new form in a homozygous condition by selection in subsequent generations.

TABLE 4.—DISTRIBUTION OF MEMBERS OF HETEROMORPHIC PAIRS OF CHROMOSOMES IN A SINGLE INDIVIDUAL OF *Trimerotropis fallax* (*Data from Eleanor Carothers*)

| Type | Expected | Observed |
|---|---|---|
| A given V (Nos. 1 or 4) | ½ × 200 = 100 | 100 |
| Only one V | ¼ × 200 = 50 | 48 |
| Two given Vs (Nos. 1 and 4) | ¼ × 200 = 50 | 46 |
| Two given Vs (Nos. 7 and 8) | ¼ × 200 = 50 | 47 |
| Any two Vs | ⅜ × 200 = 75 | 84 |
| Three given Vs (Nos. 1, 7, 8) | ⅛ × 200 = 25 | 22 |
| Three given Vs (Nos. 4, 7, 8) | ⅛ × 200 = 25 | 21 |
| Any three Vs | ¼ × 200 = 50 | 48 |
| Four Vs | 1/16 × 200 = 12½ | 8 |

The experiments also illustrate the fact that individuals may possess the same characters but may differ in the uniformity with which they transmit them, because of the different genotypes which may be included in a phenotype. In practical operations differences in transmitting ability may be ascertained by obtaining progenies of different individuals, a method known as the progeny test. Breeding ability, therefore, is disclosed by the progeny test, and selection for fixation of a desired type is most effective when based upon the progeny test, rather than merely upon the expressed characters of the individual.

### References

MORGAN: The Physical Basis of Heredity, Chapters IV and V
MORGAN: The Theory of the Gene, Chapters I–III.
SINNOTT and DUNN: Principles of Genetics, Chapter IV.

## CHAPTER XIII

# FACTOR INTERACTIONS IN DEVELOPMENT

In the hybrids which have been discussed in the preceding chapters, pairs of characters have been selected which are not only inherited independently but which also do not conflict in development, so that all possible combinations of the characters may be readily recognized in successive generations. In garden peas, for example, a plant may bear either smooth or wrinkled seeds, and these in turn may be either yellow or green in color. There is here no interference in development because the individual character pairs, smooth *vs.* wrinkled and yellow *vs.* green, affect different features of the seed. When, however, the individual characters affect the same feature of the organism, it is necessary to consider, not only the distribution of the factors in successive generations, but also the interactions which take place among them in development.

**Comb Characters in Fowls.**—A simple illustration of interaction between two pairs of allelomorphs in development is afforded by crosses between comb types in poultry (figure 46). It was first found that rose comb is dominant to single, producing, when crossed with it, rose comb $F_1$ and $F_2$ in the ratio 3 rose:1 single. Likewise, pea comb is dominant to single; $F_1$ of single × pea is pea, and $F_2$ is in the ratio 3 pea:1 single. When pea is crossed with rose, $F_1$ is walnut, a different type from either rose or pea, and $F_2$ is in the ratio 9 walnut:3 rose:3 pea:1 single.

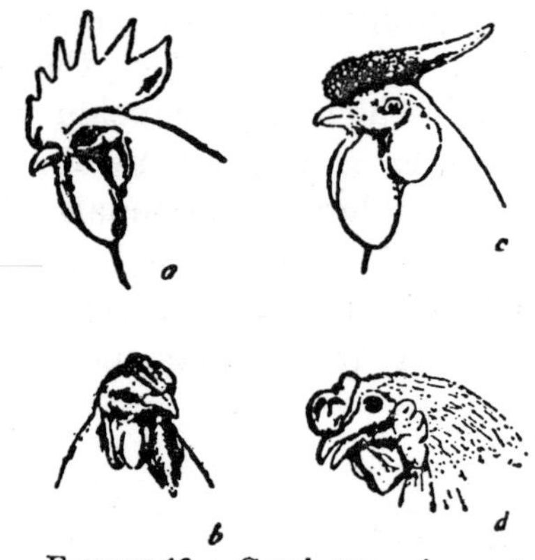


FIGURE 46.—Comb types in poultry. Single, *a*; pea, *b*; rose, *c*; walnut, *d*. (*From Morgan.*)

This cross may be analyzed in the following fashion. The cross between rose and single shows that rose differs from single in a dominant factor, $R$; so that single is $rr$ and rose, $RR$. Similarly the cross between pea and single shows that pea differs from single in a dominant factor, $P$, hence pea = $PP$ and single = $pp$. But the single-combed fowls in the two crosses were identical, so that they both must have been $rrpp$. It is now possible to complete the formulae of rose and pea. The former differs from single in a dominant factor only; hence it must be $RRpp$, and correspondingly pea comb must be $rrPP$.

These formulae may now be applied to an analysis of the results of crossing rose with pea comb from the factor standpoint. The $F_1$ is

*RrPp*, a new genotype containing both *R* and *P*. The particular expression of such a genotype cannot be predicted from preliminary results such as have been described. It can be determined only by trial. The appearance of single in $F_2$, however, is to be expected from such a cross on the basis of the previously known behavior of rose and pea with single. The distribution of factors is exactly the same as in typical cases, as may be seen by reference to figure 47. This experiment illustrates how new characters may be produced by crossing. When the two pairs of factors affect the same organ, as they do in this race, it can be predicted, on the basis of general Mendelian principles, how the factors will be distributed, but it will be possible only to guess at the character effects which they will produce by interaction in the development of the individual.

| | *RP* | *Rp* | *rP* | *rp* |
|---|---|---|---|---|
| *RP* | *RRPP* walnut | *RRPp* walnut | *RrPP* walnut | *RrPp* walnut |
| *Rp* | *RRPp* walnut | *RRpp* rose | *RrPp* walnut | *Rrpp* rose |
| *rP* | *RrPP* walnut | *RrPp* walnut | *rrPP* pea | *rrPp* pea |
| *rp* | *RrPp* walnut | *Rrpp* rose | *rrPp* pea | *rrpp* single |

FIGURE 47.—Checkerboard analysis of theoretical expectations in $F_2$ from a cross between a rose-comb *RRpp* and a pea-comb fowl *rrPP*.

**Eye Colors in Drosophila.**—A case which sometimes puzzles beginners is that of two mutant characters in Drosophila which affect the same organ. Thus the recessive eye color, purple, when crossed with wild-type (red) eye color, gives wild-type eye color in $F_1$, and 3 wild type:1 purple in $F_2$. Similarly another eye color, scarlet, when crossed with wild type, gives wild type in $F_1$, and 3 wild type:1 scarlet in $F_2$. The question arises as to how far the behavior can be predicted in a cross between purple and scarlet and what procedure can be followed in making predictions.

In any Mendelian problem of this kind, the first task is always to write the genotypic formulae of the parents. Given these formulae it is possible to set in operation the various processes by which the results of crossing them can be written out. In contrast to purple = *prpr*, wild type = *PrPr* on the basis of crosses between them. Similarly scarlet *stst* is set in contrast to wild type *StSt* on the basis of the results which they give when crossed together. The wild type in the two cases is the same; therefore, its complete formula must be *PrPrStSt*. Since purple and scarlet both differ in a single pair of factors from wild type, their formulae must be *prprStSt* and *PrPrstst*, respectively. Having assigned complete formulae to the parents, it is now possible to predict the results from crossing them. $F_1$ would be *PrprStst,* which would be wild type, and $F_2$ would give

9 wild type
3 purple
3 scarlet
1 purple-scarlet

The last class is called purple-scarlet in the Drosophila system of nomenclature because of its genotypic constitution *prprstst.* It is possible to predict its formation in $F_2$, but its appearance can be determined only by actually securing it. The designation purple-scarlet is not, therefore, descriptive of its appearance, but solely of its genetic composition. In color it is actually lighter and somewhat different in quality from scarlet, which is a transparent eye color of a bright vermilion tone.

**Allelomorphic and Non-allelomorphic Characters.**—In Chapter X a number of cases were described which may profitably be compared with those which have just been considered. It was shown, for example, that white and apricot eye color, sex-linked characters in Drosophila, both differ from wild-type eye color in a unifactorial relation, and that white crossed with apricot gives $F_1$ females having an intermediate eye color, and $F_2$ results which indicate that these two types differ from each other in a single factor. These results were explained on the assumption that wild type $= W$, apricot $= w^a$, and white $= w$ are multiple allelomorphs, so that a cross between any two members of the series gives unifactorial results. When two recessive eye colors, such as apricot and white are allelomorphic, the $F_1$ will not be wild type, because its constitution $w^aw$ is not equivalent to the wild-type condition. When two recessive eye colors are non-allelomorphic, such as purple and scarlet, $F_1$ is wild type, because each contributes to the $F_1$ the factor in which the other differs from wild type. Properly applied, this method makes it possible readily to distinguish between characters which make up an allelomorphic series and those which depend upon different pairs of factors.

**The 9 : 7 $F_2$ Ratio.**—In the preceding illustrations of factor interactions drawn from comb type in poultry and eye color in Drosophila, the typical ratio for the dihybrid, 9:3:3:1, was secured, because new, readily distinguishable classes were produced by interaction between the factors. In a number of instances, however, classes of identical appearance are produced, which depend upon different genetic constitutions.

In point of history, the first case to be analyzed satisfactorily was that of flower color in sweet peas. Bateson found in the white sweet pea, Emily Henderson, two strains, one of which had round pollen grains and the other long. He crossed these two types and found to his surprise a colored $F_1$, resembling the wild Sicilian form. The $F_2$ plants were distributed in the ratio of approximately 9 colored:7 white. The colored individuals were of six different color classes, but since these differences depend upon differences in factors other than those which determine presence or absence of color, they may be disregarded. Shape of pollen grains may also be neglected, because it depends upon a different pair of allelomorphs.

The explanation of these results depends upon a difference in genetic constitution of the two white forms which were crossed. This assumption may be developed in the following way: Assume that the two whites are genetically distinct and so related to colored that each differs from colored in a single recessive factor. Then if $c$ is one white, $C$ must be colored; and if $p$ is the other white, $P$ would be colored so far as it is concerned. Since the colored form is the same in both cases, it must contain both $C$ and $P$, and a pure-colored race will, therefore, be $CCPP$. Now since the two white races differ from colored in only one recessive factor, their complete formulae must be $ccPP$ and $CCpp$, respectively. This is equivalent to saying that color in sweet peas depends upon an interaction between $C$ and $P$: if one or both of these factors is replaced by its corresponding allelomorph, no color develops. For this reason $C$ and $P$ are called complementary factors.

Given the complete formulae of the parental types, it is now possible to proceed with the analysis of the results of crossing them. White $ccPP \times CCpp$ white gives in $F_1$ $CcPp$, which is colored because crossing has brought together the complementary factors $C$ and $P$ for color production. In the production of $F_2$, segregation, assortment, and random fertilization result in the usual distribution of two pairs of factors, so that the following classes and proportions appear:

$$
\begin{aligned}
9\ CP &= \text{colored} \\
3\ Cp &= \text{white} \\
3\ cP &= \text{white} \\
1\ cp &= \text{white}
\end{aligned}
$$

Collecting white classes, the $F_2$ ratio becomes 9 colored: 7 white.

Results such as these illustrate the flexibility of Mendelian analysis. Critics have retorted that any results may be interpretated on a Mendelian basis, if enough assumptions are made. Geneticists must be able to present a satisfactory reply to this criticism. The reply is based upon the stipulation that any Mendelian analysis must be consistent, not only with the immediate experimental data from which it was drawn, but also with results which are secured in subsequent generations with the material in question, and with those which are secured from any type of cross between the races in question. Thus in the foregoing example, if the results secured depend upon interaction between two pairs of factors, certain consequences should follow in $F_3$, which may be determined by writing out the $F_2$ genotypes as follows:

| $F_2$ phenotypes | $F_2$ genotypes | $F_3$ ratios |
|---|---|---|
| 9 *CP* = colored | 1 *CCPP* | all colored |
| | 2 *CCPp* | 3 colored:1 white |
| | 2 *CcPP* | 3 colored:1 white |
| | 4 *CcPp* | 9 colored:7 white |
| 3 *Cp* = white | 1 *CCpp* | all white |
| | 2 *Ccpp* | all white |
| 3 *cP* = white | 1 *ccPP* | all white |
| | 2 *ccPp* | all white |
| 1 *cp* = white | 1 *ccpp* | all white |

From the above analysis it should be observed that all white $F_2$ individuals should give white progeny exclusively. The colored individuals will give different types of $F_3$ progenies, depending upon their genotypes. Moreover, the different types of progenies in $F_3$ should occur in stipulated proportions; *viz.*, 1 (all colored):4 (3 colored:1 white):4 (9 colored:7 white). This trial has been made and the results agree with expectation. It is also evident that a colored $F_1$ backcrossed to a double-recessive white *ccpp* should give a progeny distributed in a ratio of 1 colored: 3 white. Here again expectation and experimental results are in agreement. When a complex analysis is followed through all its ramifications, as shown above, the agreement between expectation and actual evidence points as conclusively to the correctness of the analysis as it does in simpler cases.

**The 15:1 $F_2$ Ratio.**—As a contrast to the foregoing results may be presented the results which have been secured from studies of yellow *vs.* *white endosperm color in maize. There are at least two different genetic* types of yellow endosperm in maize which are simple dominants to white. Either one gives with white a yellow $F_1$ and 3 yellow:1 white in $F_2$. Assuming that the yellows are genetically distinct, what would be the procedure for predicting the results of crossing them?

Let *L* represent one gene for yellow, *l* its corresponding allelomorph; *Y* the other, paired with *y*. Then the results mentioned above for crosses between yellow and white are accounted for, if in one case *LL* = yellow and *ll* = white; and in the other, if *YY* = yellow and *yy* = white. But since the same white was used in both cases, its formula, considering both pairs of factors, must be *llyy*; and then the complete formula for the two yellow races, since they differ from white in a single dominant factor, must be *LLyy* and *llYY*, respectively. It is now possible to proceed with the problem. Yellow, *LLyy* × *llYY*, also yellow, gives *LlYy* in $F_1$, which is also yellow. This fact can be established only by experimentation, because the $LY$ combination might conceivably produce some other result. For production of $F_2$ the factors are distributed according to the usual scheme, as follows:

$$\begin{aligned} 9\,LY &= \text{yellow} \\ 3\,Ly &= \text{yellow} \\ 3\,lY &= \text{yellow} \\ 1\,ly &= \text{white} \end{aligned}$$

Collecting like $F_2$ phenotypes, this ratio reduces to 15 yellow: 1 white. The experimental results are in agreement with expectation. The student should be able, by writing out the $F_2$ genotypes, to determine the $F_3$ expectations for this case.

| ♀ ↓ / ♂ → | *CD* | *Cd* | *cD* | *cd* |
|---|---|---|---|---|
| *CD* → | *CD · CD* 1:0 | *CD · Cd* 1:0 | *CD · cD* 1:0 | *CD · cd* 15:1 |
| *Cd* → | *Cd · CD* 1:0 | *Cd · Cd* 1:0 | *Cd · cD* 15:1 | *Cd · cd* 3:1 |
| *cD* → | *cD · CD* 1:0 | *cD · Cd* 15:1 | *cD · cD* 1:0 | *cD · cd* 3:1 |
| *cd* → | *cd · CD* 15:1 | *cd · Cd* 3:1 | *cd · cD* 3:1 | *cd · cd* 0:1 |

FIGURE 48.—Checkerboard diagram to visualize the genetic relations in a dihybrid $F_2$ family of *Bursa bursa-pastoris* × *Heegeri*, in respect to the capsule characters. The capsules figured in each square indicate by their outline their phenotype, and by their oblique ruling their genotype, the gene $C$ being represented by lines from upper right to lower left, and $D$ from upper left to lower right. Homozygotes are densely lined, heterozygotes more sparsely. The ratios indicate the expectation in $F_3$ when a plant having the genotypic constitution indicated in the same square, is self-fertilized. (*From Shull.*)

Results of the foregoing type are not at all uncommon in genetic experimentation. The term duplicate factors has been suggested as an appropriate one for different factors which produce identical dominant character effects when contrasted with some one recessive form. A particularly good example in the common shepherd's purse, *Bursa*

(*Capsella*) *bursa-pastoris*, has been worked out very thoroughly by Shull. Here there are duplicate genes $C$ and $D$ for the common triangular capsule shape as opposed to the round or top-shaped type. Segregation of a triangular $F_1$, $CcDd$, with consequent behavior in $F_3$, is shown in figure 48. Shull has also shown that duplicate genes for certain leaf types occur in this species.

**The 13:3 F₂ Ratio.**—In maize there is a recessive white, represented by the factor $r$ which with colored $R$ gives a colored $F_1$ $Rr$, and 3 colored: 1 white in $F_2$. In contrast to this behavior there is a dominant white, represented by $I$, which with colored $i$ gives a white $F_1$ $Ii$, and 3 white: 1 colored in $F_2$. With these facts it is possible to predict the results of crossing dominant white with recessive white.

The colored type is the same in both cases; therefore, its formula must be written $iiRR$. The two whites differ from it in a single factor; therefore, they must have the formulae dominant white $= IIRR$ and recessive white $= iirr$. Dominant white $IIRR \times iirr$ recessive white then gives $F_1$, $IiRr$, white, which, on selfing, gives the following results in $F_2$:

9 $IR$ = white
3 $Ir$ = white
3 $iR$ = colored
1 $ir$ = white

Collecting like phenotypes, this $F_2$ reduces to 13 white:3 colored. For practice the student should write out the $F_2$ genotypes and determine $F_3$ expectations.

**The 9:3:4 F₂ Ratio.**—In guinea pigs, agouti crossed with some types of albino gives agouti $F_1$ and 3 agouti:1 white in $F_2$. A black guinea pig crossed with agouti gives agouti $F_1$ and 3 agouti:1 black in $F_2$. What results would be secured by crossing black with the above type of albino?

Let $A$-$a$ be the pair of factors upon which the distinction between agouti and black depends, and $C$-$c$, that responsible for colored *vs.* white. Then the genetic constitutions of the above races must be as follows, in order to yield the results described:

$AACC$ = agouti
$AAcc$ = albino
$aaCC$ = black

Then black $aaCC \times AAcc$ white would give $F_1$, $AaCc$ agouti, and an $F_2$ distributed as follows:

9 $AC$ = agouti
3 $Ac$ = white
3 $aC$ = black
1 $ac$ = white

The final results, therefore, are 9 agouti: 3 black: 4 white.

It will be observed from the foregoing analysis that it is possible to have two different homozygous white genotypes, *AAcc* and *aacc*, which may be termed agouti white and black white, respectively, although they will, of course, be indistinguishable except by appropriate breeding tests. Black whites *aacc* are readily produced by mating $F_2$ blacks together. If enough such matings are made, some of them will be of the type *aaCc* × *aaCc*, which will give rise to progenies in the ratio of 3 black:1 white, and from the parentage these whites will be *aacc* and not *AAcc*, like those at the beginning. So with other colored varieties of guinea pigs, it is possible to establish a corresponding white genotype which differs from its colored counterpart in having *c* instead of *C*. Accordingly, whites in guinea pigs, and in animals and plants generally, may exhibit a great variety of differences in genetic constitution. These differences may be unmasked by appropriate crosses with colored races, like the one described above. In practical breeding operations crosses of white with colored forms are often employed for the purpose of securing new color combinations.

**The 12:3:1 $F_2$ Ratio.**—In maize, as has been seen, white may be dominant to purple in a monohybrid relation depending on the *I-i* pair of factors. There is also a red aleurone color which is a simple recessive to purple, depending upon the *Pr-pr* pair of factors. With respect to these two pairs of factors the races under consideration have the following formulae:

*IIPrPr* = dominant white
*iiPrPr* = purple
*iiprpr* = red

When dominant white *IIPrPr* is crossed with red *iiprpr*, $F_1$ is white *IiPrpr*, and $F_2$ conforms to the ratio 12 white:3 purple:1 red, as may be seen from the following distribution:

9 *IPr* = white
3 *Ipr* = white
3 *iPr* = purple
1 *ipr* = red

Again is illustrated a very common modification of the typical dihybrid ratio, for which numerous examples might be described.

**The 9:6:1 $F_2$ Ratio.**—In swine there are breeds which have a red coat color, as for example the Duroc Jersey. There are breeds with a sandy yellowish coat color and there are white breeds, in addition to black and various mixtures which are of no concern here. The genetic relation of the colors has not been adequately established, but certain results which Wentworth has published indicate that there are two genetically distinct but phenotypically indistinguishable sandy types which bear the relations to red and white indicated by the following formulae:

$SSTT$ = red
$SStt$ = sandy
$ssTT$ = sandy
$sstt$ = white

If this analysis is correct, sandy $SStt \times ssTT$ sandy should give red $F_1$ $SsTt$ and 9 red: 6 sandy: 1 white in $F_2$, according to the following distribution:

9 $ST$ = red
3 $St$ = sandy
3 $sT$ = sandy
1 $st$ = white

This type of modification of the $F_2$ ratio is very rare. It may profitably be compared with the inheritance of yellow endosperm color in maize, which gives a 15:1 ratio.

**Aleurone Color in Maize.**—In the foregoing examples, attention has been confined to dihybrid cases of factor interactions, because they illustrate the principle involved in the simplest possible way. For the same reason, attention has been restricted to those instances where complete dominance is shown in each pair of factors. Many cases have, however, been worked out which are much more complex, as may be illustrated by considering the factors concerned in aleurone color in maize.

Among others there are six pairs of factors which bear the following relations in aleurone color production:

*I-i* — *I* is a dominant white gene which completely inhibits color production
*A-a*, *C-c*, *R-r* — Basic genes for color production, so related that *A*, *C* and *R* must all be present for color production
*Pr-pr* — *Pr* = purple, *pr* = red with *iACR*
*In-in* — *In* = dilute coloration, *in* = intense coloration in conjunction with the necessary basic color factors

The number of different kinds of homozygous races possible with these six pairs of factors is $2^6 = 64$. Of these only the following four are colored:

*iInACRPr* = dilute purple
*iInACRpr* = dilute red
*iinACRPr* = purple
*iinACRpr* = red

The other sixty genotypes are all white and indistinguishable from one another phenotypically, although readily recognizable by appropriate genetic tests. At least two of the pairs of factors are closely linked, *I-i* with *C-c*, so that it is not possible to work out all the relations among the possible races by application of the principles hitherto adduced. In fact these factors may represent a system of triple allelomorphs, so that for *I-i* and *C-c* in the above factorial tabulation it may be necessary to substitute the following:

$C^I$-*C*-*c* $C^I$ is a dominant white gene which completely inhibits color production; *C*, with *A* and *R*, is a basic gene for color production; *c*, recessive white

No better example of complex interrelations between factors could well be cited.

Although this system of factors is very complex, it has been worked out comparatively thoroughly. Doubtless not all possible crosses have been made and analyzed, but enough has been done to establish fully the correctness of the analysis. As an example of crosses which have been followed, consider the progeny of an $F_1$ of the genetic constitution *AaCcRr*. To complete its formula, *iiInInPrPr* might be added, but since $F_1$ is homozygous for these factors, all the descendants will likewise be homozygous for them, so that nothing is gained by writing them into the $F_1$ formula, even though they are undeniably present. The $F_2$ will be as follows:

| | |
|---|---|
| 27 *ACR* | = purple |
| 9 *ACr* | = white |
| 9 *AcR* | |
| 9 *aCR* | |
| 3 *Acr* | |
| 3 *aCr* | |
| 3 *acR* | |
| 1 *acr* | |

Collecting terms, the $F_2$ ratio is 27 colored:37 white. The colored individuals are of the following genotypes:

| $F_2$ genotypes | $F_3$ expectations |
|---|---|
| 1 *AACCRR* | all colored |
| 2 *AACCRr* | 3 colored : 1 white |
| 2 *AACcRR* | |
| 2 *AaCCRR* | |
| 4 *AACcRr* | 9 colored : 7 white |
| 4 *AaCcRR* | |
| 4 *AaCCRr* | |
| 8 *AaCcRr* | 27 colored : 37 white |

All white $F_2$ individuals are expected to give solely white offspring. The analysis in this case has been carried into the $F_3$, and results have been found to agree with expectation.

Conclusive demonstration of the correctness of the analysis is afforded by the isolation of three distinct races of white which may be called *a*-white, *c*-white, and *r*-white and which have the formulae *aCR*, *AcR*, and *ACr*, respectively. All possible crosses have been made between them with the following results in $F_1$:

| | *a*-white | *c*-white | *r*-white |
|---|---|---|---|
| *a*-white | white | colored | colored |
| *c*-white | colored | white | colored |
| *r*-white | colored | colored | white |

It is possible to determine the genetic constitution of unknown white races by crossing them with these three established races, for which reason they are known as aleurone testers.

**Distribution of Factors vs. Interaction in Development.**—It is possible to extend the foregoing types of cases indefinitely, but enough has been done to demonstrate adequately the principle involved in all of them. There are two distinct features which should not be confused; the first is the distribution of factors; the second, their interaction in development. In all the cases described the distribution of the factors is governed by the general principles laid down in Chapters V and XI; and hence prediction as to this part of the problem may be made, if the individual pairs of factors have been identified. The matter of interaction in development is quite another thing. Here no predictions can be made with certainty. All that can be done is to bring the factors into their various combinations by taking advantage of the operation of the hereditary mechanism. Then having secured the combinations, it may be observed how they interact in development. It is not intended to imply that an experienced investigator is entirely unable to make any predictions as to the way in which factors will interact in development. He usually can say with a fair degree of probability what the result will be, but his prediction is a guess and does not have the scientific validity of predictions as to the distribution of factors. Very often the investigator obtains very surprising results of interactions between factors, so that for certainty it is necessary to secure the combinations and observe their characters.

**The Relation between Factors and Characters.**—Some thought should be given at this time to the relation between factors and characters. What is implied, for example, by the statement that *Pr* is a factor for purple aleurone color and *pr*, an allelomorphic one for red?

In considering this problem it is helpful to go back for a moment to the facts of development. According to the conception which has been presented thus far, the chromosomes bear the factors. Obviously then the fertilized egg contains the full complement of factors upon which the development of characters in the individual depends. As has been seen development is essentially a process of multiplication in number of cells, achieved by cell division, deposition of non-living products of cellular activity, and differentiation of cells to perform specialized functions. In the process of increase in number of cells, the chromatin is always partitioned equationally between the two daughter cells, so that every cell in the body must have the entire complement of factors which were present in the original fertilized egg. So far as is known, there is no distribution of particular factors to different portions of the body where they then perform their specialized functions. On the contrary every factor is present in every cell.

The factor, therefore, is a member of a complex system, which in part provides the setting for its activities. Without this setting it would mean nothing at all. When it is said, therefore, that *Pr* is a factor for purple aleurone color and *pr* for red, it is understood that *Pr* in a certain factor complex gives rise to purple aleurone color, whereas *pr* in exactly the same complex gives rise to red. That is to say, if *R* be a symbol representing the remaining portion of the germinal system, then *R-Pr* develops into purple and *R-pr* into red. *Pr* and *pr* do not, properly speaking, produce purple and red aleurone, respectively; they merely act as differentials for purple and red. The actual production of either of these colors depends upon these factors acting in conjunction with many others. This conception is readily grasped, if certain of the known genes are considered which are present in the *R* complex. From the analytical studies of aleurone color production in maize, which have been described above, it is known that *R* must contain the factors *A*, *C*, and *R*, for the simultaneous presence of these three factors is necessary for the production of any color whatsoever. With respect to *Su-su*, the factors for starchy *vs.* sugary endosperm, it is immaterial which member is present, for these factors do not play any known differential role in color production. On the other hand the presence of one or the other of them may be absolutely essential for proper functioning of the germinal system, so that although they do not play a differential role in aleurone color production, they may nevertheless be essentially concerned in it at some stage in the developmental process. It is probably better, then, to consider all of the factors as essential to normal development; and all perhaps play some part in the production of every character in the individual. Particular factors act as differentials, which under certain conditions direct the developmental processes in certain definite ways.

From this point of view, there are some genticists who object to such statements as "*Pr* is a factor for purple," or "*Pr* is the purple factor," because *Pr* is only one of many factors necessary for the production of aleurone color, and no one of them is more essential than any other. On the same grounds, some go so far as to object to the assignment of mnemonic symbols such as *Pr* for factors, preferring to assign arbitrary symbols such as *A*, which bear no obvious relation to the characters in question. Some also go so far as to insist on writing all the known factors concerned in the production of a given character even though most of them are identical in the two races which are under immediate consideration.

It is doubtful, however, whether or not anything is gained by insisting upon these points. To most students the mnemonic system is useful, because it affords a consistent scheme of notation that recalls to their minds not only relations between factors, but also certain relations between characters under stipulated conditions. Nothing in particular is gained by the assignment of arbitrary symbols to factors, because after all their relations to characters must be described, and it is only by their relation to characters that their existence is known, so that one way or another there is just as much chance of confusion concerning the relations between factors and characters as when mnemonic symbols are employed. Similarly, writing complete formulae does not contribute to clarity of thought, because such formulae are never really complete. There still remains the large unanalyzed portion of the germinal substances, aptly termed the residual heredity, which doubtless contains numerous factors essential to the production of the characters under consideration. In working Mendelian problems, the simplest method is to write out only those factors in which the parental races differ, although the student should be fully cognizant of the essential rôle which other factors, alike in the two races, play in the production of the characters in question.

**References**

Punnett: Mendelism, 6th ed., Chapter V.
Sinnott and Dunn: Principles of Genetics, Chapter V.
Walter: Genetics, 2d ed., Chapter VII.

## CHAPTER XIV

## LINKAGE

Independent assortment applies to factors which are borne in different chromosomes, and it appears to depend upon random orientation of bivalents in the reduction division. There is, however, a very obvious limitation to this conception. In *Drosophila melanogaster*, which has only four pairs of chromosomes, in excess of 400 pairs of factors have been demonstrated. On a chromosome basis they must be distributed among the four pairs of chromosomes; hence each chromosome must contain a large number of them. Obviously then the next problem of heredity is to determine the relations among factors which are contained in the same chromosome. The development of this portion of the subject rests largely on the results of the Drosophila investigations, so that naturally most of the illustrations will be drawn from that source.

**Linkage of Sex-linked Factors.**—A natural mode of approach is to consider as a starting point factors which are known to be borne in the same chromosome. It was shown in Chapter VII that the peculiar mode of distribution of sex-linked factors is a consequence of their location in the X-chromosome. The distributional relations, therefore, which they exhibit should provide the necessary evidence on the question at issue.

In *Drosophila melanogaster* over 100 pairs of sex-linked factors have been demonstrated. Since sex-linked characters affect all portions of the fly in every conceivable way, it is self-evident that no grouping of factors in reference to chromosome location is possible on the basis of developmental effects. Some other method must be discovered.

Here again it is necessary to devise the simplest possible experiments in order to disclose the relations most clearly. This may be done in the first instance by considering the distribution of two pairs of sex-linked factors. Almost any two will serve this purpose; so the selection may include the recessive yellow body color contrasted with gray of the wild-type fly, and white eye color contrasted with red of the wild type. Inheritance of the latter has already been considered in detail; inheritance of the former follows the same scheme, if yellow be substituted for white and gray for red throughout. As formerly, let *W-w* be the pair of factors for red *vs.* white eye color, and let *Y-y* stand for gray *vs.* yellow body color. The following combinations of characters are then possible, with genetic constitutions as shown:

| | ♀♀ | ♂♂ |
|---|---|---|
| Gray red | $(YWX)(YWX)$ | $(YWX)Y$ |
| Gray white | $(YwX)(YwX)$ | $(YwX)Y$ |
| Yellow red | $(yWX)(yWX)$ | $(yWX)Y$ |
| Yellow white | $(ywX)(ywX)$ | $(ywX)Y$ |

These races may easily be established by appropriate methods, as will become clear from ensuing discussion.

Consider first a cross between a gray white female $(YwX)(YwX)$ and a yellow red male $(yWX)Y$. $F_1$ will consist of gray red females $(YwX)(yWX)$ and gray white males $(YwX)Y$. Obviously only the $F_1$ females are heterozygous for both pairs of factors; consequently attention should be confined to their hereditary behavior. What it is desired to determine are the classes of gametes which such a female produces and the proportions in which they occur. The most effective way to proceed is to mate $F_1$ females to yellow white males, the double recessive race. When this is done, the progeny fall into four character classes as follows:

| Phenotypes | Percentages | Female gametes | |
|---|---|---|---|
| Gray white | 49.5 | $(YwX)$ | 99 per cent parental combinations |
| Yellow red | 49.5 | $(yWX)$ | |
| Gray red | 0.5 | $(YWX)$ | 1 per cent recombinations |
| Yellow white | 0.5 | $(ywX)$ | |

The second column above gives the percentages of the different kinds of individuals produced, and the third column shows the kinds of gametes which the $F_1$ female must have produced in order to give these results. From the gametic series produced by the $F_1$ female, it is perfectly clear that the law of independent assortment does not apply to factors located in the same pair of chromosomes; if it applied in this instance, the four classes should have been produced in equal proportions. Instead there are two large classes $(YwX)$ and $(yWX)$ and two small $(YWX)$ and $(ywX)$. The two large classes contain the factors in the same combinations in which they occurred in the parents and are, therefore, called parental combinations. The two smaller classes represent combinations of the factors different from those of the parents and are, therefore, called recombinations. The essential feature of the above experiment is a demonstration that factors borne in the same chromosome are so related to each other that taken two by two they tend to emerge from a hybrid more often in parental combinations than in recombinations. **Linkage** is the term applied to this phenomenon. It may be defined as *a relation between factors such that they are distributed to gametes in pairs of parental combinations more often than in recombinations.*

That this statement of linkage is correct may be demonstrated by introducing the factors concerned in the above illustration in different

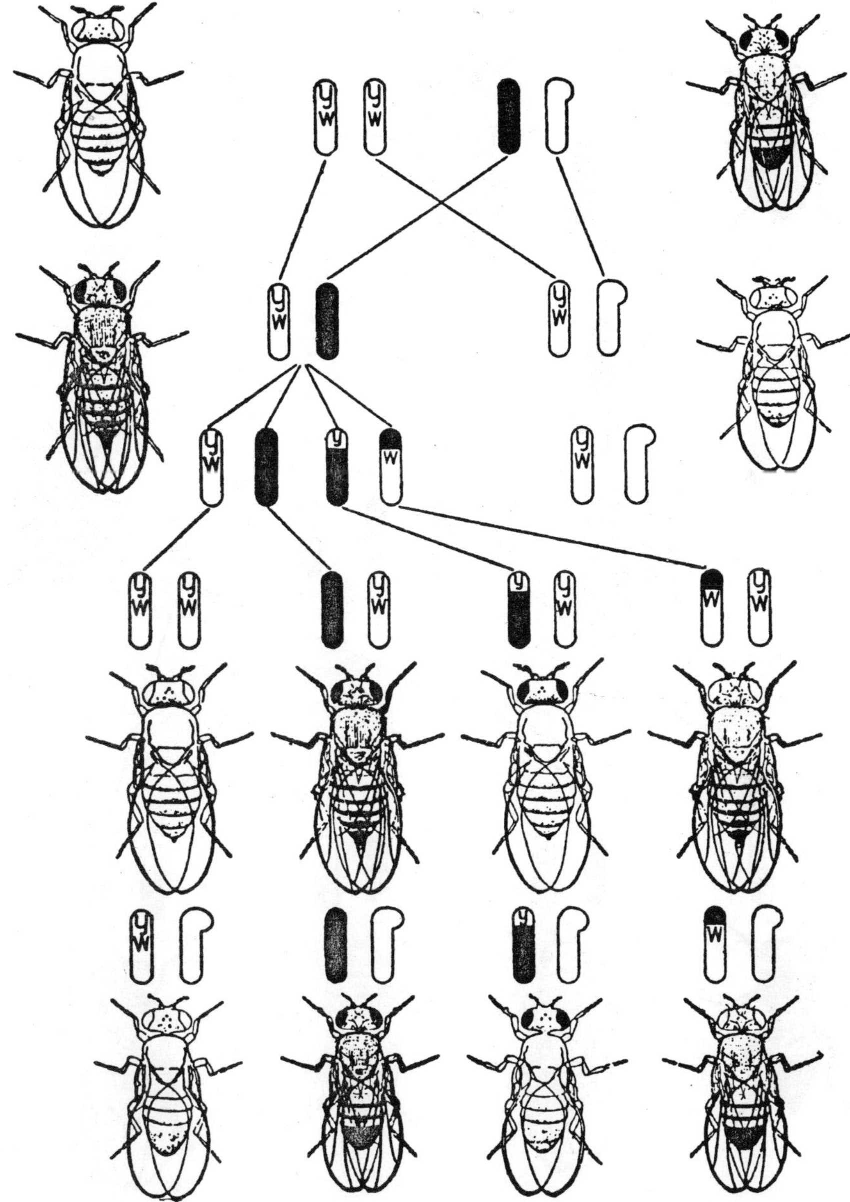


FIGURE 49.—Diagrammatic representation of the results of a cross of a yellow-white female with a gray-red male in *Drosophila melanogaster*. The $F_1$ gray-red females are mated to their yellow-white brothers for a determination of the number and proportion of gametic classes produced by the female. (*From Morgan*.)

parental combinations (figure 49). This may be done by crossing a yellow white female ($yw$X)($yw$X) with a gray red male ($YW$X)Y. $F_1$

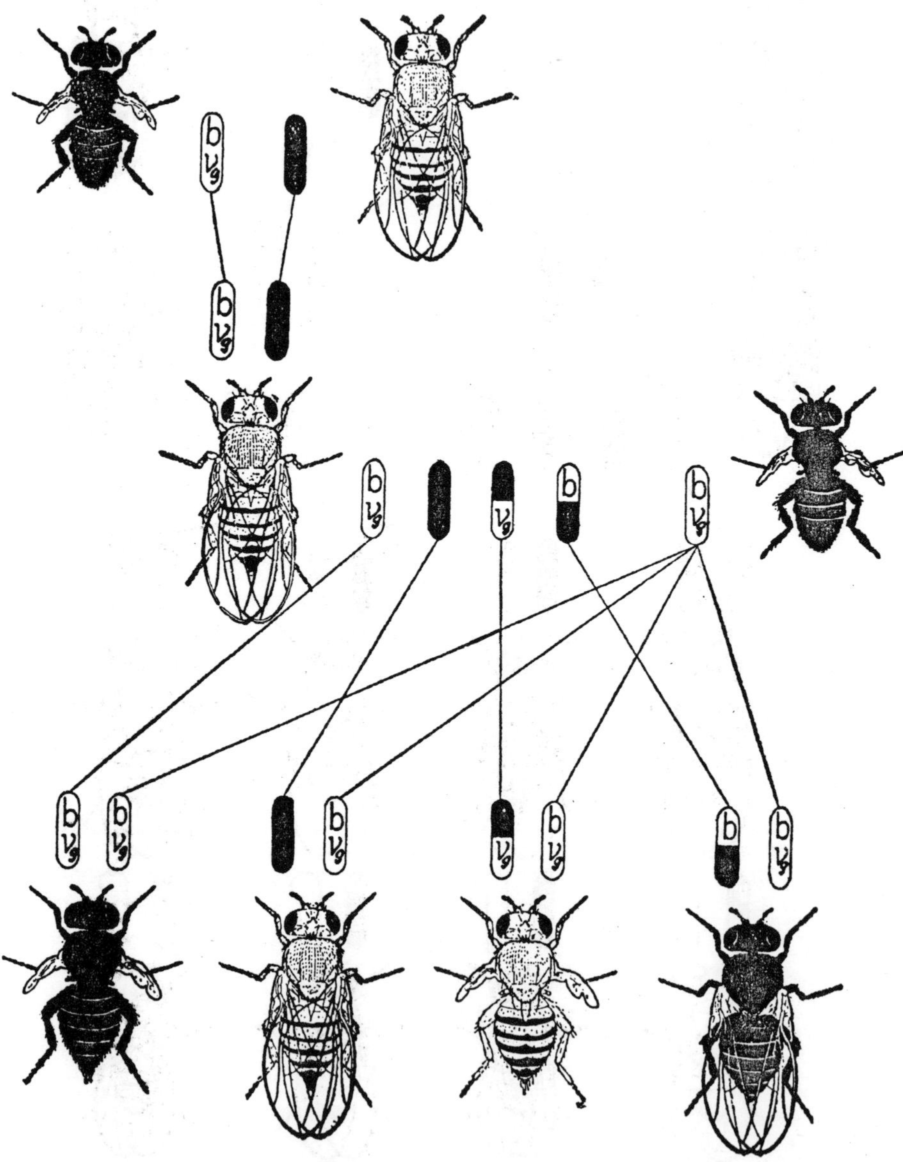


FIGURE 50.—Diagrammatic representation of the results of a cross of a gray normal female with a black vestigial male in *Drosophila melanogaster*. The $F_1$ gray normal females are mated to black vestigial males in order to determine the number and proportion of female gametic classes. (*From Morgan.*)

will then obviously consist of gray red females $(yw\mathrm{X})(YW\mathrm{X})$ and yellow white males $(yw\mathrm{X})\mathrm{Y}$. When the $F_1$ gray red females are mated to yellow white males, as before, the progeny fall into four classes as follows:

| Phenotypes | Percentages | Female gametes | |
|---|---|---|---|
| Yellow white | 49.5 | $(yw\mathrm{X})$ | 99 per cent parental combinations |
| Gray red | 49.5 | $(YW\mathrm{X})$ | |
| Yellow red | 0.5 | $(yW\mathrm{X})$ | 1 per cent recombinations |
| Gray white | 0.5 | $(Yw\mathrm{X})$ | |

The same four classes are present as in the previous experiment, but the proportions are exactly reversed. This time yellow white and gray red represent parental combinations, and yellow red and gray white, recombinations. From these two experiments, it may be stated as the first principle of linkage that the proportion of parental combinations and recombinations for any two pairs of factors is the same in the two possible types of crosses.

**Autosomal Linkage in Drosophila.**—That linkage also occurs among autosomal factors in Drosophila may be shown by considering the relation between gray *vs.* black (*B-b*) body color and normal *vs.* vestigial wings (*Vg-vg*). The results are shown graphically in figure 50. A cross between black vestigial (*bvg*) and gray normal (*BVg*) produces gray normal (*BVg*)(*bvg*) offspring of both sexes. In order to determine the gametic ratio, $F_1$ gray normal females are mated to black vestigial males, with results approximately as follows:

| Phenotypes | Percentages | Female gametes | |
|---|---|---|---|
| Gray normal | 41.5 | (*BVg*) | 83 per cent parental combinations |
| Black vestigial | 41.5 | (*bvg*) | |
| Gray vestigial | 8.5 | (*Bvg*) | 17 per cent recombinations |
| Black normal | 8.5 | (*bVg*) | |

The characteristic proportion of recombinations in this case is about 17 per cent. This evidence may be taken to indicate that these factors are borne in the same pair of autosomes (figure 50).

**Linkage in Maize.**—Linkage in other forms exhibits the same general features as those shown in Drosophila, as may be demonstrated by considering a case in maize. It has already been shown that the factor *C* for colored aleurone is a simple dominant to the factor *c* for white or colorless aleurone. Similarly non-shrunken endosperm, represented by *Sh*, is a simple dominant to shrunken endosperm *sh*. When purple non-shrunken corn is crossed with white shrunken, $F_1$ is purple non-shrunken.

The $F_1$ backcrossed to white shrunken was found to yield progeny approximately in the following proportions:

| Phenotypes | Percentages | $F_1$ Gametes | |
|---|---|---|---|
| Purple non-shrunken | 48.5 | (*CSh*) | 97 per cent parental combinations |
| White shrunken | 48.5 | (*csh*) | |
| Purple shrunken | 1.5 | (*Csh*) | 3 per cent recombinations |
| White non-shrunken | 1.5 | (*cSh*) | |

As in the illustrative case from Drosophila, parental combinations are in excess and recombinations correspondingly decreased in proportion, although the actual percentages differ from those in the previous case.

When the factors are introduced in different parental combinations, the same relations hold as in Drosophila. Colored shrunken × white non-shrunken yields a colored non-shrunken $F_1$, as before. The $F_1$ backcrossed to white shrunken (figure 51) gave the following results:

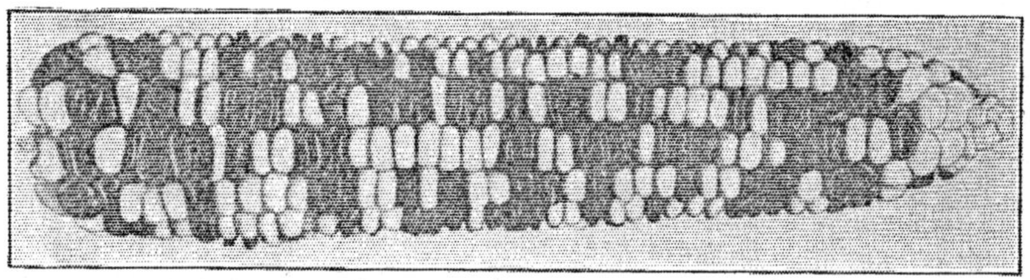

FIGURE 51.—Linkage of aleurone color and endosperm type in maize. A single ear obtained by back crossing a plant of the constitution (*Csh*)(*cSH*) with the double recessive (*csh*)(*csh*). Note occasional recombinations, colored nonshrunken and white shrunken. (*From Hutchison.*)

| Phenotypes | Percentages | $F_1$ gametes | |
|---|---|---|---|
| Colored shrunken | 48.5 | (*Csh*) | 97 per cent parental combinations |
| White nonshrunken | 48.5 | (*cSh*) | |
| Colored nonshrunken | 1.5 | (*CSh*) | 3 per cent recombinations |
| White shrunken | 1.5 | (*csh*) | |

The percentages of parental combinations and recombinations are the same in the two cases, although the actual factor combinations which represent these two categories are reversed. The experimental results upon which this analysis is based are contained in table 5. The actual values of the different classes are somewhat different in the two crosses; but this is due to the fact that they are by nature statistical and, therefore, subject to a certain amount of random fluctuation based on operation of the laws of chance.

**The Linkage Value.**—In the foregoing illustrations, the percentages of parental combinations and recombinations differ. Similarly when other pairs of linked factors are investigated, it is found that they exhibit a characteristic proportion of parental combinations and recombinations determinable only by experiment. Consequently it is necessary for two given pairs of factors, not only to ascertain whether or not they are linked but also to determine the proportion of parental combinations and recombinations which they give.

TABLE 5.—BACKCROSS TESTS FOR THE LINKAGE VALUE OF *C-sh* IN MAIZE (*Data from Hutchison*)

| Backcross | Parental combinations | | Recombinations | | Total | Linkage value |
|---|---|---|---|---|---|---|
| $\frac{Csh}{cSh} \times \frac{csh}{csh}$ | *Csh* 21,379 | *cSh* 21,096 | *CSh* 638 | *csh* 672 | 43,785 | 3.0 |
| $\frac{CSh}{csh} \times \frac{csh}{csh}$ | *CSh* 4,032 | *csh* 4,035 | *Csh* 149 | *cSh* 152 | 8,368 | 3.6 |
| Totals.... | 50,542 | | 1,611 | | 52,153 | 3.1 |

It is customary to state the degree of linkage in terms of the percentage of recombinations in the gametic series. This is called the linkage value. The linkage value for gray-yellow and red-white, usually stated as simply the yellow-white linkage value, in Drosophila is 1.1; for white with shrunken in maize, 3.1. Through a great amount of experimentation it has been found that linkage values in general range from a lower limit approaching zero to an upper limit at approximately 50 per cent. When the value lies near 50 per cent, it is, of course, difficult to determine

TABLE 6.—LINKAGE DATA FOR A NUMBER OF SEX-LINKED FACTORS OF *Drosophila melanogaster*, ILLUSTRATING DIFFERENCES IN LINKAGE VALUE CHARACTERISTIC OF DIFFERENT PAIRS OF FACTORS (*Data from Morgan and Bridges*)

| Factors | Totals | Recombinations | Linkage value |
|---|---|---|---|
| Yellow-white.............. | 81,299 | 875 | 1.1 |
| Yellow-abnormal........... | 15,314 | 299 | 2.0 |
| Yellow-vermilion.......... | 13,271 | 4,581 | 34.5 |
| Yellow-miniature.......... | 21,686 | 7,599 | 34.3 |
| White-abnormal............ | 16,300 | 277 | 1.7 |
| White-bifid............... | 23,595 | 1,260 | 5.3 |
| White-vermilion........... | 27,962 | 8,532 | 30.5 |
| White-miniature........... | 110,701 | 31,071 | 33.2 |
| Vermilion-miniature....... | 10,155 | 317 | 3.1 |
| Vermilion-bar............. | 23,522 | 5,612 | 23.9 |
| Miniature-rudimentary..... | 12,786 | 2,284 | 17.9 |

whether or not the factors are linked, for 50 per cent of recombinations normally indicates independent assortment, but special methods of experimentation have been devised to handle such cases, as will appear shortly. Table 6 contains linkage values for eleven pairs of sex-linked factors in *Drosophila melanogaster*, which illustrates the range in linkage values for these specific combinations of factors. Factors which exhibit a low linkage value, like yellow and white, are said to be closely linked, because they separate in such a low proportion of cases. When the linkage value is high, the factors are said to be loosely linked. From these facts it may be stated as a second principle of linkage that the linkage value for any two pairs of linked factors is a constant characteristic feature of their distributional relations.

**Linkage Formulae.**—Linkage is represented by formulae in various ways. In the account just given the linked factors are enclosed in parentheses. Thus $(CSh)(csh)$ represents an $F_1$ from a cross of colored non-shrunken with white shrunken; and $(Csh)(cSh)$ from colored shrunken with white non-shrunken. The same conditions are also some times indicated by *CSh-csh* or $\frac{CSh}{csh}$ and *Csh-cSh* or $\frac{Csh}{cSh}$, respectively. The fractional method of representation is usually best adapted for working out problems, for then $\frac{CSh}{csh}$ may be used to represent parental combinations and $\frac{C \mid sh}{c \mid Sh}$ recombinations. If independent assortment were found to govern the situation, the formula should be written *CcShsh*.

**Linkage Groups.**—In Drosophila any two pairs of sex-linked factors exhibit linkage relations with each other, although the particular values may lie anywhere between 0 and 50 per cent. A group of factors so related to one another that any two members will be linked is said to constitute a linkage group. It has also been found that distinct linkage groups occur among autosomal factors in Drosophila. The grouping is based upon the observed principle that if a factor $A$ is linked with two other factors $B$ and $C$, then $B$ will be linked with $C$. By application of this principle it has been found that there is a group of over 100 factors linked with factors *B-b* for gray *vs.* black body color; and a group of about the same number linked with *P-p*, the factors for red *vs.* pink eye color; and a smaller group, consisting thus far of only three pairs of factors, linked with *Bt-bt*, the factors for straight *vs.* bent wings.

By application of this principle, it is possible greatly to expedite the study of linkage. In order to demonstrate that a factor belongs to a given linkage group, it is only necessary to demonstrate that it is linked with one member of that group. Conversely, if a factor is not linked with a given member of a linkage group, it will not be linked with any other member of the group; or to state the matter another way, two factors which belong to different linkage groups will always exhibit independent assortment.

From these results a third principle of linkage may be deduced to the effect that the factors of a species belong to a limited number of linkage groups so constituted that any two members of a linkage group will exhibit linkage relations in distribution, but any two members of different linkage groups will exhibit independent assortment.

**The Number of Linkage Groups.**—In *Drosophila melanogaster* there are four linkage groups; group I, the sex-linked group; group II, those linked with black; group III, those linked with pink, and group IV, those linked with bent. So far as investigated, every factor in this species belongs to some one of these four groups. Now since linkage is evidence that factors are borne by the same chromosome, it follows that the number of linkage groups should be the same as the number of pairs of chromosomes. In *Drosophila melanogaster* the demonstration of correspondence between the number of linkage groups and the number of pairs of chromosomes has been made, for although about 300 genes have been grouped, only four groups have been discovered, and the species is known to have four pairs of chromosomes. In some other species of Drosophila additional evidence to this effect is available. *Drosophila willistoni* has three pairs of chromosomes and three linkage groups; *Drosophila obscura,* five pairs of chromosomes and five linkage groups; and *Drosophila virilis,* six pairs of chromosomes and six linkage groups. In other species of Drosophila investigation has not gone far enough to demonstrate the existence of as many linkage groups as pairs of chromosomes; but in no instance has the number of linkage groups been found to exceed the number of pairs of chromosomes.

Aside from Drosophila, linkage studies are not yet extensive enough to furnish evidence in support of the correspondence between chromosome number and linkage groups. In maize, with ten pairs of chromosomes, eight linkage groups have been demonstrated, and indications of one more have been obtained; but only a comparatively small number of the known pairs of factors have yet been grouped. In the garden snapdragon, *Antirrhinum majus,* with eight pairs of chromosomes and over 150 known pairs of Mendelian factors, a number of linkage groups have been established and the evidence obtained is in agreement with the principles stated above, but only a small fraction of the known genes have been located in a linkage system. In garden peas, *Pisum sativum,* and in sweet peas, *Lathyrus odoratus,* each with seven pairs of chromosomes, evidence has been presented for the existence of eight groups of factors. The data, however, are not extensive in either case; and since they are derived from $F_2$ distributions, it is possible that some linkage values approaching 50 per cent account for the difficulty. No acceptable demonstration, therefore, has been made that the number of linkage groups ever exceeds the number of pairs of chromosomes. It may be stated, therefore, as a fourth principle of linkage that the number of linkage groups in a given species is equal to the number of pairs of chromosomes.

**Effect of Sex upon Linkage.**—With sex-linked factors in Drosophila, recombination occurs only in females, as is obviously to be expected, because males possess only a single set of sex-linked factors, and recombination is possible only when two sets are present. Autosomal factors, however, are present in double dosage in both males and females; nevertheless it has been found that no recombination occurs in males. As an illustration, the behavior of the linked factors *B-b* for gray *vs.* black body color and *C-c* for straight *vs.* curved wings may be presented. The races pertinent to this discussion have the following genetic formulae:

| | |
|---|---|
| gray straight | $(BC)(BC)$ |
| gray curved | $(Bc)(Bc)$ |
| black straight | $(bC)(bC)$ |
| black curved | $(bc)(bc)$ |

When gray curved is crossed with black straight, both $F_1$ males and females are gray straight $(Bc)$ $(bC)$. When $F_1$ females are backcrossed to the double recessive, black curved, the following distribution is obtained:

| Phenotypes | Percentages | Female gametes | |
|---|---|---|---|
| Gray curved | 38.5 | *Bc* | 77 per cent parental combinations |
| Black straight | 38.5 | *bC* | |
| Gray straight | 11.5 | *BC* | 23 per cent recombinations |
| Black curved | 11.5 | *bc* | |

Approximately 23 per cent of recombinations occur in the progeny. If, however, $F_1$ males of the same genetic constitution are backcrossed to black curved females, the following results are obtained:

| Phenotypes | Percentages | Male gametes | |
|---|---|---|---|
| Gray curved | 50.0 | *Bc* | 100 per cent parental combinations |
| Black straight | 50.0 | *bC* | |
| Gray straight | 0.0 | *BC* | 0 per cent recombinations |
| Black curved | 0.0 | *bc* | |

Absolutely no recombinations are produced in this case. Similar studies of linked factors belonging to the other linkage groups in Drosophila demonstrate the general validity of this result, so that it may be stated with confidence that no recombination of linked factors occurs in the male

Drosophila. This result should be compared with the evidence from independent assortment, which shows no relation whatever to sex.

The absence of recombination in the male has been used in a variety of ways in Drosophila to facilitate genetic experimentation. When, for example, $F_1$ flies of the genetic constitution (*Bc*) (*bC*) are mated *inter se*, the $F_2$ progeny is distributed in the following ratio: 2 gray straight: 1 gray curved: 1 black straight: 0 black curved as may be seen by reference to figure 52. This ratio is to be compared with the distribution obtained by crossing black red flies *bbPP* with gray pink *BBpp*. $F_1$ is gray red *BbPp*, and $F_2$, obtained by mating $F_1$ flies *inter se*, is in the typical dihybrid ratio 9 gray red: 3 gray pink: 3 black red: 1 black pink.

| | ♀ | ♂ *Bc* | *bC* |
|---|---|---|---|
| 38.5 | *Bc* | 38.5 (*Bc*)(*Bc*) gray curved | 38.5 (*Bc*)(*bC*) gray straight |
| 38.5 | *bC* | 38.5 (*bC*)(*Bc*) gray straight | 38.5 (*bC*)(*bC*) black straight |
| 11.5 | *BC* | 11.5 (*BC*)(*Bc*) gray straight | 11.5 (*BC*)(*bC*) gray straight |
| 11.5 | *bc* | 11.5 (*bc*)(*Bc*) gray curved | 11.5 (*bc*)(*bC*) black straight |

FIGURE 52.—$F_2$ results of crossing gray curved with black straight. The $F_2$ ratio reduces to 2 gray straight: 1 gray curved: 1 black straight: 0 black curved.

The 2:1:1:0 ratio, obtained when the two recessive characters are introduced from opposite parents, indicates that the factors belong to the same linkage group; the 9:3:3:1 ratio, obtained under similar circumstances, indicates that they belong to different linkage groups. It is, therefore, a simple matter in Drosophila to determine the group to which a newly discovered factor belongs by submitting it to the foregoing test with convenient representatives of the known linkage groups. Other methods have also been devised which are even more convenient for this purpose, but since they are merely details of experimentation, it is unnecessary to present them here.

How widely sex has an effect upon recombination cannot be stated at present. In the silkworm, which conforms to the WZ type of sex determination, it has been found that recombination occurs only in males; so that it was previously thought that the effect of sex on recombination could be stated as a general principle that there is no recombination of linked factors in the sex heterozygote. It has since been found, however, that this statement does not possess general validity. Thus in mice and rats, the factors *C-c* for colored *vs.* albino coat and *R-r* for red *vs.* pink eye color are linked. Backcrosses of males and females heterozygous for the two factors to the double recessive prove conclusively that recombination occurs in both sexes. In mice the percentage of recombination was 16.4 for females and 13.8 for males; in rats 21.9 for females and 18.4 for males. Although these results indicate a slight difference in percentage of recombinations in the two sexes, they demonstrate that recombination does occur. Similar results have been secured for other linked factors in mice and rats.

**Recombination in Fishes.**—In fishes recombination has been demonstrated in the male for sex-linked factors, despite the fact that the inheritance of sex follows the XY scheme. In fishes red *vs.* white body color, factors *R-r*, are sex-linked in transmission, but unlike the situation in Drosophila the factors are borne by both the X- and Y-chromosomes. Consequently the genetic constitutions of red and white males must be represented as follows:

| | ♀♀ | ♂♂ |
|---|---|---|
| Red | (*R*X)(*R*X) | (*R*X)(*R*Y) |
| White | (*r*X)(*r*X) | (*r*X)(*r*Y) |

The results of reciprocal crosses between these forms have been described in Chapter VII. These results, however, were based on the assumption that the heterozygous male (*r*X)(*R*Y) produces only two kinds of gametes (*r*X) and (*R*Y). The best method of determining this point is to mate such heterozygous males to white females (*r*X)(*r*X), which has been done with the following results in one case:

| Phenotype | Numbers | Gametes |
|---|---|---|
| White females | 197 | (*r*X) |
| Red males | 253 | (*R*Y) |
| White males | 1 | (*r*Y) |
| Red females | 2 | (*R*X) |

The exceptional white males and red females in the above progeny are most reasonably explained, as further experiments carried out by Aida indicate, on the assumption that recombination has occurred between the X- and Y-chromosomes in the heterozygous males. Further experiments are necessary to establish this point, and it will be of interest to see what the final outcome of the experiments will be and how widely in other forms recombination occurs between the X- and Y-chromosomes. It cannot be expected in Drosophila, because no recombination of any kind occurs in the male sex, nor between the W- and Z-chromosomes in the silkworm, because here again no recombination occurs in the female sex; but in mammals, where recombination occurs in both sexes, it may possibly occur.

**Recombination in Plants.**—In plants recombination occurs both in the formation of pollen grains and ovules, as may be shown by the data from the Chinese primrose, *Primula sinensis,* collected in table 7. The four pairs of factors are differentials for the following characters:

| | |
|---|---|
| *B-b* | blue *vs.* non-blue flower color |
| *G-g* | green *vs.* red stigma |
| *L-l* | light red *vs.* dark red stems |
| *S-s* | short *vs.* long style |

TABLE 7.—LINKAGE IN *Primula Sinensis.* A COMPARISON OF VALUES FOR MALE AND FEMALE GAMETES DETERMINED BY BACKCROSS TESTS (*Data of Gregory*)

| Factors | Female gametes | | | Male gametes | | |
|---|---|---|---|---|---|---|
| | Total | Recombinations | Linkage value | Total | Recombinations | Linkage value |
| *S-B* | 3,047 | 270 | 7.6 | 3,286 | 403 | 12.3 |
| *S-G* | 3,209 | 1,076 | 33.5 | 3,043 | 1,228 | 40.4 |
| *S-L* | 991 | 367 | 37.0 | 1,599 | 651 | 40.7 |
| *B-G* | 6,937 | 2,151 | 31.0 | 3,974 | 1,393 | 35.1 |
| *B-L* | 2,022 | 722 | 35.7 | 1,768 | 653 | 36.9 |
| *G-L* | 1,436 | 47 | 3.3 | 2,081 | 39 | 1.9 |

An examination of this table will disclose the fact that linkage values are usually somewhat higher in pollen formation than in ovule production. The significance of this fact is not clear, and experiments in other forms have given somewhat discordant results. Moreover, it is known that linkage values are highly variable and subject to a certain extent to external influences, particularly temperature. It is, therefore, impossible at present to decide whether or not there is a real difference in the values. At any rate where recombination occurs in both sexes, the values obtained always appear to be of the same order of magnitude.

**Effect of Linkage on $F_2$ Ratios.**—In many species it is impractical to employ the method of backcrossing the $F_1$ heterozygote to the double recessive parent in order to determine the gametic ratio, because of the excessive amount of labor involved in this method. In the small grains, such as barley, oats, rice, and wheat, each operation involved in hybridization gives only a single seed at best, so that it would be necessary to make hundreds of backcrosses in order to secure a sufficient quantity of material. In sweet peas, in which linkage was first discovered by Bateson, the condition is very little better, for only five or six seeds per pod may be counted on as the result of each manipulation. Expediency demands that $F_2$ material, secured by selfing $F_1$ plants, should be utilized for determining linkage relations in such species. It is accordingly necessary to examine into the effect which linkage has upon $F_2$ distributions.

For purposes of illustration, the data for linkage of *C* and *Wx* in maize may be utilized. From a large amount of data the average recombination value for these genes has been determined to be approximately 25 per cent, so that for an $F_1$ plant of the genetic constitution (*CWx*)(*cwx*) the gametic ratio will be as follows:

0.375 (*CWx*): 0.125 (*Cwx*): 0.125 (*cWx*): 0.375 (*cwx*).

| | 0.375 (*CWx*) | 0.125 (*Cwx*) | 0.125 (*cWx*) | 0.375 (*cwx*) |
|---|---|---|---|---|
| 0.375 (*CWx*) | 0.140625 (*CWx*)(*CWx*) colored starchy | 0.046875 (*CWx*)(*Cwx*) colored starchy | 0.046875 (*CWx*)(*cWx*) colored starchy | 0.140625 (*CWx*)(*cwx*) colored starchy |
| 0.125 (*Cwx*) | 0.046875 (*Cwx*)(*CWx*) colored starchy | 0.015625 (*Cwx*)(*Cwx*) colored waxy | 0.015625 (*Cwx*)(*cWx*) colored starchy | 0.046875 (*Cwx*)(*cwx*) colored waxy |
| 0.125 (*cWx*) | 0.046875 (*cWx*)(*CWx*) colored starchy | 0.015625 (*cWx*)(*Cwx*) colored starchy | 0.015625 (*cWx*)(*cWx*) white starchy | 0.046875 (*cWx*)(*cwx*) white starchy |
| 0.375 (*cwx*) | 0.140625 (*cwx*)(*CWx*) colored starchy | 0.046875 (*cwx*)(*Cwx*) colored waxy | 0.046875 (*cwx*)(*cWx*) white starchy | 0.140625 (*cwx*)(*cwx*) white waxy |

FIGURE 53.—$F_2$ results of crossing colored starchy (*CWx*) with white waxy (*cwx*) assuming 25 per cent of recombination. The collected values are contained in table 8.

The ratio is here stated in terms of decimal fractions for convenience in working out the $F_2$ distribution. If independent assortment occurred, the classes would, of course, be in equal proportions, *viz.*,

0.25 *CWx*: 0.25 *Cwx*: 0.25 *cWx*: 0.25 *cwx*.

TABLE 8.—$F_2$ RESULTS WITH LINKAGE COMPARED WITH INDEPENDENT ASSORTMENT (See figures 53 and 54 for derivation of the $F_2$ ratios)

| Phenotypes | Linkage (*CWx*)(*cwx*) | Proportions 9:3:3:1 | Deviations | Linkage (*Cwx*)(*cWx*) | Deviations |
|---|---|---|---|---|---|
| *CWx* | 0.640625 | 0.5625 | +0.078125 | 0.515625 | −0.046875 |
| *Cwx* | 0.109375 | 0.1875 | −0.078125 | 0.234375 | +0.046875 |
| *cWx* | 0.109375 | 0.1875 | −0.078125 | 0.234375 | +0.046875 |
| *cwx* | 0.140625 | 0.0625 | +0.078125 | 0.015625 | −0.046875 |

Assuming that the gametes are produced in the same proportion in both pollen and ovule formation, the $F_2$ distribution for the above gametic series may be determined by the checkerboard method shown in figure 53. The numerical proportions in each square are obtained by multiplying together the two coefficients of the gametes which unite to form the zygote therein represented. Collecting the different phenotypes, the $F_2$ progeny is found to be distributed in the proportions shown in the second column of table 8. In the third column the proportions are stated according to the 9:3:3:1 ratio for comparison, and the fourth contains the differences between corresponding terms in the second and third columns. From these figures it may be observed that the effect of linkage on the $F_2$ ratio in this case is to increase each of the end terms, representing parental combinations of characters, and to decrease the middle terms of the ratio by an equal amount. The linkage value may be calculated from the proportions by taking the square root of the

| | 0.125 (*CWx*) | 0.375 (*Cwx*) | 0.375 (*cWx*) | 0.125 (*cwx*) |
|---|---|---|---|---|
| 0.125 (*CWx*) | 0.015625 (*CWx*)(*CWx*) colored starchy | 0.046875 (*CWx*)(*Cwx*) colored starchy | 0.046875 (*CWx*)(*cWx*) colored starchy | 0.015625 (*CWx*)(*cwx*) colored starchy |
| 0.375 (*Cwx*) | 0.046875 (*Cwx*)(*CWx*) colored starchy | 0.140625 (*Cwx*)(*Cwx*) colored waxy | 0.140625 (*Cwx*)(*cWx*) colored starchy | 0.046875 (*Cwx*)(*cwx*) colored waxy |
| 0.375 (*cWx*) | 0.046875 (*cWx*)(*CWx*) colored starchy | 0.140625 (*cWx*)(*Cwx*) colored starchy | 0.140625 (*cWx*)(*cWx*) white starchy | 0.046875 (*cWx*)(*cwx*) white starchy |
| 0.125 (*cwx*) | 0.015625 (*cwx*)(*CWx*) colored starchy | 0.046875 (*cwx*)(*Cwx*) colored waxy | 0.046875 (*cwx*)(*cWx*) white starchy | 0.015625 (*cwx*)(*cwx*) white waxy |

FIGURE 54.—$F_2$ results of crossing colored waxy (*Cwx*) with white starchy (*cWx*), assuming 25 per cent of recombination. The collected values are contained in table 8.

*cwx* term, subtracting it from 0.50, and multiplying by 2; thus $\sqrt{0.140625} = 0.375$; $2(0.50 - 0.375) = 0.25$, which is the linkage value stated as a decimal fraction.

If the $F_1$ heterozygote is of the genetic constitution (*Cwx*) (*cWx*), the coefficients in the gametic series will be reversed, as follows:

0.125 (*CWx*):0.375 (*Cwx*):0.375 (*cWx*):0.125 (*cwx*).

If a checkerboard is formed of this gametic series, like that shown in figure 54, and terms are collected as before, the proportions are obtained as shown in the fifth column of table 8. This distribution differs from the preceding one in that the end terms are reduced and the middle terms increased, but the amount of increase and decrease is different. Here again the linkage value may be determined by operating with the *cwx* class alone; $\sqrt{0.015625} = 0.125$, which multiplied by 2 gives the linkage value expressed as a decimal fraction. The reason for working with the *cwx* class may be determined by examining the corresponding $F_2$ checkerboards. The *cwx* phenotype is the only one which is represented by a single square in the checkerboard, and the coefficient of this phenotype is obviously simply the square of one of the terms in the gametic series. The other classes could be utilized, but a more complex equation would have to be solved to determine the linkage value from them.

Actual application of this method to data collected from $F_2$ material where the $F_1$ was of the type $(CWx)(cwx)$ are contained in table 9. The second column contains numbers actually obtained by counting $F_2$ populations. In the third column these are reduced to proportions by dividing each of the numbers in the first column by the total number of individuals. Six decimal places should be retained, if it is desired to calculate the linkage value to the nearest tenth of one per cent. The fourth column contains the proportions when the individuals are distributed in a 9:3:3:1 ratio. The fifth column contains the differences between corresponding values in the third and fourth columns. It will be observed that the differences are approximately equal in amount. In order to adjust them so as to obtain the best linkage value, the average of these four differences is taken, disregarding signs. The average deviation, 0.088451, is then added to the first and last terms in the 9:3:3:1 proportions and subtracted from the two middle terms to give the adjusted proportions contained in the sixth column. Computations based on the *cwx* term of the adjusted proportions give 22.2 per cent as the linkage value for *C-Wx* from these data. A seventh column obtained by multiplying the terms in the sixth column by 2,736 is added so that a comparison may be made of the calculated distribution, assuming 22.2 per cent of recombination, with the actual numbers contained in the first column.

TABLE 9.—COMPUTATION OF LINKAGE VALUE FROM $F_2$ DATA OF COLORED STARCHY × WHITE WAXY MAIZE (*Data from Bregger*)

| Phenotypes | Numbers | Observed proportions | 9: 3: 3: 1 proportions | Deviations | Adjusted proportions | Calculated numbers |
|---|---|---|---|---|---|---|
| *CWx* | 1,774 | 0.648392 | 0.5625 | 0.085892 | 0.650951 | 1,781 |
| *Cwx* | 263 | 0.096125 | 0.1875 | 0.091375 | 0.099049 | 271 |
| *cWx* | 279 | 0.101974 | 0.1875 | 0.085526 | 0.099049 | 271 |
| *cwx* | 420 | 0.153509 | 0.0625 | 0.091009 | 0.150951 | 413 |
| Totals...... | 2,736 | 1.000000 | 1.0000 | 0.353802 | 1.000000 | 2,736 |
| Average deviation....... | | ........ | ...... | 0.088451 | | |

Linkage value $= 2(0.5 - \sqrt{0.150951}) = 0.222$.

Table 10 has been constructed to facilitate the determination of the approximate linkage values shown by $F_2$ data. Use of the table requires that the data first be converted into proportions and balanced according to the methods described above. Other methods are also available for this purpose; the one which has been described is merely the simplest and it is applicable in most cases.

TABLE 10.—LINKAGE RELATIONS—RECOMBINATION IN BOTH SEXES

| $F_1$ | Percentage of recombination | Phenotypic proportions | | | |
|---|---|---|---|---|---|
| | | *AB* | *Ab* | *aB* | *ab* |
| | 1 | 0.500025 | 0.249975 | 0.249975 | 0.000025 |
| | 2 | 0.500100 | 0.249900 | 0.249900 | 0.000100 |
| | 3 | 0.500225 | 0.249775 | 0.249775 | 0.000225 |
| | 4 | 0.500400 | 0.249600 | 0.249600 | 0.000400 |
| $\frac{Ab}{aB}$ | 5 | 0.500625 | 0.249375 | 0.249375 | 0.000625 |
| | 6 | 0.500990 | 0.249100 | 0.249100 | 0.000900 |
| | 7 | 0.501225 | 0.248775 | 0.248775 | 0.001225 |
| | 8 | 0.501600 | 0.248400 | 0.248400 | 0.001600 |
| | 9 | 0.502025 | 0.247975 | 0.247975 | 0.002025 |
| | 10 | 0.502500 | 0.247500 | 0.247500 | 0.002500 |
| | 20 | 0.510000 | 0.240000 | 0.240000 | 0.010000 |
| | 30 | 0.522500 | 0.227500 | 0.227500 | 0.022500 |
| | 40 | 0.540000 | 0.210000 | 0.210000 | 0.040000 |
| *AaBb* | 50 | 0.562500 | 0.187500 | 0.187500 | 0.062500 |
| | 40 | 0.590000 | 0.160000 | 0.160000 | 0.090000 |
| | 30 | 0.622500 | 0.127500 | 0.127500 | 0.122500 |
| | 20 | 0.660000 | 0.090000 | 0.090000 | 0.160000 |
| | 10 | 0.702500 | 0.047500 | 0.047500 | 0.202500 |
| $\frac{AB}{ab}$ | 9 | 0.707025 | 0.042975 | 0.042975 | 0.207025 |
| | 8 | 0.711600 | 0.038400 | 0.038400 | 0.211600 |
| | 7 | 0.716225 | 0.033775 | 0.033775 | 0.216225 |
| | 6 | 0.720900 | 0.029100 | 0.029100 | 0.220900 |
| | 5 | 0.725625 | 0.024375 | 0.024375 | 0.225625 |
| | 4 | 0.730400 | 0.019600 | 0.019600 | 0.230400 |
| | 3 | 0.735225 | 0.014775 | 0.014775 | 0.235225 |
| | 2 | 0.740100 | 0.009900 | 0.009900 | 0.240100 |
| | 1 | 0.745025 | 0.004075 | 0.004075 | 0.245025 |

This discussion of the effect of linkage upon $F_2$ ratios is itself an argument in favor of application of the backcross method of determining linkage values. Actual computation of linkage values from $F_2$ data is not extremely difficult; but if differences occur in the percentage of recombination in the male and female series of gametes, it would be impossible to determine them from such data. The backcross experiment is the direct method of determining the gametic ratio; therefore, it is preferable whenever its employment is feasible.

**Linkage and Chromosomes.**—In terms of the chromosome theory of heredity, linkage indicates that factors are borne in the same chromosome; independent assortment, in contrast, that they are borne in different chromosomes. In linkage, therefore, a single pair of chromosomes is under consideration, in independent assortment as many pairs of chromosomes are involved as there are pairs of factors. Obviously then in a gametic series showing linkage, the emergence of parental combinations indicates that the pair of chromosomes have separated in reduction with their

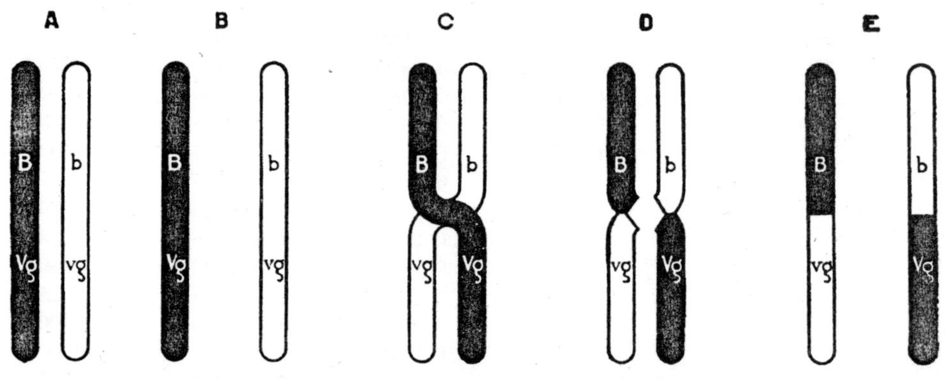


FIGURE 55.—Chromosomal relations in linkage represented diagrammatically for gamete production in an $F_1$ female of the genetic constitution, *BVgbvg*. *A*, the homologous chromosomes conjugate, and separate as in *B* without exchanging material, thus producing two classes of gametes, the parental combinations. *C*, an alternative type of conjugation, resulting in crossing-over, *D*, by which the homologues exchange equivalent portions of chromatin material. Separation of the reconstructed chromosomes, *E*, gives two classes of gametes, the recombinations. The relative frequency of the two types of behavior determines the linkage value.

original constitution unchanged, at least so far as the particular factors under consideration are concerned, and recombination indicates that the chromosomes have exchanged material, as is represented diagrammatically in figure 55. The exchange of material between homologous chromosomes is presumed to occur in the prophase of the first meiotic division, and it is called *crossing-over*. A more complete discussion of the phenomenon of crossing-over is deferred to the next chapter.

### References

CREW: Animal Genetics, Chapter IV.

MORGAN, STURTEVANT, MULLER, and BRIDGES: Mechanism of Mendelian Heredity, Chapter III.

MORGAN: Physical Basis of Heredity, Chapter VI.

SINNOTT and DUNN: Principles of Genetics, Chapter VII.

WALTER: Genetics, Chapter XI.

# CHAPTER XV

## THE ORGANIZATION OF LINKAGE GROUPS

It has been shown in the preceding chapter that the factors of a species are distributed in linkage groups corresponding in number to the number of pairs of chromosomes. Since the total number of factors greatly exceeds the number of pairs of chromosomes, it follows that in general each linkage group will contain a large number of factors. This fact has been demonstrated for *Drosophila melanogaster*, for the first three groups contain in excess of 100 known genes, although the fourth contains only three. The next practical question, therefore, is to determine the relations which exist among the members of a linkage group.

For adequate treatment of this problem, it is necessary to have extensive data for many, if not all, of the possible combinations by twos of a large number of factors belonging to the same linkage group. Such data are available only in a few species of Drosophila, particularly *Drosophila melanogaster*. To serve as a basis of discussion, table 11 has been constructed from the data of Bridges and Morgan for the linked factors of group III.

TABLE 11.—LINKAGE VALUES OF CHARACTERS OF THE FIRST RANK IN GROUP III OF *Drosophila melanogaster*

| | *ru* | *se* | *h* | *D* | *st* | *p* | *ss* | *Da* | *H* | *e* | *cd* | *ro* | *ca* | *Mg* |
|---|---|---|---|---|---|---|---|---|---|---|---|---|---|---|
| roughoid | .... | 25.7 | ? | 35.9 | 39.9 | 41.8 | ? | ? | 45.2 | 46.3 | ? | ? | ? | ? |
| sepia | 25.7 | .... | 0.5 | 13.8 | ? | ? | 26.8 | ? | 33.7 | 36.9 | ? | 42.8 | ? | ? |
| hairy | ? | 0.5 | .... | 14.2 | ? | ? | ? | ? | ? | 36.6 | ? | ? | ? | ? |
| dichaete | 35.9 | 13.8 | 14.2 | .... | 3.4 | 6.5 | 13.7 | 21.9 | 23.1 | 24.7 | 27.1 | 41.2 | ? | ? |
| scarlet | 39.9 | ? | ? | 3.4 | .... | 4.7 | 14.2 | ? | 25.4 | 28.0 | ? | ? | ? | ? |
| pink | 41.8 | ? | ? | 6.5 | 4.7 | .... | 10.4 | 18.9 | 21.2 | 23.7 | ? | 34.0 | ? | ? |
| spineless | ? | 26.8 | ? | 13.7 | 14.2 | 10.4 | .... | 7.9 | 9.9 | 12.0 | ? | 24.3 | ? | ? |
| delta | ? | ? | ? | 21.9 | ? | 18.9 | 7.9 | .... | 2.9 | 4.4 | 10.0 | ? | ? | ? |
| hairless | 45.2 | 33.7 | ? | 23.1 | 25.4 | 21.2 | 9.9 | 2.9 | .... | 1.2 | 6.2 | 21.9 | 31.5 | 34.1 |
| ebony | 46.3 | 36.9 | 36.6 | 24.7 | 28.0 | 23.7 | 12.0 | 4.4 | 1.2 | .... | 5.4 | 20.0 | 29.5 | ? |
| cardinal | ? | ? | ? | 27.1 | ? | ? | ? | 10.0 | 6.2 | 5.4 | .... | ? | ? | ? |
| rough | ? | 42.8 | ? | 41.2 | ? | 34.0 | 24.3 | ? | 21.9 | 20.0 | ? | .... | 11.6 | ? |
| claret | ? | ? | ? | ? | ? | ? | ? | ? | 31.5 | 29.5 | ? | 11.6 | .... | 5.5 |
| minute-g | ? | ? | ? | ? | ? | ? | ? | ? | 34.1 | ? | ? | ? | 5.5 | |

**The Linear Order.**—As a starting point for this discussion, select any three factors in this linkage group which give relatively low recombination values in all possible combinations taken two at a time. As a matter of fact, if two of the recombination values are low, the third must of necessity be low also. Consider the following typical cases:

| | | | |
|---|---|---|---|
| scarlet-pink | 4.7 | pink-spineless | 10.4 |
| pink-spineless | 10.4 | spineless-delta | 7.9 |
| scarlet-spineless | 14.2 | pink-delta | 18.9 |
| spineless-delta | 7.9 | delta-hairless | 2.9 |
| delta-hairless | 2.9 | hairless-ebony | 1.2 |
| spineless-hairless | 9.9 | delta-ebony | 4.4 |

In each of these four groups of three factors each, it is evident that the sum of the two smaller values is approximately equal to the largest value. Since these determinations are subject to statistical fluctuations, the actual sum of the two smaller values is not exactly equal to the largest value in any one case, but is sometimes slightly greater and sometimes slightly less. The differences are, however, slight and no more than are to be expected for statistical values. From these results it is possible to arrive at the fundamental principle of relationship among members of a linkage group, namely, that the three recombination values obtained from three factors taken two at a time are so related, when the values are low, that the sum of the two smaller values is equal to the largest value.

If now the linkage values for three factors taken two at a time are plotted, for example the scarlet-pink-spineless data, the relation stated in the above principle is evidently satisfied by spacing them along a straight line at distances proportional to their linkage values.

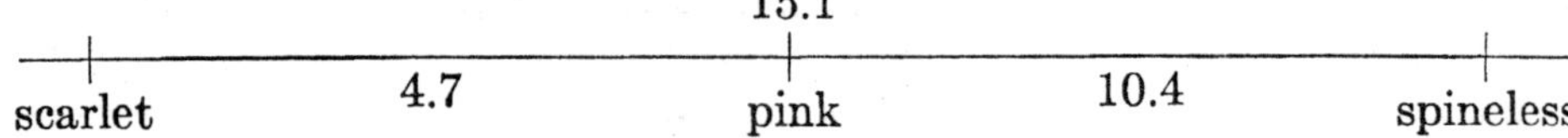


The sum of the two smaller values, 15.1, is somewhat greater than the actual determination for scarlet-spineless; but the discrepancy is not serious in view of the known variability in linkage values. Proceeding along the same lines, working always with comparatively low recombination values, every factor in the linkage group may be brought into the system, so that as the final outcome of the operation, all the factors of a linkage group may be plotted along a straight line at distances proportional to their recombination values. Such a plot of a linkage group is usually called a chromosome or linkage map. The arrangement of the factors in a line, like beads on a string, is known as the linear order of the factors.

Studies such as those described above have been carried on extensively for all four of the linkage groups of *Drosophila melanogaster* and for linkage groups in several other species of Drosophila. In figure 56 are included chromosome maps giving the relative positions of the principal factors in the four groups of *Drosophila melanogaster*.

**The Locus of a Factor.**—It follows, from the account given above, that every factor has a characteristic position or locus in its linkage group. The method of determining its locus is a comparatively simple matter. It is necessary simply to determine its recombination values with two known members of the group, provided, as in mapping the factors in

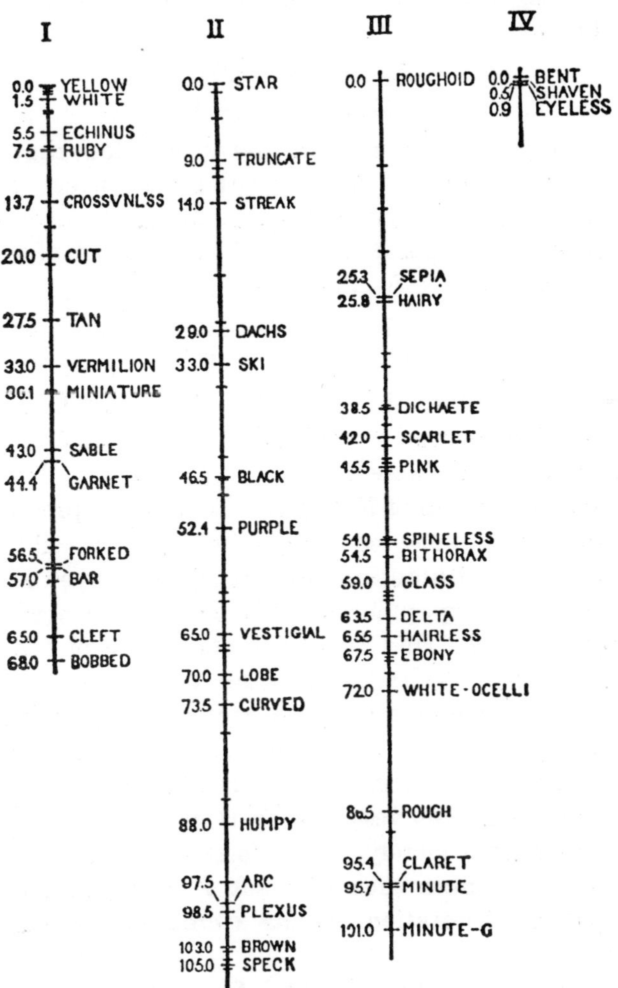


FIGURE 56.—Maps showing the loci of some of the factors in the four linkage groups of *Drosophila melanogaster*. (*From Morgan.*)

the first place, these values are comparatively low. If its value is determined with only one known, it will be impossible to say whether the factor lies above or below the known locus by the value obtained. By making a second determination, however, its position becomes definitely fixed. Once its position has been determined it is possible to predict, within certain limits, what percentage of recombinations it will show with other known members of the group.

As an illustration of the application of this method to a concrete instance, consider the factor hairless as an unknown. By reference to table 11, its linkage value with delta is found to be 2.9. The locus of delta is at 63.5. The delta-hairless value of 2.9 merely indicates that hairless lies at approximately that distance in the group from delta, but it does not fix the direction from delta. From this value the locus of hairless will be at

either $63.5 - 2.9 = 60.6$

or $63.5 + 2.9 = 66.4$

according as the locus of hairless is above or below that of delta. In order to determine which of these positions is correct, its value with some other known factor must be found. Referring again to the table, its value with ebony is given as 1.2. The locus of ebony is at 67.5; therefore, hairless must be located at

either $67.5 - 1.2 = 66.3$

or $67.5 + 1.2 = 68.7$

whence it may be seen that the locus of hairless must lie in the neighborhood of 66.35. The position of hairless in the map, 65.5, is obtained from a consideration of all available data.

**The Drosophila System of Notation.**—In order to present the remaining conceptions as to linkage most expeditiously, it is well to introduce at this point the system of notation most widely employed in stating these results in Drosophila experiments. The Drosophila system of naming phenotypes employs the wild type, the phenotype of the normal, wild representative of the species as the base of reference. The different stocks are then designated by the characters in which the individuals differ from wild type. Thus a black fly has a black body color; in other respects—eye color, wing form, etc.—it is identical with wild type. If in addition the wings are vestigial, it is called black vestigial; a fly described as vestigial, merely, is understood to have the wild-type gray body color. A cross then of black × vestigial, according to the Drosophila system of notation, means a cross of a black normal with gray vestigial.

Similarly the factor formulae take into account only those factors in which the individual differs from wild type. Thus the $F_1$ from black × vestigial is represented as $\frac{b}{vg}$, which is understood to be a wild-type fly of the constitution $\frac{b\ Vg}{B\ vg}$, according to the notation previously employed. $B$ and $Vg$ are factors appropriate to wild type; therefore, they are understood to be present in the above formula. When recombination occurs in a $\frac{b}{vg}$ female, it is represented as $\frac{b \mid vg}{\mid}$, which is understood to be equivalent to $\frac{b}{B} | \frac{vg}{Vg}$. The advantage of this system will become apparent from the discussion which follows.

**The Three-point Experiment.**—As has been seen, the recombination values obtained in experiments are subject to statistical fluctuations, so that in the case of three values obtained from three linked factors in all possible combinations, the sum of the two smaller values is rarely exactly equal to the largest value, as it should be under the stipulations of linear order. In order to avoid this difficulty, the three factors may be followed in the same experiment, so that the three recombination values will be derived from the same set of data.

The method of doing this is not difficult, although it may appear somewhat complex to the beginner. Consider the values presented for sepia-hairy-dichaete in the third group, as recorded in table 11. Taking these factors two at a time, the following results were secured:

| | |
|---|---|
| sepia-hairy | 0.5 |
| sepia-dichaete | 13.8 |
| hairy-dichaete | 14.2 |

From these data the relations would logically be plotted as follows:

```
0.0    0.5              14.3
---|-----|----------------|-----
hairy sepia           dichaete
```

But linkage values are rather notoriously variable, so that from these data it would be difficult to conclude with assurance whether hairy lay to the right or left of sepia at 0.5 unit of distance.

This difficulty is removed, however, when the three-point method of experimentation is applied. Sepia and hairy are recessive and dichaete is dominant to the corresponding wild-type allelomorphs, so that for the experiment a sepia-hairy fly $\frac{se \quad h}{se \quad h}$ was crossed with dichaete $\frac{\qquad D}{}$, producing dichaete $\frac{se \ h}{\qquad D}$ and wild type $\frac{se \ h}{}$ males and females in $F_1$, since dichaete is a heterozygous character. In order to determine the gametic ratio, $F_1$ dichaete females were backcrossed to sepia-hairy males, the triple recessive type in this case, whereupon the following results were secured:

| | | | |
|---|---|---|---|
| parental combinations $\frac{se \ h}{\qquad D}$ | sepia-hairy | 1,576 | 89.0 |
| | dichaete | 1,507 | |
| sepia-hairy recombinations $\frac{se \quad D}{h}$ | sepia-dichaete | 5 | 0.5 |
| | hairy | 12 | |
| hairy-dichaete recombinations $\frac{se \ h \mid D}{\mid}$ | sepia-hairy-dichaete | 179 | 10.5 |
| | wild-type | 184 | |

From these results it is established beyond question that the correct order is as follows:

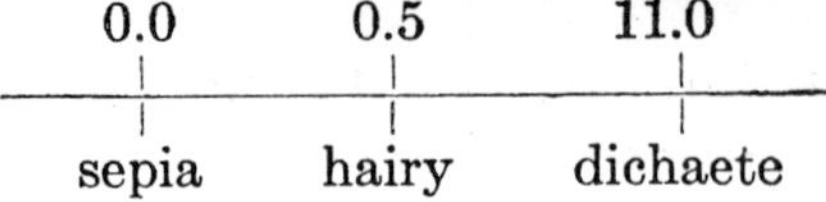


This is shown by the character of the recombination classes, all four of which represent recombinations of sepia and dichaete. If hairy were located on the left of sepia, the second two classes, sepia-hairy recombinations, would have been hairy dichaete and sepia, instead of sepia dichaete and hairy, as the student may verify for himself by assuming the constitution of $F_1$ to be $\frac{h\ se\quad}{\quad D}$ instead of $\frac{se\ h\quad}{\quad D}$. The discrepancy as to the actual location of dichaete from these two sets of data, 11.0 from sepia in one case and possibly 14.7 in the other, following the same arrangement, while large, is not unusual for data no more extensive than these. When such discrepancies occur, adjustments must be made in fitting the data into a map.

**Multiple Recombination.**—In the discussion thus far, it has been emphasized that rather low recombination values, preferably not exceeding 10 per cent, must be chosen as a basis for discovering the relations existing among the members of a linkage group and for plotting this relationship in map form. The reason for this insistence on low values is evident, when three factors giving high recombination values are considered:

| | |
|---|---|
| sepia-spineless | 26.8 |
| spineless-rough | 24.3 |
| sepia-rough | 42.8 |

From these data and from others which the student may set down by securing values from table 11, it may be seen that when the linkage values are high the sum of the two lower values regularly exceeds the largest value, and that in general the discrepancy becomes more marked the higher the values involved.

The reason for this fact may again be determined by application of the three-point experiment. Thus pink-ebony flies crossed with dichaete give dichaete and wild-type males and females in $F_1$, since dichaete is always heterozygous. $F_1$ dichaete females backcrossed to pink-ebony males gave the following results:

| Type of gamete | | Phenotypes | Numbers | Per cent |
|---|---|---|---|---|
| Parental combinations | D | dichaete | 824 | 77.7 |
| | p e | pink-ebony | 878 | |
| Single recombinations between | D \| p e | dichaete-pink-ebony | 28 | 2.7 |
| dichaete and pink | \| | wild type | 31 | |
| Single recombinations between | D \| e | dichaete-ebony | 214 | 18.3 |
| pink and ebony | p \| | pink | 187 | |
| Double recombinations | D \| p \| | dichaete-pink | 11 | 1.3 |
| | \| \| e | ebony | 18 | |

It will be observed that eight classes were obtained in this experiment, representing all possible combinations of the characters concerned. In the sepia-hairy-dichaete experiment only six classes of individuals were secured. The presence of the two extra classes is the key to the solution of the difficulty.

In determining the recombination percentages from the above data, all classes must be added together which represent recombinations between any two genes under consideration, whence

$$\begin{aligned} \text{dichaete-pink} &= 2.7 + 1.3 = 4.0 \\ \text{pink-ebony} &= 18.3 + 1.3 = 19.6 \\ \text{dichaete-ebony} &= 2.7 + 18.3 = 21.0 \end{aligned}$$

The last two classes, dichaete-pink and ebony, represent simultaneously a recombination between dichaete and pink and pink and ebony. They are, therefore, called double recombinations. In the calculations they must be added to the single recombinations of dichaete with pink and pink with ebony, because they represent changes in the relations of these factors to each other. They are not, however, included in the computation of the dichaete-ebony values, because they do not change their relations. For this reason the dichaete-ebony value is less than the sum of the two smaller values by an amount equal to twice the value of the double recombination class. Consequently when longer distances are involved, the occurrence of double and other multiple recombinations accounts for the discrepancies between values obtained by adding up the sum of smaller values and direct values obtained for the terminal factors.

Pictorially the production of the different kinds of gametes in a three-point experiment may be represented as shown in figure 57. The chromosome may separate as in *A* without interchanging material. This gives the two classes of parental combinations. If the break occurs between *D* and *p* as in *B*, the two recombination classes, representing single recombinations of the dichaete-pink type, are secured. A break between *p* and *e* gives the two classes of single recombinations of the pink-ebony type. A simultaneous break between *D* and *p* and *p* and *e* as shown in *D* would give two classes due to double recombination. It is obvious from a study of this figure that for the total dichaete-pink value it is necessary to add $B + D$; for pink-ebony $C + D$; and for dichaete-ebony $B + C$. Adding together the dichaete-pink and pink-ebony values gives $B + C + 2D$, which is in excess of the direct dichaete-ebony determination by an amount equal to $2D$, twice the value of the amount of double recombination.

**Coincidence.**—It is, of course, to be expected that a certain proportion of double recombinations will occur. Thus, if there are three factors,

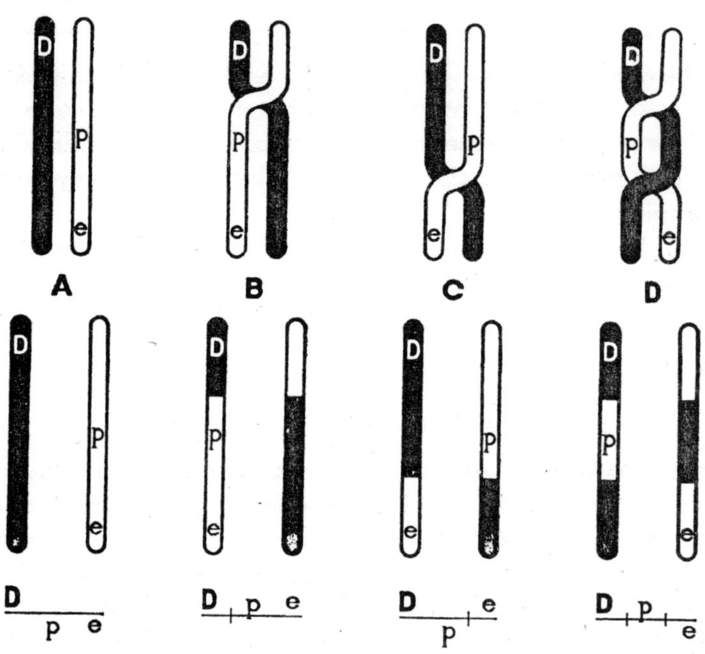


FIGURE 57.—Diagrammatic portrayal of types of recombinations secured in a three-point linkage experiment, with symbols corresponding to the four ways of gamete production, resulting in eight types of gametes.

*a*, *b*, and *c*, so spaced that *a-b* and *b-c* linkage values are each equal to 10 per cent, one would naturally expect to obtain 1 per cent of double recombinations, based upon the assumption that the two events are independent, for the expectation of simultaneous recombination in both is equal to the product of the two events taken separately; in this instance, $0.1 \times 0.1 = 0.01$ or 1 per cent. For short distances, 10 per cent or less, the proportional amount of double recombinations is negligible; for longer distances it becomes greater, to such an extent that no matter how long the map distance between two factors, considering the matter from this point of view, the direct recombination value would never exceed 50 per cent. This accounts for the fact that although map distances may be as great as 110, nevertheless the direct recombination value is never in excess of 50 per cent.

There are, however, some reservations to these assumptions, which have been determined by experiment. Consider the values obtained for the three sex-linked genes, vermilion, sable, and bar. By direct experiment the following results were secured:

| Type | Parental combinations | Single *v-s* recombinations | Single *s-B* recombinations | Double recombinations |
|---|---|---|---|---|
| Numbers......... | 5,772 | 716 | 1,015 | 21 |
| Proportion........ | 0.767 | 0.095 | 0.135 | 0.003 |

From these data the following linkage values are obtained:

vermilion-sable............................ $0.095 + 0.003 = 0.098$
sable-bar.................................. $0.135 + 0.003 = 0.138$

The expected proportion of double recombinations should equal the product of these two values, *viz.*, $0.098 \times 0.138 = 0.014$. The actual proportion of double recombinations realized, however, is only 0.003, which is only 21.5 per cent of expectation.

The term coincidence is applied to the simultaneous recombination which occurs among several factors in a series as in the preceding example. The degree of coincidence in a given case is expressed in terms of the coefficient of coincidence, a value obtained by dividing the actual proportion of coincidence by the expected proportion, assuming independent recombination. In the above example, the coefficient of coincidence has a value of 21.5 per cent as stated.

**The Mode of Interchange in Recombination.**—Extensive studies indicate that the coefficient of coincidence is low when the factors are close together and becomes progressively higher as the distance between the factors increases, attaining finally a value of 100 per cent when a certain distance is reached. These results are taken to indicate that interchange between homologous series of factors involves blocks of factors which have a certain modal length. The modal length has been found to differ in different linkage groups and in different portions of the same group. In the sex-linked group in *Drosophila melanogaster*, the coincidence value for distances less than 15 units is zero; *i.e.*, no double recombinations have been observed involving a total distance less than 15 units. The coincidence value rises gradually to 100 per cent when the total distance involved reaches 45 units. When the distance is increased beyond 45 units, a slight drop in the coincidence value again occurs, which is to be expected, if the average length of interchanged blocks is about 45 units. In the second and third groups in this species, the modal length of blocks is about 15 units in the middle of the group and 30 units at both ends.

These results indicate that, if a break occurs in a linkage series at a certain point, the adjacent regions on both sides of the break are protected to a certain extent from simultaneous recombination. The regions immediately adjacent to the break are absolutely protected, and as the distance from the break increases, the protection becomes progressively less until the distance reaches the modal length, whereupon it disappears entirely. The protection which regions adjacent to a break enjoy is termed interference, since there is a restriction of recombination in these regions. Interference is obviously merely another way of looking at coincidence.

That the interchange between homlogous series of factors involves blocks of factors, rather than single ones scattered promiscuously through the series, may be demonstrated experimentally by the use of complex stocks of linked characters containing factors so distributed that they cover a relatively long portion of the linkage group. Muller, for example, has presented data on recombination in heterozygous females containing the mutant factors for yellow, white, abnormal, bifid, vermilion, miniature, sable, rudimentary, and forked in one X-chromosome and cherry, club, and bar in the other, which may be written as follows:

$$\frac{y \quad w \quad A \quad bi \quad v \quad m \quad s \quad r \quad f}{w^c \qquad\qquad cl \qquad\qquad\qquad B}$$

Thus eleven loci in the X-chromosome were followed simultaneously in this experiment. The distances separating them may be determined by referring to figure 56. It was found that among 712 flies, 386, or 54.4 per cent, represented parental combinations; 296, or 41.7 per cent, represented single recombinations of various kinds; and only 30, or 4.2 per cent, were double recombinations. No triple recombinations occurred in this experiment, but subsequently a few were obtained in other experiments involving large numbers of individuals.

Our interest in these results lies in the types of double recombinations which were secured. These are portrayed diagrammatically in figure 58. It will be seen from this figure that in only one case was the distance between the two breaks short, and in that case it might have been as great as 20 units. This experiment amply demonstrates that recombination between homologous series of factors involves exchanges of relatively large blocks.

**Predicting Linkage Values.**—The chief usefulness of the chromosome map, lies in its value for prediction. Once the locus of a factor has been determined by application of the method described above, its approximate linkage value with any other locus in the group may be determined by simple arithmetic computations. The map distance is first obtained merely by taking the difference between the two loci. The actual linkage value may be considered a function of the map distance. For values not exceeding ten units map distances and linkage values are approximately equal, but for longer map distances a correction must be made for double and other multiple recombination. Attempts have been made to reduce the computation to a simple algebraic formula but the fact that coincidence values vary in different portions of the series limits the value of such formulae. To overcome this difficulty actual linkage values have been determined for typical factors and graphs (*cf.* figure 59) have been constructed from which the amount to be subtracted from the map distance in order to obtain the linkage value may be read off. By use of such a graph a reasonably accurate estimate of the amount of correction to be applied in other cases may be made.

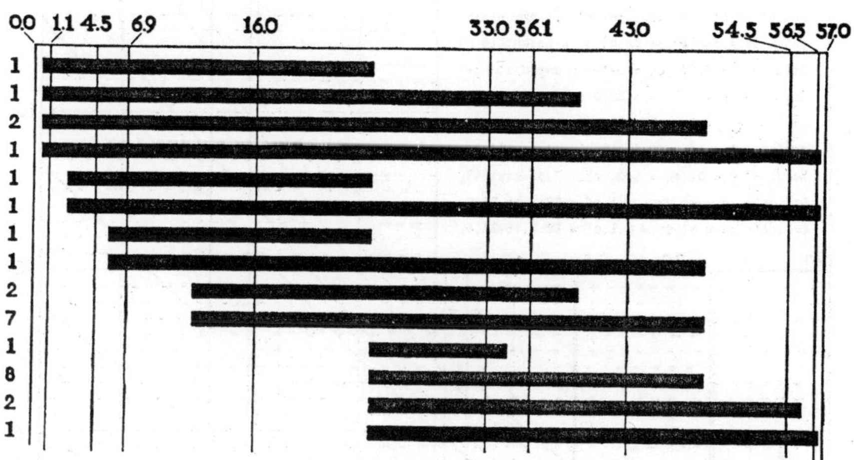


FIGURE 58.—Diagram showing lengths of the middle segment exchanged in double recombination in females of *Drosophila melanogaster* heterozygous for twelve sex-linked factors, the loci of which are represented by the vertical lines. The figures on the left indicate the numbers of times the corresponding exchange was observed in 712 cases.

**The Linear Order in Other Species.**—While the evidence on order of the factors is largely taken from *Drosophila melanogaster* and other species of Drosophila, there is no evidence extant which conflicts with its general application to animal and plant species. The only difficulty appears to be that sufficient data have not yet been accumulated to afford conclusive evidence.

In *Primula sinensis,* the Chinese primrose of commerce, Bateson reported the results of Gregory's extensive studies of a single linkage group containing four factors, the character effects of which are described in Chapter XIV. The linkage relations among these four pairs of factors in all possible combinations of twos are contained in table 7. Data are presented separately for the results from the use of heterozygous plants as male or female parents, because there seems to be a fairly consistent difference in the values obtained in the two cases. Plotting the data from female gametes, the following map is obtained:

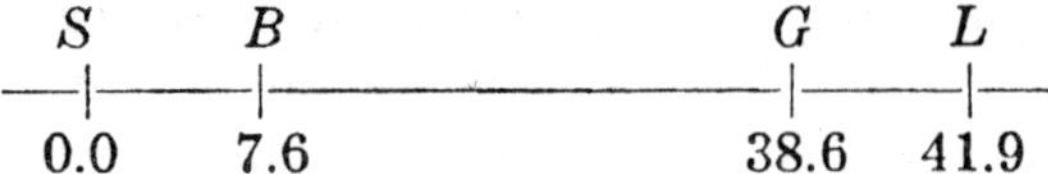


That the factors are in the linear order shown in this map and that recombination follows the same principles in Primula as in Drosophila may be determined by an examination of the data obtained from three- and four-point cxperiments. The data for the three-point experiments

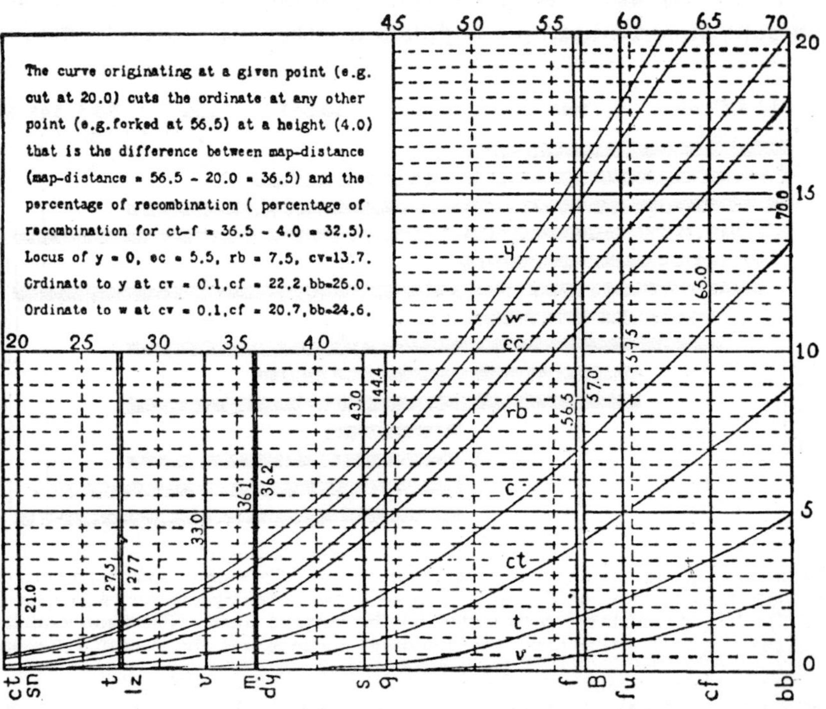


FIGURE 59.—Graph depicting the relation between map distances and linkage values for representative sex-linked factors in *Drosophila melanogaster*. (*From Morgan, Bridges, and Sturtevant, The Genetics of Drosophila in Bibliographia Genetica vol.* 2, *copyright* 1925 *by Martinus Nÿhoff. Reprinted by permission.*)

are presented in summarized form in table 12. It will be observed that double recombination occurs in Primula, just as in Drosophila, and that it exhibits the same general features as in Drosophila. Except in two cases the value of the coefficient of coincidence exceeds 100 per cent, which is probably to be expected from the rather long distances involved.

In maize Hutchison has presented extensive data for three or four pairs of linked endosperm characters. The factors concerned and their character effects are as follows:

*I-i*.......................... white *vs.* colored aleurone
*C-c*.......................... colored *vs.* white aleurone
*Sh-sh*........................ non-shrunken *vs.* shrunken endosperm
*Wx-wx*........................ starchy *vs.* waxy endosperm

The data for these four pairs of factors taken two by two are contained in table 13. It will be observed that no data are presented for linkage of *I* and *C*. This is because these two factors probably occupy the same locus, so that *I*, *C*, and *c* make up a series of triple allelomorphs. At least the factor *I*, if not an allelomorph of *C*, is so closely linked with it that it may be regarded as lying at about the same position in the linkage series. Considering *I* and *C* as allelomorphs and averaging results accordingly, a three-point map of this series may be constructed as follows:

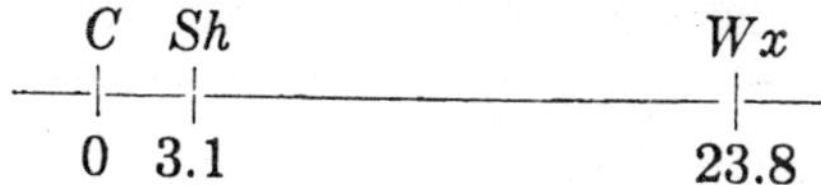


TABLE 12.—LINKAGE DATA FROM THREE-POINT EXPERIMENTS IN *Primula sinesis* (*Data of Gregory*)

| | | Types of combinations | | | | | Recombination percentages | | | |
|---|---|---|---|---|---|---|---|---|---|---|
| Loci | Sex | —— | +— | —+ | ++ | Total | +— | —+ | —+— | Coincidence |
| *S-B-G* | ♀♀ | 1,937 | 150 | 829 | 54 | 2,970 | 6.9 | 29.7 | 33.0 | 1.13 |
| *S-B-G* | ♂♂ | 1,641 | 256 | 919 | 105 | 2,921 | 12.4 | 35.1 | 40.2 | 1.21 |
| *S-B-L* | ♀♀ | 366 | 451 | 210 | 20 | 641 | 10.1 | 35.9 | 39.8 | 1.29 |
| *S-B-L* | ♂♂ | 727 | 107 | 431 | 47 | 1,312 | 11.7 | 36.4 | 41.0 | 1.19 |
| *B-G-L* | ♀♀ | 553 | 291 | 23 | 6 | 873 | 34.0 | 3.3 | 36.0 | 0.61 |
| *B-G-L* | ♂♂ | 1,006 | 574 | 20 | 7 | 1,607 | 36.2 | 1.7 | 37.0 | 0.70 |

TABLE 13.—LINKAGE DATA FOR FACTORS IN THE *I-C-Sh-Wx* GROUP IN MAIZE (*After Hutchison*)

| Factors | Total | Recombinations | Per cent |
|---|---|---|---|
| *C-Sh*.................... | 52,153 | 1,611 | 3.1 |
| *I-Sh*.................... | 19,615 | 786 | 3.0 |
| *C-Wx*.................... | 6,708 | 1,456 | 21.7 |
| *I-Wx*.................... | 10,220 | 2,713 | 26.5 |
| *Sh-Wx*................... | 14,895 | 3,019 | 20.3 |

Confirmatory data from three-point experiments are available here also. The results secured from them are contained in table 14. The figures of the two sets of results have been added together on the assumption that *I* and *C* occupy identical loci, although there may possibly be a genetic difference in the percentage of recombination in the two cases.

Whatever method of computation is used, however, is immaterial to the point at issue. The data clearly indicate a linear arrangement of factors in the order determined from the previously considered data. The coincidence values are very low, which indicates a high degree of interference over this region of the linkage group.

TABLE 14.—DATA FROM THREE-POINT LINKAGE EXPERIMENTS IN MAIZE (*After Hutchison*)

| Loci | —— | +— | —+ | ++ | Total | +— | —+ | —+— | Coincidence |
|---|---|---|---|---|---|---|---|---|---|
| *C-Sh-Wx*....... | 5,246 | 229 | 1,227 | 6 | 6,708 | 3.5 | 18.4 | 21.7 | 0.139 |
| *I-Sh-Wx*....... | 4,495 | 260 | 1,322 | 5 | 6,082 | 4.4 | 21.8 | 26.0 | 0.086 |
| Total.......... | 9,741 | 489 | 2,549 | 11 | 12,790 | 3.9 | 20.0 | 23.8 | 0.110 |

Other material, although somewhat fragmentary, might be given in support of the generality of the linear order of arrangement of factors. Sufficient, however, has been presented from widely enough separated sources to warrant the conclusion, as a general working hypothesis, that the factors in a linkage group are arranged in linear order at distances proportional to their recombination values.

**The Mechanism of Crossing-over.**—The exchange of material which occurs between homologous chromosomes is known as crossing-over. Recombination of linked factors is genetic evidence for the occurrence of crossing-over, and something of the nature of the process is disclosed by linkage studies. Crossing-over is evidently a phenomenon which involves exchange of exactly equivalent portions between homologous chromosomes. Apparently the most frequent event is the separation of chromosomes without exchanging material or with a single crossing-over, as is shown by the following approximate figures for the X-chromosome in *Drosophila melanogaster:*

| | |
|---|---|
| No crossing-over.................................. | 43 per cent |
| Single crossing-over.............................. | 43 per cent |
| Double crossing-over.............................. | 12 per cent |
| Triple crossing-over.............................. | 2 per cent |

Since crossing-over involves an exchange of material between homologous chromosomes, it seems reasonable to conclude that it must occur during the conjugation of homologous chromosomes in the prophase of the first meiotic division. The events occurring at this time have been described in detail in Chapter III, which the student should review in connection with this problem. Some evidence that crossing-over occurs at this time is afforded by Plough's experiments which disclosed a relatively marked effect of temperature differences on linkage values. It was found, by keeping females at high temperatures for short periods, that the effectiveness of the temperature appeared to be limited to eggs which were in the early maturation stages. Eggs which had passed through this stage and those which had not yet reached it, as shown by linkage values of successive broods of offspring, gave normal values.

The precise method of crossing-over has not been disclosed by cytological studies. Presumably in order to satisfy the requirements of the genetic data, crossing-over must occur in the thin thread stage when the chromosomes are spun out to the greatest possible extent. At the zygotene stage the elements of the conjugating chromosomes are arranged in linear order and accurately opposed in the two members. Moreover, these strands may be shown to cross at certain points; so that, if in the ensuing stages the conjugating chromosomes become reorganized after the scheme shown in figure 57, the genetic stipulations will be fulfilled. It is obviously a difficult task to obtain conclusive cytological evidence on such a point.

In later stages, particularly in the diplotene stage, when the chromosomes are relatively condensed, the conjugating members are often coiled about each other in various intricate ways. Some cytologists have interpreted these figures as evidence of a previous exchange of material, but it is difficult to demonstrate the correctness of this hypothesis. It is unlikely that crossing-over occurs at this period, for it is difficult to see how the genetic stipulations can be satisfied under these conditions, and the cytological studies indicate that these coils straighten out before final condensation of the chromosomes.

A further problem arises as to whether the conjugating chromosomes are in a single-strand or two-strand stage at the time of crossing-over; *i.e.*, whether the conjugating chromosomes split longitudinally before or after crossing-over. There is some evidence from cytological studies that the longitudinal split occurs during the early stages of conjugation of the chromosomes, and it appears not improbable that crossing-over occurs in the two-strand stage. If so, the exchange of material presumably involves only the two inner strands at the point of crossing, as shown in figure 60, so that the four chromatids would differ in constitution. Genetic evidence to this effect has been afforded by intricate linkage experiments on triploid females (*cf.* Chapter XXIV). Obviously, if crossing-over occurs in the two-strand stage, the distinction between reductional and equational divisions in meiosis becomes meaningless; for both divisions must then be in part reductional and in part equational.

There is still some debate, apparently gradually subsiding, concerning the interpretation of conjugation. The above discussion is based on the assumption that the chromosomes conjugate side by side (parasynapsis). If, as is contended in some cases, conjugation involves end-to-end approximation (telosynapsis), then it is difficult to see how exchange of material between homologues may occur at this stage. It is possible that some of the cytological investigations do not furnish pertinent evidence on this question, for much of it has been done upon species which have not as yet been studied genetically; but such as it is, it seems to be overwhelmingly on the side of parasynapsis. While the evidence

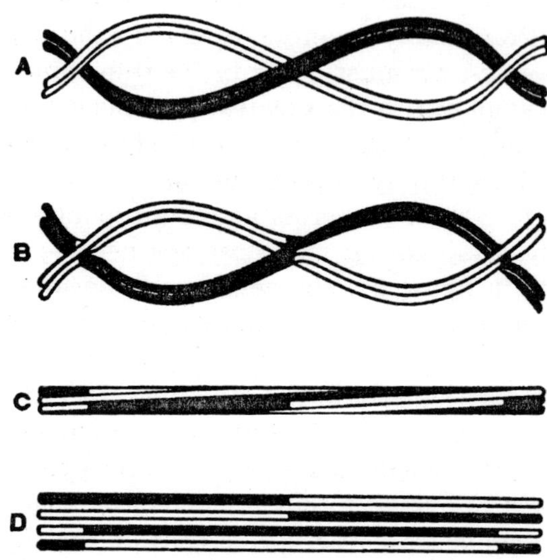


FIGURE 60.—Illustrating crossing-over on the assumption that the conjugants first split equationally. *A*, double threads twisted about each other; *B*, interchange of inner strands at the point of crossing; *C*, straightening out of the twisted strands and opening of the strands by reductional separation of the components, again giving a twisted appearance due to perspective; *D*, the four resulting chromosomes, two of which are single and two double crossovers. A progressive condensation, not represented in the figures, occurs during the process. (*Adapted from Morgan.*)

hitherto presented from cytological studies does not establish the correctness of the interpretation of crossing-over presented above, it at least is not in opposition to it.

## Reference

MORGAN: Physical Basis of Heredity, Chapters VII–XII.
See preceding chapter.

## CHAPTER XVI

## VARIATION

The discussion of variation has been deferred to this time because it is desired to emphasize genetic features and interpretation, not merely to offer a formal, descriptive treatment of the subject. Obviously these features can be understood only after a general working knowledge of the organization and operation of the germinal mechanism has been developed.

For the purpose of this discussion, variation may be defined as diversity of any kind exhibited among individuals of a given species at corresponding stages of development. These diversities may relate to a great variety of features, such as differences in size, in morphological characters, in physiological features, such as resistance to disease or response to environmental conditions; in short, in practically all features which characterize individuals of a population. So numerous are these diversities that each individual is unique; it is different in some way or other from every other individual of the species. The major problem of variation is the determination of the causes of this diversity, which implies, of course, a classification into categories of those of like import.

**Composition of the Germinal Material.**—In attacking the problem of variation, it is important to realize what general conception as to the structure and operation of the germinal substance has been evolved as a result of investigations in genetics. The foregoing portion of the text has been devoted primarily to a consideration of this matter. In *Drosophila melanogaster*, the most thoroughly investigated species, the studies have culminated in the production of a linkage map, as shown in figure 56, by the proper use of which it is possible to predict, with a reasonable degree of accuracy, the kinds and relative proportions of gametic combinations which any heterozygous individual will produce—in short to predict its breeding behavior. No doubt this feat represents one of the major triumphs of modern biological research, and its accomplishment is largely due to the efforts of Morgan and his associates at Columbia University. The general conception that the germinal material is composed of discrete elements (the factors), that the factors are arranged in linkage groups corresponding in number to the haploid number of chromosomes, and that they occupy specific loci in the linkage groups represents the point of departure for a consideration of any problem of genetics.

Mendelian studies which are in progress in other animal and plant species are directed towards the same goal. Owing, however, to the more serious technical difficulties in working with most species arising from the longer generation period and the expense of conducting investigations on a sufficiently large scale, it will be a long time before a comparable body of information shall have been obtained about them. In the meantime, however, the conceptions evolved from Drosophila experimentation have been found applicable to a great variety of animal and plant species, so that there appears to be no question as to their general validity.

Investigations in genetics have not, however, gone far enough to give an answer to some questions. It has not yet been possible to determine the number of factors contained in the germinal material, although in *Drosophila melanogaster* some rough estimates may be made. Thus if it be assumed that the minimum linkage value is 0.1 unit and that this represents the distance between two adjacent factors and if it be further assumed that the factors are spaced at this distance throughout the system, then, since the total length of the four linkage groups amounts to about 300 units, there should be about 3,000 of them in a haploid set. Estimates based on an entirely different method, namely the numbers of recurrences of mutations, give a value of about 2,000. While these estimates obviously cannot be considered very accurate, it is interesting to know that genetics has arrived at a point where first approximations may be made.

The most important unsolved problem, however, is that of the nature of the factors. They are obviously self-propagating elements of some sort; but no method has yet been devised for investigating their chemical or physical nature. If the estimates of the number of factors given above are accepted, it is possible to arrive at a rough approximation as to their size, since it is possible to measure the volume of chromatin in the cell. Here again various methods have been employed, and they give values of the same general order of magnitude, ranging from 0.010 to 0.077 micron in diameter as compared with about 0.0025 micron for such complex molecules as those of hemoglobin and casein. It is not certain, however, that the chromatin, even in the condensed condition, consists entirely of factors; and since the estimated number of factors probably represents a minimum value, these estimates of size must be regarded as maximum values. Evidently the dimensions of the factors approach those of complex molecules; but it is not necessary to conclude that the factors must be complex molecular substances.

While these matters are highly speculative, it may serve a useful purpose to call attention to them as an indication of the limitations of the present knowledge of genetics. It is self-evident that one of the major problems of cell physiology of the future lies in a determination of the composition of the factors, the manner in which they are held together in the germinal substance, and the way in which they exert their appropriate effects in development. In most instances, however, these matters are of no great concern to genetics except in so far as they indicate the complexity of the composition of the germinal substance; for problems of variation are largely problems of origin and distribution of germinal elements for a consideration of which it is necessary to proceed from the general conceptions of the mode of organization and operation of the germinal system outlined above.

**Types of Variation.**—It is self-evident that variation may arise from several distinct causes. Thus it is well known that irrespective of genetic causes individuals may differ through intervention of the environment, by reason of differences in nutrition, in surroundings, in training, etc. Since these variations arise during the course of individual development, they may appropriately be called developmental variations. Variation may also arise, as has been illustrated abundantly in the foregoing chapters of the text, as a result of difference in genetic constitution. Such diversities may be called Mendelian variations, since they depend upon differences in Mendelian factors. The very fact that differences in Mendelian factors exist raises the problem of their origin. It is known that factors, while comparatively stable, are nevertheless subject to occasional alteration. It is also known that the germinal mechanism concerned with the distribution of factors, while astonishingly regular in behavior, occasionally slips a cog, as it were, as a result of which a great variety of alterations in organization of the germinal material or in actual numbers of chromosomes may arise. These phenomena, in many instances not yet well understood, are grouped together under the term mutation. By way of a preliminary orientation, certain general characteristics of these three types of variation are discussed briefly below.

**Developmental Variation.**—It is a well-known fact that differences in the conditions surrounding an individual may profoundly affect its development. This effect is seen most clearly when groups of individuals as nearly equivalent as possible in heredity are placed under very different environmental conditions. An undernourished animal will not grow so fast as one properly fed; neither does a plant growing on poor soil develop so well as one in rich soil. When the differences are less striking, the effects naturally enough are not so clearly defined, and inasmuch as each individual in a sense develops in an environment peculiar to itself, different in some respects from that of any other individual, the differences among individuals are necessarily always in part simply developmental variations.

Obviously these differences are in a sense independent of genetic constitution. A potato tuber may be divided into two parts; one planted in rich soil under optimum conditions of moisture, temperature, etc. may develop into a large, productive plant; the other, in inferior soil, may develop into a stunted, unproductive plant. Or again, two scions may be taken from the same apple tree; one grafted on a standard crab root may grow into a large tree, whereas another grafted upon a dwarfing stock may develop into a very small tree, although both may be fruitful. In both instances genetic constitution, if not identical, is as nearly so as it is possible to have them, even though there may be such a vast difference between the resulting individuals that no one would suspect them to be of identical genetic constitution. Environment evidently modifies the course of individual development, for which reason these differences depending upon environmental differences are frequently called modifications. From this point of view heredity represents a set of potentialities which are expressed in various ways depending upon the environment, just as $H_2O$ may be a solid, liquid, or a gas, depending upon temperature and pressure.

**Mendelian Variation.**—It is hardly necessary to dwell upon Mendelian variation, for so many examples have been given in preceding chapters. Mendelian variations depend upon differences in Mendelian factors which exert their characteristic effects even when individuals develop under as nearly identical environmental conditions as it is possible to supply. The term Mendelian variation appears to be an appropriate one because they arise as a result of shuffling of factors governed by Mendelian laws; but they have also been termed combinations, because they represent different aggregations of factors.

In the preceding chapters emphasis has been placed upon the problem of organization of the factors in the germinal substance and their mode of distribution. In applying the conceptions arrived at from such investigations to variation as it occurs in nature and among domesticated animals and cultivated plants, a much more complicated situation appears than in most genetic experiments by reason of the greater number of factors involved and the attendant difficulty in classifying the individuals, and more particularly by reason of the fact that most Mendelian factors produce ill-defined character effects and affect features so highly modifiable by the usual slight environmental differences which inevitably occur that they cannot certainly be identified. As a consequence it has been necessary to develop special types of technique for dealing with such situations, although the fundamental conceptions as to germinal organization and factor distribution represent the point of departure for a consideration of the problem.

**Mutation.**—In the preceding chapters, factors have been dealt with as if they are distinct, stable elements which are distributed according to certain definite principles; and the germinal substance has been considered a definitely organized and specifically regulated mechanism. The extent to which these conceptions actually apply in living material is, of course, also a matter which should be considered, not as an article of faith, but as a problem for experimental investigation.

The very fact that Mendelian factor differences exist at once raises the question of their mode of origin. That in most instances factors are highly stable is evidenced by the results of genetic experimentation, which are inexplicable under any other assumption. Many of the variations employed in breeding experiments have been in existence for a long time, and nothing but inferential information about their origin is available. On the other hand, as exemplified particularly by the Drosophila experiments, new Mendelian variations are appearing and, when as in these experiments enormous numbers of individuals are examined, a large number of new forms may be secured in the course of time. The origin of these new Mendelian characters apparently is due to some sort of alteration in Mendelian factors, so that these factors, while relatively highly stable, are subject to occasional change.

Not only are the factors subject to occasional changes, but the mechanism itself may operate irregularly on rare occasions and thus give rise to a great variety of products differing in some respect from the normal condition. By reason of the self-propagability of germinal elements, such abnormal conditions once established are maintained in certain definite ways, so that they may be studied accurately by application of methods similar to those employed in typical Mendelian experimentation, and in many instances cytological investigation is also necessary in their elucidation.

The various phenomena involving change in germinal elements or organization are known as mutations. Evidently the term, as thus employed, is a catchall and includes many phenomena which are at present little understood. But there is no doubt that new material is continually being added in a variety of ways to the existing stock of variation.

**Problems of Variation.**—Obviously genetic investigations of variation are concerned primarily with a determination of the nature and causes of diversity. As they occur in nature and in agricultural operations, populations are subjected simultaneously to all the causes which operate to produce variation. The individuals exist under a great variety of environmental conditions which modify their development in numerous ways. Indiscriminate crossing among individuals differing in genetic constitution is the rule, rather than the exception, so that all possible combinations of existing Mendelian differences appear among them. And finally new variations, mutations of various kinds, are arising, and from time to time they become incorporated in the population. By operation of the laws of chance and by selection, natural or otherwise, as well, the composition of populations is continually changing; old factors are lost, and new factors are incorporated. Forces tending to produce change are operating in conflict with forces of conservation, as a result of which a population must be regarded as in a state of flux, even though, considered as a whole, it may exhibit a fairly constant state of equilibrium. It is this problem of variation as a whole which genetics attempts to reduce to comprehension in terms of the structure and operation of the germinal mechanism.

A number of methods of classifying variation have been proposed in order to systematize treatment, but inasmuch as they are for the most part formal, based upon characteristic features exhibited by the variations themselves rather than upon causes operating in their production, they are not very useful in genetic studies. The three foregoing categories—developmental variation, Mendelian variation, and mutation—are, however, based upon genetic causes and in most instances represent an adequate system of classification for genetic purposes. However, the actual sorting of individuals of a natural population into character classes similar to those employed in the breeding experiments described in previous chapters is rarely possible with the facility with which it can be done in experimental cultures. As a matter of fact, instead of discontinuous classes rather sharply delimited, the variation in such populations is of the continuous, intergrading type. For example, height in human beings varies over a continuous range extending from the height of the shortest to the tallest individual. Hair color, while more nearly comparable to the conditions obtaining in genetic experimentation, varies in an almost continuous series from white to black. Here there are several modes, as it were, such as black, brown, red, etc., but there are endless nuances within each class which connect them in a continuous series; so, for other characters, it is rarely possible to classify the individuals of a natural population into two distinct, alternative groups; it is more typical to find a continuous series of expressions ranging from one extreme to the other; but the frequencies of various grades of expression may differ greatly.

As a matter of fact this condition is to be expected, for many factors may affect a single feature of the organism, as has been seen. Thus, in Drosophila, there are over fifty factors concerned in eye color. If representatives of all of these were mixed in one population, they would give a continuous range of colors defying accurate classification; if in addition they were allowed to interbreed and produce all possible kinds of combinations, a condition approximating the complexity characteristic of natural populations would be set up. Add to these variations due to differences in environment and those arising occasionally from mutation, and apply the conception to every feature of bodily structure or function in order to gain an idea of the complexity of variation viewed as a whole; as compared with its simplicity in genetic experiments where the individual components have been dissected out as it were from the mass.

It is the purpose of the succeeding chapters to attempt to demonstrate the characteristic features of the different kinds of variations, to estimate the relative part which each plays in the total, in short to arrive at an accurate knowledge of the constitution of populations. In addition the discussion is designed to interpret the effects which various systems of reproduction, selection, etc. have upon the composition of populations. As the discussion proceeds, it is to be hoped that it will become clear that these matters merely represent an application in the wider sense of the conceptions of structure and operation of the germinal mechanism described in the previous chapters.

### References

Jones: Genetics, Chapter VII.
Sinnott and Dunn: Principles of Genetics, Chapter XI.
Thomson: Heredity, Chapter III.
Walter: Genetics, Chapter II.

## CHAPTER XVII

## THE STATISTICS OF VARIATION

As has been stated in the preceding chapter, many of the differences which exist among individuals of a species relate to measurable quantitative features, such as size, height, etc. In such cases accurate portrayal of the variation within the group requires that a relatively large number of measurements of individuals be made. When such measurements have been collected, it is usually discovered that no distinctly separated groups of individuals exist, but that individuals may exhibit any value in the scale of measurement within certain limits. For a character such as height in human beings, it is a familiar fact of every-day experience that a wide degree of variation exists and that a continuous series of expressions ranging from one extreme to the other is exhibited. Under such circumstances raw measurements of height lead to no definite conception of the nature of the condition, nor do they afford a readily comprehensible description of the situation. In order to overcome this difficulty, it is necessary to reduce the data. Reduction of data includes a tabulation of the measurements and a computation of certain constants which express their characteristic features in comprehensible terms. Statistical methods are employed for this purpose. "By statistical methods," says Yule, "we mean methods specially adapted to an elucidation of quantitative data affected by a multiplicity of causes." This chapter will be devoted to an exposition of the simpler statistical methods which are employed in a study of variation. The term biometry is commonly applied to the employment of statistical methods in the treatment of biological data.

**The Frequency Distribution.**—As a simple example of variation in a quantitative character, take certain measurements of length of a large number of beans representing the $F_2$ from a cross between a long- and a short-seeded bean. Johannsen measured 3,725 beans and tabulated the measurements as shown in table 15. It is important to realize what information this table contains. There are three columns which are numbered 1, 2, and 3. As shown in column 1, the measurements extend from a lower limit at 9 to an upper limit at 18 millimeters. The scale over which the measurements extend, 9 to 18 millimeters, is called the range. Column 1 contains the class boundaries in millimeters, representing equal classes, each 1 millimeter in extent, into which it has been divided. The second column contains the class centers in millimeters, representing the middle point of each class. The third column records the number of individuals—frequencies—which fall into each class. Thus there were 1,068 beans in the 13.5-millimeter class. This does not mean that all these beans measured exactly 13.5 millimeters in length, but merely that their measurements lay somewhere between 13 and 14 millimeters, the boundaries of that class. A table of this kind, usually consisting of only two items—the class centers and the corresponding frequencies—is known as a frequency distribution.

TABLE 15.—FREQUENCY DISTRIBUTION FOR LENGTH OF BEANS IN AN $F_2$ POPULATION
(*Data from Johannsen*)

| 1 | 2 | 3 |
|---|---|---|
| Class boundaries in millimeters | Class centers in millimeters | Frequencies |
| *B* | *C* | *f* |
| 9–10 | 9.5 | 2 |
| 10–11 | 10.5 | 20 |
| 11–12 | 11.5 | 136 |
| 12–13 | 12.5 | 540 |
| 13–14 | 13.5 | 1,068 |
| 14–15 | 14.5 | 1,125 |
| 15–16 | 15.5 | 636 |
| 16–17 | 16.5 | 180 |
| 17–18 | 17.5 | 18 |
| Total.................... | | $N = 3{,}725$ |

**Graphical Representation of Variation.**—The data contained in table 15 may be represented graphically in a number of ways, of which the two most commonly employed are shown in figures 61 and 62. In figure 61 a base line was laid off representing the range. The class centers were then marked on this base line in order at equal distances. At each class center a perpendicular was then erected proportional in height to the frequency of the class. The tops of successive ordinates were then connected by straight lines, forming a frequency polygon. In figure 62, a base line was laid off as before, but class limits were marked on it instead of class centers. A rectangle was then erected for each class proportional in height to the frequency of that class. Such a figure is called a rectangular histogram. In a sense it is a matter of indifference which method of representation is employed, although the rectangular histogram is most frequently used. The student should draw them both on the same base line in order to determine the relation between them.

The graphical representation of variation permits a ready grasp of certain features which are common to variation in a great many quantitative characters. In the first place the classes of greatest frequency are situated in the middle of the range. On both sides of these classes the frequencies fall off, at first slowly and then more rapidly, and the diminution in frequency from the center of the distribution is symmetrical. When very accurate measurements of a large number of individuals are available, and when the class intervals are relatively small, these features of the distribution are most pronounced.

The type of distribution which has just been described is characteristic of a great variety of quantitative characters. It is of course necessary to make a random selection of individuals for measurement, *i.e.*,

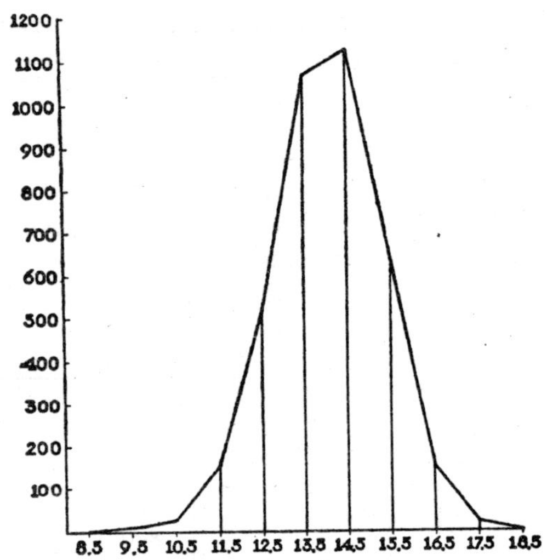


FIGURE 61.—Frequency polygon depicting the variation in length in a sample of beans. For data see Table XIV and compare this figure with figure 62.

so to choose them that all types in the general population are proportionally represented, and to avoid inclusion in the group of individuals of different stages of development, different sex, etc. In short, the material should be characterized by a multiplicity of causes, none of which has a preponderating influence in determining the expression of any subgroup of individuals.

**The Mean.**—While the mere tabulation of the data elucidates certain very characteristic features of it, a precise description of variation necessitates the computation of certain constants which express precisely significant attributes of the distribution. First, it is required to determine a constant of average type in the distribution. For this purpose the arithmetical average, or mean, is usually employed. The mean is determined very simply by computing the average value of the entire lot of individuals which make up the distribution. Its mathematical formula is as follows:

$$M = \frac{\Sigma(x)}{N}$$

where $M$ = mean, $x$ = the value of any variant, $\Sigma(x)$ indicates a summation of the values of all the variants, and $N$ = the total number of variants.

The mean is one of several measures of type and the one most frequently employed. It is the center of gravity of the distribution; *i.e.*, it represents a value such that plus and minus variations from it cancel.

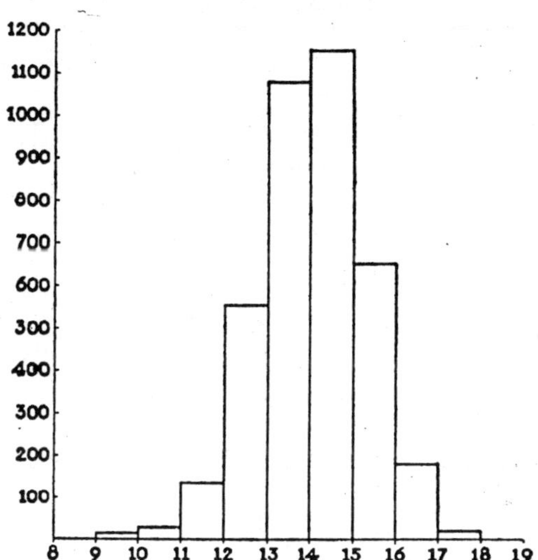


FIGURE 62.—Rectangular histogram depicting the variation in length in a sample of beans. For data see Table XIV and compare this figure with figure 61.

**The Standard Deviation.**—In the second place, it is necessary to have a measure of dispersion, *i.e.*, of distribution of individuals around the mean. The mean simply fixes the center of dispersion. Obviously two distributions may have the same mean and yet exhibit wide differences in the way in which the individuals are distributed about the means.

The customary measure of dispersion employed in biological investigations is the standard deviation $\sigma$, the formula for which is as follows:

$$\sigma = \sqrt{\frac{\Sigma(\delta^2)}{N}}$$

In this formula, $\delta$ represents the deviation of any variant from the mean. These deviations for each variant are squared, the sum of all squared deviations is taken $\Sigma(\delta^2)$, this value is divided by the total number of variants $N$, and the square root of the quotient is extracted. This gives the standard deviation, which may be defined as the root-mean-square deviation.

The standard deviation is one of many measures of dispersion which may be employed. It gives due weight to every variant in the distribution, and it has the further advantage that it is a constant of the mathematical formula of the curve of variation appropriate to the distribution. In some situations, however, it is desirable to have a relative measure of dispersion, so that variation may be compared in features expressed in different units of measurement. For this purpose the coefficient of variation $v$ is employed, *viz.*, $v = \frac{100\sigma}{M}$ per cent.

**Short Methods of Computation.**—The actual computation of the mean and standard deviation is facilitated by short-cut methods, several of which have been devised. A convenient one for use when no computing machines are available is illustrated in table 16. The principle

TABLE 16.—COMPUTATION OF MEAN AND STANDARD DEVIATION OF A FREQUENCY DISTRIBUTION BY A SHORT-CUT METHOD

| 1 | 2 | 3 | 4 | 5 |
|---|---|---|---|---|
| Class centers, millimeters | Frequencies | Deviations from arbitrary origin | Frequencies multiplied by deviations | Frequencies multiplied by square of deviations |
| $C$ | $f$ | $d$ | $f.d$ | $f.d^2$ |
| 9.5 | 2 | 0 | 0 | 0 |
| 10.5 | 20 | 1 | 20 | 20 |
| 11.5 | 136 | 2 | 272 | 544 |
| 12.5 | 540 | 3 | 1,620 | 4,860 |
| 13.5 | 1,068 | 4 | 4,272 | 17,088 |
| 14.5 | 1,125 | 5 | 5,625 | 28,125 |
| 15.5 | 636 | 6 | 3,816 | 22,896 |
| 16.5 | 180 | 7 | 1,260 | 8,820 |
| 17.5 | 18 | 8 | 144 | 1,152 |
| | $N = 3,725$ | .. | $\Sigma(f.d)$ 17,029 | $\Sigma(f.d^2)$ 83,505 |

For the mean $= M$

Arbitrary origin, $A.\,O. = 9.5$

$\bar{d} = \frac{\Sigma(f.d)}{N} = \frac{17,029}{3,725} = 4.571544$

$M = A.\,O. + \bar{d} = 14.072$ millimeters

For the standard deviation $= \sigma$

$S^2 = \frac{\Sigma(f.d^2)}{N} = \frac{83,505}{3,725} = 22.417450$

Subtract $\bar{d}^2 = 20.899014$

$\sigma_1{}^2 = 1.518436$

Subtract Shepherd's correction $(1.0 \times 1/12) = 0.083333$

$\sigma^2 = 1.435103$

$\sigma = 1.198$ millimeters

Coefficient of variation, $v = \frac{1.198 \times 100}{14.072} = 8.51$ per cent involved is to base computations on a convenient arbitrary origin, and then to correct these values in order to obtain the true values of the constants. In this table, columns 1 and 2 constitute the frequency distribution. Column 3 contains values representing the deviations of the classes from the arbitrary origin, which is here selected as the value of the lowest class, 9.5. Column 4 is obtained by multiplying corresponding numbers in columns 2 and 3 together. The sum of the numbers in column 4 divided by the total number of individuals gives the value 4.572 millimeters, which represents the average deviation of the variants from 9.5, the arbitrary origin. Since the mean is that value from which the average deviation is equal to zero, obviously

$$M = 9.5 + 4.572 = 14.072 \text{ millimeters}$$

Column 5 is obtained by multiplying together corresponding figures in columns 3 and 4. The sum of the numbers in this column divided by the total number of variants gives $S^2$, the mean square deviation from the arbitrary origin, 9.5. What is desired, however, is the mean square deviation from the mean; consequently a correction must be applied. The correction value is obtained by squaring the mean deviation from the arbitrary origin, 4.572, and subtracting it from the above value. A further correction, called Shepherd's correction, is made as a compensation for the error introduced by grouping variants into classes and basing computations on class centers. The value of this correction is one-twelfth of the class interval, which in this case is 1 millimeter. After subtraction of this value, the quantity remaining is $\sigma^2$ and extraction of the square root gives $\sigma$.

These methods are short cuts, but they give true, not approximate, values. They give exactly the same results as the unabridged method of computation, and they greatly reduce numerical work. Since long decimal fractions are eliminated, they really make for increased accuracy. The arbitrary origin from which the computations are made may be taken at any position. By one method it is taken at the middle of the range, but then plus and minus signs must be carried in the computations. Selecting the lowest class eliminates the minus signs and does not usually unduly increase the size of the figures.

**The Normal Curve.**—The two constants, the mean and the standard deviation, accurately portray the main features of variation in a population, if it conforms to the normal curve. The normal curve is a curve of the bell-shaped type shown in figure 63. The mean passes through the highest point of the curve, the point of maximum frequency, and the frequencies are distributed symmetrically about it. Its mathematical equation is of the form

$$y = \frac{N}{\sigma\sqrt{2\pi}} e^{-\frac{x^2}{2\sigma^2}}$$

in which $x$ and $y$ are coordinates, $\pi = 3.1416$ and $e = 2.7182$, the base of the natural system of logarithms, and the other terms have the same significance as heretofore. The mean is the Y-axis of the system of coordinates.

Obviously, if distributions of a given size conform to this equation, they will differ only in two respects; in the mean, which fixes the center of the distribution, and in the standard deviation, which describes the mode of dispersion of variants around the mean. By substitution of particular values of the standard deviation in the above equation, the normal curves corresponding to particular frequency distributions may be constructed. Extensive tables of ordinates and areas of the normal curves are available for this purpose, but their use is beyond the scope of this discussion.

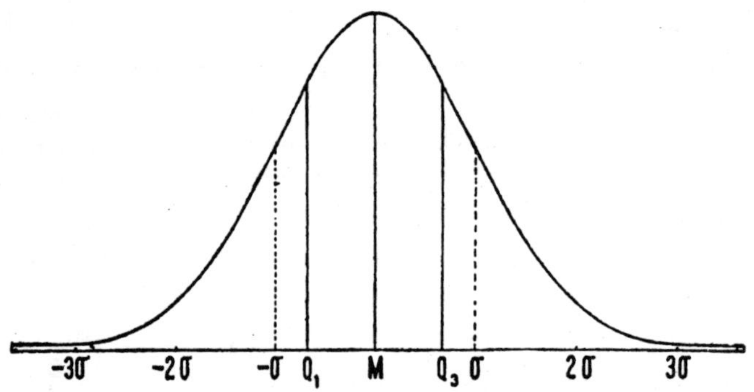


FIGURE 63.—The normal curve or curve of error. The mean intersects the maximum point of the curve. Deviations from the mean are plotted in terms of $\sigma$. The ordinates at $\pm 0.6745\sigma$ include the middle half of the area under the curve. These ordinates together with the mean divide the area under the curve into four equal parts. The proportions of the total area included within certain limits are as follows:

$$M \pm \sigma = 68.3 \text{ per cent.}$$
$$M \pm 2\sigma = 95.5 \text{ per cent.}$$
$$M \pm 3\sigma = 99.7 \text{ per cent.}$$

The normal curve exhibits several distinct features to which attention may well be called at this point. The mean is the center of gravity and the ordinate of greatest frequency. The distribution is symmetrical around the mean. The ordinates, $\pm .6745\sigma$, inclose the middle half of the area under the curve; or with the mean, they divide the curve into four equal portions. Hence these ordinates are called quartiles. The points of inflection of the curve, the ordinates which strike the curve where it changes from a concave to a convex form, are at $\pm\sigma$. The ordinates $\pm 3\sigma$ inclose 99.73 per cent of the area under the curve. In other words the range of the curve for 1,000 individuals will be about $6\sigma$. These features of the normal curve are illustrated in figure 63.

It is also helpful to consider the relation between the normal curve and the clause in Yule's definition, "elucidation of data affected by a multiplicity of causes." Suppose a multiplicity of causes are operating to determine length of bean in the illustrative example, none of the causes having a preponderant effect. They may all then be considered as equal in effect for statistical purposes. Suppose these causes are operating in such a way that some of them have plus and some minus effects on length of bean as measured from the mean, and that they are cumulative; so that the length attained by any bean is determined by algebraic addition of the various effects which are acting in a particular individual. Then, if the causes are $n$ in number and independent, the binomial $(\frac{1}{2} + \frac{1}{2})^n$ will represent their mode of distribution in a population of unit area. Expanding such a binomial for a high power of $n$ and plotting the coefficients of the terms gives us a distribution closely conforming to the normal curve, as shown in figure 64. It is thus evident that the assumption of a multiplicity of independent causes, none of preponderant effect, should give distributions of the kind which are actually obtained in biological material.

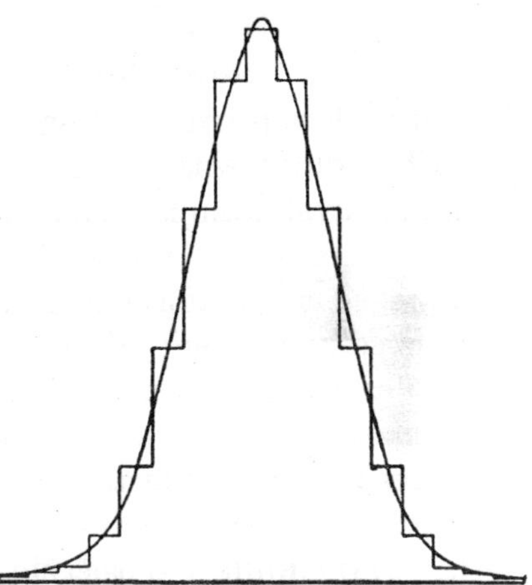

FIGURE 64.—Correspondence between the normal curve and the values obtained from expansion of the binomial $(\frac{1}{2} + \frac{1}{2})^n$, where $n = 20$.

**Other Types of Curves.**—Curves of variation may have either one maximum point, in which case they are called unimodal; or they may have two or more such points, in which case they are called multimodal. An illustration of a bimodal curve, one having two maxima, is given in figure 65. In general such curves indicate lack of uniformity in the material, or that some one or more causes of variation have a disproportionate effect.

Unimodal curves are also of several distinct types. In the first place, taking the normal curve as a basis of comparison, they may differ in symmetry. In many curves—an example is shown in figure 66—the point of maximum frequency is not the mean of the distribution. Such curves are known as skew curves. In skew curves there are three chief measures of type; the mode, which is the ordinate of greatest frequency; the median, which is the value of the middle variant in the distribution; and the mean, which as before is the average value of all the variants. In the normal frequency distribution these three measures of type coincide. Curves may also differ from the normal in the relative distribution of individuals at the mode. This feature of the curve is known as kurtosis. A curve is said to be leptokurtotic if it rises above the normal curve at the mode and platykurtotic if it does not rise as high as the normal curve at this point. The normal curve is said to be mesokurtotic. On the basis of these features of distributions, Pearson has constructed a system of curves and of criteria of different types from which a proper selection of an appropriate type may be made to fit any given distribution. Accurate fitting of curves to frequency distributions requires consideration of these features, a matter which is far beyond the scope of this treatment.

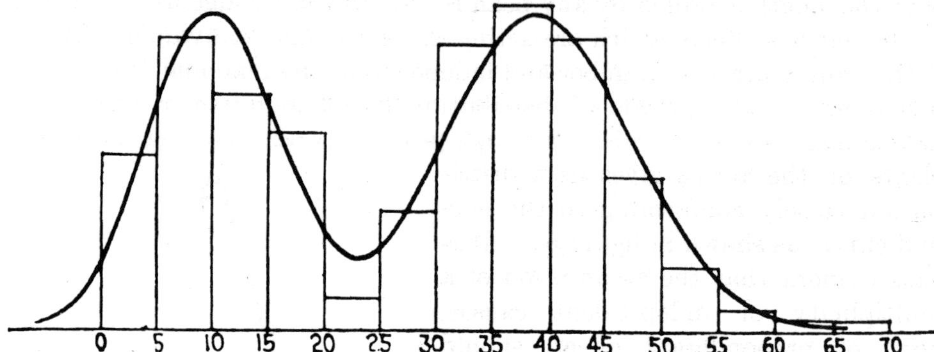


FIGURE 65.—Biomodal curve illustrating variation in percentage of aborted seeds (*Shartigkeit*) in a strain of Goldthorpe barley. (*From Johannsen, Elemente der Exakten Erblichkeitslehre, 2nd. ed., 1913, Gustav Fischer. Reprinted by permission.*)

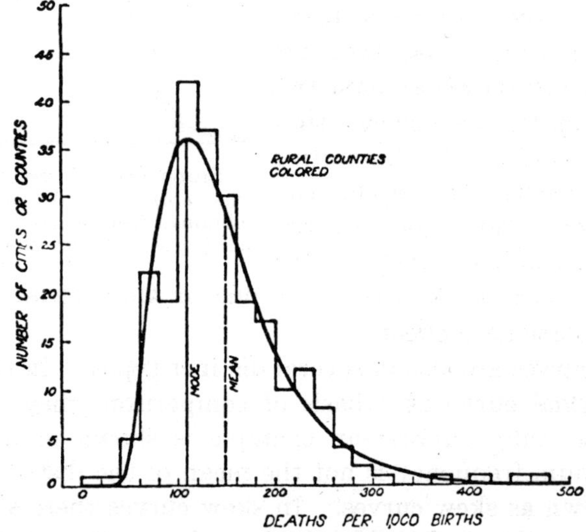


FIGURE 66.—Illustrating the characteristic features of a skew distribution. Frequency histogram and fitted skew curve for variation in the rate of infant mortality among negroes in 1918 in rural counties of the Birth Registration area. (*From Pearl, Introduction to Medical Biometry and Statistics, copyright by W. B. Saunders Company* 1923. *Reprinted by permission.*)

While most frequently distributions of biological material probably do not strictly conform to the normal type, they often differ very little from it, and no great error is introduced by assuming normality of distribution. For this reason it is usually justifiable, as a first approximation, to base computations on the assumption of normality of distribution. If a curve is obviously multimodal or skew, however, computations made according to these methods may have very little significance. If possible, multimodal curves should be split up into their components, and each component should then be treated separately.

**Correlation.**—Sometimes two characters of an organism are so related that an increase in one is attended by an increase in the other. Such, for example, is the relation between height and weight in human beings. Taller persons are, on the average, heavier than shorter ones. Or again, an increase of one character may be attended by a decrease in the other; or the two characters may vary independently. But these are problems of correlation or the degree to which characters are mutually interdependent. If an increase in one character is attended on the average by an increase in the other, the characters are said to be positively correlated; if an increase in one is attended on the average by a decrease in the other, they are said to be negatively correlated; and if there is no mutual relation between the two characters, they are said to be independent or uncorrelated. Here again, however, although it may be possible from a simple inspection of a series of pairs of values to gain some idea of the relation between them, precise treatment requires the application of statistical methods.

**The Correlation Table.**—Just as the frequency distribution expresses in a condensed form the data for some one feature of a population, so the correlation table expresses the data for two variable features. As an example, consider the data for relation between yield of grain per plant and number of culms per plant in Sixty Day oats, contained in table 17. Each entry in the body of the table represents a pair of values,

TABLE 17.—CORRELATION OF YIELD OF GRAIN PER PLANT WITH NUMBER OF CULMS PER PLANT IN SIXTY DAY OATS, ILLUSTRATING POSITIVE CORRELATION (*Data from Love and Leighty*)

| Yield in grams | Number of culms 2 | 3 | 4 | 5 | 6 | 7 | |
|---|---|---|---|---|---|---|---|
| 0–1 | 3 | | | | | | 3 |
| 1–2 | 28 | 19 | 3 | | | | 50 |
| 2–3 | 18 | 66 | 20 | 1 | | 1 | 106 |
| 3–4 | 1 | 42 | 58 | 7 | 1 | | 109 |
| 4–5 | | 7 | 59 | 11 | 3 | | 80 |
| 5–6 | | | 26 | 14 | 2 | | 42 |
| 6–7 | | | | 4 | 3 | | 7 |
| 7–8 | | | 1 | 1 | | | 2 |
| 8–9 | | | | | 1 | | 1 |
| | 50 | 134 | 167 | 38 | 10 | 1 | 400 |

*i.e.*, both yield of grain and number of culms of a given individual. Thus in the first column of figures, the entry 28 signifies that among 400 plants

28 bore two culms each and at the same time produced somewhere between 1 and 2 grams of grain each; below it the entry 18 enumerates the plants with two culms each which produced between 2 and 3 grams of grain each; and so on for the remaining entries in the table. Such a table contains all the pertinent data on yield of grain and number of culms in this population, and in addition it portrays the relation between these two variables.

Just as the tabulation of values for a single variable, without further computation, discloses a number of features with respect to it, so the construction of a correlation table discloses to some extent the relation which exists between two variables. From mere inspection of the foregoing table, it is evident that as the number of culms per plants increases, the average yield of grain per plant also increases, which indicates that the characters are positively correlated. Similarly in table 18, depicting the relation between height and number of culms in the same population, it may be seen from the table that with increase in the number of culms, the average height of plant remains approximately the same, so that these two characters are not correlated.

TABLE 18.—CORRELATION BETWEEN HEIGHT OF PLANTS IN CENTIMETERS AND NUMBER OF CULMS PER PLANT IN SIXTY DAY OATS, ILLUSTRATING ABSENCE OF CORRELATION (*Data from Love and Leighty*)

| Height in centimeters | Number of culms 2 | 3 | 4 | 5 | 6 | 7 | |
|---|---|---|---|---|---|---|---|
| 45–50 | | 1 | 1 | | | | 2 |
| 50–55 | 2 | 3 | 3 | | | 1 | 9 |
| 55–60 | 4 | 9 | 4 | 2 | 2 | | 21 |
| 60–65 | 6 | 10 | 13 | 4 | 1 | | 34 |
| 65–70 | 8 | 40 | 42 | 6 | 1 | | 97 |
| 70–75 | 15 | 41 | 53 | 12 | 2 | | 123 |
| 75–80 | 10 | 22 | 43 | 11 | 3 | | 89 |
| 80–85 | 4 | 8 | 8 | 3 | 1 | | 24 |
| 85–90 | | | | | | | 0 |
| 90–95 | 1 | | | | | | 1 |
| | 50 | 134 | 167 | 38 | 10 | 1 | 400 |

The various types of correlation are depicted graphically in figure 67. When the entries in the table are distributed in the form of an ellipse, the longer axis of which extends from the upper left corner of the table to the lower right, the characters are positively correlated. An ellipse extending from the lower left to the upper right corner of the table indicates negative correlation. If the entries are scattered uniformly in the four quadrants in the form of a circle, the characters are not correlated. In the case of positive and negative correlation, the narrower the ellipse within which the entries lie, the higher the degree of correlation; but the precise expression of the degree of correlation requires application of computations which are designed to express in one value the interrelation between them.

**The Coefficient of Correlation.**—The mathematical value employed for expressing the degree of relation between two variables is called the coefficient of correlation. It is usually represented by the letter $r$, and its formula is as follows:

$$r_{xy} = \frac{\Sigma(\delta_x \cdot \delta_y)}{N \cdot \sigma_x \cdot \sigma_y}$$

in which $r_{xy}$ is the coefficient of correlation of the variables $x$ and $y$; $\delta_x$ is the deviation of the $x$ variables from the mean value of $x$; $\delta_y$ corresponding deviations in the $y$ variables from the mean of $y$; and $\sigma_x$ and and $\sigma_y$ are the standard deviations of $x$ and $y$, respectively. The portion

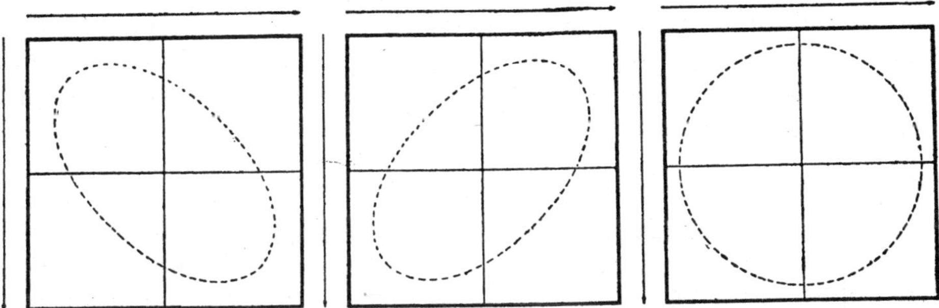

FIGURE 67.—Interpretation of the correlation table. Shape of "swarm" indicates nature and amount of correlation.

of the equation represented by $\frac{\Sigma(\delta_x \cdot \delta_y)}{N}$ is called the mean product moment. In a sense it is an expression for the mean deviation of individuals from the mean value of both variables.

The direct application of this formula is very laborious, because it requires that actual deviations from the means be employed. Since these will usually have a number of decimal places and the signs of some will be plus and others minus, which must also be taken into account, it is easier to use one of the short-cut methods of computation.

In table 19, a convenient short-cut method is illustrated. The principle involved is very simple. The computations, as in the case of the mean and standard deviation, are made from an arbitrary origin, in this case chosen as $x = 2$, $y = 0.5$, so as to eliminate minus signs. Each square in the table contains in addition to the frequency an index number, which represents the product of the deviations from the arbitrary origins for $x$ and $y$. These index numbers are multiplied by the frequencies in each square and the sum taken, as shown in the column headed $d_x \cdot d_y$. The sum of these values divided by $N$ gives what is called the mean product moment from the arbitrary origin. It must be corrected by subtracting

TABLE 19.—COMPUTATION OF THE COEFFICIENT OF CORRELATION BY A SHORT-CUT METHOD

| | 2 | 3 | 4 | 5 | 6 | 7 | $f$ | $(d_x.d_y)$ |
|---|---|---|---|---|---|---|---|---|
| 0.5 | *0*<br>3 | . . . | . . . | . . | . . | . . | 3 | 0 |
| 1.5 | *0*<br>28 | *1*<br>19 | *2*<br>3 | . . | . . | . . | 50 | 25 |
| 2.5 | *0*<br>18 | *2*<br>66 | *4*<br>20 | *6*<br>1 | . . | *10*<br>1 | 106 | 228 |
| 3.5 | *0*<br>1 | *3*<br>42 | *6*<br>58 | *9*<br>7 | *12*<br>1 | . . | 109 | 549 |
| 4.5 | . . | *4*<br>7 | *8*<br>59 | *12*<br>11 | *16*<br>3 | . . | 80 | 680 |
| 5.5 | . . | . . . | *10*<br>26 | *15*<br>14 | *20*<br>2 | . . | 42 | 510 |
| 6.5 | . . | . . . | . . . | *18*<br>4 | *24*<br>3 | . . | 7 | 144 |
| 7.5 | . . | . . . | *14*<br>1 | *21*<br>1 | . . | . . | 2 | 35 |
| 8.5 | . . | . . . | . . . | . . | *32*<br>1 | . . | 1 | 32 |
| $f$ | 50 | 134 | 167 | 38 | 10 | 1 | 400 | 2,203 |

Statistical constants for the two frequency distributions, calculated as in table 16.

$d_x = 1.5675$ $\quad d_y = 2.9575$

$M_x = 3.5675$ $\quad M_y = 3.4575$

$\sigma_x = 0.930$ $\quad \sigma_y = 1.291$

Computation of the coefficient of correlation.

$$\frac{\Sigma d_x d_y}{N} = \frac{2{,}203}{400} = 5.5075$$

$$d_x.d_y = 1.5675 \times 2.9575 = 4.6359$$

$$\frac{\Sigma \delta_x \delta_y}{N} = 0.8716$$

$$\sigma_x.\sigma_y = 0.930 \times 1.291 = 1.2006$$

$$r = 0.8716 \div 1.2006 = 0.73$$

from it the product of the two corrections which are applied to the determinations of the means, when the same arbitrary origins are used as the basis of computation. The value thus obtained, the mean product moment, is equivalent to the $\frac{\Sigma(\delta_x \cdot \delta_y)}{N}$ portion of the formula for $r$ given above, and the computation of $r$ is completed by dividing it by the product of the two standard deviations, as indicated in the formula. Here again the method of computation is a true short cut, not an approximation method.

**The Interpretation of the Coefficient of Correlation.**—The coefficient of correlation computed from the above formula always lies somewhere between $-1.0$ and $+1.0$. Plus values indicate positive correlation and minus values negative correlation. A value of $+1.0$ indicates perfect positive correlation; *i.e.*, an increase in one variable is attended by an exactly proportional increase in the other. Similarly $r = -1.0$ indicates perfect negative correlation. Values from $+0.5$ to $+1.0$ indicate high positive correlation; $r = -0.5$ to $-1.0$, high negative correlation. Values lying between $+0.3$ and $+0.5$ may be considered as moderate positive correlation, and similarly for negative values. Values from 0 to 0.3 indicate low positive correlation, and corresponding minus values indicate low negative correlation. A correlation coefficient which does not differ from zero by four times its probable error is considered as of doubtful significance; or it may be said that values of $r$ lying between the limits $0 \pm 4E_r$ indicate lack of correlation. These features may be represented graphically on a scale as follows:

| $-1.0$ | $-0.5$ | $-0.3$ | $-4E_r$ | 0 | $+4E_r$ | $+0.3$ | $+0.5$ | $+1.0$ |
|---|---|---|---|---|---|---|---|---|
| perfect | high | moderate | low | none | low | moderate | high | perfect |
| | | negative | | | | positive | | |

**Regression.**—While the coefficient of correlation expresses the degree to which two variables are interrelated, it does not give all the information which may be extracted from a correlation table. In particular, it is often useful to have some method of computing the average expected value for one variable corresponding to particular values of the other. Thus in our illustrative example, it might be desired to compute the average yield of grain per plant of that group of plants having six culms per plant. In order to do this it is necessary to express the relation between the variables in form of equations. The formulae for these are

$$\bar{d}_x = r\frac{\sigma_x}{\sigma_y}d_y$$

$$\bar{d}_y = r\frac{\sigma_y}{\sigma_x}d_x$$

The terms in these equations have the same significance as elsewhere. Particular deviations from the mean values of $x$ and $y$ are indicated by $d_x$ and $d_y$, respectively, while $d_x$ and $d_y$ indicate average values of the deviation in $x$ and $y$, respectively.

These equations obviously represent straight lines which pass through the origin, *i.e.*, the intersection of the lines representing the mean values of $x$ and $y$. It is, therefore, readily possible to plot them in a skeleton correlation table, as shown in figure 68. Now by computing the means of each column or row of squares separately, it is also possible to obtain a series of average values of one variable corresponding to particular values of the other directly from the correlation table. When these are plotted in the skeleton correlation table, it is evident that the means of the rows of squares, $x$-arrays, cluster around the line representing the first equation; and the means of the columns of squares, $y$-arrays, around the line representing the second equation. As a matter of fact, the equations of these two straight lines represent the two straight lines which most closely fit these two sets of values. The comparison of these two lines with the actual means of $x$- and $y$-arrays is also useful as an indication of the validity of the coefficient of correlation as an expression of the interrelation between the two variables. If the actual values closely approach the lines, then the lines may be considered as the best expression of their true positions, and regression is said to be rectilinear. When they fail to lie closely grouped around these lines but show an unmistakable

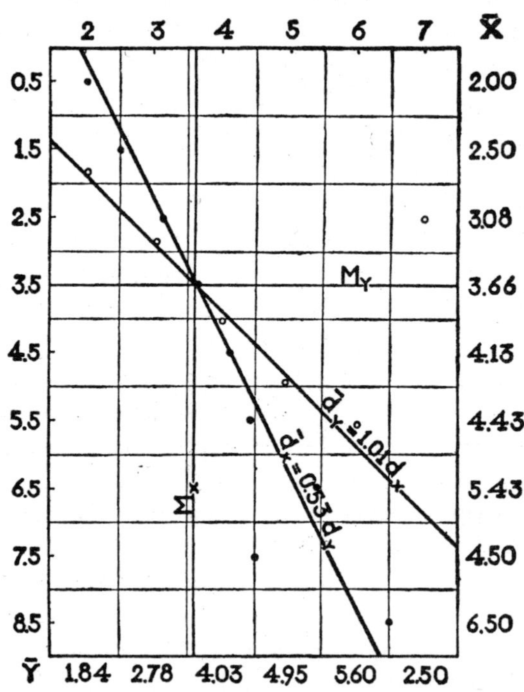


FIGURE 68.—Skeleton correlation table, representing pertinent facts of Table XVI. Black discs = average number of culms for each row of squares; circles = average weight of grain for each column of squares; note relation of these to the regression lines.

tendency to deviate in some particular way from them in certain portions of the table, then regression is non-rectilinear. The coefficient of correlation is applicable only to data exhibiting rectilinear regression; and other methods have been devised for data exhibiting non-rectilinear regression. Also, strictly, the frequency distributions of the two variables should be of the normal type to justify application of this method.

**Correlation as a Measure of Resemblance.**—Beside its use as an expression for the degree of interrelation of two variables in a population, the coefficient of correlation has an even greater value for determining the degree to which a given character is correlated in related individuals, *i.e.*, as a measure of resemblance. As a matter of fact, the correlation method was first devised to handle just such problems. A typical illustration is contained in table 26, representing the relation between height of parent and offspring in human beings. Heights of female entries are corrected by multiplying them by 1.08, since the average male height is 1.08 times the average female height. The computation of the coefficient of correlation is done in the same way as above, and the value obtained, $r = +0.449 \pm 0.026$, indicates a moderate degree of correlation between parent and offspring in height. This matter is discussed further in Chapter XIX.

**The Probable Error.**—In the examples which have been presented, it has been shown that the computation of the mean and the standard deviation from a frequency distribution makes it possible to construct a smooth curve which represents the ideal distribution of individuals in the general population. Now obviously the distribution upon which these computations depend represents a selection of a limited number of individuals from a population which may be regarded as infinite in size. The real object of the computations is to describe the characteristics of the general population, but, since the samples are limited in size, the computations which are made on the basis of the sample are subject to a certain range of fluctuation which it is necessary to determine in order to ascertain the degree of reliance which may be attributed to the computations. The degree of reliability inherent in computations is based fundamentally upon the type of distribution exhibited by the normal curve. If a single variant is selected from the general population, the probabilities are equal that it will lie either within or without the limits, $\pm 0.6745\sigma$, as shown above. Accordingly, the probable error of a single variant may be set as equal to $\pm 0.6745\sigma$. If a number of individuals are selected and averaged, the probable deviation of the mean of these individuals from the true mean of the population will decrease inversely as the square root of $N$, the number of individuals in the sample, so that the probable error of the mean is

$$E_M = \pm \frac{0.6745\,\sigma}{\sqrt{N}}$$

Similarly, the probable error of the standard deviation $E_\sigma$ is given by the equation

$$E_\sigma = \pm \frac{0.6745\,\sigma}{\sqrt{2N}}$$

Applying these formulae to the illustrative problem, the constants for length of beans, the following results are obtained:

$$E_M = \pm \frac{0.6745 \times 1.198}{\sqrt{3{,}725}} = \pm 0.0132$$

$$E_\sigma = \pm \frac{0.6745 \times 1.198}{\sqrt{2 \times 3{,}725}} = \pm 0.0094$$

Generally the values of the probable errors are computed to two significant figures, and the values of the corresponding constants are carried out to the same number of places. The mean for length of bean should then be written

$$M = 14.072 \pm 0.013$$

which signifies that it is equally probable that another $F_2$ population of the same material would give values of the mean lying somewhere between $14.072 + 0.013 = 14.085$ and $14.072 - 0.013 = 14.059$ millimeters, or somewhere outside of the limits set by these two figures. Similarly the standard deviation should be written

$$\sigma = 1.198 \pm 0.009$$

with a like interpretation.

Since the distribution of errors is assumed to be normal, the likelihood that the deviation from the true value for any constant may exceed the probable error by any stipulated amount may be computed from the known type of distribution in the normal curve. Such computations show that the following proportions of instances lie outside the limits stated:

$$\pm E = 0.5000$$
$$\pm 2E = 0.1775$$
$$\pm 3E = 0.0430$$
$$\pm 4E = 0.00698$$
$$\pm 5E = 0.000745$$

These figures show that in less than 1 per cent of instances will the true value of a constant deviate as much as four times its probable error from the computed value.

Probable errors not only indicate the degree of confidence which may be placed in the constants as computed, but they are also useful for ascertaining the significance of differences between two constants determined from similar material. Thus, for example, Johannsen presented measurements of length and breadth in two populations of beans as follows:

| Series | Mean length | Mean width |
|---|---|---|
| *E* | 12.793 ± 0.011 | 9.379 ± 0.007 |
| *GG* | 12.942 ± 0.015 | 8.152 ± 0.007 |
| Difference | 0.149 ± 0.019 | 1.227 ± 0.010 |

If the values of the constants are not correlated, the probable error of a sum or difference is computed according to the formula

$$E_S \text{ or } E_D = \pm\sqrt{E_1^2 + E_2^2}$$

where $E_1$ and $E_2$ are the probable errors of the two constants under comparison. The difference in length, 0.149 + 0.019 millimeter, is very small, but it is over seven times the probable error, so that it may safely be considered significant. Had it been less than four times its probable error, certain significance could not justifiably have been attached to it.

Extended tables are available from which probabilities corresponding to various deviations in excess of the probable error may be read off directly. It should, however, be understood that in stating that only a deviation at least four times as great as the probable error is to be regarded as significant, no absolute criterion of significance is stated. The chance at this point is better than 99 out of 100 that the deviation is significant; but if a thousand determinations of a given value are made, all from the same material, some of them will almost certainly exhibit a deviation over four times as great as the probable error, although none of them will be significant. Similarly, a deviation three times as great as the probable error is significant 95.7 times out of 100 on the average, it has no significance 4.3 times per hundred, or the odds are about 22.3 to 1 that the deviation is significant. Since an event which happens only about once in twenty-three times is not particularly unusual, it may well be said that such a deviation is not certainly significant. From what has been said, it should be apparent that statistical methods, and probable errors in particular, should not be employed in too stereotyped a fashion.

**Probable Errors of Other Constants.**—Formulae for probable errors of other constants are also available, a few of the most useful of which are given below for reference:

Of the coefficient of variation, $E_v = \pm \dfrac{0.6745\,v}{\sqrt{2N}}\left[1 + 2\left(\dfrac{v}{100}\right)^2\right]^{1/2}$

Of the coefficient of correlation, $E_r = \pm \dfrac{0.6745(1 - r^2)}{\sqrt{N}}$

Of an observed probability, $E_p = \pm 0.6745\sqrt{\dfrac{p.q}{N}}$

In the computation of the probable error of the coefficient of variation, the term within the brackets may be neglected if $v < 10$. In the probable error of an observed probability, $p$ represents the probability that a given event will occur and $q$ $(= 1 - p)$ the probability that it will not occur, expressed as decimal fractions. This formula is useful for determining the probable errors of Mendelian ratios, sex-ratios, linkage values, or other instances where mutually exclusive events occur. In these and in other cases actual labor of computation of the values may be greatly reduced by using tables which are available for the purpose.

**The Value of Statistical Methods.**—Statistical methods express in precise, comparable terms certain significant features of masses of quantitative data. Their employment is by no means restricted to genetic data, but to all data which is subject to variation and random error. As a consequence an increasing number of investigators in the biological sciences are forced to employ them, and a knowledge of them is well nigh indispensable in experimental work. A great variety of methods has been devised to deal with special situations, so that almost every contingency is provided for. Only the basic methods have been considered in this chapter. For a more extended treatment the student should consult textbooks on the subject.

In biological investigations statistical methods are employed for the most part descriptively, *i.e.*, to state in precise terms certain features of quantitative data and to determine whether or not significant differences exist in comparable sets of data. Rarely are they useful to the biologist for determining the causes of the phenomena thus disclosed. For that purpose it is usually necessary to resort to actual experimentation. By the correlation method, for example, it is possible to determine the degree of resemblance between parent and offspring; the causes for resemblance, however, should be sought in the analysis, where possible, of experimental data.

### References

Castle: Genetics and Eugenics, Chapter XI.
Fisher: Statistical Methods for Research Workers.
Jones: Genetics, Chapter VIII.
Pearl: Introduction to Medical Biometry and Statistics.
Sinnott and Dunn: Principles of Genetics, Chapter X.
Thomson: Heredity, Chapter IX.
Yule: Theory of Statistics, 6th ed.

## CHAPTER XVIII

## DEVELOPMENTAL VARIATION

Developmental variation is diversity induced in individuals in response to differences in environment. Modifiability may be defined as the ability of an individual to respond to its environment. Modifiability, thus conceived, is really not a problem of genetics at all; but since the results of genetic experimentation are based largely upon differences in characters, it is necessary to consider the influence which environment exerts in their development. A full treatment of this matter would lead far beyond the available space in this book; consequently this chapter will be devoted merely to an account of certain experiments which show unmistakably that characters may be strikingly modified by differences in specific features in the environment and to the significance which such variation has in genetics. The student who wishes to go further into the subject will find the literature of animal and plant physiology replete with instances of the same kind.

**Effect of Differences in Light.**—It is a familiar fact that potato tubers sprouting in the dark produce elongated shoots devoid of chlorophyll. Light, therefore, is a necessary condition for the development of chlorophyll in plants. Other plant characters may also vary greatly as a consequence of the action of light upon development. An instance is the so-called sunlight red pericarp in maize. The ear developing in the ordinary fashion, enclosed in husks, produces a colorless pericarp, but if the kernels are exposed to sunlight, they will develop a bright red pericarp. By enclosing the ear in a stencil, freak ears may be produced on which the kernels of the exposed portions are colored and those of the protected portions white. This instance is instructive, if considered along with the fact that races of corn exist which produce a colorless pericarp whether exposed to or protected from sunlight; and varieties also exist which produce a red pericarp under both conditions. The reaction, therefore, is one which depends upon a specific genetic constitution; but what is determined by heredity is not the production of a definite character but the power to react in a certain way to environmental conditions. The normal reaction when exposed to sunlight is the production of a red pericarp; protected from sunlight, of a colorless pericarp.

The relative length of night and day also has a profound effect upon development in certain forms. Under the long days characteristic of the summer season in New York, teosinte matures very late, if at all, so that difficulty was experienced by Emerson and his associates in utilizing it for studies of hybridization with corn, because of its frequent failure to mature. When, however, length of day was artificially shortened by placing the plants in a dark chamber early in the evening, it was found possible to bring them to maturity without difficulty and to utilize them in crossing experiments. Similarly there is a mutant form in tobacco which is characterized by an almost indeterminate habit of growth. Ordinary tobacco varieties produce a fairly definite number of leaves and then complete their growth by production of an inflorescence. In the giant varieties, however, the plants keep on increasing in height and producing new leaves throughout the growing season, and only bloom tardily if removed to a greenhouse and preserved beyond the normal season. If, however, the variety is grown under artificially shortened days, it reaches maturity at the same time as other varieties. Here again it is improper to speak of indeterminate growth habit as an inherited character. What is inherited is a certain genetic constitution which operates under some environmental conditions to produce an indeterminate habit of growth, and under others to produce plants of determinate habit. Moreover, both reactions are normal; there is no justification for considering one a normal and the other an abnormal reaction.

**Temperature Reactions.**—There is a certain variety of Chinese primrose, *Primula sinensis*, which produces white flowers if grown in a somewhat shaded greenhouse at 30°C. but in a cooler house, 15 to 20°C., it produces red flowers. A single plant may produce red and white flowers alternately by keeping it first at one temperature and then at the other. Moving it from a warm to a cool greenhouse, the opened blossoms remain white, and the first flowers which open after removal are white, but those which develop later are red. Similar reactions have been reported for other species of Primula in nature. Plants growing in warmer localities produce white flowers, and those in colder regions colored ones; white plants removed from the warmer localities to colder situations thereafter produce colored flowers. There are, however, white varieties which produce white flowers at both temperatures.

In Drosophila temperature has been shown to have outstanding effects on the development of many characters and on very important processes. Particularly extensive studies have been made of the effect of temperature upon wing development in vestigial flies and upon the development of facets in the bar-eyed fly. Vestigial flies show a marked increase in wing size with rise in temperature. At optimum temperatures the wings are mere vestiges as compared with those of wild-type flies; but as the temperature is increased they become progressively larger until, near the upper limits for the life of the fly, they actually approach the size of those of wild-type flies. The effect of temperature is to be compared with the known fact that different allelomorphs of vestigial exist which determine different grades of reduction in wing size. Here similar effects are exhibited as a consequence of differences in Mendelian factors.

In bar eye, a sex-linked dominant, only a narrow vertical strip of facets is produced. An allelomorph called ultra-bar (see Chapter XXII) produces a more extreme reduction of facet number than the normal bar mutation (figure 69). Extensive studies have been made by Krafka on the effect of temperature upon facet number in these types. Some of the details of Krafka's results are contained in table 20. The normal number in the wild-type fly is about 780 for females and 740 for males, and differences in temperature were found to have only a slight effect upon it. With bar and ultra-bar, however, a marked effect was secured over

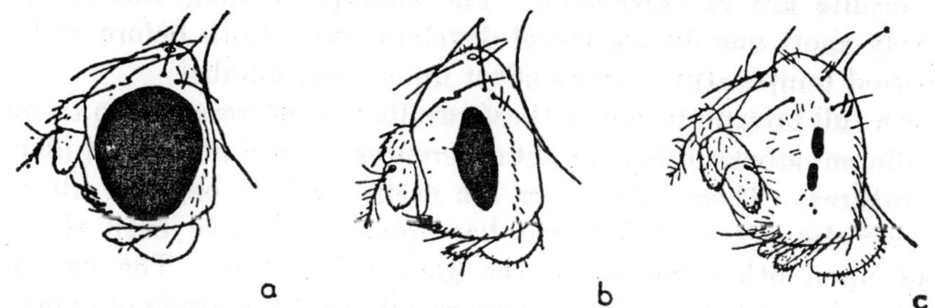


FIGURE 69.—Bar eye (*b*) contrasted with the wild type (*a*) and ultra-bar or double-bar (*c*). (*From Morgan, Bridges, and Sturtevant, The Genetics of Drosophila in Bibliographia Genetica, vol. 2, copyright* 1925 *by Martinus Nijhoff. Reprinted by permission.*)

the temperature range from 15 to 31°C., the number of facets being lowest at the highest temperatures. The data contained in the table show that males regularly exhibit a somewhat higher facet number than females, but the reduction in number runs parallel in the two sexes, and they also demonstrate that although the general effect of temperature is similar

TABLE 20.—EFFECT OF TEMPERATURE ON NUMBER OF FACETS IN LOW-BAR AND ULTRA-BAR STOCKS OF DROSOPHILA AND UPON LENGTH OF THE LIFE CYCLE
(*Data from Krafka*)

| Degrees centigrade | Low-bar stock | | Ultra-bar stock | | Life cycle, egg to adult |
|---|---|---|---|---|---|
| | ♀♀ | ♂♂ | ♀♀ | ♂♂ | |
| 15 | 189.0 | 240.2 | 51.5 | 60.8 | 31.9 |
| 16 | 158.2 | 207.2 | 43.3 | 51.1 | 22.9 |
| 17.5 | 127.3 | 164.7 | 38.6 | 45.2 | 19.2 |
| 20 | 98.9 | 122.9 | 32.6 | 37.2 | 13.6 |
| 25 | 74.3 | 103.1 | 25.2 | 27.6 | 10.4 |
| 27 | 55.1 | 79.5 | 21.3 | 23.7 | 9.2 |
| 29 | 47.4 | 65.6 | 17.2 | 19.0 | 8.3 |
| 30 | 36.6 | 57.8 | .... | .... | 9.0 |
| 31 | 28.9 | 41.7 | 14.6 | 14.2 | 9.2 |

in bar and ultra-bar, the proportional reduction between any two temperatures is not necessarily the same. Krafka also found that the effective period of the reaction, the time in embryonic development during which temperature exerted its effect, was very short. By exposing developing cultures at specific periods to given temperatures, it was found that with cultures developing at 27°C., the effective period occurred during the fourth day of development and extended over a period of about 18 hours, and with cultures developing at 15°C. the effective period occurred between the eighth and eleventh days and lasted 72 hours. By taking into account the rate of development of flies at these two temperatures, as shown in the last column of table 20, it may be seen that these results are in agreement. The effective period, therefore, is a relatively short one during larval development. Both before and after this period temperature has no effect upon facet number.

For a full interpretation of these results it is necessary also to realize that different stocks of bar exhibit differences in facet number at identical temperatures. These differences are easily seen in bar and ultra-bar; but they also occur in different bar stocks and have been shown to depend upon other factors in the germinal system. The conclusion may, therefore, be drawn from these results that the grade of expression of the character, in this instance, the number of facets, is determined by a specific genetic constitution acting under a specific set of environmental conditions.

**Nutritional Effects.**—Nutritional effects may also often be related to specific elements in the food materials. It is a well-known fact, for example, that plants grown in culture solutions totally devoid of iron salts fail to develop chlorophyll. In certain portions of Hawaii, pineapples exhibit a chlorotic condition not because there is not sufficient iron present in the soil, but because the soil conditions are such as to prevent the plant from absorbing it. The condition may be corrected by spraying plants with iron solutions. Only minute quantities of iron are necessary, but they are absolutely essential for normal development. The same statement also applies to many other elements of plant food.

In wheat wide variations in nitrogen content occur which seem to depend largely upon climatic conditions. Hard wheats which exhibit a high nitrogen content under climatic conditions prevailing in the North Central states commonly produce soft wheats low in nitrogen under California conditions. Since high gluten content is necessary to produce a superior flour for bread making, the low gluten content of California wheat necessitates importation of a certain amount of hard wheat for mixing with the local crop.

Experiments conducted by Gericke at the California Station have shown that wide variations in protein content occur, if a definite nutrient constituent, the nitrogen, is made available at different periods in development. Certain of his data are presented in table 21. All the cultures were started in the greenhouse at the same time and grown throughout under the same conditions. A definite amount of sodium nitrate was provided in each culture, but the time of application was varied. By applying sodium nitrate at the time of planting to White Australian cultures, the mature grains contained only 8.9 per cent of protein; whereas, if the time of application was delayed until 110 days after planting, nearly twice as much protein, 15.2 per cent, was contained in the final crop. Intermediate time of application gave intermediate results. With Turkey Red winter wheat, on the contrary, no significant difference in protein content with differences in time of application of the nitrate was observed, except perhaps for the last application. With Texas Red oats the results were even more striking than with White Australian wheat,

TABLE 21.—TIME OF NITROGEN APPLICATION AS RELATED TO PROTEIN CONTENT IN CEREALS (*Data from Gericke*)

| White Australian spring wheat | | Turkey Red winter wheat | | Texas Red oats | | Rye | |
|---|---|---|---|---|---|---|---|
| $NaNO_3$, days | Protein, per cent | $NaNO_3$, days | Protein, per cent | $NaNO_3$, days | Protein, per cent | $NaNO_3$, days | Protein, per cent |
| 0 | 8.9 | 0 | 14.6 | 0 | 7.5 | 0 | 9.5 |
| 17 | 9.2 | 21 | 13.8 | 19 | 8.0 | 19 | 9.0 |
| 33 | 10.6 | 36 | 14.7 | 33 | 8.5 | 28 | 9.7 |
| 48 | 11.4 | 60 | 13.4 | 48 | 9.6 | 43 | 9.8 |
| 72 | 13.0 | 81 | 14.3 | 69 | 10.8 | 74 | 10.9 |
| 110 | 15.2 | 109 | 17.9 | 90 | 12.7 | 105 | 12.0 |
| | | | | 108 | 17.2 | 133 | 14.0 |

and with rye positive results were also obtained, but somewhat less marked. Here again the same conclusion may be drawn, as in previous instances, that the expression of the character is a function both of heredity and of environment. Different wheat varieties grown under the same environmental conditions exhibit marked differences in protein content; a given variety of wheat grown under different environmental conditions exhibits even more striking differences in protein content. It would appear from these experiments that the difficulty in growing hard wheat under California conditions probably depends upon the time in development when the nitrates become available; but it does not necessarily follow that a variety of wheat might not be obtained which would produce a high protein crop under these conditions.

**Inheritance of Developmental Variations.**—A plant of the red primrose grown at 20°C. will produce red blossoms; at 30°C., white ones, as described above. If seeds are taken from two such plants and sown, the progeny will produce red-flowering plants if grown at 20°C. and white-flowering ones if grown at 30°. Or if the seeds of each plant are divided and half grown at each temperature, they will develop into red- and white-flowering plants according to the temperature at which they are grown. Obviously the particular character expression of the parent has no effect in this case upon the progeny. Similarly, in bar-eyed Drosophila, the particular modification in facet number arising from development at different temperatures is not carried over to the offspring. Krafka subjected this question to experimental investigations. Stocks of ultra-bar were grown at 15, 20, and 27°C., and these modified flies were used as parents of progenies grown at 27°C. The results as shown in table 22 indicate no significant difference in facet number in the three sets of progenies. Likewise, wheat which is soft under California conditions and hard under North Dakota conditions will develop a protein content appropriate to the conditions under which it is grown, whether the seed from which it was grown was hard or soft.

TABLE 22.—EFFECTS OF TEMPERATURE MODIFICATION ON FACET NUMBER OF PARENTS ON FACET NUMBER OF OFFSPRING IN BAR-EYED DROSOPHILA (*Data from Krafka*)

| Parents at temperature shown | | | | Progeny at 27°C. | |
|---|---|---|---|---|---|
| Degrees Centigrade | | Mean facet number | Standard deviation | Mean facet number | Standard deviation |
| 15 | ♀♀ | 50.50 ± 0.96 | 6.05 ± 0.67 | 23.67 ± 0.35 | 2.76 ± 0.25 |
| | ♂♂ | 58.94 ± 1.35 | 8.28 ± 0.95 | 24.85 ± 0.28 | 2.53 ± 0.20 |
| 20 | ♀♀ | 35.50 ± 0.54 | 4.53 ± 0.38 | 20.85 ± 0.24 | 2.61 ± 0.18 |
| | ♂♂ | 40.60 ± 0.48 | 4.29 ± 0.34 | 23.12 ± 0.24 | 2.80 ± 0.17 |
| 27 | ♀♀ | 20.80 ± 0.34 | 3.02 ± 0.24 | 20.50 ± 0.29 | 2.39 ± 0.20 |
| | ♂♂ | 22.96 ± 0.12 | 2.43 ± 0.08 | 22.68 ± 0.31 | 2.56 ± 0.22 |

These results show that it is improper to speak of the inheritance of a character, even though it may often be convenient to do so. Characters are not inherited. A specific germinal constitution is inherited, which enables the individual to react in a certain definite way to environmental conditions. Shortly stated, modifications are not inherited but modifiability is. It is also necessary to point out that our ideas of normal types are often faulty. It is just as normal for a Chinese primrose to produce white flowers at 30°C. as red flowers at 20°C., for wheat to be soft under California conditions as to be hard under North Dakota cultivation, or for bar-eyed Drosophila to have a low facet number at high temperatures as to have a high facet number at low temperatures. What is called normal development is usually an average development under an arbitrary set of environmental conditions under which it is customary to see the individual developing.

It should also be evident that in studies of variable characters, it is often possible to secure more definite results by growing the individuals under conditions most favorable to a full expression of the character. Observations on segregation of red and white in Primulas obviously should be carried out at low temperatures, for although the same genetic segregation occurs at the higher temperatures, it is impossible to demonstrate it on account of the effect of temperature on development.

**Heredity and Environment in Development.**—In the foregoing examples, instances have been considered which have been placed upon an experimental basis, and specific environmental effects have been considered under controlled conditions. But in considering the development of a character in all its manifestations, a different point of view must be taken, and a conception of the environment as a whole must be developed. From the beginning, the zygote is surrounded by environmental conditions which exert a constant molding effect upon it as development proceeds. Now one feature of the environment, now another, plays a conspicuous role, and the final expression of any single character may have been modified by innumerable environmental influences, a change in any one of which might have altered the final expression of the character. It is perhaps no exaggeration to say that all the environmental influences under which an individual develops have played some part, direct or indirect, in the expression of every one of its characters.

It has been pointed out in numerous places that characters are not inherited, but that factors are the things received by an individual from its parents. These factors probably represent so many chemical substances contained in the nucleus. Together they form what might be styled a reaction system; *i.e.*, the various factors are intimately interacting with one another in a harmoniously balanced fashion, and the products of the operation of this system, development and functional activity in all their manifestations, are to be considered as the outcome of the activities of the reaction system as a whole. In the expression of every character then, in the last analysis, every factor must be considered as having played some essential role, and a change in any one of them may be expected to alter in some way the final outcome of its activities.

Now in considering the total play of causative agencies which are actively concerned in the production of characters, environment has to be considered in the same way; in fact in the development of every character both heredity and environment, every factor in the heredity and every feature of the environment, must be considered. As Holmes and others have stated, the expression of a character is a function both of the heredity and of the environment, a conception which may be represented equationally thus:

$$C = f(HE)$$

If the heredity is changed and environment is kept constant, differences will appear in the characters, and they will also respond, if heredity is kept constant and environment altered.

Viewed from this angle the old debate as to the relative importance of heredity and environment in the development of characters becomes completely meaningless. There is no way of arriving at a decision on this question save by cutting the Gordian knot and declaring them both equally important. What is needed is an experimental consideration of particular cases. If it is observed that individuals in a population differ, it may be ascertained by experimental methods whether the differences are due to differences in heredity or to differences in environment or to both, and if due to both, the relative part which each plays in the situation may be determined. It is necessary to realize that a given heredity may give a variety of developmental products depending upon environmental conditions. It is a matter of experimentation to determine what those products are under specified conditions and what the range of variation is under different conditions. As a matter of fact, all the improved methods of tillage utilized in crop production, of nutrition and training in animal production, and of these same factors in human life are based upon the knowledge that organisms are modified and that in order to realize the fullest potentialities in a given instance the greatest possible attention must be paid to environment; but this fact does not conflict with the conception that the best possible result is secured when attention is given both to heredity and environment.

### References

Morgan, Sturtevant, Muller, and Bridges: Mechanism of Mendelian Heredity, Chapter X.

Morgan: Physical Basis of Heredity, Chapter XIX.

## CHAPTER XIX

# QUANTITATIVE CHARACTERS

It is a familiar fact that many variable features, commonly expressed in quantitative terms, such as weight, skeletal measurements, etc., in human beings exhibit a graduated series extending over a considerable range. The same thing is true for quantitative variations in general in animal and plant populations. In part this variation is obviously developmental, due to the action of environmental influences which call forth plus and minus responses in developing organisms, but it is equally obvious that heredity exerts an influence in most cases, as for example in height among human beings.

The part that heredity plays in such instances is, however, more clearly shown by differences in quantitative features among established breeds of animals and varieties of plants. A survey of existing varieties of wheat, for instance, will disclose great differences in yielding capacity, nitrogen content, stooling ability, number of grains per head, time of maturity, etc., features expressed on a quantitative scale. It is well known that no amount of forcing will make a bantam fowl grow to the size of standard breeds or a Shetland pony to the size of a draft horse, or will make a beef cow produce the quantity of milk of a good dairy cow. Numerous other illustrations immediately suggest themselves of differences in quantitative features which have become more or less fixed in definite races. The existence of such races, or their production in species in which they have not already become fixed, makes it possible to subject them to experimental analysis by approximately the same method as has been employed in dealing with qualitative features in the same forms.

**Simple Quantitative Variations.**—By way of introduction, it may be stated that variations in quantitative characters are frequently determined by single factor differences, just as are variations in qualitative characters. Mendel's case of tall and dwarf in peas is a classical example of such an instance. Here the two characters are discontinuous; *i.e.*, there is a gap between the tall and dwarf races, so that the differences may be treated as a qualitative one, and plants in a segregating population may be classified into tall and dwarf categories without recourse to measurement. There is, of course, variation in height both in the tall and dwarf classes; but the shortest tall plants are distinctly taller than the tallest dwarfs. Similarly, tall and dwarf strains dependent upon a single factor difference have been reported in tomatoes, sweet peas, beans, maize, snapdragons, and many other plants. In beans, Emerson reports that the difference between tall and dwarf races is in the type of growth; the tall are indeterminate, the apex keeps on growing indefinitely, and the flowers are produced laterally; while the dwarf varieties are determinate, and the axis of stems and branches terminate in an inflorescence after production of from four to eight internodes (figure 70).

In sweet peas there are several distinct dwarf forms which exhibit simple Mendelian relations to the tall form. One of these is the low-

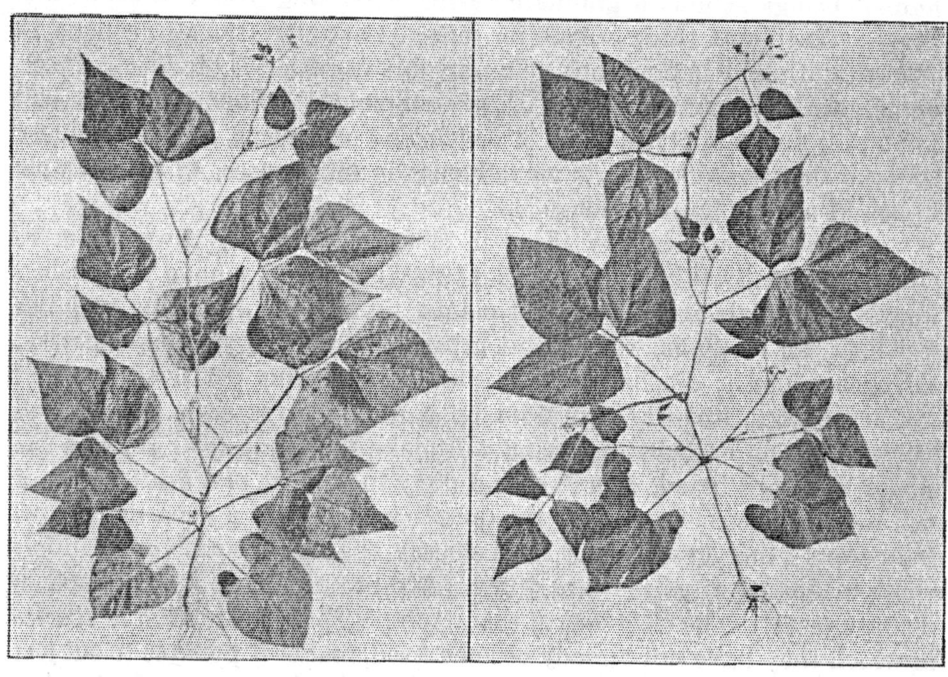

FIGURE 70.—Young bush bean plant (right) showing determinate habit of growth. The axis is terminated by a flower cluster. Young pole bean plant (left) showing indeterminate habit of growth. The flowers are all in the axils of the leaves. (*From Emerson.*)

growing, prostrate cupid sweet pea which attains a height of only about 6 inches. Originally it arose as a mutant form from Emily Henderson, a tall white variety, and apparently it has since arisen independently in a few other instances as a mutation from other tall varieties. When crossed with tall varieties, it gives a tall $F_1$ and the $F_2$ progenies consist of 3 tall : 1 cupid. Evidently it is a simple recessive to tall. Another is the bush sweet pea, which produces a bushy mass of short, erect branches. It also arose as a mutation from tall, and in crosses with tall, it has proved to be a simple Mendelian recessive. If *C-c* stands for tall *vs.* cupid and *B-b* for tall *vs.* bush, it is evident from the above described relations that the genetic formulae of the three types must be as follows:

| | |
|---|---|
| Tall | *BBCC* |
| Bush | *bbCC* |
| Cupid | *BBcc* |

When cupid is crossed with bush, therefore, $F_1$, *BbCc*, is tall, and $F_2$ exhibits the following distribution:

9*BC* = tall
3*Bc* = cupid
3*bC* = bush
1*bc* = bush-cupid

The double-recessive class represents a new combination of factors establishing a new phenotype. In appearance bush-cupid plants are very dwarf, but they differ from cupid in having erect instead of prostrate branches.

In the above instances the dwarf forms are recessive to the normal type. Occasionally giant or enlarged types are met with which are recessive to normal; for example, the mammoth character in tobaccos. Normal tobaccos have a determinate habit of growth. They produce a fairly definite number of internodes and then an inflorescence; but in mammoth tobaccos the stem continues to grow through a much longer period and consequently produces many more leaves than normal forms. This character is recessive to the normal type, and $F_2$ progenies exhibit a distribution of approximately 3 normal: 1 mammoth, although the proportion of mammoth plants in $F_2$ is often far below expectation, apparently because of their lower viability as seedlings. Definite Mendelian variations in size may, therefore, take place in both directions from normal, a fact which is also confirmed by the Drosophila studies. Two smaller mutant forms of *Drosophila melanogaster* are known; dwarf (*dw*) at III-51.0 and (*dw-b*) III-12, and a recessive giant (*gt*-III) form has also been located at III-64.0.

FIGURE 71.—Average flowers of two varieties of *Nicotiana longiflora* with an average flower of the $F_2$ from a cross between them in the middle. (*From East.*)

**Flower Size in Nicotiana.**—Usually, however, crosses between races differing in a quantitative character produce $F_2$ progenies which exhibit a continuous series of expressions extending over a considerable range. A typical illustration is afforded by crosses made by East between two varieties of *Nicotiana longiflora* which differ strikingly in flower size (figure 71). Flower size in tobacco is a particularly favorable object for studies of inheritance of a quantitative character, because it has been found to be relatively constant under different environmental conditions. Plants of a pure line grown under widely different conditions of nutrition produce flowers of very nearly the same size. Moreover, in some species established varieties exist which differ greatly in this respect, although they may be hybridized readily and produce perfectly fertile hybrids. For his investigations East selected two varieties of *N. longiflora,* one with an average corolla length of about 40 millimeters and the other, of 93 millimeters. Both parent varieties varied within narrow limits, so that they may be assumed to have been homozygous for flower size. The data which were secured from crosses between them are contained in table 23. The results may be summarized as follows:

1. $F_1$ was intermediate in size. Its variability, measured by the standard deviation, was approximately the same as that of the parents.

2. $F_2$ was also intermediate. It exhibited about the same mean as $F_1$. It differed, however, in exhibiting a marked increase in variability as shown by the higher standard deviation and a great increase in the range. $F_2$ progenies from different $F_1$ plants were identical.

3. $F_3$ progenies from different $F_2$ plants exhibited differences in mean and standard deviation, as contrasted with behavior of $F_2$ progenies from different $F_1$ plants. In general the standard deviations of $F_3$ progenies were smaller than that of $F_2$.

4. In subsequent generations races were isolated which exhibited different mean corolla lengths and gradual attainment of constancy as shown by a reduction in the standard deviation until it approximated that of the original parental races. Among others, races were isolated which exhibited mean corolla lengths approximately identical with those of the parents.

It is obvious from a consideration of quantitative differences which have been subjected to a definite Mendelian analysis that if the number of factor differences in a given cross were increased, a point would soon be reached where it would be quite impossible to distinguish $F_2$ classes. This is the conception which is applied to such instances as the Nicotiana cross described above. It is assumed that the differences in size of flowers in the two parental races is due to a difference in so many Mendelian factors, each of which has so small an effect upon flower size, that it is impossible to recognize distinct $F_2$ classes, a conception known as the multiple-factor hypothesis.

In the application of this conception to the Nicotiana cross described above, assume for the sake of simplicity that the parental means are 40 and 90 millimeters, respectively, and that the two parental races differ in five pairs of Mendelian factors. Then variety A with a mean of 40 milli-

TABLE 23.—FREQUENCY DISTRIBUTIONS FOR COROLLA LENGTH IN A CROSS BETWEEN VARIETIES OF *Nicotiana longiflora* (*After East*)

| Designation number | Year | Generation | Parent size | Class centers in millimeters | | | | | | | | | | | | | | | | | | | | | | |
|---|---|---|---|---|---|---|---|---|---|---|---|---|---|---|---|---|---|---|---|---|---|---|---|---|---|---|
| | | | | 34 | 37 | 40 | 43 | 46 | 49 | 52 | 55 | 58 | 61 | 64 | 67 | 70 | 73 | 76 | 79 | 82 | 85 | 88 | 91 | 94 | 97 | 100 |
| 383 | 1911 | .. | .. | .. | 13 | 80 | 32 | | | | | | | | | | | | | | | | | | | |
| 383 | 1912 | .. | .. | 1 | 4 | 28 | 16 | | | | | | | | | | | | | | | | | | | |
| 383 | 1913 | .. | .. | .. | 4 | 32 | 1 | | | | | | | | | | | | | | | | | | | |
| 330 | 1911 | .. | .. | .. | .. | .. | .. | .. | .. | .. | .. | .. | .. | .. | .. | .. | .. | .. | .. | .. | .. | 6 | 22 | 49 | 11 | |
| 330 | 1912 | .. | .. | .. | .. | .. | .. | .. | .. | .. | .. | .. | .. | .. | .. | .. | .. | .. | .. | .. | .. | 2 | 16 | 32 | 6 | 1 |
| 330 | 1913 | .. | .. | .. | .. | .. | .. | .. | .. | .. | .. | .. | .. | .. | .. | .. | .. | .. | .. | .. | .. | 5 | 7 | 10 | 2 | |
| 383 × 330 | 1911 | $F_1$ | .. | .. | .. | .. | .. | .. | .. | .. | 4 | 10 | 41 | 75 | 40 | 3 | | | | | | | | | | |
| (383 × 330)1 | 1912 | $F_2$ | 61 | .. | .. | .. | .. | .. | .. | 1 | 5 | 16 | 23 | 18 | 62 | 37 | 25 | 16 | 4 | 2 | 2 | | | | | |
| (383 × 330)2 | 1912 | $F_2$ | 61 | .. | .. | .. | .. | .. | .. | 2 | 4 | 2 | 24 | 37 | 31 | 38 | 35 | 27 | 21 | 5 | 6 | 1 | | | | |
| (383 × 330)1-1 | 1913 | $F_3$ | 72 | .. | .. | .. | .. | .. | .. | .. | .. | .. | 4 | 20 | 25 | 59 | 41 | 19 | 2 | | | | | | | |
| (383 × 330)1-2 | 1913 | $F_3$ | 46 | .. | .. | .. | 1 | 4 | 26 | 44 | 38 | 22 | 7 | 1 | | | | | | | | | | | | |
| (383 × 330)1-3 | 1913 | $F_3$ | 50 | .. | .. | .. | 6 | 20 | 53 | 49 | 15 | 4 | | | | | | | | | | | | | | |
| (383 × 330)1-4 | 1913 | $F_3$ | 60 | .. | .. | .. | 2 | 3 | 9 | 25 | 37 | 70 | 19 | 10 | | | | | | | | | | | | |
| (383 × 330)2-1 | 1913 | $F_3$ | 77 | .. | .. | .. | .. | .. | 1 | 0 | 1 | 1 | 1 | 2 | 16 | 33 | 43 | 34 | 20 | 6 | 1 | | | | | |
| (383 × 330)2-3 | 1913 | $F_3$ | 81 | .. | .. | .. | .. | .. | .. | .. | .. | .. | 1 | 1 | 8 | 16 | 20 | 32 | 41 | 17 | 3 | 3 | 1 | | | |
| (383 × 330)2-4 | 1913 | $F_3$ | 80 | .. | .. | .. | .. | .. | .. | .. | .. | .. | 2 | 8 | 14 | 21 | 39 | 39 | 32 | 10 | 1 | | | | | |
| (383 × 330)2-5 | 1913 | $F_3$ | 50 | .. | .. | .. | .. | 7 | 25 | 55 | 55 | 18 | | | | | | | | | | | | | | |
| (383 × 330)2-6 | 1913 | $F_3$ | 82 | .. | .. | .. | .. | .. | .. | .. | .. | .. | .. | .. | 3 | 5 | 12 | 20 | 40 | 41 | 30 | 9 | 2 | | | |
| (383 × 330)1-2-1 | 1914 | $F_4$ | 44 | .. | .. | 8 | 42 | 95 | 38 | 1 | | | | | | | | | | | | | | | | |
| (383 × 330)1-3-1 | 1914 | $F_4$ | 43 | .. | .. | 2 | 23 | 122 | 41 | 1 | | | | | | | | | | | | | | | | |
| (383 × 330)2-6-1 | 1914 | $F_4$ | 85 | .. | .. | .. | .. | .. | .. | .. | .. | .. | .. | .. | .. | .. | 4 | 9 | 38 | 75 | 59 | 6 | 3 | 1 | | |
| (383 × 330)2-6-2 | 1914 | $F_4$ | 87 | .. | .. | .. | .. | .. | .. | .. | .. | .. | .. | .. | 4 | 5 | 6 | 11 | 21 | 33 | 41 | 29 | 8 | 5 | 1 | |
| (383 × 330)1-3-1-1 | 1915 | $F_5$ | 41 | 3 | 6 | 48 | 90 | 14 | | | | | | | | | | | | | | | | | | |
| (383 × 330)2-6-2-1 | 1915 | $F_5$ | 90 | .. | .. | .. | .. | .. | .. | .. | .. | .. | .. | .. | .. | .. | 2 | 3 | 8 | 14 | 20 | 25 | 25 | 20 | 8 | |

meters may be written as *aabbccddee* and variety B with a mean flower length of 90 millimeters, *AABBCCDDEE*. Let it be further assumed that each factor has an equal effect upon length of flower and that they act cumulatively, so that each capitalized symbol represents an increment of 5 millimeters to the basic flower length of 40 millimeters characteristic of variety A. Then $F_1$, *AaBbCcDdEe*, obviously will produce flowers 65 millimeters long, midway between the parental forms. In $F_1$ each pair of factors will segregate as usual, so that the $F_2$ population may be represented by $(AA + 2Aa + aa)(BB + 2Bb + bb)(CC + 2Cc + cc)(DD + 2Dd + dd)(EE + 2Ee + ee)$. Since each factor is assumed to have the same effect on size, instead of laboriously multiplying this expression out, the binomial $(a + b)^{10}$ may simply be expanded, coefficients of the terms representing frequencies, powers of $a$ the number of recessive factors, and of $b$ dominant factors, thus:

| | $a^{10}$ + | $10a^9b$ + | $45a^8b^2$ + | $120a^7b^3$ + | $210a^6b^4$ + | $256a^5b^5$ + | $210a^4b^6$ + | $120a^3b^7$ + | $45a^2b^8$ + | $10ab^9$ + | $b^{10}$ |
|---|---|---|---|---|---|---|---|---|---|---|---|
| Classes | 40 | 45 | 50 | 55 | 60 | 65 | 70 | 75 | 80 | 85 | 90 |
| Frequencies | 1 | 10 | 45 | 120 | 210 | 256 | 210 | 120 | 45 | 10 | 1 |

From these assumptions, therefore, it follows that $F_2$ will exhibit the same mean value as $F_1$, but the range will extend from the lower parental value to the higher, and the distribution will be of the normal type. Parental values will reappear only rarely, one of each kind in 1,024 individuals with five pairs of factors; the rest of the individuals will represent intermediate grades of expression, the greatest frequency occurring in the middle class. In $F_3$ and subsequent generations, variability will be reduced, because some of the factors will become homozygous; and eventually homozygous races will be established for intermediate expressions such as *AAbbccddee* = 50 millimeters, *AABB ccddee* = 60 millimeters, etc. Evidently on the assumptions which have been made, all the conditions stipulated by the foregoing experimental results are satisfied.

It is not, however, intended that this explanation should represent the actual factor conditions in the above-described experiment. In the first place, there is no way of determining accurately in how many factors the parental varieties differ. Very likely they differ in more than five pairs of factors. Parental values were not recovered in $F_2$. If the two $F_2$ populations are added together and the expectancy of obtaining parental values is computed from the standard deviation on the assumption of normal distribution, it is found that the range separating the two parental values amounts to about 40. On this basis the chances of recovering parental values are about 4 in 10,000, which would indicate a difference in about six pairs of factors. But this is a very rough method of estimate and should not be taken too seriously.

The explanation which has been offered above is only a formal one. The hypothesis was presented in its simplest form in order to emphasize the essential features of the multiple-factor hypothesis. In the first place it is improbable that the factor differences have equal weight as size determiners. If, however, they are numerous and have comparatively small effects individually, the statistical relations would not be significantly changed by slight differences in their effects. In the second place, the assumption of a strictly cumulative effect is probably unjustifiable. It is probably just as true of size factors as of qualitative ones that their interactions in development are complex and not conformable to a simple scheme such as has been described. However, even if these assumptions do prove incorrect, when the number of factors is large the interactions among them will no doubt produce individuals which will conform to the normal distribution.

It has also been assumed that size factors exhibit no dominance, but that the heterozygous condition produces an expression intermediate between that of the parents. This assumption is based on the observed fact that crosses between two races differing in a quantitative character often give rise to an intermediate $F_1$. This intermediacy of $F_1$, however, is really not incompatible with complete dominance in every pair of factors, for it may simply be an expression of interaction between dominant factors some of which tend to increase and some to decrease size. When parental races differ in two or more factors affecting the same character, judging as to dominance from the character expression of $F_1$ is exceedingly uncertain, and nothing but a complete factor analysis will yield dependable results.

It is intended in this discussion merely to emphasize the fact that quantitative factors, like qualitative ones, no doubt exhibit endless idiosyncrasies which can be discovered only if some method can be devised for identifying them and studying them individually. But whatever their idiosyncrasies, they would not seriously affect the principal conclusions which have been reached above.

**Size Inheritance in Fowls.**—A further instance of investigations of size inheritance is afforded by Punnett's studies of crosses between Silver

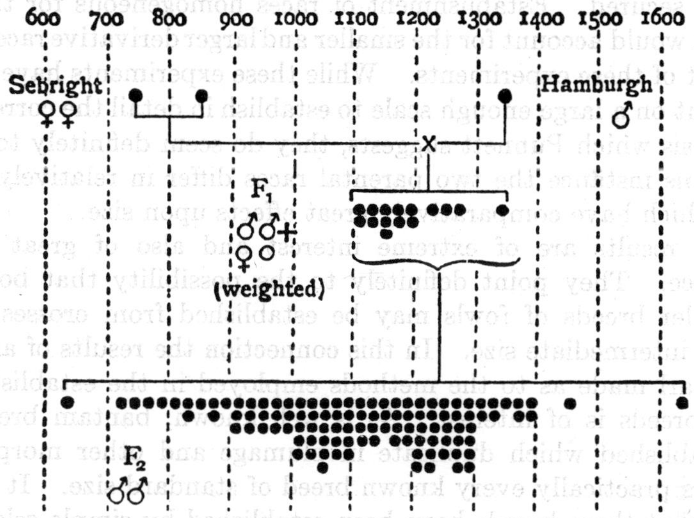


FIGURE 72.—Scheme illustrating inheritance of weight in a cross between Sebright bantam ♀ and Gold-penciled Hamburgh ♂. The figures at the top indicate weight in grams. The ♀♀ of the parental and $F_1$ generations are weighted, *i.e.*, their actual weights are × 5/4 to bring them into line with those of the ♂♂. In $F_2$ the weights of the ♂♂ alone are given. (*From Punnett, Heredity in Poultry, copyright by Macmillan and Company Ltd.*, 1923. *Reprinted by permission.*)

Sebright bantams and Golden Hamburgh fowls. A portion of these results is set forth diagrammatically in figure 72. Since males are on the average somewhat larger than females, female weights have been multiplied by 1.25 in order to convert them into comparable male values. Sebright bantams exhibit a mean weight of about 800 grams and Hamburghs, 1,350 grams. The mean weight of $F_1$ was intermediate but nearer the larger parent, and the range of variation was small. The modal value in $F_2$ was about the same as that of $F_1$, but the range extended from values lower than that of the Sebright to values exceeding that of the Hamburgh. In subsequent generations derivative races have been established which breed true for sizes smaller than that of the Sebright and larger than that of the Hamburgh as well as for intermediate expressions. This illustration has been introduced to show that segregation for size factors may extend the range of values beyond the limits set by the parental races, a phenomenon known as transgressive variation.

Punnett suggests tentatively that a four-factor difference may exist between the two parents and that these factors may have unequal effects. The Sebright may then be represented by the formula *aabbccDD*, and the Hamburgh by *AABBCCdd*, capitalized symbols as before standing for factors which increase size. The $F_1$ would then be heterozygous for all four pairs of factors, *AaBbCcDd*, and in $F_2$ some individuals of the formula *aabbccdd*, representing a smaller size than the Sebright, and some *AABBCCDd* and *AABBCCDD*, representing sizes larger than the Hamburgh, would be secured. Establishment of races homogeneous for these constitutions would account for the smaller and larger derivative races secured as a result of these experiments. While these experiments have not been carried out on a large enough scale to establish in detail the correctness of the analysis which Punnett suggests, they do seem definitely to indicate that in this instance the two parental races differ in relatively few size factors which have comparatively great effects upon size.

These results are of extreme interest and also of great practical significance. They point definitely to the possibility that both larger and smaller breeds of fowls may be established from crosses between breeds of intermediate size. In this connection the results of an inquiry which Pearl made as to the methods employed in the establishment of bantam breeds is of interest. As is well known, bantam breeds have been established which duplicate in plumage and other morphological characters practically every known breed of standard size. It might be thought that these breeds have been established by simple selection for smaller size in larger breeds, but Pearl found that they have invariably originated from crosses between bantams and normal fowls. They have, therefore, been the result of combining size factors of bantams with the factors for plumage type and other morphological features characteristic of the parallel standard breeds.

**Number of Leaves in Tobacco.**—A very instructive instance of the outcome of multiple-factor conditions is afforded by results which Hayes has reported for leaf number in tobacco. The results in $F_1$ and $F_2$ are shown in table 24. The two parent varieties in this cross show no significant difference in number of leaves; $F_1$, obtained by crossing them, exhibited the same mean leaf number and, if anything, less variation than the parental varieties. $F_2$ exhibited approximately the same mean leaf number as the parents and $F_1$, but the range was greatly extended and the standard deviation was over twice as great. Assuming that the $F_2$ variation is here due to genetic segregation and combination, these results indicate clearly that the parent races, although exhibiting the same mean leaf number, must have differed in their genetic factors for leaf number. As a consequence of production of new genetic combinations for leaf number in $F_2$, new types appeared with higher and lower numbers than that of the parents. These results are, of course, to be expected on the multiple-factor hypothesis, for if many factor differences for size exist, then obviously many of the races which exhibit the same expression of the character may still differ in their factors.

TABLE 24.—FREQUENCY DISTRIBUTION OF NUMBER OF LEAVES PER PLANT IN CROSSES BETWEEN HAVANA AND CUBAN VARIETIES OF *Nicotiana tabacum* (*From Hayes*)

| Generation and year | Leaf numbers | | | | | | | | | | | | | | | | | | | | Total | $M$ | $\sigma$ |
|---|---|---|---|---|---|---|---|---|---|---|---|---|---|---|---|---|---|---|---|---|---|---|---|
| | 14 | 15 | 16 | 17 | 18 | 19 | 20 | 21 | 22 | 23 | 24 | 25 | 26 | 27 | 28 | 29 | 30 | 31 | 32 | 33 | | | |
| Cuban, 1910 | .. | .. | 1 | 7 | 16 | 37 | 36 | 35 | 12 | 4 | 1 | 1 | .. | .. | .. | .. | .. | .. | .. | .. | 150 | 19.9 ± 0.082 | 1.50 ± 0.059 |
| Cuban, 1911 | .. | .. | .. | .. | 3 | 16 | 34 | 46 | 20 | 5 | .. | .. | .. | .. | .. | .. | .. | .. | .. | .. | 124 | 20.6 ± 0.066 | 1.09 ± 0.047 |
| Havana, 1910 | .. | .. | .. | 3 | 22 | 44 | 42 | 22 | 10 | 6 | 1 | .. | .. | .. | .. | .. | .. | .. | .. | .. | 150 | 19.8 ± 0.076 | 1.38 ± 0.054 |
| Havana, 1911 | .. | .. | 1 | 10 | 11 | 23 | 34 | 31 | 16 | 12 | 3 | 2 | .. | .. | .. | .. | .. | .. | .. | .. | 143 | 20.3 ± 0.101 | 1.80 ± 0.072 |
| $F_1$, 1910 | .. | 1 | 1 | 3 | 8 | 39 | 60 | 30 | 7 | .. | .. | 1 | .. | .. | .. | .. | .. | .. | .. | .. | 150 | 19.8 ± 0.067 | 1.21 ± 0.047 |
| $F_2$, 1911 | 3 | 4 | 8 | 8 | 20 | 18 | 30 | 24 | 25 | 17 | 16 | 5 | 4 | 3 | 1 | 1 | 1 | 1 | 2 | 1 | 192 | 20.9 ± 0.161 | 3.31 ± 0.114 |

**Hybrid Vigor.**—It has frequently been observed that when two varieties of animals or plants are crossed, the $F_1$ is distinctly more vigorous than either parental form, or as in the case of fowls described above, larger than the average of the two parental varieties. This matter of vigor may be expressed in a variety of ways: in increased size, greater resistance to disease, greater ease of vegetative reproduction, lessened susceptibility to the action of unfavorable environmental conditions, greater fertility, etc. It is often a complex of effects difficult to express on any accurate quantitative scale, and the general term hybrid vigor is frequently used to describe it.

East has presented data for height of plant in crosses between two varieties of *Nicotiana rustica* which illustrate the general course of the phenomena in such cases. It will be observed from the data presented in table 25 that the mean height of $F_1$ obtained from crossing *scabra* with *brasilia* exceeds that of the taller parent, and the variability of $F_1$ as

TABLE 25.—FREQUENCY DISTRIBUTION FOR HEIGHT OF PLANT IN CROSSES BETWEEN *Nicotiana rustica* VAR. *scabra* AND *N. rustica* VAR. *brasilia* (*From East*)

| Variety or cross | Class centers in inches | | | | | | | | | | | | | | | | | | | | | | Total |
|---|---|---|---|---|---|---|---|---|---|---|---|---|---|---|---|---|---|---|---|---|---|---|---|
| | 15 | 18 | 21 | 24 | 27 | 30 | 33 | 36 | 39 | 42 | 45 | 48 | 51 | 54 | 57 | 60 | 63 | 66 | 69 | 72 | 75 | 78 | |
| Brasilia | .. | .. | .. | 4 | 10 | 22 | 14 | 7 | .. | .. | .. | .. | .. | .. | .. | .. | .. | .. | .. | .. | .. | .. | 57 |
| Scabra | .. | .. | .. | .. | .. | .. | .. | .. | .. | 2 | 1 | 5 | 11 | 16 | 17 | 6 | .. | .. | .. | .. | .. | .. | 58 |
| Scabra × brasilia $F_1$ | .. | .. | .. | .. | .. | .. | .. | .. | .. | .. | .. | .. | 1 | 3 | .. | 5 | 5 | 6 | 1 | 1 | .. | .. | 22 |
| Brasilia × scabra $F_1$ | .. | .. | .. | .. | .. | .. | .. | .. | .. | .. | .. | .. | .. | .. | 3 | 5 | 2 | 4 | 6 | 5 | 1 | 2 | 28 |
| Brasilia × scabra $F_2$ | .. | .. | .. | 2 | 3 | 6 | 2 | 7 | 12 | 8 | 12 | 13 | 12 | 12 | 4 | 15 | 4 | .. | 1 | .. | 1 | .. | 114 |
| Brasilia × scabra $F_3$ | 4 | 9 | 32 | 43 | 17 | 12 | 2 | .. | .. | .. | .. | .. | .. | .. | .. | .. | .. | .. | .. | .. | .. | .. | 119 |
| Brasilia × scabra $F_5$ | .. | .. | .. | .. | .. | .. | .. | .. | 3 | 4 | 6 | 10 | 9 | 7 | 28 | 13 | 10 | 8 | 3 | .. | .. | .. | 101 |

shown by the distribution is not significantly different from that of the parental varieties. In $F_2$ the mean value drops considerably below that of the $F_1$ and the variability is greatly increased. The two $F_3$ populations exhibit the familiar outcome of such investigations, establishment of different mean values and reduction of variability as compared with $F_2$.

Hybrid vigor may also be accounted for on a multiple-factor basis, for it is certainly reasonable to assume that many factors contribute to such a complex feature of development. Assuming that numerous dominant factors tend to augment vigor and the corresponding recessive allelomorphs to depress it, it is reasonable to suppose that each variety will possess some dominant factors and some recessive factors for vigor, and that consequently hybridization, which brings all the dominant factors together, will sometimes produce an $F_1$ superior in this respect to either parent. Multiple-factor segregation in $F_2$, giving rise to various combinations of dominant and recessive factors, accounts for the depression in average vigor and the wide extension of the range. It will be observed that under these assumptions $F_1$ represents a maximum expression and that the $F_2$ range extends from this condition as an upper extreme to a lower limit lying below the range of the inferior parent as is shown by the data. More complete details as to the interpretation of hybrid vigor will appear in subsequent chapters.

**Parent-offspring Correlation.**—In many instances, when it is impossible or inadvisable to employ the experimental method of procedure, it is possible to secure some evidence of the mode of inheritance by use of the correlation coefficient. Thus in a human population, height, or almost any other quantitative character, exhibits a graduated expression over a wide range. Morever, in general it is found that average height of offspring increases regularly with height of parents; that is, tall parents have tall offspring and short parents, short offspring. Considering the evidence merely in this general way it is evident that variation in height in human beings must rest in part on a genetic basis; since, if variation in height were due entirely to environmental influences, there should be no marked positive correlation between parents and offspring.

The matter is placed upon a more precise basis by resort to statistical methods. In table 26, certain data collected by Galton on relation of parental to offspring heights are presented in somewhat condensed form. The data for women in this table have been multiplied by 1.08 to bring them to a height comparable to men, a somewhat doubtful procedure statistically, but introducing no great error for these purposes. The constants determined from the data are as follows:

| | | |
|---|---|---|
| For parents | $M = 68.06$ | $\sigma = 1.853$ |
| For offspring | $M = 68.09$ | $\sigma = 2.583$ |
| Correlation | $r = 0.449 \pm 0.026$ | |

From these constants the regression line for height of offspring on parents, referred to intersection of means, is

$$\bar{d}_x = 0.625d_y$$

in which $\bar{d}_x$ is the average deviation of offspring corresponding to particular parental deviations $d_y$ from the mean. On the average the offspring, therefore, exhibit a deviation from the mean only approximately two-thirds as great as their parents. These figures express in precise terms the average relationship of parents and offspring in height. They may reasonably be interpreted to indicate that hereditary diversity for height exists in human populations; and in other populations positive parent-offspring correlations may usually be taken as indicative of genetic diversity. In some instances it is necessary to guard against permanent differences in environment acting in like manner upon parent and offspring; but such considerations are part of the general warning that statistical methods must be employed properly with adequate realization of the various elements which enter into a particular situation.

TABLE 26.—CORRELATION BETWEEN HEIGHT OF PARENT AND HEIGHT OF OFFSPRING IN HUMAN POPULATIONS. FEMALE HEIGHTS MULTIPLIED BY 1.08 TO MAKE THEM COMPARABLE TO MALE HEIGHTS (*Data from Galton*)

| Height of parents in inches | Height of offspring in inches | | | | | | | | |
|---|---|---|---|---|---|---|---|---|---|
| | 60.7 | 62.7 | 64.7 | 66.7 | 68.7 | 70.7 | 72.7 | 74.7 | |
| 64 | 2 | 7 | 10 | 14 | 4 | ... | .. | .. | 37 |
| 66 | 1 | 15 | 19 | 56 | 41 | 11 | 1 | .. | 144 |
| 68 | 1 | 15 | 56 | 130 | 148 | 69 | 11 | .. | 430 |
| 70 | 1 | 2 | 21 | 48 | 83 | 66 | 22 | 8 | 251 |
| 72 | .. | .. | 1 | 7 | 11 | 17 | 20 | 6 | 62 |
| 74 | .. | .. | ... | ... | ... | ... | 4 | ... | 4 |
| | 5 | 39 | 107 | 255 | 287 | 163 | 58 | 14 | 928 |

$$r = 0.449 \pm 0.026$$

The correlation coefficient or more properly the regression lines have been used by some investigators as a measure of the intensity of heredity. That is to say, in the instance just described, the regression lines show that a parental deviation of a certain amount from the mean is on the average attended by a deviation in the offspring of the group of about two-thirds that amount from the mean, the remaining one-third being the regression. But regression may also be due partly or wholly to heredity, as a moment's consideration will indicate, for if individuals are heterozygous for factors determining height, segregation may give rise to larger and smaller individuals. If such segregation produces a disproportionate number of individuals more nearly approaching the mean of the race, then regression will occur. Also if parental deviations are partly due to environmental causes, the effect of which is eliminated in the offspring, a regression toward the mean of the race will occur. These facts must be considered in judging the significance of parent-offspring correlation.

**Types of Heredity.**—It was at one time thought that different types of heredity occurred, as is reflected in Galton's classification of heredity into three categories: alternative, particulate, and blending. Eye color in human beings was considered a typical example of alternative inheritance, inasmuch as individuals of a population may be grouped into relatively few classes; black, brown, blue, etc. Such inheritance is now

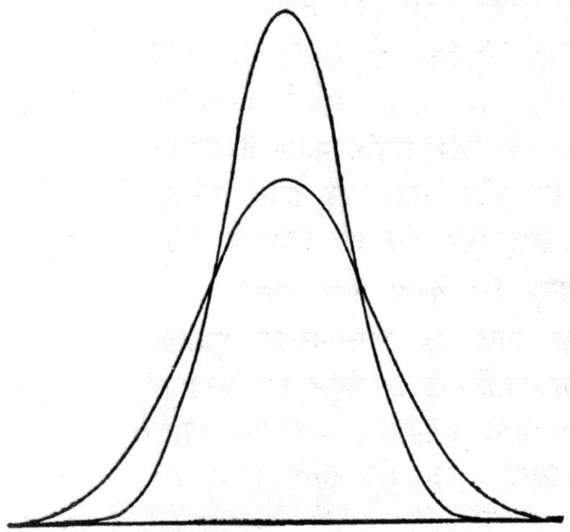

FIGURE 73.—Theoretical distribution curves for five and ten pairs of factors, assuming that the range covered by the two curves is equal.

known to be typical of Mendelian situations in which relatively few factor differences are involved. Particulate inheritance is well illustrated by certain crosses between white and black breeds of swine which give a spotted hybrid. The term particulate was applied to such instances because it was thought that the hybrid exhibited the character of one parent in some and the other parent in the remaining portions of the body. The Mendelian interpretation is that white races may bear factors for spotting, which exert their effect only in association with factors for the colored condition. Hence these two categories of heredity are readily reduced to a Mendelian formulation.

Blending inheritance presented more serious difficulties. A typical example is afforded by crosses between rabbits differing greatly in ear length. $F_1$ was intermediate and $F_2$ was also intermediate and exhibited only a slight increase in variability as compared with $F_1$. Apparently

$F_1$ established a new mean condition, a blend, which, thereafter, remained as constant as the parental conditions. If, however, the multiple-factor hypothesis is applied to such instances, the difficulty vanishes, for it may be shown that, as the number of factors concerned in a given range of variation increases, the population will be concentrated more strongly around the mean value. As a consequence, if a very large number of factors contribute to ear length in rabbits, a limited $F_2$ population will consist almost entirely of individuals exhibiting very nearly the average condition, so that segregation will appear to be lacking. Under such conditions, so large $F_2$ progenies would be necessary for the recovery of parental types as to make it impracticable to grow them. As shown heretofore, with five pairs of factors, parental combinations of each kind are recovered only once among 1,024 individuals, but when ten pairs of factors are involved, 1,048,576 individuals are necessary for their reappearance. Comparisons as to the relative distributions of populations of 1,024 individuals over a given range when five and ten pairs of factors are involved are shown in figure 73. Evidently blending inheritance also is susceptible of Mendelian interpretation on the multiple-factor basis, so that all three of these proposed categories reduce to aspects of Mendelian heredity.

**The Composition of Populations.**—The matters discussed in this chapter permit some important inferences to be drawn concerning the composition of populations. As will be seen shortly, however, some attention must be given to modes of reproduction in order to reach accurate conclusions. For this reason, assume the existence of an ideal population, mating completely at random, as the basis of discussion. The foregoing discussion has shown that numerous Mendelian factors, each having such a slight effect as to be unrecognizable, are usually concerned in the expression of qualitative characters, in particular, but no doubt in others also. Now, if a given pair of factors $A$-$a$ exists in the proportions $m_1A$ and $n_1a$ in the entire population where $m_1 + n_1 = 1$, then the gametes will obviously be produced in these same proportions $m_1A + n_1a$; and under conditions of random mating, the composition of the population with respect to this pair of factors will obviously be $(m_1A + n_1a)^2 = m_1^2AA + 2m_1n_1Aa + n_1^2aa$. These same remarks apply to any other pair of factors, so that for $B$-$b$ the result is $m_2^2BB + 2m_2n_2Bb + n_2^2bb$; for $C$-$c$, $m_3^2CC + 2m_3n_3Cc + n_3^2cc$; and so on. Now then, under conditions of equilibrium, the total composition of the population in terms of all these factors is evidently the continued product of these various expressions; *viz.*,

$$(m_1^2AA + 2m_1n_1Aa + n_1^2aa)(m_2^2BB + 2m_2n_2\,Bb + n_2^2bb)$$
$$(m_3^2CC + 2\,m_3n_3Cc + n_3^2cc)\ .\ .\ .$$

Expansion of this expression would then give all the combinations occurring in such an ideal population with the frequencies appropriate to each.

Linkage, of course, would disturb these relations somewhat, but not sufficiently to invalidate the general conclusions. The student should be able to verify for himself that under the stipulated conditions the composition of the population would remain unchanged from generation to generation.

In order to grasp the significance of this conclusion, assume that the members of each pair of factors are present in equal proportions and that each pair of factors is equally effective in the production of the character in question, assumptions which are justifiable when the number of factors concerned is fairly large. Then the general form of distribution in a population of unit size may be obtained by expansion of the binomial $(\frac{1}{2} + \frac{1}{2})^n$, where $n$ equals the number of pairs of factors involved, which, as has been shown, gives a normal curve. If, however, the number of factors involved is large, it may well be that populations of moderate size, such as are met with in nature and in practical operations, will not contain all the possible combinations. Consequently the actual population presumably represents merely a random sample of the theoretical population, and, because of concentration around the mean, the range of expression would be restricted by reason of elimination of extreme classes of low frequency. The purpose of animal and plant breeding under such circumstances is to produce populations uniform for a desirable expression of a character, or rather, combination of characters, a result which may be achieved only by adoption of proper methods; and it is important to realize further that extreme combinations lying entirely outside the range of variation exhibited by a population of limited size may be secured.

### References

CASTLE: Genetics and Eugenics, Chapter XXVIII.

MORGAN, STURTEVANT, MULLER, and BRIDGES: Mechanism of Mendelian Heredity, Chapter IX.

SINNOTT and DUNN: Principles of Genetics, Chapter X.

WALTER: Genetics, Chapter VIII.

# CHAPTER XX

## PURE LINES AND CLONES

In the preceding chapter it was shown that differences in numerous factors, each of which has a relatively slight effect, is responsible for much of the variation existing in populations. It is also self-evident that slight differences in environment produce plus and minus deviations from the mean of a similar character. It is important to devise means of distinguishing between these two types of variability and of determining the part played by each in a given situation.

**Johannsen's Experiments.**—The first clear-cut experimental demonstration of this distinction was the outcome of certain studies made by Johannsen on various quantitative characters in the Princess bean, particularly weight of seed. He observed that a great deal of variation occurred in unselected populations, and he was able to show that by dividing the seed into classes according to weight the smaller beans in general produce small-seeded and the larger ones larger-seeded offspring. But he also observed that seeds in a given class produced progeny of rather widely different weights and this observation led him to compare average size of beans in progenies from different mother plants. The progenies thus separated still exhibited variability in seed weight, but to a lesser extent than that in the original population; and moreover the average seed weight of these progenies exhibited striking differences. Johannsen established nineteen such strains by growing separately the progenies from as many mother plants from the original mixed crop. These nineteen strains, called pure lines, were grown separately, with careful attention to prevention of admixture, and their characteristic features were studied in great detail and compared with those of the original population.

In the first place Johannsen found that each of the pure lines exhibited a characteristic mean weight of seed, ranging from 35.1 centigrams for line 19, the smallest seeded one, to 64.2 centigrams for line 1, which produced the largest seeds. He concluded from this fact that the original commercial variety was a mixture of pure lines. He defined a pure line as the progeny of a single self-fertilized individual of homogeneous factorial composition.

There was still, of course, variability within each pure line, for each of them exhibited curves of distribution for weight of the normal type. It remained necessary, therefore, to determine the meaning of this variation. For this purpose the beans in each pure line were segregated into size classes and grown separately. It was found that the different classes of beans within a pure line produced progeny of the same average weight, as is shown in table 27, which should be studied carefully in this

TABLE 27.—RESULTS OF PLANTING BEANS OF SELECTED SIZES IN THE NINETEEN PURE LINES OF JOHANNSEN ON MEAN SEED WEIGHT OF THE PROGENY

| Pure line number | Weight in centigrams of mother beans | | | | | | Mean weight of pure line |
|---|---|---|---|---|---|---|---|
| | 20 | 30 | 40 | 50 | 60 | 70 | |
| 1 | .... | .... | .... | .... | 63.1 | 64.9 | 64.2 |
| 2 | .... | .... | 57.2 | 54.9 | 56.5 | 55.5 | 55.8 |
| 3 | .... | .... | .... | 56.4 | 56.6 | 54.4 | 55.4 |
| 4 | .... | .... | .... | 54.2 | 53.6 | 56.6 | 54.8 |
| 5 | .... | .... | 52.8 | 49.2 | .... | 50.2 | 51.2 |
| 6 | .... | 53.5 | 50.8 | .... | 42.5 | .... | 50.6 |
| 7 | 45.9 | .... | 49.5 | .... | 48.2 | .... | 49.2 |
| 8 | .... | 49.0 | 49.1 | 47.5 | .... | .... | 48.9 |
| 9 | .... | 48.5 | .... | 47.9 | .... | .... | 48.2 |
| 10 | .... | 42.1 | 46.7 | 46.9 | .... | .... | 46.5 |
| 11 | .... | 45.2 | 45.4 | 46.2 | .... | .... | 45.5 |
| 12 | 49.6 | .... | .... | 45.1 | 44.0 | .... | 45.5 |
| 13 | .... | 47.5 | 45.0 | 45.1 | 45.8 | .... | 45.4 |
| 14 | .... | 45.4 | 46.9 | .... | 42.8 | .... | 45.3 |
| 15 | 46.9 | .... | .... | 44.6 | 45.0 | .... | 45.0 |
| 16 | .... | 45.9 | 44.1 | 41.0 | .... | .... | 44.6 |
| 17 | 44.0 | .... | 42.4 | .... | .... | .... | 42.8 |
| 18 | 41.0 | 40.7 | 40.8 | .... | .... | .... | 40.8 |
| 19 | .... | 35.8 | 34.8 | .... | .... | .... | 35.1 |
| Mean....... | 44.0 | 44.3 | 46.1 | 49.0 | 51.9 | 56.1 | 47.9 |

connection. Thus to take line 13 as an instance, beans weighing 30, 40, 50, and 60 centigrams were planted separately. The average weights of the progeny were 47.5, 45.0, 45.1, and 45.8 centigrams, respectively. Clearly within a pure line large-sized beans do not tend to produce large-sized progeny and small-sized ones small progeny, but all tend to reproduce the mean seed weight characteristic of the pure line to which they belong. In other words the variation within a pure line is purely developmental; it depends upon slight differences in environmental conditions of all kinds acting upon the individuals, and it has no genetic significance.

This conclusion was submitted to further verification by subjecting each pure line to continuous selection in a plus and minus direction for a number of years. Each year for 6 successive years a number of the largest beans were selected for continuing the line in a plus direction and the smallest ones for the minus series. The data for two of the pure lines, line 1 which produced the heaviest seeds and line 19 the lightest, are shown in tables 28 and 29. The selection, although carried out on a sufficiently large scale, failed in either case to establish a significant difference between the plus and minus series during a 6-year period. A further statistical demonstration of the same fact was made by determining the parent-offspring correlation for seed weight within a pure line.

TABLE 28.—SELECTION EFFECT DURING SIX GENERATIONS IN LINE 1 OF THE PRINCESS BEANS (*From Johannsen*)

| Harvest years | Total number of beans | Mean weight of mother beans of the select strains | | Difference $b - a$ | Mean weight of progeny seeds of select strains | | Difference $b - a$ |
|---|---|---|---|---|---|---|---|
| | | $a$-minus | $b$-plus | | $a$-minus | $b$-plus | |
| 1902 | 145 | 60 | 70 | 10 | 63.15 ± 1.02 | 64.85 ± 0.76 | +1.70 ± 1.27 |
| 1903 | 252 | 55 | 80 | 25 | 75.19 ± 1.01 | 70.88 ± 0.89 | −4.31 ± 1.35 |
| 1904 | 711 | 50 | 87 | 37 | 54.59 ± 0.44 | 56.68 ± 0.36 | +2.09 ± 0.57 |
| 1905 | 654 | 43 | 73 | 40 | 63.55 ± 0.56 | 63.64 ± 0.41 | +0.09 ± 0.69 |
| 1906 | 384 | 46 | 84 | 38 | 74.38 ± 0.81 | 73.00 ± 0.72 | −1.38 ± 1.08 |
| 1907 | 379 | 56 | 81 | 25 | 69.07 ± 0.79 | 67.66 ± 0.75 | −1.41 ± 1.09 |

TABLE 29.—SELECTION EFFECT DURING SIX GENERATIONS OF LINE 19 OF THE PRINCESS BEANS (*From Johannsen*)

| Harvest years | Total number of beans | Mean weight of mother beans of the select strains | | Difference $b - a$ | Mean weight of progeny seeds of select strains | | Difference $b - a$ |
|---|---|---|---|---|---|---|---|
| | | $a$-minus | $b$-plus | | $a$-minus | $b$-plus | |
| 1902 | 219 | 30 | 40 | 10 | 35.83 ± 0.44 | 34.78 ± 0.38 | −1.05 ± 0.58 |
| 1903 | 200 | 25 | 42 | 17 | 40.21 ± 0.65 | 41.02 ± 0.43 | +0.81 ± 0.78 |
| 1904 | 590 | 31 | 43 | 12 | 31.39 ± 0.29 | 32.64 ± 0.21 | +1.25 ± 0.36 |
| 1905 | 1,657 | 27 | 39 | 12 | 38.26 ± 0.16 | 39.15 ± 0.17 | +0.89 ± 0.23 |
| 1906 | 1,367 | 30 | 46 | 16 | 37.92 ± 0.22 | 39.87 ± 0.16 | +1.95 ± 0.27 |
| 1907 | 594 | 24 | 47 | 23 | 37.36 ± 0.30 | 36.95 ± 0.21 | −0.41 ± 0.37 |

The data for pure line 13 contained in table 30 show a correlation coefficient of −0.018 ± 0.038, which is not significantly different from zero. This result should be compared with that obtained from table 31, which represents the parent-offspring correlation for seed weight in the mixed population consisting of all nineteen pure lines. Here the correlation coefficient, +0.336 ± 0.008, indicates a positive relationship between parent and offspring. The former value shows that no genetic difference exists among the individuals of a pure line. The latter result is due to the fact that the mixed material consisted of a number of distinct pure lines differing in mean seed weight. The positive correlation coefficient here obtained is not inconsistent with absence of correlation within each pure line as shown by table 27, which represents a segregation of this same material into its component pure lines.

TABLE 30.—PARENT-OFFSPRING CORRELATION IN PURE LINE 13 OF THE PRINCESS BEANS (*Data from Johannsen*)

| Seed weight of parents | Seed weight of offspring in centigrams: 17.5 | 22.5 | 27.5 | 32.5 | 37.5 | 42.5 | 47.5 | 52.5 | 57.5 | 62.5 | Total |
|---|---|---|---|---|---|---|---|---|---|---|---|
| 27.5 | .. | .. | 1 | 5 | 6 | 11 | 4 | 8 | 5 | .. | 40 |
| 32.5 | .. | .. | .. | 1 | 3 | 7 | 16 | 13 | 12 | 1 | 53 |
| 37.5 | .. | 1 | 2 | 6 | 27 | 43 | 45 | 27 | 11 | 2 | 164 |
| 42.5 | 1 | .. | 1 | 7 | 25 | 45 | 46 | 22 | 8 | .. | 155 |
| 47.5 | .. | .. | 5 | 9 | 18 | 28 | 19 | 21 | 3 | .. | 103 |
| 52.5 | .. | 1 | 4 | 3 | 8 | 22 | 23 | 32 | 6 | 3 | 102 |
| 57.5 | .. | .. | 1 | 7 | 17 | 16 | 26 | 17 | 8 | 3 | 95 |
| Total....... | 1 | 2 | 14 | 38 | 104 | 172 | 179 | 140 | 53 | 9 | 712 |

$r = -0.018 \pm 0.038$

TABLE 31.—PARENT-OFFSPRING CORRELATION IN A MIXED POPULATION OF PRINCESS BEANS (*Data from Johannsen*)

| Seed weight of parents | Seed weight of offspring in centigrams: 10 | 20 | 30 | 40 | 50 | 60 | 70 | 80 | 90 | Total |
|---|---|---|---|---|---|---|---|---|---|---|
| 20 | .. | 1 | 15 | 90 | 63 | 11 | .. | .. | .. | 180 |
| 30 | .. | 15 | 95 | 322 | 310 | 91 | 2 | .. | .. | 835 |
| 40 | 5 | 17 | 175 | 776 | 956 | 282 | 24 | 3 | .. | 2,238 |
| 50 | .. | 4 | 57 | 305 | 521 | 196 | 51 | 4 | .. | 1,138 |
| 60 | .. | 1 | 23 | 130 | 230 | 168 | 46 | 11 | .. | 609 |
| 70 | .. | .. | 5 | 53 | 175 | 180 | 64 | 15 | 2 | 494 |
| Total...... | 5 | 38 | 370 | 1,676 | 2,255 | 928 | 187 | 33 | 2 | 5,494 |

$r = +0.336 \pm 0.008$

**Genetic Significance of Pure Lines.**—Johannsen correctly ascribed these results to the homozygous condition present in the Princess beans, as a result of long-continued self-pollination. As a matter of fact, Mendel first showed that starting with a hybrid *Aa*, by continued self-pollination in successive generations the population soon comes to consist of equal proportions of *AA* and *aa* plants with only a negligible proportion of *Aa* remaining. Mathematically expressed, as the student may readily verify for himself, after *n* generations of self-fertilization the proportions of the various genotypes will be as follows:

$$AA = \frac{2^n - 1}{2^{n+1}} \qquad Aa = \frac{1}{2^n} \qquad aa = \frac{2^n - 1}{2^{n+1}}$$

The proportion of heterozygous individuals decreases one-half in each successive generation. After six generations only 1.6 per cent of heterozygous individuals remain; and after ten generations, less than one-tenth of 1 per cent.

For one pair of factors the total proportion of homozygotes $AA + aa$ is evidently $\frac{2^n - 1}{2^n}$. If the original hybrid is heterozygous for a number of factors, each of them will follow the same formula. If there are $m$ such pairs of factors, the proportion $x$ of completely homozygous plants after $n$ generations of self-fertilization will be given by the equation:

$$x = \left(\frac{2^n - 1}{2^n}\right)^m$$

This formula does not take into account linkage or selective survival, but it gives a basis of judgment as to the effect of long-continued self-pollination in a population with numerous factor differences. With ten pairs of independent factors and five generations of self-fertilization, 73 per cent of the population will be completely homozygous; and even with 100 pairs of independent factors, obviously more than can exist in most animal and plant species as evidenced by their chromosome numbers, and ten generations of self-pollination no less than 91 per cent of the individuals will be completely homozygous. Since linkage may be demonstrated mathematically to hasten reduction to a homozygous condition, it is evident that in even the most extreme cases, self-pollination will rapidly reduce the population to a homozygous condition.

The end result will be a homozygous but not a homogeneous population. A homozygous population consists entirely of homozygous individuals; while a homogeneous population is one which consists of genetically identical individuals. The population will not be homogeneous because it will contain many different kinds of homozygous plants. If there is only a single pair of factors, there will be two kinds of homozygous individuals $AA$ and $aa$; if, however, there are $m$ pairs of factors, there will be $2^m$ different kinds of homozygous individuals. If $m = 10$, then there will be $2^{10} = 1{,}024$ different kinds of homozygotes; if $m = 20$, not an inconceivable figure, over a million kinds will be possible.

**Clones.**—In agricultural operations it is a common practice to propagate many plant varieties by asexual methods. Thus potatoes are propagated from tubers, sugar cane from offsets, grapes from cuttings, many herbaceous plants from divisions of the crown, fruit trees and ornamental shrubs and trees by budding, grafting, etc., lilies, daffodils, tulips, etc., from bulbs, and other methods of asexual propagation immediately present themselves. A variety propagated asexually from a single original individual is known as a clone.

There are certain other methods of reproduction which are equivalent genetically to the foregoing methods of asexual propagation. In citrus fruits, in the avocado, and in other less familiar cases a single seed may produce several seedlings of which only one, and sometimes none, is the product of sexual reproduction. The others arise from adventitious embryos having their origin in somatic cells of the mother plant, a phenomenon known as polyembryony. Obviously, polyembryony is a form of vegetative multiplication, although it may not occur except under the stimulus of pollination. In certain species the production of embryos from some vegetative cell in the mother or from unreduced egg cells is not uncommon. Usage is somewhat confused as to terminology in these instances; but the essential feature from the genetic point of view, among the higher plants at least, is that these species produce seeds without fertilization from diploid cells of the mother plant, a phenomenon known in general as apogamy. The stimulus of pollination may be necessary for development in some instances, but no true fertilization takes place. Offspring thus produced are identical genetically with the mother plant.

Apogamy is apparently not a common phenomenon among plants habitually grown from seeds; but it has been demonstrated in certain species. Thus in many species of Hieracium, the hawkweed, most of which are exceedingly variable in nature, seed progeny uniformly reproduce the mother type; and viable seeds are produced even when pollination is prevented. Apogamy has also been reported in Alchemilla, Antennaria and Taraxacum, the last being fairly close to Hieracium; but in Crepis, a genus delimited from Hieracium with difficulty, apogamy has not been discovered in over fifty species under investigation. Certain wild roses, particularly those of the *Caninae* group, have been found to produce apogamous seed and the same phenomenon probably also occurs in blackberries and raspberries, as shown by attempts at hybridization which yield maternal plants either exclusively or in part.

In animals, especially among lower forms, parthenogenesis is not uncommon. Diploid parthenogenesis, development of unreduced egg cells without fertilization, occurs naturally in aphids and some other insects. In most of the animal forms with which agriculturists have to deal, however, reproduction is restricted to the normal sexual process.

All the forms of reproduction which have been described above are equivalent to vegetative reproduction. They give rise to clones which are genetically homogeneous but not necessarily homozygous; that is, the individuals are all identical genetically, but they may be and usually are highly heterozygous. Thus a variety of apple, say the Ben Davis, propagated by grafting, comes true to type, but if seedlings are grown they are highly diverse. Reproduction by sexual means is necessary in clones in order to determine the genetic constitution.

**Variation in Clones.**—A clone is equivalent to a pure line in its genetic homogeneity, but variation occurs in clones, just as in pure lines, particularly in respect to quantitative features. The different trees of a given variety of apples will differ widely in size, in fruit production, and sometimes even in habit, but these differences are usually entirely due to environmental influences, of which soil conditions and differences in the stocks upon which the scions are grafted have the greatest effect. This fact may be demonstrated by growing comparatively large numbers of trees from mother trees which have differed greatly in yield. Such experiments have shown no significant differences in yield between trees grown from scions selected from high-yielding trees and those grown from scions taken from low-yielding trees of the same variety.

The most satisfactory demonstration of the absence of genetic differences in clones has been obtained in aphids. The oat aphid, *Aphis avenae*, reproduces during the summer entirely by diploid parthenogenesis, and many generations may be obtained in a comparatively short time. Ewing conducted extensive selection experiments for a number of variable characters such as length of cornicles, length of antennae, and body length. The results of selection for body length are shown in figures 74 and 75. In each generation the largest individuals were selected as parents for the next, and length of body of their progeny was compared with that of unselected progenies. In each generation mean body length of progeny from selected individuals did not differ significantly from that of the unselected stock. Regression was complete as in pure lines; and the conclusion was reached that the variability in these clones was entirely developmental. A considerable portion of it was found to depend upon fluctuations in temperature. When temperature was held constant, less marked fluctuation in body length occurred but selection was still ineffective as a means for establishing a race with a greater body length than that normal for the unselected stock.

In some instances an apparent diversification into distinct lines has been demonstrated in clones. In certain varieties of potatoes high- and low-yielding lines have been established by hill selection. A more thorough study of these lines, however, has shown that the differences are due to obscure diseases, such as mosaics, which become established in certain lines and are transferred to the next generation through infected tubers. Although such cases may simulate inheritance, they are not really exceptional, for disease effects merely represent one of the sources of modification.

**Artificial Production of Pure Lines.**—Pure lines exist normally in self-pollinated plants as a consequence of automatic reduction to a homozygous condition in such species. Consequently in peas, beans, barley, oats, wheat, and other plants normally self-pollinated, the production of pure lines is readily accomplished by simply growing the progeny of selected individuals separately. In normally cross-pollinated plants, however, the individuals are prevailingly highly heterozygous, but if it is possible to subject them to self-pollination for a number of consecutive generations, pure lines may be produced from them. Maize, for example, is almost wholly cross-pollinated under normal conditions; but by protecting the silks and applying pollen collected under bag from the

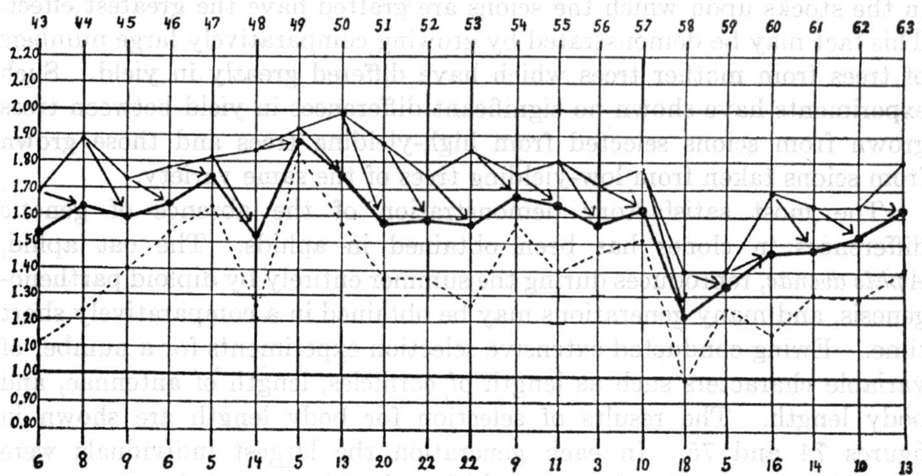


FIGURE 74.—Chart illustrating the effects of selection for increased body length during twenty generations in the oat aphis. The heavy line represents the fraternal mean; the light solid line, fluctuations of the largest variant; the light broken line, fluctuations of the shortest variant. Temperature uncontrolled. (*After Ewing from Morgan, The Physical Basis of Heredity, copyright* 1919 *by J. B. Lippincott Company. Reprinted by permission.*)

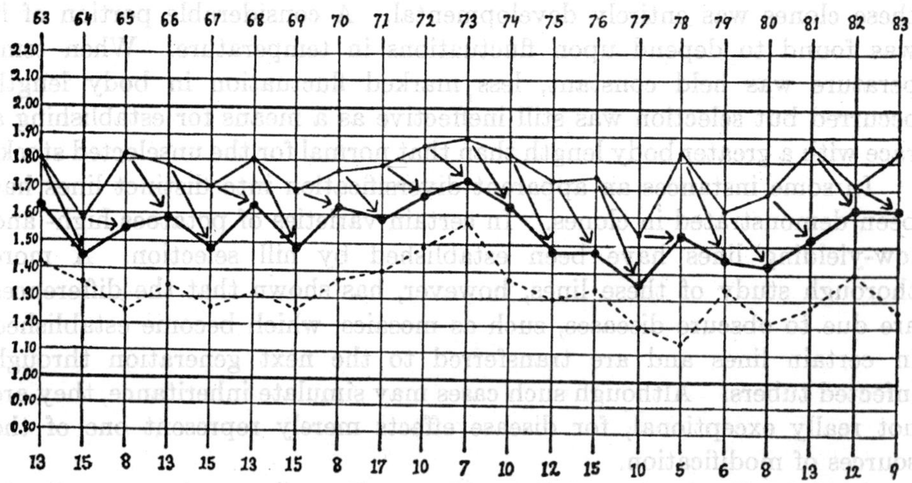


FIGURE 75.—Continuation for twenty succeeding generations of selection experiment illustrated in figure 74. Temperature controlled at 65°F. (*After Ewing from Morgan, The Physical Basis of Heredity, copyright* 1919 *by J. B. Lippincott Company. Reprinted by permission.*)

same plant, it is possible to self-pollinate it, and to create artificially the conditions necessary for the production of pure lines.

Maize has been studied so extensively from this point of view that it affords an excellent illustration of the phenomena which attend application of the pure-line conception to cross-pollinated species. When individual plant selections are made in any commercial variety and self-pollination is practiced thereafter continually within them, very characteristic results are secured. The most notable effect, perhaps, is a rapid and conspicuous decline in vigor and productiveness. The rate of decline is most rapid in the first few generations, then becomes less marked, and finally after seven or eight generations ceases altogether. The decline in vigor is accompanied by a differentiation in morphological characters, so that the different inbred lines become distinct in this respect. Moreover, the amount of decline in vigor and productiveness is not the same in different inbred lines; but each ultimately reaches an equilibrium at a point which is just as characteristic for it as its distinctive morphological characters. Lines thus established exhibit the characteristic uniformity in every respect of pure lines.

**Vigor and Productivity.**—The ultimate uniformity of these inbred lines reasonably may be ascribed to the fact that they become homozygous through enforced self-pollination. This result is to be expected on the theoretical grounds amplified in an earlier portion of this chapter. But whence comes the reduction in vigor and productiveness? At one time this latter feature was ascribed to inbreeding. Inbreeding was considered of itself a harmful process. But this explanation is hardly consistent with conditions in normally self-pollinated plants which are vigorous despite long periods of self-pollination. The correct explanation can be arrived at only by considering in detail phenomena attending the reduction in vigor and fertility. The inbred lines during the early generations produce many defective, abnormal individuals. Some produce albinos, which by reason of their inability to synthesize starch die in the seedling stage. Dwarf plants and plants with defective anthers or pistils also appear. These defectives segregate out in ratios characteristic of Mendelian recessives; and some have been studied and shown to depend upon such factors. The conclusion is reached, therefore, that individuals of the general variety are heterozygous for numerous recessive factors which have an adverse effect upon vigor and fertility and that some of them become homozygous in the inbred lines. The decline in vigor and productiveness is, therefore, purely a Mendelian phenomenon dependent upon the original genetic constitution of the crossbred variety; and it reaches an equilibrium after seven or eight generations because the inbred lines have then become homozygous. Theoretically it might be possible to produce an inbred strain with the vigor and productiveness of the original crossbred variety; but practically it is exceedingly difficult to do so because of the number of adverse recessive factors which exist in it and because of linkage between them and favorable factors. This matter is treated more fully in Chapter XXXII.

This conclusion is further verified by results of crossing different inbred strains together. Jones, for example, crossed two such strains together, one which had been inbred seventeen successive generations and the other sixteen. One parent gave an average yield of 19.5 bushels per acre and the other 19.6 over a 6-year period. The $F_1$ produced an average of 101.2 bushels per acre. Many other crosses have confirmed these results. The $F_1$ from two inbred strains usually shows an immediate return to the vigor and fertility of the original variety, and it may even exceed it in these respects. Subjection of such an $F_1$ to self-pollination repeats the course of events shown in self-pollination of selections from the original variety, as shown by the data in table 32.

TABLE 32.—THE PRODUCTION OF GRAIN IN BUSHELS PER ACRE OF TWO INBRED STRAINS OF MAIZE AND THEIR HYBRID, FROM THE $F_1$ TO THE $F_8$ GENERATION, SUCCESSIVELY SELF-FERTILIZED (*Data from Jones*)

| Year grown | Generations | | | | | | | | | |
|---|---|---|---|---|---|---|---|---|---|---|
| | $P_1'$ | $P_1''$ | $F_1$ | $F_2$ | $F_3$ | $F_4$ | $F_5$ | $F_6$ | $F_7$ | $F_8$ |
| 1917 | 5.5 | 21.5 | 64.5 | 56.0 | | | | | | |
| 1918 | 23.5 | 26.9 | 120.6 | 128.4 | 15.1 | | | | | |
| 1920 | 27.8 | 16.2 | 127.6 | 47.6 | 34.8 | 29.0 | 9.9 | | | |
| 1921 | 13.1 | 19.6 | 72.8 | 54.5 | 49.2 | 32.7 | 15.3 | 23.1 | | |
| 1922 | 26.1 | 20.0 | 160.4 | 83.2 | 73.7 | 67.5 | 48.8 | 35.7 | 22.7 | |
| 1923 | 21.0 | 13.2 | 61.0 | 45.0 | 40.5 | 47.0 | 15.8 | 23.0 | 26.2 | 27.2 |
| Mean..... | 19.5 | 19.6 | 101.2 | 69.1 | 42.7 | 44.1 | 22.5 | 27.3 | 24.5 | 27.2 |

Now if the reduction in vigor and fertility of inbred lines depends upon production of Mendelian combinations homozygous for harmful recessive factors which existed, mostly masked by the favorable dominant allelomorphs, in the original population, then obviously the reduction in vigor in different inbred strains would depend upon different combinations of these recessive factors. When, therefore, two such strains are crossed. the harmful recessive factors of one will be largely masked by favorable dominant allelomorphs from the other and *vice versa*. In other words the vigor and fertility of $F_1$ is a consequence of interaction among a series of dominant complementary factors, some introduced from one parent and others from the other, a conclusion in harmony with the preliminary remarks on hybrid vigor contained in the preceding chapter. A purely Mendelian explanation is sufficient to account for all the results secured from these extensive and illuminating investigations.

The pure-line method of analysis has produced similar results when applied to other normally cross-fertilized species. In almost every instance, reduction in vigor and fertility has occurred as a result of enforced inbreeding. Wild species normally cross-pollinated exhibit the same phenomena. Collins has inbred strains of *Crepis capillaris* and has demonstrated loss of vigor and fertility in consequence, and crosses between such inbred strains exhibited vigor and fertility comparable to that of normally crossbred populations. Baur has found that wild species of Antirrhinum as well as the cultivated garden snapdragon follow the same course of events. It has been suggested, and observation seems to support the idea, that the reduction in vigor and fertility is roughly proportional to the degree of cross-pollination occurring in the species.

**Pure Lines in Animals.**—Normally, of course, pure lines do not exist in animals, the separation of the sexes promoting a degree of cross-breeding equivalent under ideal conditions to random mating. As a consequence unselected animal populations may be expected to exhibit genetic variability and heterozygosis comparable to that in cross-pollinated plants. It is possible, however, by practicing brother-sister matings over a long period of successive generations, to obtain strains of animals homozygous for their genetic factors. Considering the problem purely from a mathematical point of view, seventeen generations of brother-sister mating are approximately equivalent in effectiveness to six generations of self-fertilization in attainment of a homozygous condition. Farther degrees of relationship up to first cousins likewise tend gradually to increase the proportion of homozygosis, but less rapidly in proportion to the degree of remoteness of the relationship.

The results which have been secured by the application of this method to animals are well illustrated by Wright's analysis of guinea-pig experiments conducted by the Bureau of Animal Industry of the U. S. Department of Agriculture. Sixteen families were established by brother-sister matings and continued for thirty successive generations. Most of them showed marked reduction in vigor and fertility as compared with the crossbred control stock. They were not, however, equal in these respects. All the families were remarkably uniform. The results, as with maize, demonstrate that there are many factor differences affecting vigor, fertility, and vitality, and that different combinations of them become fixed by inbreeding. $F_1$ hybrids from inbred parents exhibited marked increase in these respects comparable to that secured in maize, so that these two sets of experiments are in agreement in every respect. They indicate clearly that normally crossbred species, whether animal or plant, consist of individuals prevailingly heterozygous for numerous factors, among them many recessive factors which have a detrimental effect when homozygous. Dominant factors of this type are not to be expected as normal permanent constituents of such populations; because they would be eliminated by selection; but harmful recessives may persist for a long time masked by the corresponding favorable dominant allelomorphs. The effects produced by resorting to inbreeding are purely a consequence of the constitution of the original crossbred stock.

**Fluctuating Variability.**—Investigations on pure lines demonstrate that developmental variations under ordinary conditions are distributed in the form of a normal curve. The form of this curve does not differ from that of the distribution of variations determined by numerous Mendelian factors having minute effects; but in the latter case the variations are heritable, and in the former they are not. As a matter of fact both forms of variation are ordinarily present in populations; it is only in pure lines and clones that, by reason of genetic homogeneity, only developmental variation occurs.

A moment's consideration will disclose the reason for distribution of developmental variations in the form of a normal curve. Under ordinary conditions the environmental features which affect individual development are numerous and have relatively slight effects. In a plant population, for example, the soil varies slightly in fertility, texture, etc., from place to place in the field; the moisture conditions also differ; spacing of plants, exposure to light, insect and fungus injury, etc., are also unequal, and so for numerous other external agents. A given plant is affected favorably by some of these features and unfavorably by others; and its final development represents the balance struck by these various effects. Only very rarely are all the contributing conditions which affect a plant uniformly favorable or unfavorable. From this point of view the distribution of environmental effects, assuming that they are numerous and each comparatively slight, may again be represented by the binomial $(\frac{1}{2} + \frac{1}{2})^n$, which on expansion gives a normal curve.

The plus and minus variation exhibited by individuals in response to environment may appropriately be termed fluctuating variability, and particular expressions may be called fluctuations. These terms have, however, also been so employed as to include all quantitative variations which exhibit a normal distribution, whether due to genetic factors or environmental effects or both. Its restriction to developmental variations which exhibit a normal distribution, as suggested above, would be highly desirable. Fuctuating variability in this sense is important in practical operations, principally because it may easily be confused with Mendelian variation unless proper methods are adopted to discount its effects.

**Composition of Populations.**—In the preceding chapter it was shown that the distribution of factors in a population mating at random may be represented by a continued expression of the following type:

$$(m_1^2AA + 2m_1n_1Aa + n_1^2aa)(m_2^2BB + 2m_2n_2Bb + n_2^2bb)(m_3^2CC + 2m_3n_3Cc + n_3^2cc) \dots$$

In such a population most of the individuals are obviously highly heterozygous. In a self-fertilized species, however, it has been shown that the population consists of a mixture of pure lines; and, following the same notation, the distribution of factors under such circumstances may be represented thus:

$$(m_1AA + n_1aa)(m_2BB + n_2bb)(m_3CC + n_3cc) \ . \ . \ .$$

If, as before, the factors are numerous and have slight effects, this expression for all practical purposes reduces to the binomial $(\frac{1}{2} + \frac{1}{2})^n$, which, together with the fluctuating variability exhibited by each pure line, accounts for the normal distribution shown by mixtures of pure lines.

Random mating and self-fertilization represent two extreme methods of reproduction. In view of their important genetic consequences, it is convenient to distinguish species reproducing mainly by cross-fertilization as allogamous; those reproducing solely by self-fertilization, or nearly so, as autogamous. Animal forms in general belong to the allogamous category by reason of separation of sexes. Among plants, all diœcious species, such as asparagus, hemp, and hops; all self-sterile species or varieties such as rye, radishes, and many crucifers (not invariably, however); monœcious plants, such as maize, squashes, etc.; wind-pollinated plants; in short, all species which are prevailingly cross-pollinated belong to the allogamous group. The autogamous category includes such important agricultural species as barley, oats, wheat, and most legumes, all of which are prevailingly fertilized before the flower opens. In nature, however, most allogamous plants may be self-pollinated to a limited extent; most autogamous plants do cross occasionally; and all intergradations between the two extreme conditions are represented. Under such circumstances the composition of the population lies somewhere between the two extremes indicated above, depending on the relative proportion of self- and cross-pollination. The importance of this distinction from a practical standpoint is considered in detail in other chapters; here it is sufficient to point out that individual selection in autogamous species immediately establishes uniform races, because it isolates pure lines; whereas a long period of selection is necessary in allogamous species, because of the heterozygous constitution of the individuals.

It is further necessary to distinguish between the theoretical composition of a population of an autogamous species and its actual composition. By reason of the enormous number of pure lines which are possible with a large number of factors, only a comparatively small proportion of them will be represented in any given population, and by mere operation of the laws of chance some will be eliminated from time to time. If any of them are handicapped in any way, by reason of low viability, low fertility, etc., they will tend to be eliminated in the course of time, and the population eventually may come to consist of a comparatively small number of those pure lines which are best fitted for given condition. Despite these tendencies toward elimination, however, the actual number of pure lines in such populations is usually large.

**References**

WALTER: Genetics, Chapter VI.
See general texts.

## CHAPTER XXI

## SELECTION

In general, animal and plant improvement has been effected by selection. That is to say, when an animal or a plant breeder desires to improve his forms, experience has shown him that the logical method of procedure lies in choosing in each generation for breeding stock those individuals which approximate most closely to his ideal. This practice continued consistently generation after generation almost invariably leads to a marked improvement in the desired direction and to increasing uniformity. It is this method which has led to the differentiation and establishment of the numerous breeds of animals and varieties of plants which are now utilized by farmers. Selection is responsible for the high-producing dairy cattle of the present times, for the speed of our race horses, the strength of the draft horse, the high egg-laying capacity of our breeds of fowls, the wool production of sheep, etc. Selection has transformed the common beet with an average sugar content of about 10 per cent into the sugar beet with an average field run of 15 per cent, and an average content of over 20 per cent in highly selected stocks. It has lengthened the fiber of cotton and increased its strength, and it has operated in many other ways to augment the value of animal and plant forms for practical and ornamental purposes. These practical results have built up a great body of evidence on the effectiveness of selection in molding the characters of animals and plants to man's uses.

In the domain of nature a similar condition is supposed to prevail. The Darwinian theory of evolution holds that species have become gradually modified through the course of ages by natural selection; that is to say, certain individuals which have proved best fitted to survive live to produce offspring, and those less fitted are eliminated in the struggle for existence. If those which survive differ genetically in some respect from those which fail, a gradual modification in a more adaptable direction may be expected, a process analogous to the selection which occurs in agricultural operations, as Darwin so clearly pointed out.

In addition to the body of observational evidence which has been accumulated from studies of evolution and from agricultural practice, a large number of experiments have been performed in order to determine the extent to which selection is effective and the reason for its effectiveness. It is this last point which is of most interest at this time, and it is to this aspect of the problem that the present chapter is largely devoted.

**Chemical Composition in Maize.**—One of the most famous series of selection experiments is that of the Illinois Station on chemical composition in maize. These experiments arose from the observation, originally made by Hopkins, that ears of corn taken from an established variety and analyzed individually differed in chemical composition, particularly in oil and protein content. As a foundation stock for the experiment, 163 ears of Burr's white corn were individually analyzed for oil and protein content with the results shown in table 33. Protein content varied from 8 to 14 per cent, and oil from 3.8 to 6.0 per cent. The average protein content was 10.93 per cent; oil 4.69 per cent.

TABLE 33.—DISTRIBUTION OF PROTEIN AND OIL, IN 163 EARS OF BURR'S WHITE CORN WITH WHICH THE ILLINOIS STATION MAIZE-BREEDING INVESTIGATIONS WERE STARTED (*Data from L. H. Smith*)

| Oil, per cent \ Protein, per cent | 8.0 | 8.5 | 9.0 | 9.5 | 10.0 | 10.5 | 11.0 | 11.5 | 12.0 | 12.5 | 13.0 | 13.5 | 14.0 | |
|---|---|---|---|---|---|---|---|---|---|---|---|---|---|---|
| 3.8 | .. | .. | .. | .. | .. | .. | .. | 1 | .. | .. | .. | .. | .. | 1 |
| 4.0 | .. | .. | 2 | 2 | .. | .. | 2 | 3 | 1 | 1 | 1 | .. | .. | 12 |
| 4.2 | .. | .. | .. | 3 | 2 | 3 | 3 | 2 | 3 | .. | 2 | .. | .. | 18 |
| 4.4 | .. | .. | .. | 3 | 4 | 2 | 7 | 3 | 2 | .. | .. | .. | .. | 21 |
| 4.6 | .. | .. | 2 | 1 | 1 | 8 | 5 | 8 | 4 | 3 | .. | .. | .. | 32 |
| 4.8 | 1 | 1 | .. | 5 | 6 | 3 | 7 | 2 | 4 | 5 | .. | .. | .. | 34 |
| 5.0 | .. | 1 | .. | 2 | 1 | 3 | 3 | 5 | 4 | .. | 1 | .. | .. | 20 |
| 5.2 | .. | .. | .. | 2 | 3 | 2 | 2 | 1 | .. | 3 | 1 | .. | .. | 14 |
| 5.4 | .. | .. | .. | .. | .. | 2 | .. | .. | .. | .. | .. | .. | .. | 2 |
| 5.6 | .. | .. | .. | 1 | 1 | 1 | 1 | .. | 1 | .. | 1 | .. | .. | 6 |
| 5.8 | .. | .. | .. | .. | .. | .. | 1 | .. | .. | .. | .. | .. | 1 | 2 |
| 6.0 | .. | .. | .. | .. | 1 | .. | .. | .. | .. | .. | .. | .. | .. | 1 |
| | 1 | 2 | 4 | 19 | 19 | 24 | 31 | 25 | 19 | 12 | 6 | .. | 1 | 163 |

Selection was carried on in four directions: (1) high protein, (2) low protein, (3) high oil, (4) low oil. In each generation twenty-four ears were selected, mainly on the basis of chemical composition, but with some secondary consideration of yield and morphological type, to perpetuate each of the four selection lines. The selected ears were planted in isolated plots, a row from each ear, and after the ninth generation alternate rows were detasseled in order to prevent too close inbreeding. The methods have been varied somewhat during the course of the experiments, but they have remained essentially the same so far as the problem at issue is concerned.

The results which have been secured are shown in table 34. In the protein selection strains, starting with an average content of 10.93 per cent, the high selection line has gradually moved up to an average of over 15 per cent, and the low line has decreased to an average of approximately 7.5 per cent, about half as great as that of the high line. Despite the existence of marked seasonal fluctuations, there has been a steady progress in the direction of selection as is shown by the increasing divergence of the two selection lines. The results for oil content have been more striking and less subject to seasonal fluctuations. The high line average has risen from 4.69 per cent to 9.35, and the low line value has decreased to 1.87 per cent, approximately a fifth of that of the high line after a period of 23 years. In both the protein and oil series, it is to be noted that selection has established strains which have average values for the entire crop lying outside the limits of the range of values in the foundation stock and while the progress has been less marked in the latter years of the experiments, there is evidence that the limit has not yet been reached.

TABLE 34.—RESULTS OF SELECTING MAIZE FOR HIGH AND LOW PROTEIN AND HIGH AND LOW OIL AT THE ILLINOIS EXPERIMENT STATION. AVERAGE PER CENT OF PROTEIN AND OIL IN EACH GENERATION (*Data from L. H. Smith*)

| Year | High protein | Low protein | Difference | High oil | Low oil | Difference |
|---|---|---|---|---|---|---|
| 1896 | 10.93 | 10.93 | 0.00 | 4.69 | 4.69 | 0.00 |
| 1897 | 11.10 | 10.55 | 0.55 | 4.73 | 4.06 | 0.67 |
| 1898 | 11.05 | 10.55 | 0.50 | 5.15 | 3.99 | 1.16 |
| 1899 | 11.46 | 9.86 | 1.60 | 5.64 | 3.82 | 1.82 |
| 1900 | 12.32 | 9.34 | 2.98 | 6.12 | 3.57 | 2.55 |
| 1901 | 14.12 | 10.04 | 4.08 | 6.09 | 3.43 | 2.66 |
| 1902 | 12.34 | 8.22 | 4.12 | 6.41 | 3.02 | 3.39 |
| 1903 | 13.04 | 8.62 | 4.42 | 6.50 | 2.97 | 3.53 |
| 1904 | 15.03 | 9.27 | 5.76 | 6.97 | 2.89 | 4.08 |
| 1905 | 14.72 | 8.57 | 6.15 | 7.29 | 2.58 | 4.71 |
| 1906 | 14.26 | 8.64 | 5.62 | 7.37 | 2.66 | 4.71 |
| 1907 | 13.89 | 7.32 | 6.57 | 7.43 | 2.59 | 4.84 |
| 1908 | 13.94 | 8.96 | 4.98 | 7.19 | 2.39 | 4.80 |
| 1909 | 13.41 | 7.65 | 5.76 | 7.05 | 2.35 | 4.70 |
| 1910 | 14.87 | 8.25 | 6.62 | 7.72 | 2.11 | 5.61 |
| 1911 | 13.78 | 7.89 | 5.89 | 7.51 | 2.05 | 5.46 |
| 1912 | 14.48 | 8.15 | 6.23 | 7.70 | 2.17 | 5.53 |
| 1913 | 14.83 | 7.71 | 7.12 | 8.15 | 1.90 | 6.25 |
| 1914 | 15.04 | 7.68 | 7.36 | 8.29 | 1.98 | 6.31 |
| 1915 | 14.53 | 7.26 | 7.27 | 8.46 | 2.07 | 6.39 |
| 1916 | 15.66 | 8.68 | 6.98 | 8.50 | 2.08 | 6.42 |
| 1917 | 14.44 | 7.08 | 7.36 | 8.53 | 2.09 | 6.44 |
| 1918 | 15.48 | 7.31 | 8.17 | 9.35 | 1.87 | 7.48 |

Although these four strains have been derived from the same foundation stock, they have become distinctly different in morphological features. This result of the experiments is analogous to the distinctive types of conformation which have been obtained in race horses, as a result of selection for speed; in dairy cows, as a result of selection for milk production; and in fowls, as a result of selection for high egg production. So in maize it may be shown that the differences in morphological features of the four strains bear a more or less close relation to features of chemical composition for which selection has been practiced. All the strains are less productive than the original variety, so much so that they are of doubtful value for commercial planting (table 35).

TABLE 35.—DATA ON PRODUCTION IN BUSHELS PER ACRE OF SELECTED STRAINS OF MAIZE COMPARED WITH THE UNSELECTED VARIETIES (*After East and Jones*)

| Year | High protein | Low protein | High oil | Low oil | Control |
|---|---|---|---|---|---|
| 1913 | 30.2 | 35.5 | 31.0 | 23.9 | 39.6 |
| 1914 | 36.2 | 43.3 | 37.4 | 48.7 | 55.2 |
| 1915 | 42.4 | 57.2 | 45.2 | 49.9 | 53.5 |
| 1916 | 14.6 | 29.6 | 16.7 | 19.8 | 28.2 |
| 1917 | 48.9 | 56.3 | 55.9 | 51.3 | 63.9 |
| 1918 | 38.8 | 47.8 | 46.6 | 58.2 | 62.8 |
| Average..... | 35.2 | 45.0 | 38.8 | 42.0 | 50.0 |

It is impossible from these data alone to state with assurance what the effects of selection depend upon, but a highly probable explanation can be presented for them, when all circumstances attending the experiments are taken into account, and particularly when the results of other investigations, conducted from a different point of view, are compared with them. An explanation essentially in accord with Mendelian principles starts from the assumption that the original, highly heterogeneous population consisted of individuals heterozygous for various combinations of factors affecting protein and oil content, and that selection has gradually sorted out those combinations most favorable for the expression of the desired type.

Perhaps there is some evidence in favor of this explanation from the analysis which Surface made of the pedigrees of ears chosen after ten generations of selection for planting the eleventh generation. He showed that in the high-protein line all twenty-four were descended from one of the original high-protein ears; in the low-protein line, only two of the original ears were represented; in the high-oil line three; in the low-oil line, two. Not much weight can, however, be placed upon this analysis in the absence of computations of the mathematical probabilities of survival on chance grounds alone, for obviously much elimination would occur on this account. Moreover, this analysis disregards paternal lines in the ancestry which were manifestly not controlled in these experiments.

East and Jones, however, have shown, especially in the case of protein content, that although the conditions of inheritance are exceedingly complex and protein content is peculiarly susceptible to environmental effects, the logical treatment points to a Mendelian explanation of the kind stated above. This conclusion is particularly well substantiated by certain experiments which they performed on selection for high protein content in self-fertilized lines. Their studies of inheritance of protein content definitely showed that it must depend upon a large number of Mendelian factors, a few of which might have comparatively great results. Moreover, they showed that selection in self-fertilized lines led to a much more rapid increase in protein content than that shown in the Illinois experiments. After 5 years a Stadtmueller's Leaming strain reached a high level of 17.07 per cent, as compared with a maximum of 15.66 in the Illinois experiments in 23 years. An Illinois high-protein strain derived from the sixteenth generation of those experiments gave rise after five generations of self-fertilization with selection to a high level of 18.69 per cent. These experiments show that the method of self-fertilization, which rapidly reduces populations to a homozygous condition, combined with selection, is a much more effective means of establishing high-protein lines than simple mass selection without control of pollination. Moreover, they also show that the Illinois high-protein strain is still heterozygous for factors for protein content, which may be quickly reduced to a homozygous condition by a few generations of self-pollination.

East and Jones also showed that crosses between different high inbred strains gave rise to comparatively vigorous hybrids of fair yielding capacity which still retained a comparatively high protein content. One such cross averaged 15.33 per cent protein, another 14.66, and others ranged between these values and 13.71, the lowest average percentage. Selfed ears from these populations were grown and crossed, whereupon one lot averaged 15.81, another 14.99, and others lower. From these, by selfing and simple selection, four lots averaging 17.68, 17.25, 15.84, and 16.22 per cent of protein were secured. One selfed ear gave 20.14 per cent of protein. These results indicate that it is possible to combine factors for high protein from different inbred lines, thereby further increasing protein content. The results are in accord with Mendelian expectation, and they show how it is possible to apply Mendelian principles and secure a more rapid approach to the desired ideal.

**The Hooded Pattern in Rats.**—Castle has shown that the hooded pattern in rats behaves as a simple recessive in crosses with self-colored or Irish pattern (white belly). Self-colored crossed with hooded gives self-colored $F_1$ and 3 self-colored: 1 hooded in $F_2$. Castle further found that the hooded character is highly variable in extent of pigmentation as shown in figure 76. Some of the rats have the pigment confined to the head; others show an extension of a stripe of variable width down the back.

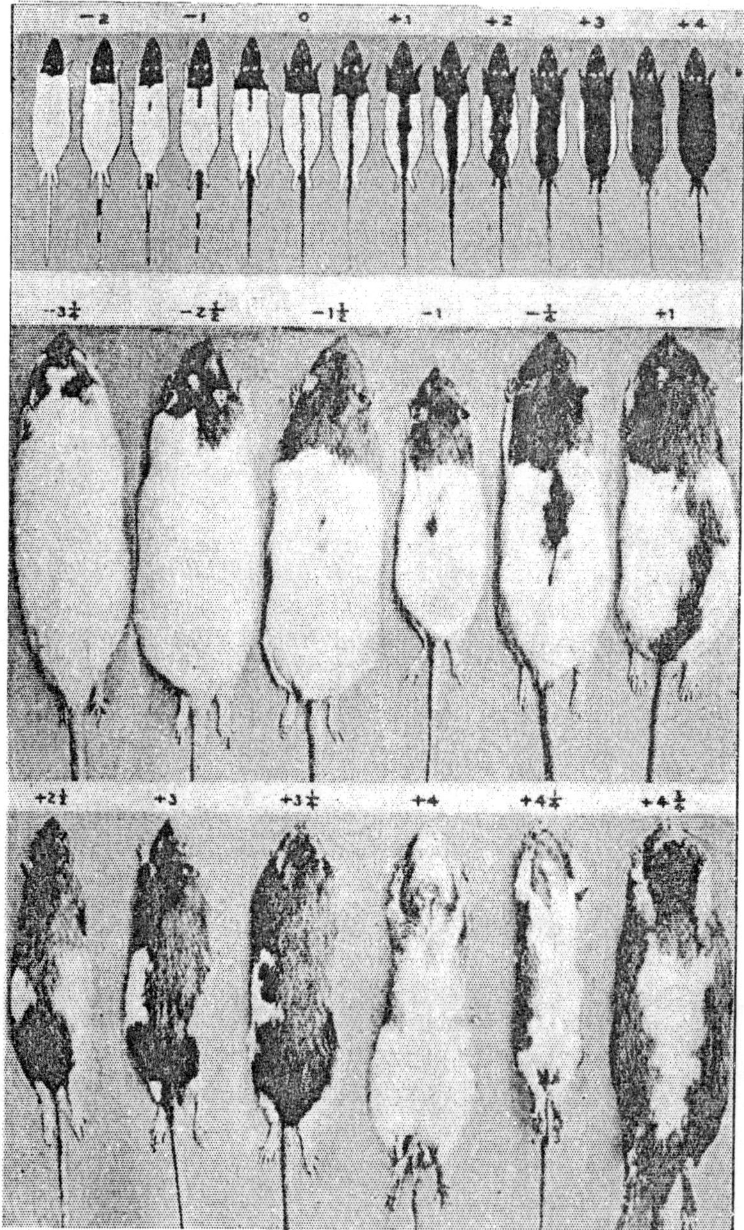


FIGURE 76.—Top row, a set of arbitrary rades used in the classification of hooded rats. Middle and bottom rows, skins of rats graded as indicated by the numerals above each skin. The animals +4, +4¼, +4¾, being entirely dark above, are shown in ventral view. (*From Castle and Phillips.*)

On the basis of these plus or minus variations in the degree of pigmentation, Castle conducted a series of experiments designed in the minus direction to decrease the pigmented area and in the plus direction to increase it. The plan of the experiments was comparatively simple. An arbitrary scale of values shown in figure 76 was made for the purpose of recording expeditiously the degree of pigmentation of an animal. A variable population was used as the foundation stock. In the plus series individuals were selected for increased pigmentation; in the minus direction for decreased area of pigmentation. Selection was continued for twenty successive generations and comparatively large numbers of individuals were grown; a total of 17,716 in the minus series and 17,250 in the plus. In the minus series the average grade of pigmentation decreased from −1.00 to −2.74 in the twenty-first generation. On Castle's scale of values −2.74 represents an animal with only slight pigmented areas confined to the head. In the plus series, on the other hand, the degree of pigmentation increased from 2.05 to 4.61. Individuals graded as 4.61 have the unpigmented area confined to a small

TABLE 36.—RESULTS OF PLUS AND MINUS SELECTION IN HOODED RATS CONTINUED THROUGH TWENTY SUCCESSIVE GENERATIONS (*Data from Castle*)

| Generation | Plus selection | | | Minus selection | | |
|---|---|---|---|---|---|---|
| | Mean grade of parents | Mean grade of offspring | Number of offspring | Mean grade of parents | Mean grade of offspring | Number of offspring |
| 1 | 2.51 | 2.05 | 150 | −1.46 | −1.00 | 55 |
| 2 | 2.52 | 1.92 | 471 | −1.41 | −1.07 | 132 |
| 3 | 2.73 | 2.51 | 341 | −1.56 | −1.18 | 195 |
| 4 | 3.09 | 2.73 | 444 | −1.69 | −1.28 | 329 |
| 5 | 3.33 | 2.90 | 610 | −1.73 | −1.41 | 701 |
| 6 | 3.52 | 3.11 | 861 | −1.86 | −1.56 | 1,252 |
| 7 | 3.56 | 3.20 | 1,077 | −2.01 | −1.73 | 1,680 |
| 8 | 3.75 | 3.48 | 1,408 | −2.05 | −1.80 | 1,726 |
| 9 | 3.78 | 3.54 | 1,322 | −2.11 | −1.92 | 1,591 |
| 10 | 3.88 | 3.73 | 776 | −2.18 | −2.01 | 1,451 |
| 11 | 3.98 | 3.78 | 697 | −2.30 | −2.15 | 984 |
| 12 | 4.10 | 3.92 | 682 | −2.44 | −2.23 | 1,037 |
| 13 | 4.13 | 3.94 | 529 | −2.48 | −2.39 | 1,006 |
| 14 | 4.14 | 4.01 | 1,359 | −2.64 | −2.48 | 717 |
| 15 | 4.38 | 4.07 | 3,690 | −2.65 | −2.54 | 1,438 |
| 16 | 4.45 | 4.13 | 1,690 | −2.79 | −2.63 | 1,980 |
| 17 | 4.81 | 4.48 | 351 | −2.86 | −2.70 | 868 |
| 18 | 4.80 | 4.46 | 420 | −3.09 | −2.84 | 330 |
| 19 | 4.66 | 4.49 | 280 | −3.10 | −2.89 | 13 |
| 20 | 4.66 | 4.61 | 92 | −2.81 | −2.78 | |
| 21 | .... | .... | ...... | −2.58 | −2.74 | 3 |
| Total....... | .... | .... | 17,250 | ...... | ...... | 17,71 |

portion of the belly. Briefly it may be stated that starting with a common foundation stock, selection in the minus direction practically eliminated the colored areas, whereas in the plus direction a race was produced which was almost self-colored. The effects of selection were at first comparatively rapid and then became slower (table 36); but obviously the final point reached represents about the limit which can be obtained.

The interpretation of these experiments at one time raised an interesting and spirited controversy in genetic circles. Castle contended, on the basis of the behavior of hooded in crosses with self-colored and Irish pattern, that the hooded character was a Mendelian unit and that these experiments showed that selection may alter such units. In effect he contended that the hooded factor $h$ might exist in a number of forms, $h_1, h_2, h_3 \ . \ . \ . \ h_n$, which determined different degrees of pigmentation, and that this series of factors was variable and might be changed from one average grade to another progressively by selection. The adherents of the multiple-factor hypothesis, on the contrary, held that the factor $h$ simply is necessary for the development of the hooded pattern, but that the degree of pigmentation is determined by numerous other genes acting in conjunction with it, some of which tend to increase and others to decrease the degree of pigmentation. They assumed that the original stock was highly heterogeneous and that selection had merely sorted out in the minus direction combinations of factors which decreased the pigmented area and in the plus direction combinations which increased it.

Castle himself eventually demonstrated the correctness of the multiple-factor hypothesis. He argued that, if the changes effected by selection were due to subsidiary factors cooperating with $h$ in the determination of different grades of pigmentation, then replacement of those factors which had accumulated in the plus and minus strains by selection by a set of such factors from another source should wipe out the differences between the two strains. He, therefore, crossed the two strains with a non-hooded race, and in $F_2$ selected the hooded segregants and again crossed them with the same non-hooded race. The cross was repeated three times. Each time it was observed that the hooded segregants were less divergent, and in the third extraction some of the hooded individuals descended from the minus race exhibited pigmented areas as great as some of those descended from the plus race. Obviously the effect of this successive crossing to a non-hooded race is gradually to supplant the subsidiary factors in the plus and minus races by those present in the non-hooded race. The evidence indicates that completion of this process, which would require a number of generations, would wipe out entirely the differences in degree of pigmentation between the two strains. Castle, therefore, concluded that his earlier views were incorrect, and that the multiple-factor interpretation represented the most probable explanation of the results.

**Modifying Factors.**—The foregoing experiments on hooded rats demonstrate that the particular expression of the hooded character is dependent on the factor *h* interacting in development with numerous independent factors which affect the degree of pigmentation. Without *h* these subsidiary factors are totally ineffective. The factor *h* then may be regarded as the main factor for the hooded condition, and the subsidiary factors which cooperate with it to determine the degree of pigmentation are commonly called modifying factors. The number of these and their specific quantitative effects were undetermined, as is true in most experiments involving modifying factors; but the fact that they do exist was adequately demonstrated. In the Drosophila investigations, Bridges has been able to show that it is possible not only to demonstrate the existence of modifying factors, but also to study them individually and to show that they differ in no essential respect from other factors.

Eosin eye color is a member of the series of multiple allelomorphs of white described in Chapter X. From time to time eosin flies have been discovered which differed in eye color from the original eosin, and stocks were established from them. These stocks varied in color from darker than eosin by progressive steps to white, indistinguishable in appearance from the white determined by the factor *w* as follows:

eosin, dark
eosin
eosin, pinkish
eosin, cream *c*
eosin, cream *b*
eosin, cream *a*
eosin, cream III
eosin, cream II
eosin, whiting

In each of these cases the eosin factor $w^e$ was present associated with some other mutant factor which reduced or increased the degree of coloration by a specific amount. It was also found that these modifying factors of eosin had no effect upon other eye colors such as vermilion, sepia, pink, etc., nor, what is stranger still, did they have any effect upon the other allelomorphs of the white series, such as coral, cherry, and blood. For this reason Bridges called them specific modifiers of eosin. Some of them were definitely located by the usual methods adopted in Drosophila experimentation. The factor for pinkish was found to be located in group II at 106.0; cream *b* in group II at 22.5; and cream III in group III at 33.5. In this instance the modifying factors have been studied individually by the usual experimental methods, a procedure which was possible by reason of their occurrence in different stocks. Had they all been mixed in a single stock in the first place, the phenomena of modification of eosin eye color and progressive alteration by mass selection might well be as striking as that of hooded in rats.

The existence of modifying factors explains many at first sight anomalous possibilities. A sugary corn may be made sweeter by crossing with a starchy variety and isolating sugary segregants in $F_2$ and subsequent generations. A dwarf variety of peas may be made shorter by crossing with a tall variety; a bantam fowl may be made smaller, as Punnett actually showed, by crossing with a larger breed; and the results, properly understood, are no more inexplicable than the modification of flower colors, for example, which are sometimes obtained as a result of crossing colored and white varieties.

**Mass Selection.**—By mass selection is simply meant selection in each generation of a number of individuals most nearly approaching the desired type for progenitors of the next generations. Such selection, as has been shown, is ordinarily effective over a comparatively long period of time. It has also been found in some instances that a conspicuous variation in the character still remains after long-continued selection, and that in the absence of selection the race may subsequently lose some of the ground gained by the process; in other words that selection under such circumstances is often not permanently effective but must be continued in order to keep the race at its new level.

A moment's thought as to the implications of the multiple-factor hypothesis will suggest an explanation for this widely observed fact. If there are numerous pairs of factors affecting the character expression, a particular expression of the character for which selection is being practiced may depend upon various combinations of factors which will consequently be mixed in the selected population. Crosses between these factor combinations will produce by segregation other combinations of factors for different expressions of the character, and theoretically no amount of selection according to this method will produce a genetically homogeneous race. In the absence of selection, such a stock may obviously show a partial retrogression toward the level of the original unselected stock.

This difficulty may be obviated by combining selection with inbreeding. The selection has the effect of isolating combinations of factors favorable for the desired expression of the character; inbreeding has the effect of producing stocks homozygous for a particular combination of factors. The operation of this system as compared with mass selection is well illustrated by the results of East and Jones on protein content as compared with those obtained at the Illinois Station. Isolation of several such selected inbred lines, followed by crossing and further application of the method to the crossed progeny, theoretically and apparently practically also, is the method best calculated to produce the most rapid and effective isolation of desired types.

**Selection and the Composition of Populations.**—In the preceding chapters it was shown that populations of allogamous species are highly heterogeneous and that they consist largely of mixtures of different combinations of highly heterozygous individuals. Autogamous species, on the other hand, are highly heterogeneous, but they consist of mixtures of homozygous individuals of different combinations. The results of selection described in this and the preceding chapter are in agreement with these conceptions. In allogamous species the effects of selection are gradual and continuous over a long period of time, and theoretically at least the limits are not reached until the population becomes homogeneous for a single homozygous combination of factors. Reduction to such a condition obviously requires a considerable period of time, If, however, selection in an allogamous species is accompanied by inbreeding, this limit is approached much more rapidly by reason of the automatic attainment of a homozygous condition under such circumstances. In autogamous species, on the contrary, as shown in the preceding chapter, individual selection is immediately followed by creation of a pure line, which is no longer subject to modification by selection inasmuch as it is completely homozygous. Mass selection, however, in autogamous species may give continuous results over a period of years superficially resembling those obtained from allogamous species; for under such a system a single pure line is not immediately isolated, but rather a mixed sample of superior pure lines; and the selection may be effective until the material finally is composed entirely of the best pure line. In both cases the results of selection are obviously in accord with Mendelian principles operating under specific methods of reproduction.

**References**

Jones: Genetics, Chapter IX.
Walter: Genetics, Chapter VI.
General texts.

## CHAPTER XXII

## FACTOR MUTATION

In the preceding chapters an attempt has been made to show that the heritable diversity which occurs so abundantly in animal and plant species is almost entirely due to Mendelian variation. The qualitative variations which have been employed for the purpose of demonstrating Mendelian principles belong clearly enough to this category; and the more numerous variations in size and other quantitative features also appear to do so, even though they are not so readily demonstrable. In variable animal and plant species, therefore, numerous sets of allelomorphs exist which are evidently distributed according to Mendelian principles. This demonstration has afforded a clear conception of the nature of variation; but before the problem can be considered finally solved, it will be necessary to know not only of the existence of Mendelian allelomorphs and of laws governing their distribution but also of their mode of origin. This last problem is obviously fundamental; but unfortunately it is one of the most obscure features of genetics. Nevertheless many interesting facts have been determined about it, and an attempt will be made to outline some of them in this chapter.

In many instances nothing is known as to the origin of the Mendelian factors which exist in animal and plant species. Mendel, for example, simply made an extensive collection of varieties of peas and found that they differed in certain well-defined contrasted characters; tall *vs.* dwarf, round *vs.* wrinkled, yellow *vs.* green, etc. He did not know how those differences came into existence; in fact he was not concerned with this problem. But he did demonstrate that the visible differences in characters are dependent upon the existence of allelomorphic factors. So also in the great majority of instances, Mendelian studies have simply disclosed the factor relations which are responsible for diversity, but they have not shown how the factors came into existence.

**The Mutation Theory of de Vries.**—Thought and opinion on the origin of variations prior to the advent of Mendelian methods necessarily suffered from a lack of knowledge of the fundamental basis of variation. But nevertheless the interest aroused in scientific circles by the enunciation of the doctrine of evolution led to numerous speculations and some investigation on the source of variations. At about the same time that experiments were in progress destined to lead to the rediscovery of Mendel's paper, one of these investigators, the Dutch botanist de Vries, was also conducting studies on the origin of species. He observed that the evening primrose, *Oenothera Lamarckiana*, although for the most part constant, produced a small percentage of well-defined new types when grown in large numbers. The data contained in table 43 (Chapter XXVI) representing eight generations of self-fertilized progenies, show that among 54,343 individuals, 834, approximately 1.5 per cent, were distributed among seven distinct types, so different from the parental form as to lead de Vries to consider some of them as valid new species and to suggest that new species arise, not by gradual accumulation of minute differences, as Darwin contended, but by sudden departures from the parental type, a process which he called mutation.

Subsequent research has shown that the so-called Oenothera mutations are probably connected with a complex genetic situation which makes it difficult, even at the present time, to offer a satisfactory interpretation of them, as is shown in Chapter XXVI. They are mentioned here because they led to a greater emphasis upon the significance of the occasional, sudden departure from type, often called "sports," which are familiar to animal and plant breeders and had been observed by investigators prior to the time of de Vries.

**Mutation in Drosophila.**—Although the Oenothera mutations may no longer be considered as typical examples of the manner in which new forms usually come into existence, acceptable evidence is now available from numerous other sources. The Drosophila investigations, in particular, are unique in the fact that some information is available about the origin of practically all the forms which have been studied experimentally. In Drosophila it has been observed that in experiments involving examination of large numbers of individuals, an occasional individual, or sometimes comparatively large numbers, exhibit some new character, the origin of which does not depend upon the ordinary processes of segregation and assortment. From such individuals new stocks are readily established which uniformly exhibit the new character, provided it is viable as a homozygote.

The consideration of a few actual cases will illustrate some of the conclusions which may be drawn from observations on the appearance of mutations. Muller describes the case of a female heterozygous for a number of known second-group factors, which produced 139 females and 132 males of the expected eye colors, and 9 males which had a new, very pale eye color, ivory. Tests of this eye color proved it to be a sex-linked recessive, allelomorphic to white; therefore, its origin must have been due to change of the factor $W$, which all of the flies in this stock contained, to $w^i$, a new factor for ivory eye color. This is the process which is called factor mutation, a definite, discontinuous alteration in a single element or factor of the germinal substance.

The female which produced these ivory males was herself $WW$ in constitution. If she had been heterozygous for ivory $Ww^i$, half her sons should have been ivory instead of only 9 in 141, as the counts show. The small number of ivory males which she produced indicates that at some comparatively late stage in development, a cell of the germ plasm had changed from $WW$ presumably to $Ww^i$. Since $Ww^i$ cells by reduction would produce gametes, half $W$ and half $w^i$, the change must have occurred early enough so that approximately 13 per cent of the eggs arose from such cells. Muller suggests that, since this value is so nearly equal to one-eighth, the mutation may have occurred in one of the first eight germ cells.

Another instance described by Muller is also of interest. In a stock heterozygous for known group factors, a single male was discovered with one red and one white eye. Mated to the red-eyed females of the same stock, a red-eyed progeny was secured and these mated *inter se* gave red-eyed females and males half of which had red and half white eyes. Subsequent tests showed that this new white eye was probably identical with the previously discovered white; therefore, since this stock was $WW$ in constitution, it must have arisen by a mutation from $W$ to $w$.

In this instance, however, the mutation must have arisen at an early stage in the male which had one red and one white eye. This male must have started as a $(WX)Y$ zygote, and the red eye must have been produced by unaltered cells of this type. The white eye, however, must have been produced by a change in constitution to $(wX)Y$, and this change occurred early enough in ontogeny so that the germ cells were all produced by $(wX)Y$ tissue, since all of the daughters of this male were heterozygous for white. From these and other observations, the conclusion is drawn that factor mutations may occur in either females or males and that they may occur either early or late in development; in fact, most of the evidence indicates that mutation is equally likely to affect any cell at any stage in development. To be transmissible, however, the mutation must occur in cells of the germ track; otherwise it obviously has no opportunity to become incorporated in the gametes.

**Appearances of Mutants.**—The mode of appearance of mutant forms is in complete agreement with Mendelian principles, as is evidenced particularly by data obtained from the Drosophila and Antirrhinum experiments. In Drosophila recessive sex-linked mutants first appear either in a few males or in half the male progeny of a given female. The appearance of a few males indicates that the mutation occurred in the germ track of the mother; and the number of mutant individuals which appear may be regarded as evidence as to the time of mutation. A single mutant male might indicate that mutation had occurred during or immediately preceding the meiotic divisions; a greater number of males would indicate its occurrence earlier in development, so that the mutated cell produced a group of cells by division prior to the production of gametes. The circumstance that males receive their only X-chromosomes from their mother obviously accounts for the fact that sex-linked recessive mutants appear at first in male progeny and may appear in the immediate progeny of a female which contains mutated cells in her germ track. If, however, half of the sons of a given female are of a mutant type, then the female must herself have been heterozygous for the mutant factor, which leads to the conclusion that the mutation must have occurred in the germ cells of the preceding generation, either in her mother or father. Obviously also occurrence of a sex-linked recessive mutation in a male must necessarily lead to production of some heterozygous females among his progeny, and half of their sons would be of the new mutant type and would represent its first appearance.

Autosomal recessive mutants on the other hand first appear in about one-fourth of the offspring of a given pair of flies. A moment's consideration will show why this should be so. If an autosomal recessive mutation occurs in a given individual, it will produce some gametes containing the new recessive mutant factors. These gametes obviously will be fertilized by unmutated gametes from an individual of the opposite sex, thus producing a certain proportion of heterozygous flies in the progeny. If now, by chance, two heterozygous flies are mated together, one-fourth of their progeny, being homozygous for the new factor, will be of the new mutant type. Obviously, therefore, the appearance of a new autosomal recessive form simply indicates that the two parents were heterozygous; the actual mutation did not occur in the parents, but may have occurred in one of the grandparents or in an earlier generation. In animals, in particular, several generations may elapse between the actual mutation and the production of the mutant form by reason of the fact that the latter event usually depends upon chance mating of two heterozygous individuals.

Dominant mutants, whether sex-linked or autosomal, usually appear in a few individuals of either or both sexes. Obviously if a dominant mutation occurs in the germ cells of an individual, those of its progeny which receive mutated gametes will exhibit the corresponding character.

In plants circumstances of reproduction alter the details of the phenomena. A mutation to be transmissible must be incorporated in the germ track, which in higher plants is equivalent to saying that it must either occur in the hypodermal layer of cells in the primary meristem or in some way become incorporated therein. Since both pollen grains and ovules arise from this layer (*cf.* Chapter XXX), a recessive mutant in plants may appear in the immediate progeny of a self-fertilized plant in which the mutation has occurred. It may even happen that the mutation, occurring comparatively late in ontogeny, will affect a single branch or only a few branches. These branches will then behave as heterozygotes segregating for the new recessive type, whereas unaffected branches will give nothing but descendants normal to the particular strain. In most instances, however, recessive mutations apparently arise so late in ontogeny that only a few gametes of the mutant type are produced. The chances are then overwhelmingly in favor of a mutant gamete meeting with a normal one, rather than with another mutant gamete. The result would be the production of a few heterozygous individuals among the progeny of the plant in which the mutation occurred; and these would segregate for the mutant character in the next following generation according to Mendelian principles. Dominant mutants, of course, would immediately make their appearance manifest, but they, as in Drosophila, are apparently far less common than recessives.

In plants also the circumstance that each bud is tipped with a primary meristem which produces both gametes and somatic cells makes it possible for the dominant mutations immediately to manifest themselves in a change in the character of a particular branch on a plant. Such a phenomenon is known as bud mutation and is comparable to the case above described of a Drosophila fly with one eye red and one white. The question of bud mutation, or somatic mutation, raises special problems which will be discussed in Chapter XXX. Here it is sufficient to point out that this phenomenon occurs in homozygous plants only when a dominant mutation appears. If, however, a plant is heterozygous for a pair of factors *Aa*, and mutation changes *A* to *a* or to some other allelomorph of *A*, the change may be immediately shown as a bud mutation in the plant in which it occurs.

These remarks on the appearances of mutants are simply made to point out that a mutant factor is governed by Mendelian laws from the moment of its origin. It is also perfectly obvious that at times it is difficult to distinguish between the origin of a new form by mutation and its appearance simply as a result of segregation in a mixed stock. Appearance of a new character is not enough, for as has been shown in Chapter XIII, new characters may arise as a result of new combinations of factors in stocks to all appearances of the same genetic type. In order to be positive that a new form has arisen by mutation, it is necessary to be able to demonstrate that somewhere in its ancestry a definite change has occurred in a specific factor, producing a new factor responsible for the appearance of the new type.

**The Frequency of Mutation.**—Deliberate attempts to determine the frequency of mutation experimentally have been made in a few instances, notably by Muller in Drosophila and by Baur in Antirrhinum. Such studies present numerous difficulties, so that it becomes necessary to adopt special methods of technique for the purpose.

In the studies reported by Muller and Altenburg particular attention was devoted to the occurrence of sex-linked lethal factors, because the existence of such factors in females is shown immediately by a ratio of 2♀♀:1♂ in their progeny. These investigators tested 1,062 females known to be the progeny of females which were not themselves heterozygous for lethal factors, and found that twenty of them gave 2:1 sex-ratios, representing a frequency of about 1 in 53 for the X-chromosome. Similar studies made by Muller on special III-group stocks gave about thirty mutations in 5,000 chances, a rate of 1:160. Since these two chromosomes make up about three-fifths of the germinal substance, this frequency corresponds to about 1.4 per cent for the entire germinal substance. On this basis about 0.7 per cent, or 1 in 140, of the gametes produced by an individual contain a single mutant factor.

As in Drosophila so also in Antirrhinum most mutations are recessive; consequently it is necessary to grow not only the immediate selfed progeny of the plant, but also selfed progenies from each of its offspring in order to determine whether some of them may not be heterozygous for a recessive mutant factor derived from the mother. Or what amounts to the same thing, numerous selfed progenies from a single homozygous line may be grown in successive generations and the proportion of these which exhibit segregation for a new character may be determined. Plants lend themselves well to this form of experimentation because self-fertilization is the most expeditious method of obtaining homozygous races.

From 384 selfed progenies of a single line, Baur obtained sixteen new mutant types. This represents a frequency of nearly 5 per cent for the entire germinal substance. However, the new forms here taken into account were comparatively easily recognizable types. Baur emphasizes the fact that slight, easily overlooked mutations may also occur and that mutations may occur which require a special genetic constitution in order to make them recognizable. This feature may be illustrated by the specific modifiers of eosin described in Chapter XXI, which produce no visible effect except in eosin flies; consequently, if they occur in other stocks, they would not be detected until they were incorporated with eosin by crossing. Relations of this type indicate the difficulty which is met with in obtaining satisfactory evidence of the rate of mutation. Baur assumed that at least as many, probably more, mutations having slight or unrecognizable effects occurred in his experiments. Consequently about 10 per cent of the plants in homozygous stocks must be heterozygous for a single mutant factor, or expressed in terms of gametes, the mutation frequency would be approximately 5 per cent.

While seven-tenths of 1 per cent may appear to be a relatively high rate of mutation, when translated into terms of the individual factor it gives an extremely low frequency. Thus, if it is assumed that there are 3,000 factors in a haploid chromosome complex in Drosophila, as estimated by the methods described in Chapter XVI, then an average mutation rate of only one in about 400,000 is indicated. This method of estimation is, however, open to serious objections, and the value stated is probably too high.

It might seem possible to obtain some information on this matter from observations on repeated mutations of easily recognizable types. Thus in the Drosophila investigations in which literally millions of flies have been examined, a mutation rate as high as this should be reflected in a number of instances of reappearance of known mutant types. As a matter of fact such data as are available for easily recognizable types, when equated against the total number of flies which have been examined, indicate a much lower rate, even for those which have reappeared most frequently. Among the most frequent mutants are those recorded for the white locus, about twenty-five times; cut, sixteen; yellow, vermilion, and rudimentary about fifteen times each; forked, twelve; pink, eleven; and ebony and lozenge, ten each; but the number of flies counted is estimated to be in the neighborhood of twenty million. Evidently the studies which have been made indicate that factors have a very high order of stability; but possibly some are more frequently subject to mutation than others.

In Muller's investigations, lethal mutations were found to occur about six times as frequently as those giving rise to a new morphological type. In both Antirrhinum and Drosophila recessive mutants appeared more frequently than dominant ones. Moreover, just as factors having very minute effects are more numerous than those producing conspicuously visible effects, so mutations of comparatively small effect are more frequent; in fact it is possible that mutations having no recognizable character effect are most frequent of all. Under these circumstances, the difficulty of arriving at an accurate estimate of the rate of mutation is obvious.

External conditions appear to have little if any effect upon the rate. In certain experiments Muller and Altenburg observed a somewhat higher rate under increased temperature; but the difference was hardly sufficiently great to be significant statistically. Thus far no method of increasing mutability of factors is known.

**Mutation and Allelomorphism.**—Mutation evidently is the method by which allelomorphic factors come into existence, for if, for example, $W$ mutates to $w$ in Drosophila, giving rise to the white-eyed type, the mutant factor $w$ occupies the locus in the X-chromosome formerly filled by $W$. Similarly, if $W$ mutates in other ways, as it has, to $w^e$, eosin, or $w^a$, apricot, etc., these mutant factors will all occupy the same locus as $W$ thus giving rise to a system of multiple allelomorphs. As a matter of fact, factor mutation is the only known method of setting up allelomorphic relations. From this point of view any Mendelian analysis provides presumptive evidence for the occurrence of mutation at some previous time. The numerous series of Mendelian allelomorphs which exist in species in nature and under domestication presumably have had their origin in factor mutation.

**Mutation in the Bar Locus in Drosophila.**—Zeleny has shown that the dominant sex-linked mutant, bar, reverts to wild type in about 1 in 1,600 offspring of homozygous bar females and produces a more extreme form, called ultra-bar, in about 1 in 28,000 individuals. Ultra-bar in turn reverts to wild type and bar in the ratios of about 1 in 1,700 and 1 in 2,900, respectively. The bar allelomorphs, therefore, represent the most mutable factors discovered thus far in Drosophila (figure 77).

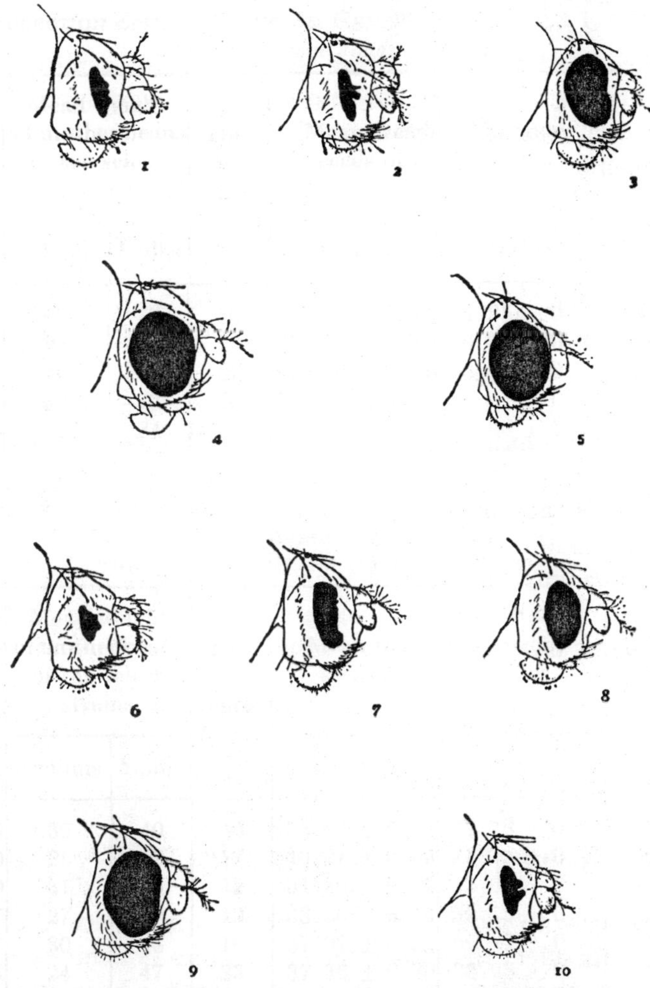


FIGURE 77.—Eye types in the bar series in *Drosophila melanogaster* compared with the normal condition. (1) $\frac{B}{B}$ ♀; (2) $\frac{B}{Y}$ ♂; (3) $\frac{B}{+}$ ♀; (4) reversionary wild type ♀; (5) reversionary wild type ♂; (6) $\frac{BB}{Y}$ ♂; (7) $\frac{B^i}{B^i}$ ♀; (8) $\frac{B^i}{Y}$ ♂; (9) $\frac{B^i}{+}$ ♀; (10) $\frac{B^iB^i}{Y}$ ♂; (*From Sturtevant.*)

Subsequent investigations have shown that mutation in bar and ultra-bar is confmed to females, since males of these types do not produce reversionary offspring. Moreover, Sturtevant and Morgan have shown that the phenomenon is connected with crossing over at or near the bar locus, for females homozygous for bar and heterozygous for forked and fused, $\frac{f\,B}{B\,fu}$, produced reversionary offspring all of which were crossovers between forked and fused. On the basis of these results, they suggest that mutation in bar is due to unequal crossing-over by which a female of the above constitution by inaccurate juxtaposition of the factors, symbolically $\frac{f\,B}{B\,fu}$, by crossing-over gives $\frac{f\,B \mid B\,fu}{\mid}$; *i.e.*, one X-chromosome with two bar factors, the other with none, the former in the homozygous condition giving ultra-bar and the latter wild-type reversion.

Another allelomorph, infra-bar, $B^i$, less extreme than bar, exhibits the same phenomena. It reverts to wild type and produces a more extreme form, called double-infra-bar, $B^iB^i$, about as frequently as the corresponding phenomena in bar. Moreover, females heterozygous for these two factors $\frac{B^i}{B}$ also revert to wild type and produce as a corresponding extreme form either bar-infra-bar $BB^i$, or infra-bar-bar $B^iB$, two forms which are alike in appearance but which may be shown by linkage tests to have the factors in different order. These two types break down in the heterozygous condition into their components, bar and infra-bar, with a frequency of about 1 in 1,000, which is about the same as the frequency of reversion of ultra-bar to bar and double-infra-bar to infra-bar under similar conditions.

A rather puzzling outcome of these experiments arises from a comparison of facet number of ultra-bar and double-infra-bar heterozygotes with homozygous bar and infra-bar, respectively, as shown by the following figures:

$$\frac{B}{B} = 68.1 \pm 1.1 \qquad \frac{BB}{+} = 45.4 \pm 0.2$$

$$\frac{B}{B^i} = 73.5 \pm 1.2 \qquad \frac{BB^i}{+} = 50.5 \pm 0.4$$

$$\frac{B^i}{B^i} = 320.4 \pm 14.2 \qquad \frac{B^iB^i}{+} = 200.2 \pm 8.6$$

These results seem to indicate that two factors in the same chromosome produce a greater effect than the same two factors in homologous chromosomes, a phenomenon for which no explanation has been offered.

Naturally the question immediately arises as to how far these phenomena are typical of factor mutation. The original mutations from the wild-type factor to bar and from bar to infra-bar may represent true qualitative changes, which is assumed to be the characteristic feature of factor mutations. The derivative types, however, according to the above explanation, merely represent quantitative changes associated with crossing-over. Inasmuch as it has been shown that factor mutation may occur in somatic cells in males and in females of special constitution which exhibit no crossing-over in certain sections of the chromosomes, it is impossible that this explanation can be applied as a general interpretation. As a matter of fact, the demonstration that unequal crossing-over gives rise to wild type as one of its products would appear to indicate that the original appearance of bar was due to incorporation of a new factor in the germinal material, rather than qualitative change in one

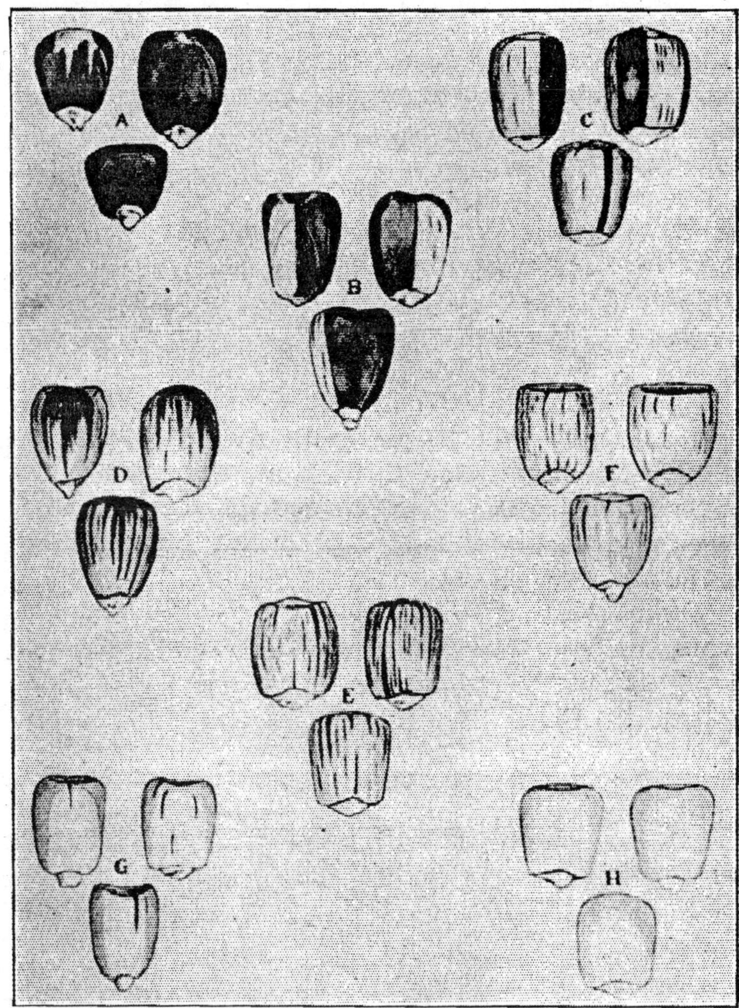


FIGURE 78.—Variation in distribution of color in variegated maize. *A*, self-colored and near self; *B*, more than half self; *C*, less than half self; *D*, dark-crown variegated; *E*, medium variegated; *F*, light variegated; *G*, very light variegated; *H*, colorless. (*From Emerson.*)

already present, and that possibly this addition is responsible for inequalities in alignment resulting in unequal crossing-over. The phenomena of the bar locus must, therefore, be regarded as unique; but they are of interest because they indicate that factor mutation, as it is now recognized, may arise from diverse processes.

**Variegation in Maize.** In the calico types of maize, the pericarp consists of colorless material more or less conspicuously striped with red. Variegation is a simple allelomorph of self-colored and colorless pericarp, recessive to the former and dominant to the latter. There is also evidence that different types of variegation, distinguishable by the fineness of the colored stripes depend upon different factors for variegation. Emerson was able to show that medium variegation $V^m$ was dominant to very light variegation $V^l$, and that in backcrosses of heterozygous plants $V^mV^l$, approximately half the plants were of the medium variegated and half of the light variegated type. Further evidence indicates that this series will have to be extended, for apparently there are many grades of variegation which depend upon differences in the factor for variegation.

Considering, however, the series of multiple allelomorphs, $S$ self-colored, $V^m$ medium variegation, $V^l$ light variegation, and $W$ colorless or white pericarp, some interesting and important observations have been made upon mutability. It has been found that variegated ears whether produced from homozygous $VV$ or heterozygous $VW$ plants exhibit marked variation in expression, as shown in figure 78. Occasionally they produce kernels which are entirely self-colored or self-colored with a light crown. Sometimes the kernels have dark crowns, as shown in $D$ (figure 78). Occasionally ears are produced which have considerable areas of self-colored dark-crown kernels. Various degrees of striping with self-color or variegation were found as shown in the figure.

Emerson segregated the kernels of homozygous and heterozygous ears from plants which had been backcrossed to a colorless strain and grew them separately. Table 37 contains the data obtained from homozygous variegated ears cross-pollinated with colorless. It will be observed

TABLE 37.—PROGENIES OF DIFFERENT CLASSES OF MAIZE SEEDS FROM EIGHTY HOMOZYGOUS VARIEGATED EARS CROSS-POLLINATED BY COLORLESS MAIZE

*(Data from Emerson)*

| Type of seeds planted | Progenies | | | | |
|---|---|---|---|---|---|
| | Number of plants | | | Percentages | |
| | Self-colored ears | Variegated ears | Total | Self-colored ears | Variegated ears |
| Self and near-self...... | 71 | 72 | 143 | 49.65 | 50.35 |
| More than half-self..... | 43 | 67 | 110 | 39.09 | 60.91 |
| Less than half-self...... | 16 | 150 | 166 | 9.64 | 90.36 |
| Dark-crown variegated. | 1 | 528 | 529 | 0.19 | 99.81 |
| Medium variegated.... | 2 | 607 | 609 | 0.33 | 99.67 |
| Light variegated....... | 2 | 125 | 127 | 1.57 | 98.43 |
| Very light variegated... | 1 | 269 | 270 | 0.37 | 99.63 |
| Colorless............. | .. | 18 | 18 | .... | 100.00 |

that self and near-self kernels produced approximately 50 per cent each of self-colored and variegated ears. Kernels more than half-self produced 39.09 per cent of self-colored ears, and kernels less than half-self produced 9.64 per cent of self-colored ears. The remaining classes, dark-crown variegated, medium, light, and very light variegated, and colorless produced an occasional self-colored ear, the rest of the ears being variegated.

The homozygous variegated plants $VV$ should produce only variegated progeny $VW$, when crossed back to colorless. The appearance of self-colored ears in 50 per cent of the progeny of self-colored kernels indicates that one of the $V$ factors had changed to $S$, so that the constitution of these kernels had become $SV$, instead of $VV$. Hence crossing to $WW$ gave $SW$, self-colored, and $VW$, variegated in approximately equal numbers. In those kernels more than half and less than half self-colored, a lower proportion of self-colored progenies indicates that the self-colored stripes are due to a change from $VV$ to $SV$ which has occurred so late in ontogeny that only a portion of the kernel is affected, for if the change occurs late, there is a chance that it will not affect the particular area from which the germ cells arise. This chance will obviously be less as the width of the stripe diminishes; consequently the data are in agreement with expectation in this respect.

The dark-crown variation illustrates a different situation. These kernels evidently are $VV$ so far as heredity is concerned, although the dark-crown kernels are almost as conspicuous as the self-colored ones. They probably arise in the same way, by conversion of $VV$ cells into $SV$ at some late stage in ontogeny. The failure of inheritance is apparently accounted for on histological grounds. The change from $VV$ to $SV$ in self-colored kernels occurs in the hypodermal layer of cells, *i.e.*, the layer just below the epidermis. The change responsible for dark-crown variegation occurs in the epidermal layer. Now it is well known that the gametes arise from the hypodermal layer, and this fact accounts for inheritance of self-colored changes and lack of inheritance of dark-crown changes, although fundamentally the same process is doubtless responsible for both phenomena.

These results, therefore, show that the factor for variegation has a comparatively high degree of mutability from $V$ to $S$. Inasmuch as the colorless seeds proved to be $VV$, apparently $V$ does not change to $W$, or at least very infrequently. Moreover, the fact that the self-colored kernels were always $SV$ and not $SS$ indicates that the change occurs in only one of the two $V$ factors. Consequently, the conditions responsible for the change must be so highly localized in the cell that only one of two identical factors is affected by them. These remarks apply especially to the factor for medium variegation $V^m$. Studies of the factor for very light variegation $V^l$ gave similar results, but the frequency of change was found to be very much lower. Studies of mutation in heterozygotes of the type $V^mV^l$ indicated that the change to self-color invariably converted $V^m$ into $S$, so that the constitution became $SV^l$. Some studies were also made of the comparative mutability of $V$ in $VV$ and $VW$ plants. The frequency should be twice as great in the former since two $V$ factors are present, but as a matter of fact it was approximately the same in both. These results indicate that the frequency of mutation in a given factor may be changed by the conditions under which it exists. One instance was obtained which indicated a change from $V^m$ to $V^l$, and it was further observed that a strain of self-colored arising from $V^m$ produced a very small proportion of variegated kernels, from which variegated strains have been established. On these and other grounds it is apparently established that this series of allelomorphs exhibits marked differences in mutability. Self-colored and colorless races are exceedingly constant. Then there is a self-colored type which exhibits a comparatively low frequency of mutation to variegation. The very light variegated type stands next in order, mutating occasionally to self-colored. Medium variegation exhibits a higher ratio of change to self-colored, and a very dark variegation is recognized which produces numerous self-colored or nearly self-colored kernels.

At first sight it might appear possible to relate these results with those which have been described in the bar-eyed series in Drosophila. But there is a fundamental difference in the two cases which makes any analogy impossible. The changes in the bar series occur during the meiotic process, while those in maize occur in meristematic cells which are simply undergoing mitotic divisions. It is, however, possible that these phenomena represent a special category of mutation.

Eyster has made some interesting studies on orange pericarp which gives striking evidence of mutability. Orange pericarp varies in intensity of coloration from cherry red to colorless. It also produces variegated ears which exhibit a graded series of intensity from very coarse variegation to very fine variegation, almost equivalent to colorless. These different types were found to change from one to another with different frequencies. Eyster interprets these results as indicating that the factor for orange pericarp consists of pigment-producing and non-pigment-producing elements, and that the intensity of orange pericarp depends upon the relative proportion of these elements. Those of medium intensity produce a progeny exhibiting a graded series of intensities ranging from nearly colorless to dark orange or red in a distribution approximating a normal curve. Progeny from light- or dark-orange ears exhibit a similar graded series with the distribution skewed toward the parental type. Eyster assumes that random distribution of factor elements accounts for this result, and further that segregation of the elements in somatic tissue is responsible for variegation. If this conception is correct, it will be necessary to revise current conceptions of mitosis; for the characteristic feature of this process has been presumed to lie in its exact partition of germinal elements to the two daughter nuclei, so that each of them contains the full complement of elements contained in the mother nucleus.

**Variegation in Antirrhinum.**—In Antirrhinum, Baur has described and studied two types of variegation, one of which is responsible for the production of striped flowers, the other for the production of flaked leaves, which have irregular areas of green, chlorophyll-bearing tissue interspersed in tissue devoid of chlorophyll. Both are inconstant. They vary in type of variegation from few large, colored areas through a graded series down to many small, colored areas. Both produce occasional self-colored shoots.

Taking the green-flaked type as a basis of discussion, Baur assumes that the zygotes are *zz*. The factor *z* is incapable of producing chlorophyll, but it mutates comparatively frequently to *Z*. If the change from *z* to *Z* occurs early in ontogeny, the flaking produced is of the coarse type few, but large green areas. If the change occurs relatively late in ontogeny, the fine-flaked type is produced. If a branch arises in a *Zz* sector, it is wholly green and exhibits Mendelian segregation for flaking. Flaked branches produce variable proportions of green and flaked descendants. It will be observed that Baur does not assume that there is, properly speaking, a factor for variegation, but rather that an unstable factor mutating during the course of ontogeny is responsible for the production of variegated patterns. He also presents evidence that the factor *z* may exist in different conditions which may determine whether it mutates early in ontogeny to produce coarse variegation or late to produce finer types.

As far as the evidence hitherto presented is concerned, it seems impossible to determine which conception of the ontogenetic processes concerned in variegation is correct. It is, however, worthy of note that variegations in plants commonly exhibit the range of phenomena which have been described in maize and snapdragons. Striped flowers are characteristic of many horticultural varieties, and almost invariably they exhibit marked inconstancy in type of striping and occasionally produce self-colored blossoms. These variations evidently depend upon genes which are peculiarly prone to mutate. In so far as they are concerned, selection, whether in clones or pure lines, may be expected to give positive results. The ineffectiveness of selection under ordinary conditions depends upon a high stability in the genes. If some of them are highly mutable, the conditions for this ineffectiveness disappear. As Baur points out, it is possible in the flaked Antirrhinum to secure positive results in clonal selection in the type of flaking, and to change from one type to another at will by continuing the process.

There is, however, one feature of the phenomena which throws some doubt upon Baur's interpretation. If, in variegated maize, mutation from *V* to *S* is responsible for production of variegation, it would seem reasonable to expect a comparatively high frequency in the production of self-colored progeny from variegated plants. Actual results, however, indicate a frequency of only about 1 per cent. On this basis it would seem impossible to assume that *VV* areas are white and that colored ones represent a change to *SV*. It is more probable that both colored and colorless areas are *VV* but that production of color in some and lack of it in others is a phenomenon of ontogenetic differentiation, unattended by genetic change.

**The Nature and Cause of Mutation.**—From the foregoing discussion it is evident that factor mutation represents a definite, qualitative change in a factor to a new level, usually as stable as the original condition, comparable to change in composition of chemical compounds. Nothing is really known, however, concerning the actual nature of the process, and it is probable that bar in Drosophila and variegation in plants represent special categories.

It was at one time widely believed that recessive mutation merely represented loss of the corresponding dominant factor from the germinal material, hence the term loss-mutation applied to the origin of recessives. That not all recessive mutation is due to loss is clearly shown by the origin of systems of multiple allelomorphs by mutation; for example, the eleven known recessive allelomorphs of *W* in Drosophila, and numerous analogous series; although in other instances, as in miniature, which are of frequent recurrence, apparently only a single type has been produced. It is of course possible that factor mutation may sometimes simply represent loss of a factor but it is difficult to prove the contention.

There is evidence of reversibility in the mutation process in some cases. Thus eosin in Drosophila originally arose from white and has since been observed to arise directly from wild type. Eosin in turn has been observed to mutate back to both red and white. White, which arose directly from red, has not certainly been shown to revert back to red.

Concerning the causes of mutation, nothing is known. If the change in the factor is due to action of some agent external to the factor itself, then it must be exceedingly localized in its effect; for it has been shown that mutation in the first instance is due to a change in genetic constitution of a single cell, which is later multiplied by the ordinary process of cell division; and not only that but of two identical factors in a given cell, only one usually mutates, thus giving rise to a heterozygous cell from one previously homozygous. Numerous attempts have been made to produce mutations by x-ray treatment, subjection to radium emanations, high and low temperatures, and administration of various toxic substances, but apparently without success. Unquestionably the problem of the nature and cause of factor mutation is one of the most important unsolved problems of genetics.

**Mutation and Breeding.**—It has been argued that mutation is so rare that it may be left out of account in breeding methods. The results obtained by Muller and Baur, however, do not support this conclusion, for they indicate that mutations are occurring frequently enough to provide an abundant source of genetic diversity. The experience in the improvement of such species as the sweet pea, the Chinese primrose, and numerous others, which apparently has depended fundamentally on the appearance and utilization of factor mutations likewise contradicts this assumption. The numerous mutations which have appeared in the various species of Drosophila during the recent period of extensive genetic exploitation also illustrates the type of diversity which may be secured by mutation. Even if most mutations are of little value for practical purposes, when animals and plants are grown in sufficiently large numbers, new forms of practical value may be expected to appear occasionally.

The practical utilization of mutations requires resort to hybridization in order to bring them into various combinations with existing factors. The appearance of a single new mutation of value, followed by its utilization by hybridization, doubles the potential number of varieties or combinations of factors which may be secured in the species. Thus the appearance of the cupid type of sweet pea by mutation makes it possible to produce by hybridization a complete series of cupid varieties having all the colors and types of blossoms hitherto existing in the tall varieties. With ten mutations, $2^{10} = 1{,}024$ combinations may be produced by hybridization; with twenty, in excess of a million. A few favorable mutations properly utilized may provide a basis for the creation of numerous new varieties of economic value.

**Mutation and Selection.**—The remarks which have been made in the previous chapters on the effect of selection in pure lines, clones, and mixed populations must be modified to take account of mutation. Mutations have been observed in pure lines, in clones, and in highly inbred strains derived from mixed populations. If such mutations affect the expression of the feature for which selection is being practiced, some progress may still be expected in clones or under sexual reproduction in lines which have been reduced to a homozygous condition. The general experience, however, indicates that mutation under such conditions so rarely provides material for further progress that it is an uneconomical method of procedure in practical operations in comparison with other methods which may be employed.

### References

Castle: Genetics and Eugenics, Chapter XXX.

Morgan: Theory of the Gene, Chapters V and VI.

Muller: Mutation, *in* Eugenics, Genetics and the Family, **1**: 106–112.

# CHAPTER XXIII

## PARALLEL VARIATION

Related species of animals and plants often exhibit a striking similarity as respects the distinctive features of their varieties. Wheat, rye, barley, and oats, for example, all have bearded and beardless varieties; barley and rye have similar series of grain colors, and so on through a wide array of characters. *Antirrhinum majus*, the garden snapdragon, includes an extensive collection of varieties differing in flower color, form, and numerous other features, almost all of which may be found more or less faithfully duplicated in some species of Linaria. Corn, teosinte, and coix exhibit a striking similarity in grain color. In rodents, as is well known, mice, rats, guinea pigs, and rabbits exist in distinct breeds, chiefly characterized by differences in coat color, which are practically identical in the four forms. Breeds of pigeons and of fowls present many similarities in plumage color and other features. So also the same kind of remarkable correspondence may be shown to exist among many related species, and in general it may be stated that the more closely related the species, the more nearly identical are the series of varieties which characterize them. It is to this feature of diversity that the term parallel variation is applied; and it is proposed to devote this chapter to a discussion of some of its aspects.

**Color Variation in Rodents.**—In the familiar pet rodents, mice, rats, rabbits, and guinea pigs, a great variety of breeds has been established, chiefly characterized by differences in coat color. Because of their rapidity in breeding and the ease of rearing large numbers inexpensively, they have been thoroughly studied genetically. Among mammals no better illustrations of parallel variation can be obtained. In the tabulation presented herewith (table 38) an attempt has been made to indicate the extent to which homology exists among some of the known Mendelian factors basic to the production of the various coat color forms of fanciers and investigators.

A number of allelomorphic conditions of the color factor $C$ have been demonstrated. Their characteristic effect is exhibited in dilution of the black and yellow pigments; but, as Wright has shown, a given factor does not necessarily reduce the two pigments proportionally. $C$ represents the factor for full pigmentation found in all four species and dominant to other members of the series in all. Slight dilution $c^k$ found only in guinea pigs and full dilution $c^d$ mainly reduces the intensity of the yellow pigment. In ruby dilution $c^r$, which is apparently represented in all four, the eyes are dark red, but yellow in the fur is practically changed to white, and black is somewhat diluted. In the Himalayan albino of rabbits and guinea pigs $c^h$, the eyes are pink and the fur is prevailingly white, except for the colored extremities, but even white portions may give evidence of slight sepia pigmentation. Finally, the complete albino $c$, found in mice, rats, and rabbits, develops no pigment in coat or eyes.

TABLE 38.—PARALLEL VARIATION IN RODENTS

(An x entered in a column indicates occurrence of the factor in the corresponding species; a – sign, that it has not yet been found)

| Factors and character effects | Mouse | Rat | Rabbit | Guinea pig |
|---|---|---|---|---|
| $C$ = colored | x | x | x | x |
| $c^k$ = slight dilution | — | — | — | x |
| $c^d$ = full dilution | x | — | — | x |
| $c^r$ = ruby-eyed dilution | x | x | x | x |
| $c^h$ = Himalayan dilution | — | — | x | x |
| $c$ = complete albinism | x | x | x | — |
| $E$ = extension | — | x | x | x |
| $E^d$ = dominant black | — | x | x | — |
| $e^b$ = brindle (or tortoise) | — | — | x | x |
| $e^y$ = yellow | — | x | x | x |
| $A$ = agouti | x | x | x | x |
| $A^y$ = dominant yellow | x | — | — | — |
| $A^l$ = light-bellied agouti | x | — | — | x |
| $a^t$ = black-and-tan | — | — | x | — |
| $a$ = black | x | x | x | x |
| $B$ = black agouti | x | — | x | x |
| $b$ = brown agouti | x | — | x | x |
| $P$ = intense pigmentation | x | x | — | x |
| $p$ = pink-eyed dilution | x | x | — | x |
| $R$ = intense pigmentation | — | x | — | — |
| $r$ = red-eyed dilution | — | x | — | — |
| $I$ = intense pigmentation | x | — | — | x |
| $i$ = dilute pigmentation | x | — | — | x |

The factor $E$ has a less extensive series of allelomorphs. The factor $E^d$ for dominant black or steel gray occurs in rats and rabbits. The factor $e^b$ for brindle or yellow-spotted coat gives the black-and-yellow "Japanese" rabbit and the tortoise guinea pig. Finally, $e^y$ is responsible for production of recessive yellow in rats, rabbits, and guinea pigs. Curiously enough no member of the series has been demonstrated in mice.

The agouti factor $A$, responsible for the ordinary gray or agouti coloration in rodents, determines the production of yellow in the middle of each hair and black at the tip and base. In the white-bellied agouti type in mice, the hairs of the belly are white throughout, but those of the rest of the body have the characteristic distribution of black and yellow in bands. This factor is found in mice, rabbits, and guinea pigs, but not in rats. The dominant yellow $A^y$, a form found only in mice, is peculiar in that it can exist only in the heterozygous condition. Genetic experimentation and obvious degeneration of embryos in matings of yellow × yellow indicate that homozygous yellow is a lethal condition. The factor $a^t$ of black-and-tan rabbits determines a black, slightly ticked coloration of the back with yellow or white belly. It probably should be considered a parallel of light-bellied agouti in mice and guinea pigs. Finally, black coat $a$ is found in all four rodents.

The factor $B$, responsible for black pigmentation, has an allelomorph $b$ for brown pigmentation in coat and eyes, which gives the brown-eyed brown agouti type. It has been found in the mouse, rabbit, and guinea pig but not in rats.

Pink-eyed dilution $p$, a variation from $P$ found in mice, rats, and guinea pigs, reduces black to pale, slaty brown but does not affect yellow, so that the form may be called pink-eyed yellowish gray. In rats a red-eyed dilution is also known which is distinguishable from pink-eyed dilution only by eye color, and in mice and rabbits a dilution factor $i$ as a variation from intense pigmentation $I$ reduces black to a slaty blue and yellow to cream, the maltese type of dilution characteristic of blue mice and rabbits.

With this illustration in mind, it is well to turn to the problem of the meaning of parallel variation. Genetically it is desirable to restrict the term to the possession by related species of factors which are homologous elements of the germinal system. From this point of view related species are to be considered as having germinal systems which are either identical or very similar in organization. But in the case of species such as these, it is doubtful whether it is justifiable to conclude merely from similarity of character effect that two factors are identical. It has already been shown in numerous instances that different factors within a species may produce identical character effects, and it may be that chemical desiderata are such that only a limited number of color variations may be produced. The parallels which have been drawn in the above discussion, therefore, should be considered merely as suggestive of true homology.

When as in the $C$, $E$, and $A$ series of factors above described a similar set of multiple allelomorphs is shown to exist in the different species, the probability that the factors are truly homologous is somewhat strengthened. The presumption is made even stronger when linkage relations are taken into account. In rats the factors $C$-$R$-$P$ belong to a single linkage system, probably in the order indicated. In mice, *R* has not yet been demonstrated; but *C* and *P* are linked and give recombination values comparable to those for the same factors in rats. If a red-eyed yellow type were to be discovered in mice and found to give values with *C* and *P* comparable to those in rats, the evidence for true homology of the factors would be very strong. In guinea pigs, however, the *P-p* pair of factors apparently is not linked with the *C* series of allelomorphs.

**Parallel Variation in Cavia Species.**—In the above instances it is quite impossible conclusively to demonstrate homology, because the different rodents cannot be crossed. When species are closely enough related so that they may be crossed, it is possible to determine whether or not they have factors in common. Such crosses have been made among Cavia species: *C. porcellus*, the common guinea pig; *C. cutleri*, a wild Peruvian cavy; and *C. rufescens*, the Brazilian cavy. Crosses between the Peruvian cavy and the guinea pig seem to show that the agouti factor *A* is the same in these two species; but the Brazilian cavy has a darker form of agouti which has been shown to be allelomorphic to *A* of the other species and recessive. This is the tick-bellied agouti $A^r$ of table 38. It is probable, however, that both *A* and $A^r$ occur in the Brazilian cavy.

Further investigations made by Detlefsen and others on crosses between the guinea pig and the Brazilian cavy have shown that the factors *C*, basic for color production, *B* for dark eye and coat, *E* for extended pigmentation, *P* for dark-eyed as contrasted with *p* for pink-eyed brown agouti, and *r* for smooth coat as contrasted with *R* for rough occur in the Brazilian cavy. Similarly Castle and others have shown that the Peruvian cavy also possesses these same factors. Here the evidence for true identity of factors is based upon actual transfer of specific factors from one species to the other and by observations on effects of the factors in segregation following hybridization.

**Comparative Genetics in Antirrhinum.**—In Antirrhinum, the section Antirrhinastrum contains a group of well-characterized species which may be crossed without difficulty, all of which have eight pairs of chromosomes. This group includes the garden snapdragon, *A. majus*, with which Baur has made such extensive genetic investigations, and a number of variable wild species distributed in the Mediterranean region of Europe.

Baur has crossed eight of these wild species with *A. majus* and has grown $F_2$ populations in excess of a thousand plants in each of them. The results in general are similar. $F_2$ consists of a great variety of types, so numerous as to defy accurate analysis, and the conclusion is drawn that the species differ in a large number of factors which have comparatively minor effects, so that, as in the case of quantitative characters, the results cannot readily be subjected to a specific Mendelian analysis. The cultivated garden races of *A. majus*, however, when crossed together give results which indicate that they differ in comparatively few factors which have marked character effects, so that at least the major differences which characterize varieties of *A. majus* may readily be subjected to Mendelian analysis.

In *A. majus* Baur has demonstrated the existence of numerous well-defined recessive characters which in various combinations are responsible for the major portion of the noticeable diversity in these species. He has introduced these variations into the species crosses and finds that there also they behave as simple Mendelian recessives. Thus eosin-red *majus* × *latifolium* gives fuchsin $F_1$ and segregation into 3 fuchsin: 1 eosin-red in $F_2$ superimposed upon the normal complex segregation in minor factors characteristic of this species cross. Proceeding along similar lines, he has shown that the wild species contain the dominant allelomorphs corresponding to almost all the recessive factors of *A. majus*. It might be concluded from these results that these species may be expected to produce a series of mutant forms comparable to those of *A. majus*, if subjected to intensive investigation.

**Parallel Variation in Drosophila.**—The most significant studies on parallel variation hitherto reported are those of Sturtevant on *D. melanogaster* and *D. simulans*. These two species are so nearly identical in morphological features that for a long time they were not distinguished; in fact their separation dates from the discovery of the difficulties met with in crossing them and the fact that the $F_1$ hybrids are completely sterile. Since this discovery it has been found that they are characterized by numerous minute differences, of which certain morphological features of the genitalia are apparently most striking. They may be crossed, but the $F_1$ hybrids of both sexes are sterile both *inter se* and when mated to the parents. The gonads of these hybrids have been shown to be rudimentary. The fact, however, that $F_1$ hybrids are produced makes it possible to test parallel variations in them and to determine homology without question.

Some difficulty has been experienced in crossing the two species and results of reciprocal crosses differ as follows:

1. *melanogaster* ♀ × *simulans* ♂ = $F_1$ ♀ ♀, no $F_1$ ♂ ♂
2. *simulans* ♀ × *melanogaster* ♂ = few $F_1$ ♀ ♀, $F_1$ ♂ ♂

In the first cross the hybrid progeny are all females; in the second they are almost all males. Since many larvae die, there is reason to believe that the missing individuals simply fail to reach maturity. Obviously, in the case of sex-linked factors, tests for homology must be made by mating *melanogaster* females to *simulans* males; for the sex-linked characters of hybrid males are received from the mother as in trials within the species; but for autosomal characters, it is immaterial how the cross is made; convenience may be permitted to dictate.

Under laboratory study *simulans* has produced a series of variations comparable to those in the more thoroughly studied *D. melanogaster*. For purposes of illustration, consider, for the moment, the group of sex-linked characters in each. In *simulans* yellow body color, prune eye color, ruby eye color, cross-veinless wings, dusky wings, vesiculated wings, singed bristles, rudimentary wings, garnet eye color, forked bristles, fused wing veins and bobbed have been shown to be parallel characters according to the following system:

1. *melanogaster* yellow ♀ × *simulans* wild type → $F_1$ wild type ♀ ♀
2. *melanogaster* wild type ♀ × *simulans* yellow → $F_1$ wild type ♀ ♀
3. *melanogaster* yellow ♀ × *simulans* yellow → $F_1$ yellow ♀ ♀

In the case of recessive characters, cross 3 may be considered adequate evidence of true homology of factors. Factors which have been demonstrated to be truly homologous in two species are called isomorphs. Where the factors are not isomorphic, even though the character effects are identical, three such crosses give different results as may be illustrated by *melanogaster* miniature and *simulans* dusky, two similar mutant forms with small wings:

1. *melanogaster* miniature ♀ × *simulans* wild type → $F_1$ wild type ♀ ♀
2. *melanogaster* wild type ♀ × *simulans* dusky → $F_1$ wild type ♀ ♀
3. *melanogaster* miniature ♀ × *simulans* dusky → $F_1$ wild type ♀ ♀

Here cross 3 yielding wild-type $F_1$ ♀ ♀ may be taken as evidence that *melanogaster* miniature and *simulans* dusky are not homologous, although similar enough to be considered identical, were it impossible to secure evidence from crossing. Since a *melanogaster* dusky is also known, crosses were made with *simulans* dusky which yielded dusky $F_1$, thus establishing homology of these two characters. Two tiny bristled forms have been discovered in *simulans*, but crossing tests show that neither is isomorphic with tiny bristles of *melanogaster*.

With the above crossing test, homology may be determined beyond question in these two species. But the experiments have gone far beyond this point, particularly into the problem of the comparative organization of the chromosomes in the two species. Cytological examination discloses the fact that the two species have identical chromosome groups. The genetic experiments on linkage indicate that for the most part the factors which have been shown to be homologous occupy corresponding loci in the two systems, as is shown comparatively in figure 79. For isomorphs which have been studied from this point of view, there are differences in detail in the two species. It is probable that some of them will disappear when the investigations on *simulans* approach those on *melanogaster* in extent.

One case of real disagreement, however, has been discovered in the third group of the two species. Scarlet, peach, and deltoid of *simulans* have been shown to be isomorphs of scarlet, peach, and delta, respectively, in *melanogaster*. Linkage studies, however, give not only strikingly different values but indicate that the sequence differs in the two species, as is also shown in figure 79. These results are very significant, because they do apparently demonstrate a difference in chromosomal organization in the two species. They also indicate that isomorphic factors do not necessarily exhibit identical sequence of genes in the chromosomes. Obviously this demonstration increases the difficulty of drawing acceptable conclusions from species which cannot be crossed.

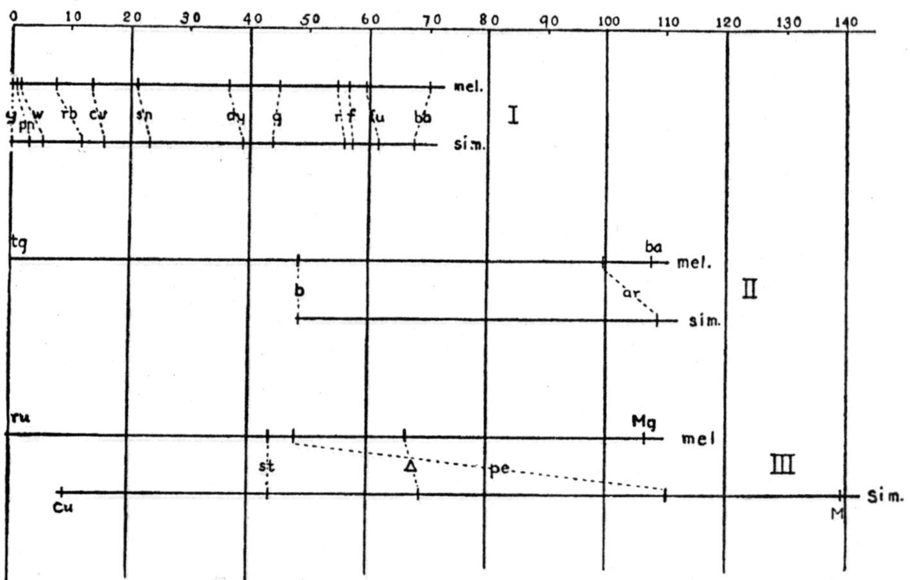


FIGURE 79.—Chromosome maps showing loci of corresponding factors of *Drosophila melanogaster* and *D. simulans*. (*From Morgan, Bridges, and Sturtevant, The Genetics of Drosophila in Bibliographia Genetica vol.* 2, *copyright* 1925 *by Martinus Nijhoff. Reprinted by permission.*)

**Comparative Genetics in Drosophila.**—The species of Drosophila exhibit characteristic differences in their chromosome groups, both as respects number and shape, as shown in figure 80. Obviously an interesting problem in evolution is raised by the occurrence of these chromosome patterns; a problem upon which it may be possible to obtain some information by a thoroughly comparative study of the genetics of representative species. Such investigations are now in progress; but they have not yielded sufficient dependable data as yet to furnish more than a slight insight into the significance of the cytological differences. *D. willistoni* with three, *D. melanogaster* and *simulans* with four, *D. obscura* with five, and *D. virilis* with six pairs of chromosomes have been most thoroughly investigated as to sex-linked characters.

Since a comparison has already been made of *simulans* with *melanogaster*, the former may be disregarded and attention may be devoted to the relations of *melanogaster*, *virilis*, *willistoni*, and *obscura*, particularly as respects sex-linked characters. No attempt will be made to present a complete account, which would require a rather extensive review of the literature. It is immediately evident from such a comparison that the sex-linked groups of these four species taken as a whole contain numerous characters of striking similarity, much greater than that of any autosomal group of *melanogaster* with the sex-linked characters of any of the other three; so that it is fully justifiable to conclude that the sex chromosomes of the four species are to some extent similar in genetic content.

When the examination descends to details of linkage values and sequence, the data are far less convincing. As a basis of discussion only

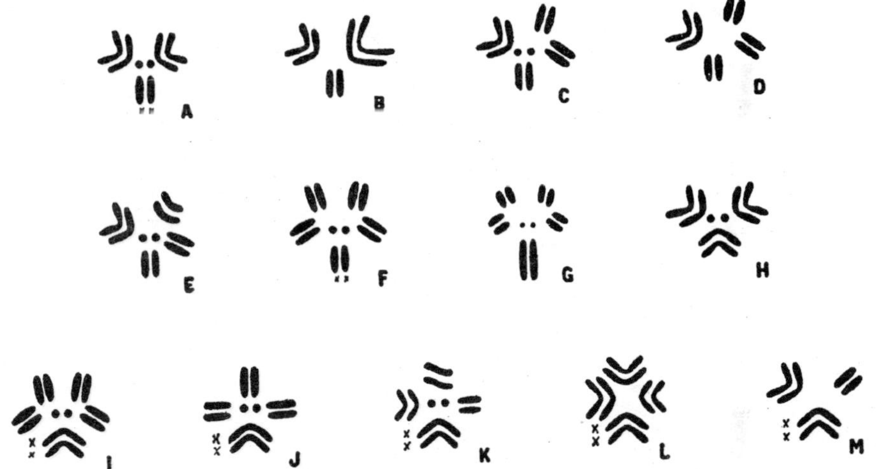


FIGURE 80.—Types of chromosome groups of species of *Drosophila*. *A*, *melanogaster* and *simulans*; *F*; *virilis*; *J* *obscura*, *M*, *willistoni*. (*From Metz and Moses.*)

those characters will be considered which appear to give the best evidence of correspondence in the four species, as shown in figure 81. It will be observed that yellow has been found in all four species; and that in *melanogaster* and *virilis* its position is terminal or nearly so, whereas in *willistoni* and *obscura*, it is near the middle of the linkage group. In a sense this fact is in harmony with the cytological findings, which have demonstrated the existence of rod-shaped X-chromosomes in the two former and V-shaped ones in the two latter. It is tempting to conclude that the X-chromosome in *willistoni* and *obscura* consists of one arm corresponding to the rod-shaped X-chromosome of *melanogaster* and *virilis*, and another representing added material; but further comparison does not wholly support this idea, and the recent evidence that the spindle fiber is not attached at the yellow end of the X-chromosome in *melanogaster* is also opposed to it.

Comparisons of map distances for corresponding factors at once disclose striking similarities and differences; yellow and scute are close together in all four, but yellow and vermilion range from 2.5 to 33.0. Miniature and dusky are close together in *melanogaster*, as are also diminutive and wee, two similar characters in *virilis*. Differences in sequence, however, are probably more significant than differences in linkage values, and such differences also exist, as may be seen from the following comparison:

| | |
|---|---|
| *melanogaster*........................... | yellow-scute-lozenge-vermilion |
| *virilis*................................. | yellow-scute-vermilion-rugose |
| *willistoni*.............................. | morula-scute-yellow-vermilion |
| *obscura*................................ | glazed-scute-yellow-vermilion |

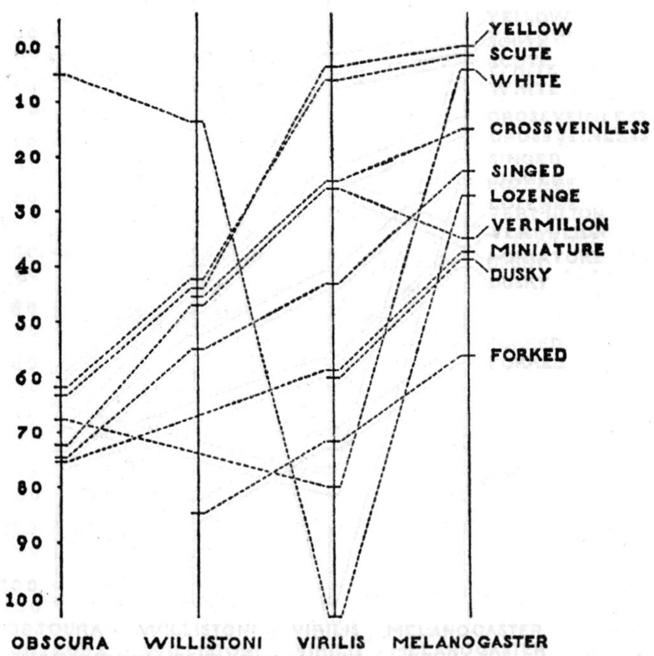


FIGURE 81.—Comparative maps of sex-linked factors in four species of Drosophila. Suspected instances of parallels, based on similarity of character effects are connected by dotted lines.

It may be possible that these discrepancies indicate rearrangement of components in the X-chromosomes of these species; or on the other hand, they may merely indicate that supposed correspondences are at fault. The fact that it is impossible to cross the species imposes a severe restriction on the value of the investigations.

**Parallel Variation and Mimicry.**—Mimicry is a term applied to resemblance of one species to some other distinct species. The phenomenon has been most thoroughly studied in moths and butterflies, and many exceedingly striking instances have been disclosed. In the typical cases, the species which serves as a model is generally protected from its enemies, presumably by an unpalatable taste or poisonous properties. In order, as it were, to take full advantage of this feature, these species are often conspicuously colored or marked, the so-called warning coloration thus advertising their disagreeable properties. The mimetic species, in typical cases, are usually palatable and otherwise unprotected; but in some way they have developed in form and coloration so striking a resemblance to their corresponding models that even experienced collectors are sometimes deceived into grouping them together. A distinction should be drawn between mimicry and protective resemblance. The latter term is used to designate those cases in which species resemble closely some feature in the environment, as for example the resemblance to leaves of some species of butterflies when at rest with wings closed. The discussion below applies only to true mimicry as above defined.

Vavilov has described an interesting instance from agricultural practice, which, because of its simplicity, may throw light upon the fundamental basis of mimicry. The cultivated lentil, *Ervum lens*, includes a great number of diverse varieties which differ strikingly in seed color and also in size of seed, time of maturity, etc. The common vetch, *Vicia sativa*, which also is highly diverse, often occurs as a weed in lentil fields. Now the forms of vetch present a parallel series of variations in shape, color, and size of seed to those in the lentil. In the growing of lentils it is customary to sort the seed before planting, rejecting all seeds which do not conform to the varietal type as well as all weed seeds of all kinds. This has led to the elimination from each variety of lentils of all vetch types except the particular one which closely mimics the lentil variety in which it grows. The mimicry extends beyond seed characters, for the vetches flower and ripen at the same time as the corresponding lentil varieties. Obviously this example of mimicry may be interpreted in the following way. Parallel variation accounts for the existence of a corresponding diversity in the two species. Selection operating in lentil varieties has led to the rejection of all vetch forms occurring in them as weeds except the particular one which so closely resembles the corresponding lentil variety as to pass undetected by the methods of sorting utilized in the cultivation of lentils. The accident of resemblance brought about by parallel variation has permitted the survival of certain vetch forms as weeds in lentil varieties.

Returning to butterflies, the instances of mimicry which are perhaps most illuminating are those in which a single species exists in a number of different forms, some of which bear mimetic relations to other species. *Papilio dardanus*, an African butterfly, is a classical instance. The male is non-mimetic and apparently subject only to slight variation over its entire range; the female, however, is extremely variable and apparently the different forms are more or less coextensive in distribution with the corresponding model species. Punnett has described the case fully and has presented an interesting colored plate of models and mimics.

In South Africa three forms of female are found: *cenea* resembling *Amauris echeria; hippocoon, niavius; trophonius, Danais chrysippus.* In the last case, *Danais chrysippus* has a variant form, *dorippus,* which corresponds to *dorippoides* of *Papilio dardanus.* Northwards *planemoides* closely resembles *Planema poggei.* However, other forms of the female which are apparently non-mimetic also exist, such as *dionysius* and *trimeni,* and in Abyssinia two mimetic forms, *niavioides* and *ruspina,* in addition to a non-mimetic form resembling the male. On Connor's island is also found the form *Humboldti,* on Madagascar, *meriones,* both non-mimetic and resembling the male. In resumé then, in this remarkable and complicated instance, corresponding to a single male type, no less than eleven distinct female forms exist, some of which provide notable examples of mimicry to distasteful species, more or less coextensive in range.

According to the theory of natural selection, mimetic forms are supposed to have come into existence through a gradual accumulation of rather minute differences, leading eventually to a close resemblance between model and mimic. If this theory is correct, genetic evidence from breeding experiments should give evidence of complex, multiple-factor differences among the distinct forms. As a matter of fact, however, such genetic evidence as is available indicates a comparatively simple state of affairs. In the case of *Papilio polytes,* described in Chapter XI, it was shown that the three female types might be explained on the basis of two pairs of Mendelian factors, limited in their determinative power to females. Here again the relation is one of mimicry, for the *polytes* female resembles *Papilio aristolochiae* and *romulus, P. hector,* both presumably distasteful species. *Papilio dardanus* unfortunately has not been worked out so completely; but in so far as evidence is available, it points to a similar conclusion.

The genetic evidence, therefore, indicates that these distinct forms have arisen by mutation, not by gradual accumulation of minute differences, and that natural selection has operated merely to preserve those forms which happened to bear a close resemblance to some distasteful species. Parallel variation with selection, as in the vetches occurring as weeds in lentils, probably accounts for these interesting instances of mimicry.

In general, parallel variation in related species results in the production of homologous series of varieties which resemble each other most closely in species which are closely related taxonomically. This, of course, is to be expected from the Darwinian theory of evolution, for, if two species have evolved from some common ancestral form, they may be expected to have not only many factors in common, but also a marked similarity in germinal organization; and manifestly these similarities should be more marked the more closely related the species.

In certain instances the phenomenon may be accounted for on the basis of Mendelian variation. Vavilov has called attention to the fact that varieties of *Hordeum vulgare* and *H. distichum* present an identical series of characters. But these two groups of varieties are distinguished as species on purely formal grounds, *H. vulgare* representing the six-rowed and *H. distichum* the two-rowed barleys. The two-rowed and six-rowed condition in barley is a simple Mendelian difference determined at most by a difference in one or two pairs of factors. It is better to consider all cultivated varieties of barley as a single species, for all may be crossed without difficulty, the hybrids are fully fertile, and the differences may readily be reduced to a Mendelian formulation. The existence of parallel series of six- and two-rowed barleys arises, therefore, largely through recombinations of characters between the two groups. So also within limits annual teosinte in Mexico has been found to exhibit variations identical with those found in maize growing in the same region, a circumstance which is accounted for by intercrossing between the two with the production of fertile hybrids. Instances such as these should not be included under the term parallel variation, for they represent merely examples of the operation of Mendelian recombination. The term should be confined to the independent origin of parallel series of variations in distinct species.

Thus delimited, the study of parallel variation offers some interesting possibilities. It may, for example, be possible to predict what types of diversity are most likely to appear in a species from a knowledge of variation in another, as Vavilov has shown in a number of instances. Similarity in germinal organization may enable an investigator to make more rapid progress in linkage studies within a species, if some information on this question has already been obtained in a related species. Where the germinal organization is obviously different, as in related species which differ in chromosome number, comparative studies of genetics may throw some light upon the fundamental features of the differences and may thereby lead to a more intelligent conception of the processes which have been responsible for the divergence of species. Future progress in this field may well disclose similarities in germinal composition and organization as impressive as those which have been discovered in comparative studies of development and anatomy.

A wide field of research, as yet but little exploited, here lies open for exploration, particularly among economic forms in which a high degree of diversity exists. Rye and the various species of wheat and of Aegilops, in which such an impressive series of homologies in variation exist, may be crossed, and it should, therefore, be possible to obtain satisfactory evidence of the existence of isomorphic characters according to the method used by Sturtevant in studies of parallel variation in *Drosophila melanogaster* and *D. simulans*. The evidence at present available, based merely upon similarity in characters, does not permit us to draw the conclusion that the several species belonging to this group present these series because they contain certain factors in common; such a conclusion must stand or fall on the basis of further and more searching investigations.

### References


MORGAN: Theory of the Gene, Chapter VII.

MORGAN, BRIDGES, and STURTEVANT: Genetics of Drosophila, Chapters XVIII–XX.

VAVILOV: The law of homologous series in variation, *Jour. Genetics*, **12**: 48–89, 1922.

# CHAPTER XXIV

## CHROMOSOMAL VARIATION

On the whole, chromosome number in animal and plant species is a highly constant feature of their germinal organization. As a matter of fact, the precision with which Mendelian distribution is effected stands as evidence for a remarkable regularity in the behavior of the chromosome mechanism both in meiosis and in mitosis, for as has been seen, the phenomena of heredity merely represent the workings of factors which are distributed in a certain way because they occupy definite loci in particular chromosomes.

Nevertheless there are well-established instances of aberrant behavior in both mitosis and meiosis, some of which have led to the production of strains or races which exhibit a difference in chromosome number from that normal for the species. As interest has been aroused in the relation between cytological phenomenon and genetic behavior, numerous such instances have come to light and some of them have been studied in great detail. Moreover, it has been discovered that important groups of economic plants often exhibit variations in chromosome number, a knowledge of which is necessary in order properly to interpret their genetic behavior and also in order properly to utilize them in plant-breeding operations.

Comparative cytological studies, such as Metz and his associates have made in Drosophila, clearly show that evolutionary processes have at times gone in the direction of reorganization of the germinal material. The mere existence of species of Drosophila with three, four, five, and six pairs of chromosomes points to this conclusion. Similar studies made within numerous other genera, both animal and plant, have demonstrated that difference in chromosome number among the species of a genus is a phenomenon of widespread occurrence. Particularly impressive instances have been disclosed in plant genera. Sometimes, as for example in Crepis, $n$ = 3, 4, 5, 6, 8, 9, 11, 20, and in Carex, $n$ = 9, 15, 16, 19, 24, 25, 26, 27, 28, 29, 31, 32, 33, 34, 35, 36, 37, 38, 40, 41, 42 and 56, the numbers run more or less consecutively. In other instances, as in Triticum (wheat) $n$ = 7, 14, and 21, in Avena (oats), $n$ = 7, 14, and 21, in Rosa, $n$ = 7, 14, 21, and 28, and in Chrysanthemum $n$ = 9, 18, 24, 36, and 45, the numbers appear to fall into an arithmetical progression, successive multiples of some common base number. In still other instances both types of seriation occur in the same genus, as in Nicotiana, $n$ = 9, 10, 12, 16, and 24. If these differences in number are the outcome of evolutionary processes, as in fact they must be, then it should be possible *a priori* to obtain evidences of variation in chromosome number.

As a matter of fact, numerous instances of variation in chromosome number have been discovered in genetic experiments. In order to discuss the matter in an orderly way, a tentative classification may be presented of the types of variation which have been found. In this discussion it is well to adopt a definite system of terminology. The chromosomes of a given kind may be called a chromosome set. Normally the number of sets of chromosomes is equal to the haploid number of chromosomes; and in somatic tissues each set consists of two members. Variation may occur in single sets or in the entire chromosome complex. Variations in number which affect certain chromosome sets and leave the others undisturbed constitute the unbalanced type of variation in chromosome number, for in such cases the proportion of certain elements in the germinal system is increased or decreased in relation to the sum total of elements normally present. Variations which affect the entire chromosome complex constitute the balanced type, for in such instances the proportions of elements remain unchanged. In the unbalanced types, several different kinds of variation have been discovered. Thus a single chromosome set may have only one member, in which case it is known as a monosome; two, the normal condition, a disome; three, a trisome; four, a tetrasome; etc. In the balanced type the number of chromosomes present is a multiple of the haploid number; and a series of forms has been discovered which are designated haploid, diploid, triploid, tetraploid, etc., according as they have one, two, three, or four haploid complexes, respectively, in their somatic cells. The term polyploidy is often applied to this phenomenon.

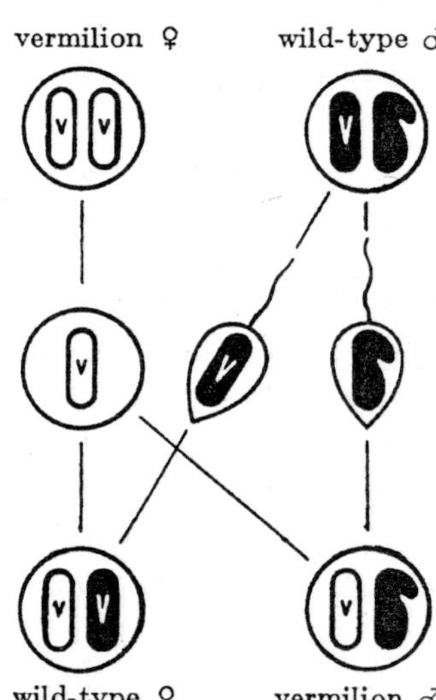


FIGURE 82.—The normal result of crossing a vermilion female with a wild-type (red) male in Drosophila. The straight figures represent X-chromosomes and the hooked ones, Y-chromosomes. (*Adapted from Bridges.*)

It should be understood that variations in chromosome number within a species are here dealt with, not variations from species to species within a genus. Lumping the two categories together cannot be countenanced because the homologous chromosomes in different species may differ greatly in their elements and mode of organization. This feature greatly complicates the situation in interspecific hybridization; consequently a discussion of variation in chromosome number arising from crosses between species is deferred to Chapter XXVII.

**Non-disjunction in Drosophila.**—In Drosophila the normal result of crossing a vermilion-eyed female with a red-eyed male is the production of an $F_1$ consisting of red-eyed females and vermilion-eyed males in approximately equal numbers, as shown in figure 82. When, however, numerous cultures are reared and enumerated, it is found that about one individual in 2,000 is exceptional, either a vermilion female or a red male. The red exceptional males are invariably sterile; but the vermilion females are of normal fertility. When mated to normal red males, however, they do not give results normal for such a mating, but yield progenies consisting of approximately 92 per cent of regular offspring, red females and vermilion males, and 8 per cent of exceptional offspring, vermilion females and red males. The vermilion exceptional females

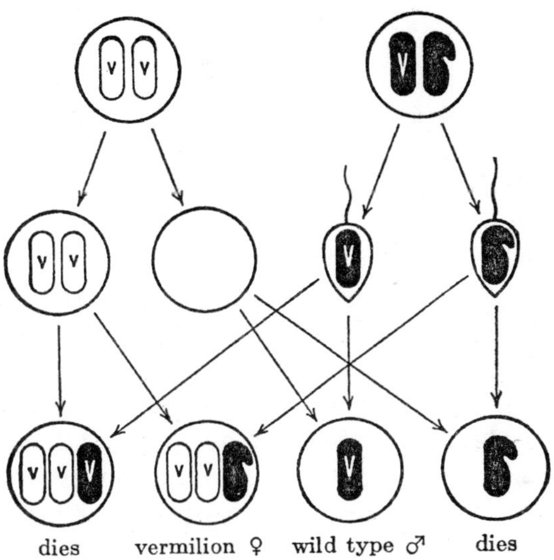


FIGURE 83.—Diagram of the production of exceptional individuals, vermilion females and wild-type males, through primary nondisjunction from matings of vermilion female by wild-type male. (*Adapted from Bridges.*)

repeat the behavior of the mother; the red males, however, although exceptional, are fertile and behave like normal red males in genetic tests.

These results led Bridges to conclude that the exceptional vermilion females arise from an irregularity in the distribution of the X-chromosomes in meiosis, as shown diagrammatically in figure 83. If the two X-chromosomes in certain rare instances fail to separate in the reduction division and pass to the same pole, eggs will be produced which have either two X-chromosomes or none, according as the two X-chromosomes are included in the egg nucleus or pass out into the polar body. This phenomenon is termed primary non-disjunction because of the failure of X-chromosomes to disjoin, as they normally do, in the reduction division. Four possibilities arise in fertilization of these eggs. They may be fertilized by X-bearing sperm, in which case two types of zygotes are produced: XXX, which almost always die, or XO, sterile males with red eye color because they have received their X-chromosome from the father, instead of from the mother, as is normally the case. If they are fertilized by Y-bearing sperm, two types of zygotes are again produced: XXY, exceptional vermilion females, because they receive both X-chromosomes from the mother, and a lethal combination which fails to appear.

The exceptional vermilion females produce further exceptional offspring on account of irregularities in distribution arising from the presence of the Y-chromosome. If in reduction in XXY females two chromosomes pass to one pole and one to the other by chance, three equally frequent types of distribution are possible:

X – XY
X – XY
XX – Y

Since the first two give identical results, two-thirds of the eggs should be X and XY and one-third XX and Y. As a matter of fact, the breeding experiments show that approximately 92 per cent of the eggs are of the first two types and only 8 of the second two; so that there is evidently a preferential distribution in favor of X-chromosomes passing to opposite poles in reduction. The breeding results interpreted on this basis give rise to the classes of individuals to be expected on the assumption that the exceptional vermilion females are actually XXY in constitution (*cf.* figure 84), and Bridges by following the problem in all its numerous ramifications was able to show agreement at every point between theory and observation. Moreover, cytological examination disclosed the presence of a supernumerary Y-chromosome in suspected XXY females and the absence of a Y-chromosome in sterile males produced by primary non-disjunction. The presence of XYY males was also demonstrated genetically and cytologically, and crosses of XXY females with XYY males produced some females with two extra Y-chromosomes XXYY, as expected.

These experiments show that an abnormal chromosomal content is not automatically rectified but is preserved unaltered by somatic divisions, and that it eventually gives rise to an irregular distribution of chromosomes in reduction. The irregularity in distribution is paralleled at all points by a corresponding deviation in distribution of factors presumed to be located in the chromosomes in question; consequently primary and secondary non-disjunction, far from being real exceptions to the chromosome theory of heredity, constitute the strongest possible evidence in favor of it.

The Y-chromosome on genetic grounds has been considered a neutral body performing no positive function in development. For the most

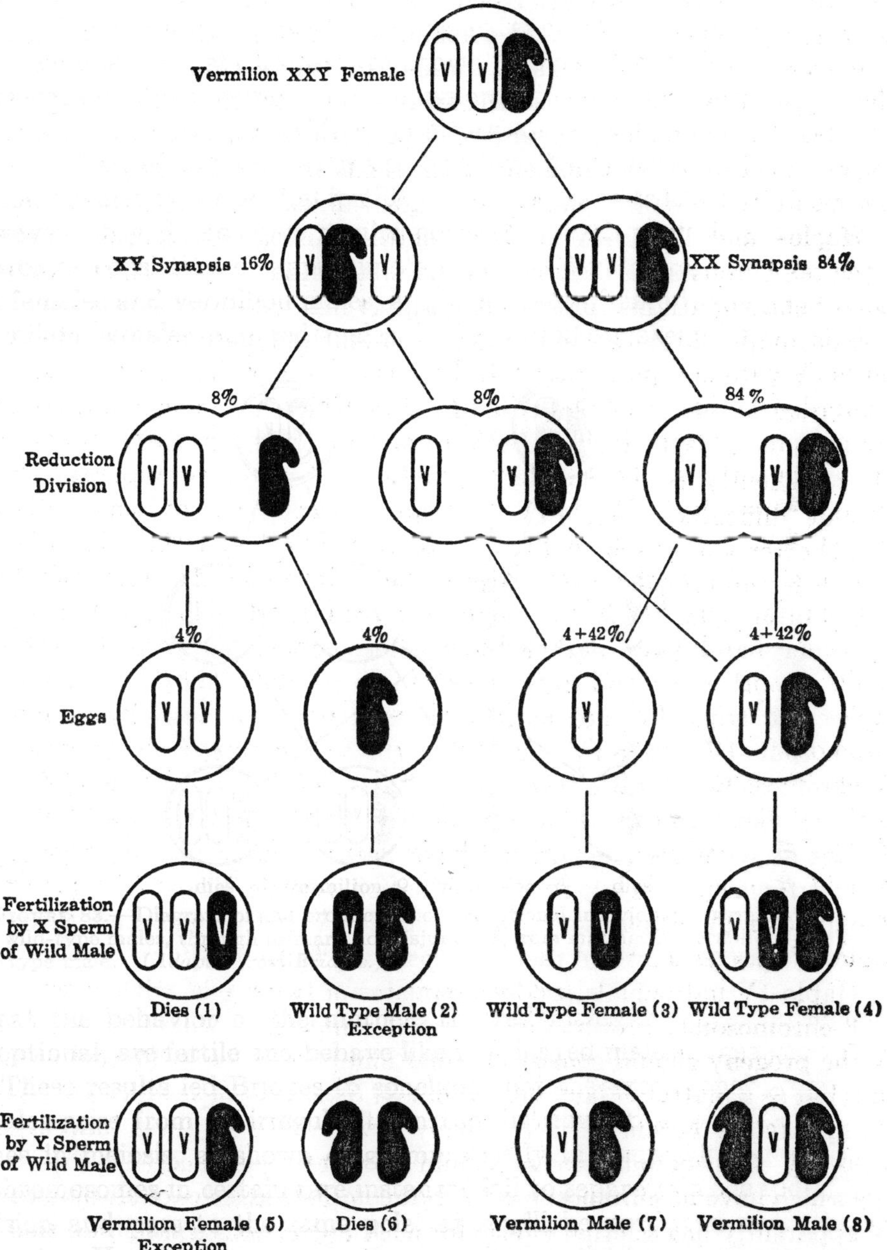


FIGURE 84.—Secondary nondisjunction in the female. Diagram showing the constitution of an exceptional vermilion female, the two types of synapsis, reduction, and the four classes of eggs produced. Each kind of egg may be fertilized by either of the two X and Y kinds of sperm of the wild male, giving the eight classes of zygotes shown. (*From Bridges.*)

part these investigations bear out this conclusion, for XXY or XXYY flies are females, indistinguishable in appearance from normal XX females; and XO and XYY flies are males, indistinguishable in appearance from normal XY males. Evidently the Y-chromosome has no effect upon the development of the zygote; but a positive role is indicated in males, for XO males are sterile. The sterility appears to be due to a general effect upon development of the testis and spermatogonial tissues, as a result of which few sperm are produced and they are non-motile.

**Haplo- and Triplo-IV in Drosophila.**—Historically non-disjunction of the sex chromosomes represents the first instance of irregular chromosomal behavior studied in Drosophila; but it is not in some ways typical of variation in chromosome number because the supernumerary chromosomes (Y-chromosomes) are neutral bodies. A more instructive instance is afforded by haplo-IV and triplo-IV flies, which have one and three IV-chromosomes, respectively. Both forms differ only slightly from normal, but sufficiently so to enable a practiced observer to separate them without difficulty. Haplo-IV flies are characterized by smaller body size; shorter and slenderer bristles; paler body color; darker thorax pattern; larger and rougher eyes, with smaller hairs; reduced or absent aristae; and blunter, cloudier wings, slightly spread apart. Triplo-IV flies on the other hand have darker body color; suppressed trident pattern; small, smooth eyes; and narrow and more pointed wings. Presumably haplo- and triplo-IV flies both arise as a result of non-disjunction of chromosome-IV, whereby eggs or sperm are produced either with no IV-chromosome or with two. As a matter of fact, however, although haplo-IV flies are produced in a frequency of about 1 in 2,000, such triplo-IV flies as have been recognized have been obtained in the progeny of triploid females. Cytological examination reveals clearly the presence of only one IV-chromosome in the haplo-IV individuals and three in triplo-IV flies as shown in the accompanying figure (85).

Haplo-IV individuals produce gametes of two kinds, with or without a IV-chromosome, presumably in equal numbers. Mated to a normal fly the progeny should consist of equal numbers of haplo-IV and normal flies; but as a matter of fact, only about 34.5 per cent of the flies are haplo-IV, apparently because of a relatively low viability of these individuals. Haplo-IV flies mated *inter se* should produce some nullo-IV flies, *i.e.*, flies which have no chromosome-IV, but complete absence of chromosome-IV apparently has a lethal effect; for such individuals are not secured.

With these facts, it is possible to consider the genetic consequences of the haplo-IV condition. Three mutant characters, bent, shaven, and eyeless, from breeding experiments with flies of normal chromosomal constitution, have been assigned to chromosome-IV. They are so closely linked that no double recessives had been secured; consequently no data as to order and linkage values were available. When an eyeless female is mated to a haplo-IV male with normal eyes, offspring are of two types, wild type and haplo-IV eyeless. The haplo-IV individuals are eyeless, because they have but one IV-chromosome received from the mother and it bears the factor for eyeless. Theoretically the numbers of these two types should be equal; but actually the combination of a single IV-chromosome with the IV-group mutant factor eyeless so greatly diminishes the viability of this class that only about 12.6 per cent of haplo-IV eyeless flies are produced in matings of this type. Evidently, however, genetic results and chromosomal condition run parallel.

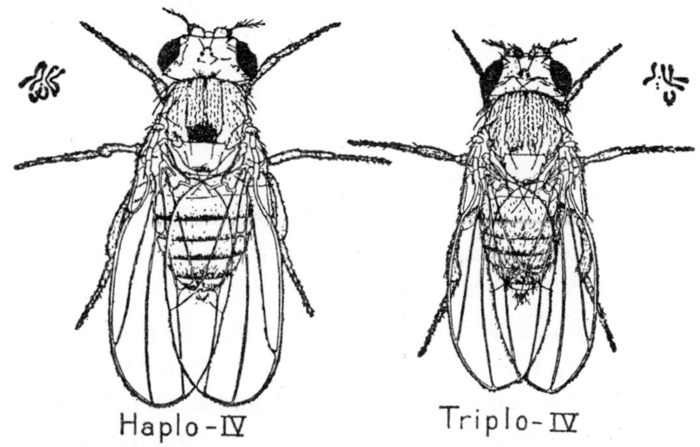


FIGURE 85.—The haplo-IV female contrasted with the triplo-IV female with their respective chromosome groups. *(From Morgan, Bridges, and Sturtevant, The Genetics of Drosophila in Bibliographia Genetica, vol. 2, copyright 1925 by Martinus Nyhoff. Reprinted by permission.)*

The possession of haplo-IV flies led to the employment of a unique method for determining linkage relations between the IV-group factors, bent and eyeless. Bent and eyeless flies were crossed giving an $F_1$ wild type $\frac{bt}{\quad ey}$. These $F_1$ heterozygotes were then mated to haplo-IV flies whereupon they produced wild-type flies, and haplo-IV flies of four classes. *viz.*, non-crossovers, bent and eyeless; crossovers, wild type and bent eyeless. An equivalent of seventeen crossovers among 1,987 individuals indicates a linkage value of $0.86 \pm 0.14$ for these two factors. The method is strictly parallel to that employed in determining linkage values of sex-linked factors in Drosophila, for the male, as a consequence of the neutral behavior of the Y-chromosome in development, is virtually a haplo-X fly.

The connection between fourth-group factors and chromosome-IV has also been strikingly confirmed from studies of triplo-IV flies. Triplo-IV flies produce gametes, half of which have only one fourth chromosome and half have two. As a consequence, when mated to wild-type flies, half the descendents are wild type and half triplo-IV. Mated *inter se*, they should theoretically give flies in the ratio 1 tetra-IV: 2 triplo-IV : 1 diplo-IV (wild type). As a matter of fact tetra-IV flies do not appear. Evidently the presence of two extra chromosomes-IV in a zygote is incompatible with development.

A diplo-IV eyeless fly crossed with a triplo-IV normal fly gives triplo-IV and diplo-IV offspring, both with normal eyes. The triplo-IV $F_1$ flies are *EyEyey* in constitution on account of the fact that they have three chromosomes-IV, two of which come from the triplo-IV normal parent. The distribution of these three chromosomes may be effected in three ways, *viz.*,

$$EyEy - ey$$
$$Eyey - Ey$$
$$Eyey - Ey$$

so that the complete gametic ratio is

$$EyEy + 2Eyey + 2Ey + ey$$

If now such an $F_1$ fly is mated to a diplo-IV eyeless fly, the offspring will obviously be distributed thus:

$$EyEyey + 2Eyeyey + 2Eyey + eyey$$

or 3 triplo-IV normal: 2 diplo-IV normal: 1 diplo-IV eyeless. All the triplo-IV progeny have normal eyes, and among the diplo-IV two have normal eyes to one eyeless. Backcrossing an ordinary heterozygote, of course, gives a ratio of 1 normal eyes: 1 eyeless. When the triplo-IV heterozygotes are mated *inter se* the progeny are in a ratio approximating: 18 triplo-IV normal: 8 diplo-IV normal: 1 diplo-IV eyeless. The nine missing members in the series represent individuals with four chromosomes-IV which fail to survive. This ratio is to be compared with the 3 normal: 1 eyeless ratio obtained by mating ordinary heterozygotes together.

Just as results with haplo-IV individuals bear witness to the fact that factors of the fourth group must be borne by the small chromosome-IV, so also these results with triplo-IV fully confirm the supposition. No better support for the chromosome theory of heredity could be expected, for these experiments definitely connect the group-IV of factors with the small chromosome-IV.

**Datura Trisomics.**—Perhaps the most remarkable series of non-disjunctional products hitherto discovered is that of the Jimson weed, *Datura stramonium.* The normal haploid chromosome number of this species is twelve. From time to time individuals have been found which differ from the normal type of the species in a complex of characters, relatively slight differences in capsule form, leaf shape, habit, etc. All are less viable than the normal form, all produce a certain percentage of aborted pollen grains, and all exhibit a peculiar mode of inheritance.

Twelve primary mutants of this class have been discovered. They have been given names from the most conspicuous feature which distinguishes them from normal, *viz.*, globe, poinsettia, cocklebur, ilex,

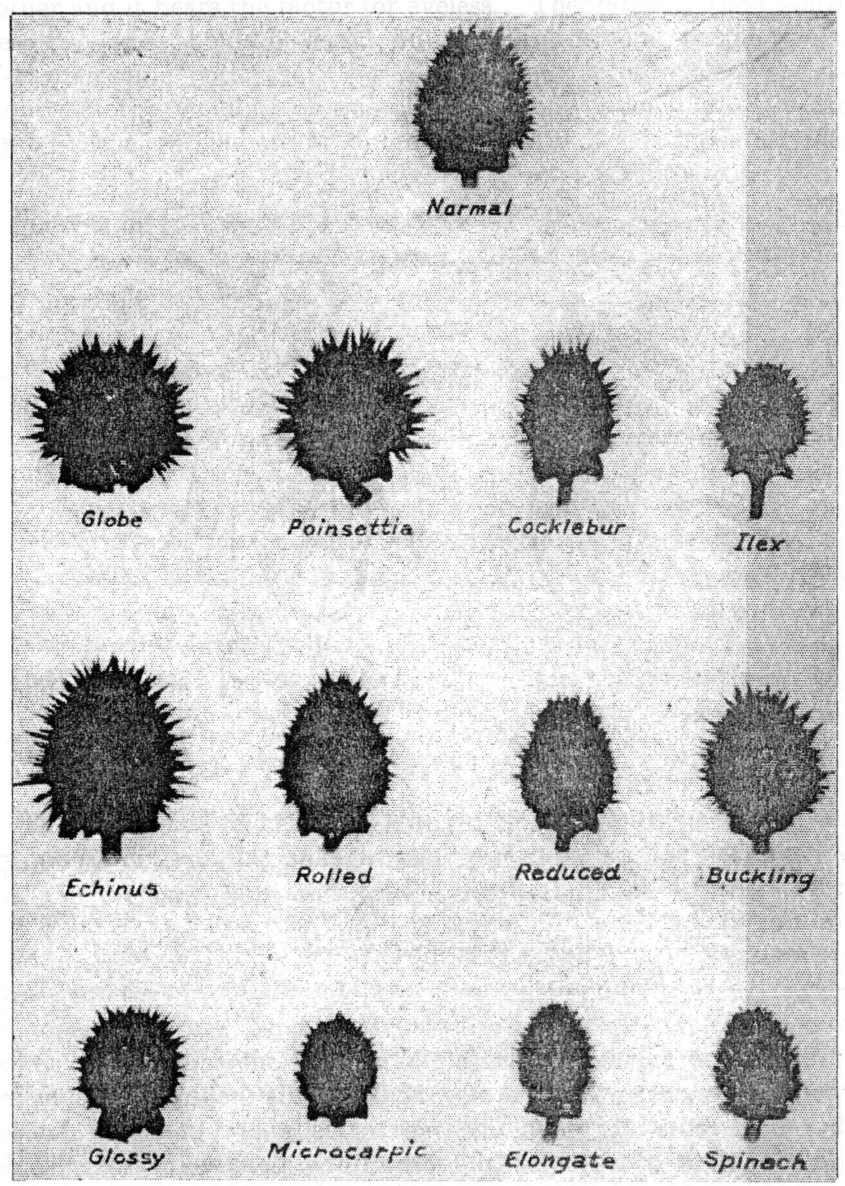


FIGURE 86.—Capsules of the twelve primary trisomics and of the normal type in *Datura stramonium*. (*From Blaskeslee and Belling, Courtesy the Journal of Heredity.*)

echinus, rolled, reduced, buckling, glossy, microcarpic, elongate, and spinach (?). Distinctive capsule characteristics of these forms are illustrated in figure 86. Cytological studies have shown that each of the twelve types has a single extra chromosome ($2n + 1$).

The mode of inheritance of these trisomic characters is highly distinctive. Consider the trisomic, poinsettia, as an example. The following data are reported by Blakeslee and Farnham:

| Type of progeny | Normal | Poinsettia | Per cent poinsettia |
|---|---|---|---|
| Poinsettia selfed | 5,316 | 2,098 | 28.30 ± 0.35 |
| Poinsettia ♀ × poinsettia ♂ | 3,041 | 1,384 | 31.28 ± 0.47 |
| Poinsettia ♀ × normal ♂ | 2,864 | 1,304 | 31.29 ± 0.48 |
| Normal ♀ × poinsettia ♂ | 1,126 | 0 | 0.00 |

From these records it is evident that the poinsettia complex is transmitted exclusively through egg cells. Theoretically, of course, half the gametes, both egg cells and pollen grains, should bear an extra chromosome. While it has not been determined accurately for what reasons the actual results differ from the theoretical, yet it has been shown that in all twelve of these types, transmission through ovules amounts to an average of about 25 per cent, and through pollen grains, to not more than 2 or 3 per cent. Buckholz and Blakeslee have shown that low ratio of transmission through pollen grains may possibly be due to relative rate of growth of pollen tubes of normal grains and those having an extra chromosome. The latter grow more slowly; consequently with an excess of pollen very few of them reach the ovules before all have been fertilized by normal pollen.

The poinsettia character complex, therefore, is determined by the presence of an extra chromosome. Similarly the other eleven all depend upon the presence of a single additional chromsome. It is natural to conclude that each type of trisomic is due to the presence of an extra member in some specific chromosome set, a conclusion which is to a certain extent confirmed by cytological study. Belling has shown that there are considerable differences in size among the chromosomes of this species. In the reduction division through conjugation of homologous chromosomes the trisomes exhibit eleven bivalents and one trivalent. In the trisomic, rolled, the mean product of length and breadth of members of the trivalent was 54.9 ± 0.9; in echinus, 32.1 ± 0.3; in microcarpic 30.9 ± 0.3; and in globe, 20.9 ± 0.4.

In Datura a number of pairs of ordinary Mendelian characters are known, among which *P-p* for purple *vs.* white flower color and *A-a* for spiny *vs.* smooth capsules have been studied extensively in relation to trisomic types. Thus poinsettia has been found to give ratios for purple *vs.* white which demonstrate that the factors for this pair of characters *P-p* must be borne by the members of the chromosome set concerned with the production of poinsettia. To illustrate, consider the results obtained from crosses of purple poinsettia with white normal. $F_1$ consists of approximately 30 per cent of purple poinsettias and 70 per cent of purple normals. The $F_1$ purple poinsettias must be $PPp$ on the above assumptions, since two chromosomes bearing $P$ are received from the mother, and the homologous chromosome from the male parent bears $p$. The $F_1$ purple normals are obviously $Pp$ and give the usual results. The $F_1$ purple poinsettias have been analyzed in various ways with the results shown in table 39. An $F_1$ purple poinsettia $PPp$, assuming random

TABLE 39.—TRISOMIC TRANSMISSION OF PURPLE VS. WHITE IN RELATION TO POINSETTIA IN DATURA (*Data from Blakeslee*)

| Parentage | Purple poinsettia | White poinsettia | Purple normal | White normal |
|---|---|---|---|---|
| $PPp \times PPp$ observed | 692 | 0 | 1,826 | 220 |
| 8: 1 calculated | 692 | 0 | 1,819 | 227 |
| $PPp \times pp$ observed | 409 | 0 | 645 | 281 |
| 2: 1 calculated | 409 | 0 | 617 | 309 |
| $pp \times PPp$ observed | * | * | 1,958 | 1,040 |
| 2: 1 calculated | 0 | 0 | 1,999 | 999 |
| $Ppp \times Ppp$ | 870 | 223 | 1,287 | 995 |
| 5: 4 7:2 | 850.1 | 242.9 | 1,267.8 | 1,014.2 |
| $Ppp \times pp$ | 206 | 84 | 219 | 436 |
| 2: 1 1: 2 | 193.3 | 96.7 | 218.3 | 436.7 |
| $pp \times Ppp$ | * | * | 1,166 | 2,392 |
| 1: 2 | 0 | 0 | 1,186 | 2,372 |

* Not determined but probably none since poinsettia is not transmitted through pollen grains.

assortment of chromosomes in the trivalent set, will give the following three types of distribution:

$$Pp - P$$
$$Pp - P$$
$$PP - p$$

The gametic ratio, therefore, should be

$$PP + 2Pp + 2P + p$$

From this ratio it is possible easily to determine theoretical compositions for the populations in the table, but in making these determinations correction should be made for the ratio of transmission of the extra chromosome in ovules and pollen grains. It is most convenient to consider poinsettia and normal classes in progenies separately and to deal with purple-white ratios within these classes. The results may all be deduced from the checkerboard in figure 87 for $F_1$ purple poinsettia selfed. In this instance all the poinsettia plants will be purple; and among the normals the ratio will be 8 purple: 1 white. From $F_1$ poinsettia purple $\times$ normal white a progeny should be secured corresponding to the four squares in the right-hand column of the checkerboard. Again all the poinsettias should be purple and the normals should be distributed in the ratio of 2 purple:1 white. Finally from white normal ♀ × $F_1$ purple poinsettia, since there is no transmission of the poinsettia complex through pollen grains, the results as shown in the two lower squares of the checkerboard should be all normal in the ratio of 2 purple:1 white. Table 39 shows that the calculated frequencies based on these ratios are in close agreement with observed results.

In a similar way it has been shown that the members of the chromosome set concerned in the production of the trisomic, cocklebur, bear the factors *A-a* for spiny *vs.* smooth capsules. With all other primary trisomics, the normal Mendelian ratios are secured. These results should be compared with those described above for triplo-IV Drosophila. The presence of an extra member in the chromosome set concerned with transmission of a particular pair of factors is shown to give rise to the expected type of irregularity in the distribution ratios.

| ♂ | 2*P* | *p* |
|---|---|---|
| ♀ *PP* | 2 *PPP* purple poinsettia | *PPp* purple poinsettia |
| 2*Pp* | 4 *PPp* purple poinsettia | 2*Ppp* purple poinsettia |
| 2*P* | 4*PP* purple normal | 2*Pp* purple normal |
| *p* | 2*Pp* purple normal | *pp* white normal |

FIG. 87.—Checkerboard depicting results of self-fertilizing a purple poinsettia $F_1$ from purple poinsettia ♀ × white normal ♂. Owing to the low ratio of transmission of poinsettia, the two groups should be considered separately.

**Balanced Variations in Chromosome Number.**—In the balanced series of chromosomal mutants, forms have been discovered with *n*, 3*n*, and 4*n* chromosomes in their somatic cells as compared with the normal 2*n* condition. Of these forms those with even multiples of the haploid number, 2*n* and 4*n*, may be called even balanced forms; whereas the *n* and 3*n* types are conveniently distinguished as odd balanced types.

In general features balanced mutants do not differ strikingly from each other, except for the fact that odd members are highly sterile, as shown among plants by the fact that many of their pollen grains are shriveled and devoid of contents, and they produce very few seeds. The most conspicuous difference is in size, the different types showing an ascending series in the order *n*, 2*n*, 3*n*, 4*n*, as shown particularly well in Datura, in which all four types are available for comparison. In other respects, save for fertility, they differ but little from normal. It seems at first paradoxical that the differences occasioned by adding a single chromsome to the normal complex are more striking than those which arise from doubling the entire chromosome number; but in the former case the normal proportions of the germinal elements are altered; in the latter they remain unchanged.

**Haploidy.**—Haploidy is a normal condition in the gametophyte in plants, and in the male in rotifers and hymenopteran insects. In these cases, however, reproductive cells are formed without reduction. Haploidy as a mutant condition of the sporophytes in higher plants has been studied only in a few species, notably in *Datura stramonium*, wheat (Gaines and Aase), the ten-weeks stock (Frost and Lesley, unpublished), and the common tobacco, *Nicotiana tabacum*.

In those instances which have been studied, the characteristic features and behavior are very similar. The haploid sporophytes, as compared with the normal diploid representatives of the corresponding species, are somewhat dwarfed and less vigorous but otherwise exhibit comparatively slight differences in morphological characters from normals. They exhibit, however, a high degree of sterility as is shown by their pollen grains, which are mostly shriveled and devoid of contents.

The reason for the high sterility of haploids has been shown by cytological examination. In pollen formation, they exhibit meiotic divisions which are very irregular in appearance. This is due to the fact that the unpaired chromosomes are drawn into the spindle and distributed by chance to the poles. In the haploid of *Datura stramonium* with 12 chromosomes, the chromosomes are distributed 6 < > 6, 5 < > 7, 4 < > 8, etc. The second division is equational. The pollen cells thus formed may theoretically have from 0 to 12 chromosomes; but the great majority will have intermediate numbers such as 4, 5, 6, 7, 8. Apparently these diminutive pollen cells soon abort. Rather frequently, however, instead of the pseudoreduction above described, the pollen mother cell divides but once, equationally, giving two cells, each with the full haploid chromosome complement. These apparently develop normally and constitute the only functional pollen grains.

The Datura haploid self-fertilized gives diploid offspring only, which probably points to the fact that only gametes containing a full haploid complement of chromosomes may function. The offspring thus produced are of interest genetically because, in consequence of their derivation by self-fertilization from a haploid parent, they must be complete homozygotes.

Progeny from the *Nicotiana tabacum* haploid have been obtained only from crosses of diploid ♀ × haploid ♂. Like the offspring of the Datura haploid they are almost entirely normal diploids, but about 2 per cent of the plants appear to have less than the full diploid complement of chromosomes. This indicates that Nicotiana haploids may produce some viable pollen grains and ovules with less than the full haploid complement of chromosomes. It may be that in this species, with $n = 24$ chromosomes, a gamete containing one or two less than the normal number of chromosomes may function.

Theoretically, if all gametes of a haploid proved viable, a cross between a haploid and a diploid should give an entire range of forms from haploid to diploid. Assuming, as shown by the lack of viability of nullo-IV in Drosophila, that zygotes which do not have a complete germinal system are inviable, none of these forms would be constant or would give rise to any constant form except the normal diploid. However, it should be possible to establish from such a mating a complete set of monosomics ($2n$-1), which might then be employed for Mendelian analysis in the same way as monosomic-IV in Drosophila. These should also furnish interesting material for comparison with the corresponding trisomics.

**Triploidy.**—Triploidy has been discovered as a mutant condition in numerous species of plants and also in some animals. Most complete studies of the phenomenon have been made in *Oenothera Lamarckiana* var. *semigigas, Datura stramonium* and *Drosophila melanogaster*. The Oenothera triploid is involved with a complex genetic situation which makes it difficult to utilize it as a demonstration of triploidy uncomplicated with other phenomena.

Triploids may arise in at least three different ways: (1) by omission of the reduction division, thus giving diploid gametes which give triploid zygotes on union with normal haploid gametes; (2) by production of diploid gametes by a tetraploid group of cells, which have arisen through doubling of the somatic chromosome complex; and (3) by crossing tetraploids with diploids. In Datura the triploid condition has been obtained only from crosses of tetraploid ♀ × diploid ♂. The triploid plants have the same general appearance as diploids on an enlarged scale. They are. however, highly sterile as shown by the fact that a large proportion of pollen grains are aborted.

Cytological studies of reduction in triploids show that there are twelve sets of trivalents instead of twelve sets of bivalents as in normal diploids. In each trivalent, one member goes to one pole and two to the other in the first division, and the second is equational. Since the assortment in different trivalents is at random, nuclei are produced with chromosome numbers ranging from 12 to 24 in frequencies approximating those stipulated by chance, as shown by Belling for actual determinations of assortment in 84 pollen-mother cells which gave the following results:

| Type of distribution | 12/24 | 13/23 | 14/22 | 15/21 | 16/20 | 17/19 | 18/18 |
|---|---|---|---|---|---|---|---|
| Number observed | 1 | 1 | 6 | 13 | 17 | 26 | 20 |
| Number calculated | 0.04 | 0.5 | 2.7 | 9.0 | 20.3 | 32.5 | 19.0 |

In some instances lagging of chromosomes was observed, as a consequence of which one or more chromosomes would fail to be included in the four nuclei and would give rise to diminutive cells, or microcytes.

The triploid Datura produced progeny both by self-fertilization and from crosses of triploid ♀ × diploid ♂. The cytological studies show that all types of gametes are formed according to the laws of chance, but the high percentage of aborted pollen grains indicates that many of these fail to function. In progeny from selfed triploids determinations of chromosome numbers of 87 plants disclosed the presence of 30 normal diploids ($2n$), 43 simple trisomics ($2n + 1$), 11 double trisomics ($2n + 1 + 1$), and 3 tetraploids ($4n$). From crosses of triploid ♀ × diploid ♂, 697 plants were classified, 215 normal diploids, 381 simple trisomics, and 101 double trisomics. These results indicate that the viable gametes are limited to $n$, $n + 1$, $n + 1 + 1$, and possibly $2n$. The intervening numbers in the series, which should, on the basis of probability be much more frequent, are lacking. No figures are presented for diploid ♀ × triploid ♂ which should provide evidence on constitution of viable pollen grains; but the close agreement in results between triploid selfed and triploid ♀ × diploid ♂ indicates that viable pollen grains are almost entirely $n$, with possibly an occasional functional $2n$ pollen grain. Among the trisomics, eleven of the twelve types occurred in approximately equal frequencies. The twelfth type, spinach, occurred only twice, which may indicate that it is a secondary trisomic and that the corresponding primary has not been discovered (see Chapter XXV).

Similar studies made by van Overeem on the triploid mutant of *Oenothera Lamarckiana* var. *semigigas* indicate that in this form, which has $3n = 21$ chromosomes, the viable ovules represent all possible chromosome numbers in approximately the comparative frequencies stipulated by chance. The viable pollen grains, however, were clearly of only two types, $n$ and $2n$. It is possible that species differ widely as respects viability of unbalanced chromosomal combinations; but the method of mating triploids reciprocally provides a method of determining accurately which combinations are viable.

In *Drosophila melanogaster* triploid females have been discovered, but there are no corresponding males. When they are mated to normal males, a variety of chromosomal types is secured, mostly representing variations in the numbers of X-chromosomes in relation to the number of total sets of autosomes, as shown in table 40. Apparently the normal ratio of X-substance to A-substance (this representing the complete autosomal complex) is 1:1 in females and 1:2 in males. A ratio lying between these two values, such as 2:3 in XX + AAA flies gives an intersexual form, whereas extreme ratios such as 3:2 in XXX + AA gives an extreme, sterile female type and 1:3 as in X + AAA an extreme, sterile male type. The idea involved is that the X-substance has a net female tendency as opposed to a net male tendency in the autosomal substance, and that the balance struck by these opposing tendencies determines sexual development. This hypothesis of factor balance has received additional support by the discovery of tetraploid flies, XXXX + AAAA, which are normal females and by the further demonstration that the haploid condition X + A gives a female type of development. Conspicuous by their absence in the progeny of triploids are individuals unbalanced for chromosomes II and III, *i.e.*, containing one of the former and two of the latter or *vice versa*. Evidently such an unbalance is incompatible with development. On the other hand unbalance in the IV-chromosome may be obtained, presumably in conjunction with all the types listed in the table.

Table 40.—Relation of Sex to Factor Balance in Offspring of Triploid ♀ × Diploid ♂ in Drosophila (A = complete set of autosomes)

| Chromosomal formula | Ratio X:A | Sex type |
|---|---|---|
| XXX + AA | 1.5 | superfemale |
| XXX + AAA | 1.0 | triploid female |
| XX + AA | 1.0 | normal female |
| XX + AAA | 0.67 | intersex |
| X + AA | 0.50 | normal male |
| X + AAA | 0.33 | supermale |

Genetic evidence has been obtained on a number of interesting points from triploid females. In linkage experiments with normal diploids two large parental types of gametes are obtained because of distribution of the two chromosomes in each set to gametes without crossing-over. In triploids, three large parental classes are obtained because three chromosomes are present in each set. Moreover, from very complex and ingenious experiments it has been shown that chromosomes may be built up from parts of all three of the chromosomes in a set; *i.e.*, all three members of a chromosome set may be involved in crossing-over. It has also been shown from the character of classes obtained that crossing-over occurs in the two-strand stage; *i.e.*, each chromosome splits equationally before undergoing crossing-over. Here again there are interesting examples of parallel investigations in the field of cytology and genetics.

**Tetraploidy.**—Complete doubling of chromosome number, giving rise to tetraploids, has been observed in a number of plant species and in some lower forms of animals. Particularly pertinent studies have been made in plant species, of which the most extensive are those with the tetraploid form of *Datura stramonium*.

Such information as is available seems to indicate that the doubling of chromosome number which is responsible for the occasional appearance of tetraploid plants occurs immediately or soon after fertilization, rather than through the occasional production of unreduced gametes by diploid plants. The product of the latter process would almost inevitably be a triploid individual; since the chances of a diploid gamete meeting another of the same constitution are practically nil. Observations also show that occasional branches on diploid plants may be tetraploid, indicating that mutation to the tetraploid condition has occurred in a somatic cell. The process may be easily visualized as one in which a nuclear division is arrested after the chromosomes split, whereupon a single nucleus is formed with $4n$ instead of $2n$ chromosomes.

Tetraploid individuals are not conspicuously different from the corresponding diploid forms. They are somewhat enlarged in size and are usually less fertile than diploids; but the gametes do not show a high percentage of abortion, and the progeny are for the most part uniformly tetraploids. Crossed with diploids, which in some instances is attended with considerable difficulty, triploids are formed which are usually highly sterile as has been shown. Tetraploid forms consequently bear much the same relations to the parental form as many distinct species bear to each other.

Detailed studies of reductional phenomena of the tetraploid *Datura stramonium* have been made by Belling. He finds that the chromosomes in this form, 48 in number, associate in twelve groups of fours, quadrivalents. Usually two members of each quadrivalent go to each pole, so that most of the gametes have $2n = 24$ chromosomes. In about 3 per cent of cases a 3< >1 distribution occurs. As a consequence less than 75 per cent of figures show 24< >24 distribution; and distributions of 23< >25, 22< >26, and 21< >27 have been observed in descending frequency in the order named, as shown below:

| 24< >24 | 23< >25 | 22< >26 | 21< >27 |
|---|---|---|---|
| 73.2 | 24.4 | 2.1 | 0.2 |
| 342 | 114 | 10 | 1 |

Occasionally also elimination of chromosomes occurs, and rarely non-reduction with formation of giant pollen grains having $4n$ chromosomes.

The genetic behavior of the factors in tetraploids constitutes a distinct problem since there are four chromosomes in each set. This problem has been investigated most fully in Datura for the factor pairs *A-a* for spiny *vs.* spineless capsules and *P-p* for purple *vs.* white flowers. The data for the latter character pairs are somewhat more extensive because the color difference may be determined in young seedlings, whereas the former necessitates growing the plants to the stage of capsule formation. When purple tetraploid is crossed with white tetraploid, $F_1$ is purple; but since the gametes are diploid, the $F_1$ is of the constitution $PPpp$. Assuming random distribution in the quadrivalent, the following types of distribution should occur:

$$\begin{array}{ll} PP' & pp' \\ Pp & P'p' \\ P'p & Pp' \end{array}$$

The gametic series is, therefore,

$$PP + 4Pp + pp$$

Self-fertilization should give a ratio of 35 purple:1 white, and backcrosses to white a ratio of 5 purple:1 white. Actual results of these experiments are in agreement with expectation (*cf.* table 41), and

TABLE 41.—INHERITANCE OF MENDELIAN CHARACTERS IN TETRAPLOID DATURAS
(*Data from Blakeslee*)

| Parentage | Dominant | Recessive |
|---|---|---|
| $PPpp \times PPpp$ | 9,199 purple | 225 white |
| Calculated 35:1 | 9,162.2 | 261.8 |
| $PPpp \times pppp$ | 546 | 122 |
| Calculated 5:1 | 556.7 | 111.3 |
| $AAaa \times AAaa$ | 3,383 | 118 |
| Calculated 35:1 | 3,403.7 | 97.3 |
| $AAaa \times aaaa$ | 518 | 137 |
| Calculated 5:1 | 545.8 | 109.2 |

numerous other genetic tests upon which Blakeslee and his associates have presented data, both for the flower and plant color factors $P$–$p$ and for those responsible for spiny or smooth capsules $A$–$a$, have demonstrated essential agreement of observation and expectation. The dihybrid ratio, if it may be called such, in tetraploids has also been studied for the $A$-$a$ and $P$-$p$ pair of factors. $F_1$ here is $AAaaPPpp$, and $F_2$ is in close agreement with the expected ratio, *viz.*,

1,225 purple spiny:35 purple smooth:35 white spiny:1 white smooth.

An interesting case of tetraploidy under experimentally controlled conditions has been reported in mosses. In mosses the gametophyte is the dominant phase in the life history, the sporophyte being restricted to a slender stipe surmounted by a capsule. The Marchals have shown that sections of the stipe or capsule placed under proper conditions will grow and produce protenema which will give rise to typical leafy shoots like those of the normal gametophyte. Since this regeneration, however, has occurred from sporophytic tissue with the $2n$ chromosome number, these gametophytes have the diploid number of chromosomes. and the corresponding sporophytes which they produce are tetraploid. Regeneration from these tetraploid sporophytes in turn gave rise to $4n$

gametophytes, but they were sterile and failed to produce octoploid $8n$ sporophytes.

**Modified Tetraploids.**—The 3<>1 type of distribution shown to occur occasionally in tetraploids obviously provides a mechanism for the production of a group of modified tetraploids showing both plus and minus deviations from the $4n$ chromosome number. Actual counts of 62 plants from a self-fertilized tetraploid Datura showed 55 of $4n$, 1 of $(4n - 1)$, 5 of $(4n + 1)$, and 1 of $(4n + 1 - 1)$. These modified tetraploids in a sense are parallels to the modified diploids previously described; but although not such thorough studies have been made of them, the indications are that a greater variety of forms may be obtained than of modified diploids. Of these the most readily handled and doubtless most easily secured are those with $4n - 1$ or $4n + 1$ chromosomes. Twelve of each of these are possible, as among modified diploids; but only five trisomics and eight pentasomics have been recognized.

In the pentasomic globe $(4n + G)$ in Datura, the characteristic features are the same as in the trisomic globe $(2n + G)$ but less extreme in expression, doubtless because the chromosomal unbalance is less marked in the previous case. Also the transmission ratios for pentasomic globe appear to be higher, 60 per cent in selfed progenies and 14 per cent through pollen grains. Addition of further chromosomes to this set has given $(4n + GG)$ and $(4n + GGG)$ globes which show a progressive accentuation of globe characters. Evidence as to transmission of factors when five or more members are present in a chromosomal set has not been published, nor details of the breeding behavior of modified tetraploids. Modified tetraploids of the types $(4n + 1 - 1)$ and $(4n + 1 + 1 - 1 - 1)$ have also been discovered. They have the same chromosome number as normal tetraploids but some chromosome sets have three and some five members. Mere count of chromosome number, therefore, is an insufficient basis for classification. Evidently the constancy of the tetraploid condition is a relative matter depending on the superior viability of the $4n$ condition and the reversion of modified tetraploids to this type.

**Resumé.**—A review of these instances of variations in chromosome number reveals a number of interesting things. In the first place the question of interchromosomal balance in development may be considered. A change in the number of chromosomes, so long as it does not disturb the normal proportions of elements as seen in the balanced series of forms, haploid, diploid, triploid, and tetraploid; does not greatly alter morphological characters but does increase size with ascending multiples of the haploid number. The results from mosses, however, seem to set a limit to the extent to which this process may be carried.

On the other hand disturbance of the normal proportions of the germinal elements, as seen in monosomics, trisomics, etc., does alter, sometimes strikingly, the characters of the individual and its vigor and fertility. Moreover, the alterations are specific; *i.e.*, a change in the number of members in a given chromosomal set has a particular character effect. Consequently, as has been surmised from genetic studies, each chromosome must bear a particular set of factors. The set of factors borne in each chromosome also must have manifold effects, reaching into every part of the body, for the character effects produced by these unbalanced chromosomal types are complex and affect every feature of the plant.

There are also evidently limits to the extent to which unbalance in chromosome number may go. This is shown particularly well in the Daturas where a fringe of unbalanced types cluster around the diploid and tetraploid types as centers. In the case of modified diploids, all except one have more than the diploid number. Apparently subtraction of a chromosome is a more serious matter than addition of one. Addition of successive chromosomes has a cumulatively adverse effect upon vigor. Moreover, addition of two chromosomes of a kind, as in tetrasomic globe Datura, has such a deleterious effect upon vigor and fertility that this method of permanent increase in chromosome number is almost unthinkable. The further fact that the four chromosomes of such a set conjugate as a quadrivalent and show a certain proportion of 3<>1 distribution is a further argument against non-disjunction as a method of permanent increase in chromosome number.

The most significant feature in studies of chromosomal variations lies in the parallel which may be drawn between cytological observation and results of genetic experiments. It is becoming evident that each specific type of chromosomal variation presents characteristic breeding features by which a skilled investigator may detect it, and actual cytological examination may be restricted to a verification of the conditions presumed to exist from a study of genetic phenomena. This is particularly true when associated Mendelian factors are available; for evidently the disturbances in transmission are characteristic of the conditions existing in the set of chromosomes governing the situation. When individual chromosomes may be recognized by size or other morphological features, these investigations present a method by which particular chromosomes may be determined to contain specific groups of Mendelian factors.

### References

Morgan: Physical Basis of Heredity, Chapter XIII.

Morgan: Theory of the Gene, Chapters VIII–XII.

Morgan, Bridges, and Sturtevant: Genetics of Drosophila, Chapters XI–XV.

# CHAPTER XXV

## SECTIONAL VARIATION

A number of instances have come to light which indicate that a portion of a chromosome has become lost or has been moved to a new location in the germinal system. To these instances may be added others in which chromosomes have been split into two sections, or two chromosomes have become permanently joined, or an entire chromosome has suffered some type of rearrangement of materials. These cases are included together under the head of sectional variation, not, however, a very satisfactory term for all of the phenomena. Many of the phenomena here described have not been fully investigated; but nevertheless it is evident that they promise to yield interesting results. For the sake of completeness of treatment, it is hardly possible to do less than present a general account of them, although future research may necessitate a thoroughgoing revision of present conceptions.

**Deficiency.**—The studies on haplo-IV in Drosophila described in the preceding chapter show that loss of one chromosome-IV produces characteristic alterations in somatic characters by which it is possible to separate haplo-IV from normal flies by visual examination. The character differences peculiar to haplo-IV as compared with normal flies are slight; but they affect many portions of the body. It has also been observed that many mutations have appeared from time to time in Drosophila which belong to this same category; *i.e.*, they exhibit slight but numerous character differences as compared with wild type. They also exhibit other features in common. Like haplo-IV they are heterozygous characters incapable of existence in a homozygous condition. They are transmitted as dominants to approximately half the offspring when mated to wild type, or to two-thirds of the offspring when mated *inter se*, although in some instances they are defective in viability. Linkage determinations show that they may occur in almost any portion of the germinal system. Bridges has called them minutes, on account of their slight effects and has suggested that like haplo-IV they have arisen as a consequence of actual loss of a small portion of the germinal material. In all cases hitherto investigated, the portion lost from a given chromosome is so small that cytological demonstration has been impossible; reliance must be placed on genetic data. Loss of a section of a chromosome is known as deficiency.

The sex-linked dominant, notch-8, which has been studied on a large scale by Mohr may be considered here as typical of the phenomenon of deficiency. Notch flies are characterized by production of wings scalloped at the tips as shown in figure 88. Notch females mated to normal males give progeny as follows:

1 notch ♀ : 1 wild type ♀ : 1 wild type ♂

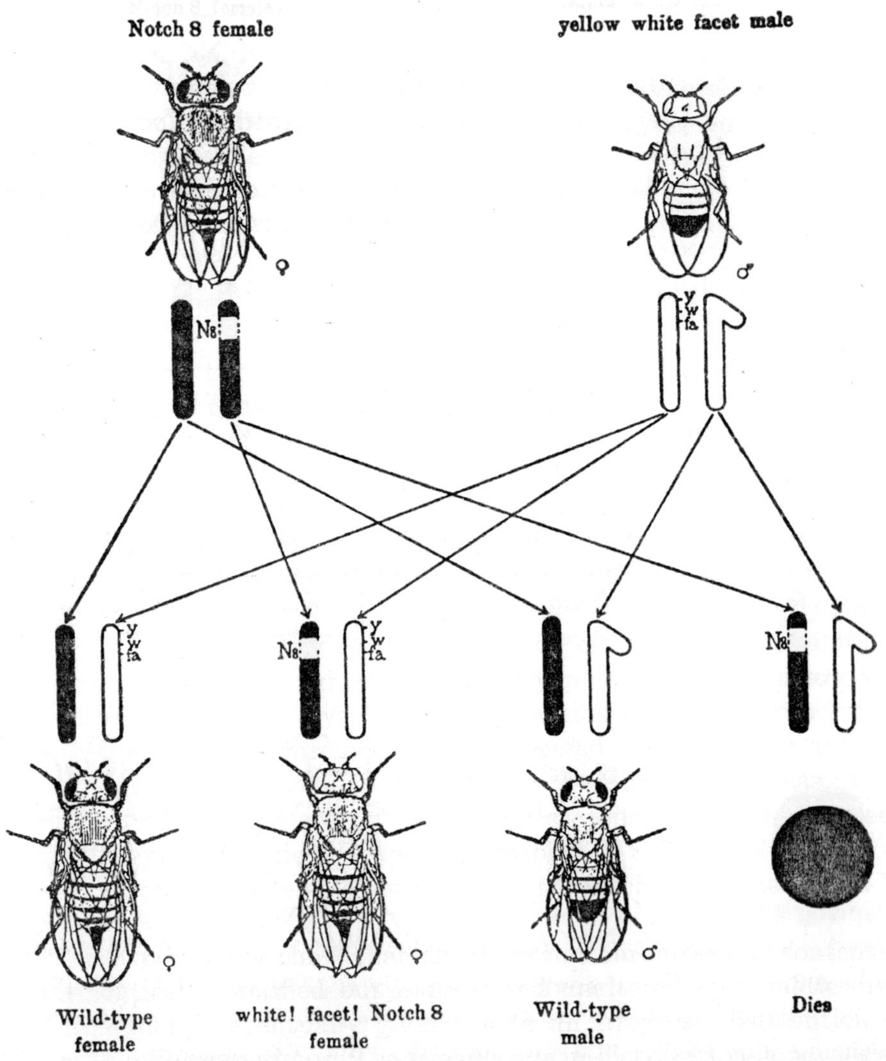


FIGURE 88.—Operation of inheritance involving a deficiency. Notch-8 females, heterozygous for deficiency represented by a blank space in one of the X-chromosomes, crossed with yellow-white facet males gives wild type ♀♀, white facet ♀♀ and wild type ♂♂ in approximately equal numbers. The expected notch males die on account of the lethal action of the deficiency. The $F_1$ notch females are white facet because the deficiency covers this section; they are not yellow, because the factor for yellow lies outside the deficient section. Facet, a roughened-eye character, is not shown in the diagram. (*From Mohr.*)

Evidently notch males die, indicating that there is a recessive lethal effect associated with the character.

In an attempt to determine the locus of notch, numerous crosses were made of notch females with males of various sex-linked mutant types, some of which gave surprising results. Thus notch ♀ × white ♂ gave

1 notch white ♀ : 1 wild type ♀ : 1 wild type ♂

Similarly, notch ♀ × apricot ♂ gave

1 notch apricot ♀ : 1 wild type ♀ : 1 wild type ♂

In this last instance the eye color of the notch flies was considerably lighter than apricot, in fact, lighter than apricot-white compound.

Since the notch flies were not heterozygous for white, in which case they should have produced occasional white-eyed flies by crossing-over, it was assumed from these results that notch was due to loss of a portion of the X-chromosome including the locus of *W*. The operation of this conception may be seen by reference to figure 88. This led to experiments designed to determine the extent of the lost section. Crosses were made with numerous mutant types the factors for which lie in the X-chromosome near *W*, such as yellow, scute, broad, prune, facet, abnormal, echinus, etc. With white and its allelomorphs, and with facet and abnormal, the $F_1$ notch female exhibited the mutant characters in addition to notch, indicating that the deficient section covered these loci; with the others, mutant characters did not appear in association with notch in $F_1$ females; consequently these must lie outside the deficient section. These data together with the linkage tests showed that the deficient section extended from about 1.5 to 5.3; consequently its length must have been about 3.8 units.

The parallelism of these phenomena with those of haplo-IV, in which the actual absence of one IV-chromosome was demonstrated by cytological examination, justifies the supposition that both are phenomena of the same kind. Just as there is pseudo-dominance for IV-chromosome recessive characters in association with haplo-IV, so there is pseudo-dominance of certain sex-linked factors in association with notch. Similarly just as character effects of group-IV factors are exaggerated in haplo-IV, so are those included in notch deficiency. Further interest attaches to the case from the fact that numerous recurrences of notch deficiency have been observed which affect the same portion of the chromosome, but the deficient region varies in extent. Three of them did not cover the white locus but did include facet and abnormal.

Not all the instances of supposed deficiency, however, exhibit dominant character effects by which they may be followed. A case in point is bar deficiency. An eosin ♀ crossed with a white bar male should produce eosin-white bar ♀♀ + eosin ♂♂, but one $F_1$ female in such a culture was eosin-white but not bar. It was presumed in this instance that the white-bar chromosome had lost a section including the bar locus.

Outcrossed to a wild type ♂, this female gave a 2♀♀ : 1♂ sex-ratio, demonstrating that the deficient chromosome had a lethal effect; and the deficiency was maintained in stock by the simple expedient of selecting cultures which gave this sex-ratio.

The bar locus (57.0) is 0.2 unit to the right of forked (56.8), a recessive character. Crosses of deficient-bearing females with forked males gave a female progeny half of which were forked (!) and half wild type. The lost fragment, therefore, included the forked locus, but tests with rudimentary at 55.1 and fused at 59.5 showed that the loss was not extensive enough to cover these two loci. The length of the section must, therefore, be *more than 0.2 and less than 4.4 units. Linkage data* indicated that it probably had a length of about 0.7 unit. Although no observable morphological difference between normal and heterozygous bar-deficient females was discovered, it was found that bar-deficient females were less fertile and lower in viability than wild-type females.

It will be recalled that most of the dominant mutations are lethals in the homozygous condition, and the suggestion has been made that they are really deficiencies for a small section of the chromatin material. Some support for this hypothesis has been obtained in certain instances. Mohr has found that the dominant gull located at 10 in chromosome-II crossed with fat, a group-II recessive, gives fat-gull and wild-type offspring. This would indicate that gull is deficient for the section in which fat is located. Bridges has pointed out that delta and hairless, two group-III dominants of this class, exhibit character effects which are approximately opposite. Crossed together, the $F_1$ delta-hairless combinations exhibit both characters in a comparatively less extreme form than that of the types. The supposition is made that every section of the germinal material contains factors which modify the expression of numerous features of the body, some exaggerating and some diminishing the expression of specific characters. They may be called plus and minus modifiers. If a small section is removed, it may be unbalanced for these modifiers, containing an excess of plus modifiers of some characters and of minus for others. This results in an upsetting of the normal balance with a production of a specific character complex as a result. If now two deficiencies which are unbalanced in the opposite sense are crossed, the *$F_1$ double deficiencies* may be less extreme because the two neutralize each other.

That the numerous minutes which have been observed in Drosophila are deficiencies has been presumed from their lethal effect when homozygous and their slight but numerous differences from wild type. The conclusion has been strengthened somewhat by the fact that one of them, minute-1, located in the right end of the II-chromosome, is deficient for plexus and arc, two recessives occupying loci about one unit apart.

**Duplication.**—In Drosophila a number of instances are known in which a section of a chromosome has become detached from its normal position and joined onto some other portion of the chromosome, a phenomenon known as duplication. Thus in vermilion sable duplication, apparently a piece of an X-chromosome containing the normal allelomorphs of vermilion and sable became attached to the zero end of the X-chromosome. Since the distance from vermilion to sable amounts to approximately 10 units, tests were made in order to determine whether the duplicated section covered this entire region. These tests, however, indicated that the duplication was restricted merely to the vermilion and sable portions of the region.

Some highly interesting genetic phenomena were obtained from a study of this duplication. For purposes of convenience, the X-chromosome with the duplicated section may be designated as X-*VS*. Males of the constitution (*vs* X-*VS*)Y were found to be wild type. Females of the constitution (*vs*X)(*vs*X-*VS*) were found to be vermilion sable in appearance, and (*vs*X-*VS*)(*vs*X-*VS*) females were wild type. A wild type of the latter constitution crossed with a vermilion sable male gave vermilion sable female and wild-type male $F_1$ offspring. Linkage experiments with the duplicated race may be so arranged as to indicate a locus of zero for vermilion and sable instead of the normal values, 33 and 43 respectively.

**Attachment.**—In *Drosophila melanogaster*, a form known as double yellow has the two X-chromosomes permanently attached. In the double yellow stock, females are XXY and males are normal wild type. Each of the X-chromosomes of the female carries a factor for yellow so that the formula may be represented (*y*X-*y*X)Y. Inheritance then proceeds in the stock according to the following scheme:

| | | | | | | |
|---|---|---|---|---|---|---|
| | double yellow ♀ | | | | normal ♂ | |
| | (*y*X − *y*X)Y | | × | | (*Y*X)Y | |
| gametes | (*y*X − *y*X) | Y | | | (*Y*X) | Y |
| | (*y*X − *y*X)(*Y*X) | + | (*y*X − *y*X)Y | + | (*Y*X)Y | + YY |
| | triplo-X ♀ ♀ usually die | | double yellow ♀ ♀ | | wild type ♂ ♂ | dies |

Gray females with the three X-chromosomes occasionally survive but they are highly abnormal in appearance and are weak and sterile; consequently they have no effect upon the normal behavior of the stock which consists in each generation of yellow females and normal males as a consequence of permanent attachment of the two X-chromosomes in females. The normal transmission of yellow is completely transformed as a consequence of this alteration in the chromosome mechanism.

Cytological examination in this instance confirmed conclusions drawn from genetic data. Double yellow females have a long V-shaped chromosome and a Y-chromosome in place of the two rod-shaped X-chromosomes characteristic of females.

The two attached X-chromosomes occasionally break apart and thus give rise to a small percentage of offspring which do not conform to the above-described scheme. Investigations of the phenomena consequent upon the breaking of the attachment, however, show that the point of fracture is not always exactly at the original point of union of the two X-chromosomes. As a consequence X-chromosomes may be produced which are deficient and others which have duplicated portions.

**Translocation.**—In certain instances in Drosophila genetic evidence indicates that a piece of one chromosome has become detached from its normal position and removed to a new position in another chromosome. Only fragmentary accounts of this phenomenon have been published; consequently it is impossible to go into details concerning it. An interesting instance is that of pale-translocation. Here the right end of chromosome-II covering the loci for a number of known mutant factors has been transferred to the middle of chromosome-III. Races have been built up which contain normal chromosomes-II and chromosomes-III, each of which contains this pale-section. Such individuals are reported to be only slightly different from normals in appearance. However, individuals deficient for the pale-section in both chromosomes-II and with normal chromosomes-III die. The addition of this section to the normal chromosomal complement has little effect upon development; but its subtraction evidently removes vital materials so that such individuals die.

The interesting feature of this phenomenon is that sections of chromatin material may under certain conditions be transferred from one chromosome to another, thus setting up entirely new linkage relations. Despite this fact, however, the developmental effects of the factors remain unchanged; *i.e.*, for development the important desideratum is the presence of the factors; their position in the genetic system appears to be of little consequence. The same conclusion may be drawn from studies of duplication.

**Secondary Trisomics in Datura.**—Certain of the primary trisomics in Datura occasionally but regularly produce other trisomic types which exhibit only a portion of the features of the corresponding primaries. The secondaries thus produced in turn regularly give a small proportion of primaries in their offspring. As shown in table 42, six of the primaries have been found to give such secondary types and Belling has found apparently that the trisomics prone to give secondaries are those which involve chromosome sets having large members.

The exact relationship between primaries and secondaries has not yet been determined, and there is evidence that the same phenomena are not concerned in all. From chromosome measurements, Belling has

TABLE 42.—OFFSPRING OF PRIMARY AND SECONDARY TRISOMICS IN DATURA (*Data of Blakeslee*)

| Type | Per cent primaries | Per cent secondaries | Total |
|---|---|---|---|
| Poinsettia | 27.86 | 0.23 | 30,933 |
| Wiry | 0.73 | 18.31 | 2,998 |
| Cocklebur | 22.95 | 0.01 | 8,860 |
| Wedge | 1.45 | 19.06 | 2,477 |
| Echinus | 25.35 | 0.00 | 718 |
| Mutilated | 1.38 | 13.99 | 2,688 |
| Nubbin (?) | 3.25 | 30.50 | 400 |
| Rolled | 18.16 | 0.14 | 11,140 |
| Sugarloaf | 2.96 | 20.90 | 5,230 |
| Polycarpic | 0.81 | 2.44 | 123 |
| Buckling | 33.57 | 0.03 | 3,101 |
| Strawberry | 3.80 | 30.31 | 947 |
| Maple | 0.20 | 21.27 | 1,984 |
| Elongate | 5.50 | 0.00 | 1,206 |
| Undulate | 0.68 | 23.65 | 592 |

found that in some secondaries the three members of the trivalent chromosome set are of the same size, but the types of union exhibited during reduction are different from those of the primaries. A comparative study of these figures has led to the suggestion that some secondaries have arisen by a reversed type of crossing-over which has given rise to a chromosome which has lost half of its genetic material but has duplicated the other half (figures 89 and 90).

Some of the genetic phenomena bear out this supposition. Thus wedge, a secondary of cocklebur, gives 3:1 $F_2$ ratios for spiny *vs.* smooth capsules, whereas cocklebur, as has been seen, gives trisomic ratios. This difference may be explained by loss in wedge of that portion of the chromosome which bears the factor $A$, and duplication of the other half of the chromosome. In that case the extra chromosome in wedge would have no effect upon expression of the spiny-smooth characters because of deficiency for that portion of the chromosome, nor upon transmission because in effect the wedge chromosome almost always acts as the odd member.

In three instances, rolled, buckling, and poinsettia, and possibly also in echinus, complementary secondaries exist; pairs of secondaries one member of which exhibits certain of the peculiarities of the corresponding primary in a lessened degree, and the other the complementary peculiarities. In two instances it has been shown that one of the secondaries gives disomic and the other trisomic ratios for factors borne in that chromosome set. This observation is obviously in accord with the theory that secondaries have arisen by duplication of a half-chromosome.

In wiry, a secondary of poinsettia, one member of the trisomic set is only approximately half as large as the other members. This has led to the suggestion that wiry is a ($2n + \frac{1}{2}$) mutant. Wiry, however, gives trisomic ratios with the *P-p* pair of factors; consequently the half-chromosome present must bear these factors. In one group Blakeslee and his associates apparently have found corresponding to a given primary

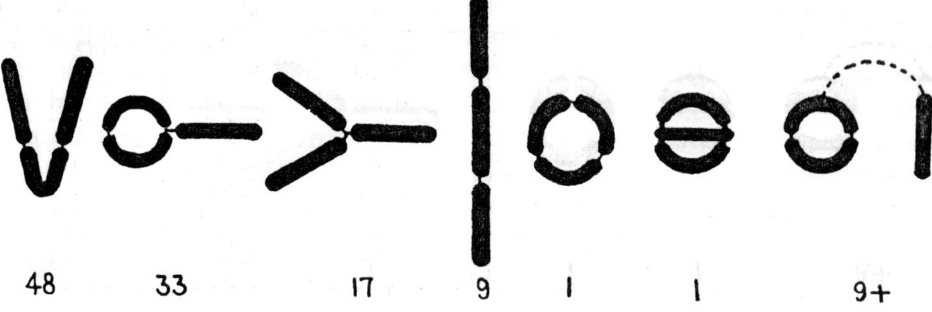


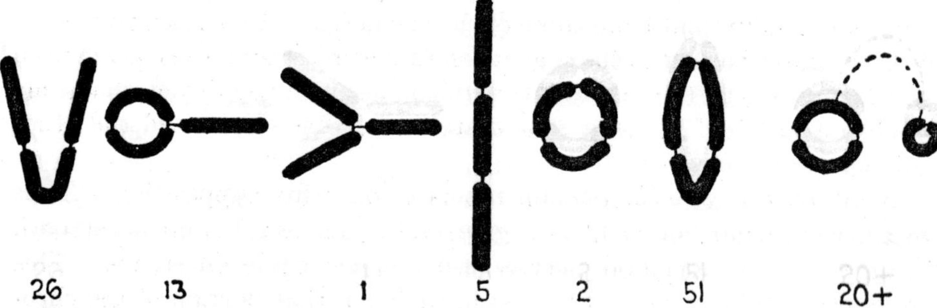


FIGURE 89.—Comparative configurations of the trivalent chromosomes in primary and secondary ($2n + 1$) daturas. Frequencies are indicated beneath the corresponding types of configuration. See figure 90 for significance. (*From Belling.*)

two complementary secondaries with extra double half-chromosomes and two complementaries with extra half-chromosomes.

Evidently the occurrence of more than twelve ($2n + 1$) mutants in Datura does not invalidate the assumption that the addition of a particular chromosome has a specific effect due to alteration of the proportions of the materials present in that chromosome in relation to the sum total of elements present in the system. As a matter of fact, it extends the idea to include varying proportions of portions of chromosomes as well. There is evidence that the types of secondaries above mentioned do not exhaust the possibilities which may arise in connection with internal rearrangement or fragmentation of chromosomes, but these additional instances have not been investigated sufficiently to permit a logical treatment at this time.

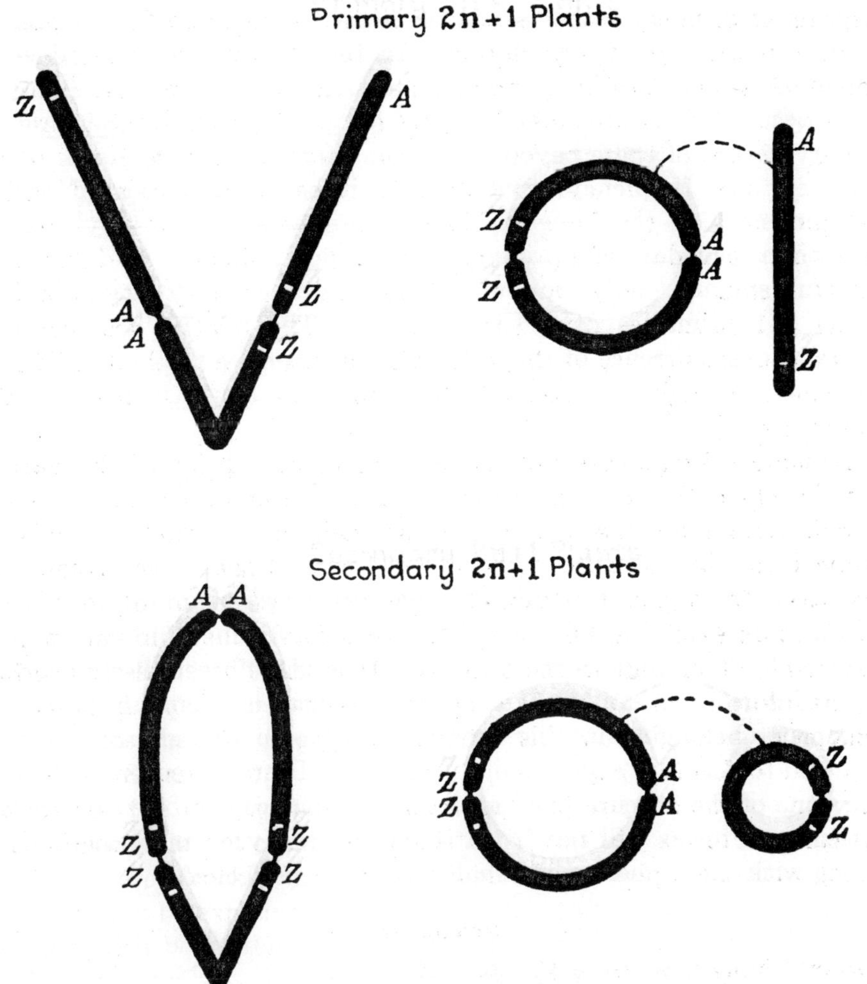


FIGURE 90.—Diagram to illustrate hypothetical interpretation of certain characteristic types of configuration in trivalents of primary and secondary $(2n + 1)$ daturas. The extra chromosome in secondaries has become reorganized so as to have duplicated ends. (*From Belling.*)

Cytological demonstrations of the existence of small extra chromosomes such as occur in wiry have also been made in stocks, rye and Oenothera. In the stock, *Matthiola annua*, a number of definite mutant types have been found to be associated with their presence. In rye the normal chromosome number is seven; but certain strains have eight pairs, the extra chromosome pair being smaller than any pair of the normal chromosomal complex. Studies of genetic phenomena have not been made in this instance.

Where primary and secondary $(2n + 1)$ mutants have been discovered, it is a matter of some importance to be able to distinguish between them. Blakeslee and his associates have shown that the $(2n + 1)$ offspring of triploids either selfed or crossed with diploids provides an accurate method of determination. In the offspring of triploid ♀ × diploid ♂ which has been studied most extensively, eleven $(2n + 1)$ forms occur with approximately equal frequency, thus establishing the primary nature of these beyond question. The twelfth, spinach, occurs in a very low frequency; consequently it has been indicated with a question mark in the foregoing lists of primaries, for it may possibly represent a secondary of a primary as yet undetermined. Aside from the doubtful spinach, only one unquestionable secondary was obtained among 381 primaries from such a cross. The relative frequency of spontaneous occurrence of the forms also represents a method of distinction, since, as might be expected, the primaries occur in distinctly higher frequencies.

**Resumé.**—Fragmentary as is the present knowledge of the various ways in which the internal organization of chromosomes may become altered, there is no reason to doubt that many and diverse methods are involved in the instances which have come under observation. In most cases the adequate study of these rare types of mutation necessitates a more extensive knowledge of the normal genetic situation than is available at present in most forms. It is therefore well-nigh certain that as information about the genetics of species is extended, numerous phenomena belonging in this category will come to light and will be subjected to accurate experimental analysis. There is reason to believe that some of the obscure phenomena of inheritance exhibited by certain horticultural forms will not be satisfactorily analyzed until methods of dealing with these phenomena shall have been perfected.

#### References

MORGAN: Theory of the Gene, Chapter XII.

MORGAN, BRIDGES, and STURTEVANT: Genetics of Drosophila, Chapters XVI–XVII.

# CHAPTER XXVI

## THE OENOTHERA INVESTIGATIONS

The Oenothera investigations began in 1886 with a study by de Vries of a population of the evening primrose, *Oenothera Lamarckiana*, found growing wild in an abandoned potato patch in Hilversum near Amsterdam (figure 91). During the subsequent forty years de Vries has continued these investigations, and many other investigators, attracted by

FIGURE 91.—*Oenothera Lamarckiana.* (*From de Vries, Gruppenweise Artbildung, Gebrüder Borntraeger,* 1913. *Reproduced by permission.*)

the novelty and complexity of the Oenothera phenomena, have engaged in studies not only on this species and its derivatives but also on the numerous other species and forms of the section Onagra of Oenothera. These studies have been concerned both with genetic and cytological behavior, and as a result of activity in this field the literature on Oenothera is more extensive than that of any other plant genus. Despite this fact, however, so many complications have been met with in the investigations that for a long time they were regarded as outstanding exceptions to the universal application of Mendelian principles. Recent investigations, however, have shown that the Oenothera phenomena are highly diverse and that a proper combination of cytological and genetic study is necessary for their complete elucidation.

**Mutation in Oenothera.**—When *Oenothera Lamarckiana* is grown in large pedigreed cultures, it is found to produce, as de Vries first demonstrated, about 1.5 per cent of variant individuals, or mutants. These derivatives are distributed among comparatively few types, many of which show a rather high frequency of recurrence, as shown in table 43.

TABLE 43.—PEDIGREE OF A FAMILY OF OE. LAMARCKIANA, 1886–1899 (*Data of de Vries*)

| Generation | Gigas | Albida | Oblonga | Rubrinervis | Lamarckiana | Nanella | Lata | Scintillans |
|---|---|---|---|---|---|---|---|---|
| I | | | | | 9 | | | |
| II | | | | | 15,000 | 5 | 5 | |
| III | | | | 1 | 10,000 | 3 | 3 | |
| IV | 1 | 15 | 176 | 8 | 14,000 | 60 | 73 | 1 |
| V | | 25 | 135 | 20 | 8,000 | 49 | 142 | 6 |
| VI | | 11 | 29 | 3 | 1,800 | 9 | 5 | 1 |
| VII | | | 9 | | 3,000 | 11 | | |
| VIII | | 5 | 1 | | 1,700 | 21 | 1 | |
| Total....... | 1 | 56 | 350 | 32 | 53,509 | 158 | 229 | 8 |

Thus among 54,343 offspring recorded in the table 350 were of the form called *oblonga*, 229 *lata*, 158 *nanella*, and others in smaller numbers. Subsequent investigations have greatly added to the list of mutants, but by far the greater number still fall within relatively few classes. Some of the forms like *gigas* are found only rarely, and one type at least, *brevistylis*, was found only in a single individual in the original wild population. Moreover, the derivatives in turn exhibit a rate of mutability for the most part equivalent to that of *Lamarckiana*, although some of them may show characteristic differences in the types of mutations which they produce.

Many of the derivatives of *Lamarckiana*, however, differ in important respects from the factor mutations discussed in Chapter XXII. In particular most of them do not differ in some specific limited feature from the parental form, as is true for most factor differences. On the contrary they commonly exhibit general differences from the parental form in leaf shape, size and texture, in habit, etc., and they are distinguished and named on the basis of these character complexes. In the second place they usually exhibit a peculiar mode of inheritance when tested with the parental type. The whole character complex is inherited as a unit; but $F_1$ and $F_2$ behavior is extremely individualistic.

Naturally in view of the extensive investigations with the species, during the progress of which careful search has been made for variant individuals, it would be expected that every conceivable category of mutation would be found represented among them. Such, in fact, is the case, and it is now possible to present as a basis for discussion a classification of the various mutant types, based primarily upon the chromosomal situation, and secondarily upon genetic behavior. In this classification, a scheme suggested by de Vries is followed mainly.

*Derivatives of Oenothera Lamarckiana* ($2n = 14$)

*A.* 14-chromosome derivatives:
- 1. Factor mutations: *brevistylis, laevifolia.*
- 2. Recombination products of chromosome-I:
  - *a.* Alethal derivatives: *blandina, decipiens, deserens, tarda, fragilis.*
  - *b.* Sesquiplex derivatives: *simplex, secunda, compacta, elongata, favilla, linearis.*
  - *c.* Balanced lethals: *nanella.*

*B.* 15-chromosome derivatives:
- 1. Primaries: *lata, scintillans, cana, pallescens, liquida, spathulata, pulla.*
- 2. Secondaries:
  - *a.* Lata group: *semilata, albida, flava, delata, subovata, sublinearis.*
  - *b.* Scintillans group: *oblonga, aurita, auricula, nitens, distans, diluta, militaris.*
  - *c.* Cana group: *candicans.*
  - *d.* Pallescens group: *lactuca.*

*C.* Polyploid derivatives:
- 1. 21-chromosome: *semigigas.*
- 2. 28-chromosome: *gigas.*

The forms enumerated above represent by no means all of the derivatives which have been obtained from *Lamarckiana*, but they do give a fair conception of the different categories found among them. An attempt will be made in the following pages to discuss some of the distinctive phenomena which characterize these categories.

**14-chromosome Derivatives.**—Among the 14-chromosome derivatives only the two listed above as factor mutations exhibit typical Mendelian relations to the parental form. *Brevistylis* differs from *Lamarckiana* almost solely in its short styles. Crossed with long-styled forms, as for example, *Lamarckiana*, $F_1$ is a long-styled form, and $F_2$ shows typical Mendelian segregation of approximately 3 long-styled: 1 short-styled. Crossed with some other mutant type, segregation for style length is superimposed upon segregation for other characters, indicating independent assortment. *Brevistylis* was originally discovered in the wild populations from which de Vries secured his material, and he has not obtained it as a mutant form. Its factor relations to *Lamarckiana*, however, leave little reason to doubt that it arose as a factor mutation. Perhaps also the form *laevifolia*, which has the smooth instead of the roughened leaves of *Lamarckiana* and presents other associated differences in size and shape of leaves and petals, is to be included here. It also was found by de Vries in the original wild population and has never appeared as a mutation in his cultures.

The remaining 14-chromosome mutants exhibit peculiar relations to the parent which have led to the conclusion that they represent segregation products consequent upon a heterozygous condition in the parent type. Many complications enter into the breeding results, and so often has it become necessary to revise statements concerning them that even yet it is difficult to be assured of the correctness of the conclusions.

As an illustration, take the type *rubrinervis*, which arises comparatively frequently as a derivative of *Lamarckiana*. It is particularly characterized by increased red coloration, brittle stems, and narrow hairy leaves. At first it was considered constant, but later investigations have revealed it as a heterozygote which produces a closely similar homozygous form, *deserens*. *Deserens* is slightly weaker than *rubrinervis*, but it produces no empty seeds, whereas from 25 to 50 per cent of those of *rubrinervis* are empty.

*Deserens*, crossed with *Lamarckiana* gives results conforming to the following scheme:

$P_1$ *deserens* × *Lamarckiana*

$F_1$ *rubrinervis* + *lucida*

$F_2$ *deserens rubrinervis* *deserens lucida*

*Rubrinervis* crossed with *Lamarckiana* gives somewhat more complex results, according to most recent reports, *viz.*,

$P_1$ *rubrinervis* × *Lamarckiana*

$F_1$ *rubrinervis* + *lucida* + *Lamarckiana*

$F_2$ *deserens rubrinervis* *deserens lucida* *Lamarckiana*

If *deserens*, as seems probable, is a homozygous form, the results of the first cross indicate that *Lamarckiana* is heterozygous and produces two classes of gametes which are to be considered as differing in a complex of factors rather than in a single factor. These factor complexes are called *gaudens* and *velans* (*laeta* and *velutina*, respectively, of de Vries). In terms of gametic composition *Lamarckiana* is *gaudens-velans*. Its constancy depends upon the fact that *gaudens-gaudens* and *velans-velans* combinations fail to survive. This factor-complex theory, first proposed by Renner and since accepted by the majority of Oenothera investigators, including de Vries, has led to a consistent interpretation of most of the puzzling phenomena exhibited within the genus.

The origin of *deserens* and *rubrinervis* in terms of factor complexes may now be diagrammed as in figure 92. A *gaudens* complex lost its lethal factor and suffered some other changes including incorporation of factors for brittle stems and deeper red coloration. This new factor complex *deserens* in combination with unmodified *velans* gives *rubrinervis,* and *rubrinervis* produces *deserens* by segregation as shown in the same figure.

A further diagram (figure 93) will illustrate how this conception may be applied to interpret the results of a cross between *rubrinervis* and *Lamarckiana* as outlined above. Some discrepancies occur in statements of the outcome of this cross which are apparently due in part to

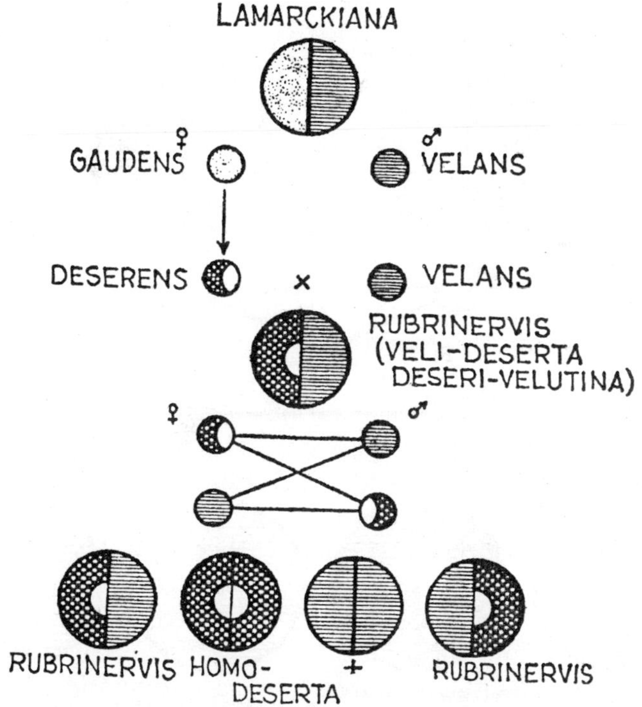


FIGURE 92.—Origin of *rubrinervis* and *deserens* from *Lamarckiana* in terms of the factor complex theory. (*After Lehmann, Die Theorien der Oenotheraforschung,* 1922, *by Gustav Fischer. Reprinted by permission.*)

failure to distinguish between the closely similar *lucida* and *Lamarckiana* on the one hand and *rubrinervis* and *deserens* on the other. In many cases, also, a form called *subrobusta,* which is intermediate between *Lamarckiana* and *rubrinervis* and which does not have the brittle stems of the latter, replaces *rubrinervis* in these crosses. Apparently this is due to differences in the *Lamarckiana* plants which have been employed, some of which produce *velans* gametes with a factor for tough stems.

Similarly each of the other alethal derivatives listed above has arisen through a characteristic modification of one or the other of the factor complexes of *Lamarckiana,* which has resulted in the production of a heterozygous form from which the homozygous type has been produced by segregation. Thus *gaudens* converted into *decipiens*, a factor complex similar to *deserens* but with a factor for tough stems, gave the combination *decipiens-velans*, which is *erythrina*, a form almost identical with *subrobusta*, described above. *Erythrina* then produced the form called *decipiens*, through production of *decipiens-decipiens* combinations. Of particular interest are *blandina* and *fragilis*. *Blandina* is *blandina-blandina* in terms of factor complexes; and *fragilis* is *perficiens-perficiens*. The *blandina* factor complex is merely a *velans* complex minus the lethal factor, and *perficiens* bears a similar relation to *gaudens*. *Blandina* × *fragilis* gives an $F_1$ almost indistinguishable from *Lamarckiana* but producing practically no empty seeds. By resort to this conception of factor complexes, a formal and consistent interpretation of most of the breeding results with these 14-chromosome derivatives may be offered, but some details still remain unsolved.

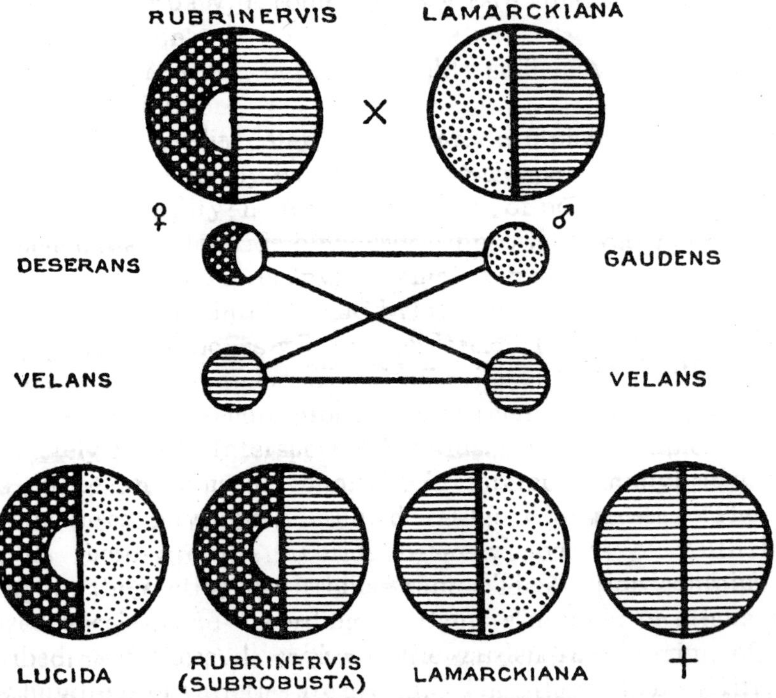


FIGURE 93.—Results of crossing *rubrinervis* with *Lamarckiana*. (*After Lehmann, Die Theorien der Oenotheraforschung*, 1922, *by Gustav Fischer. Reprinted by permission.*)

**The Factor Complexes of Oenothera.**—It remains to determine what the factor complexes of Oenothera represent in terms of chromosome organization and behavior, and what significance is to be attached to the changes in these factor complexes such as occur when *gaudens* is converted into *deserens*, *decipiens*, etc., and *velans* into *blandina*, etc.

Assuming that reductional phenomena in Oenothera follow the same course as in other plants, for which there seems to be adequate evidence at present, the logical interpretation is that *gaudens* and *velans* represent a pair of chromosomes which may be called chromosome-I. The two members of this pair of chromosomes must differ in many factors; and in *Lamarckiana* and some of its derivatives each of the two chromosomes bears a recessive lethal factor. If these two chromosomes differ in numerous factors, then the enforced heterozygous condition of *Lamarckiana* has the effect of preventing numerous recessive factors from coming to light. Crossing-over between the two chromosomes would give rise to viable homozygous combinations of certain of these factors; and crossing-over at different levels would naturally give rise to different combinations of the factors. The derivatives of this class, therefore, are assumed to represent simply recombination products of chromosome-I.

To give a concrete illustration of this conception, the *gaudens* chromosome appears to contain a factor $n$ for dwarf stature. Due to the lethal effect of the *gaudens-gaudens* combination, dwarfness does not appear as a segregant from *Lamarckiana*. *Gaudens* may be represented as $(nl_1L_2)$; *velans* as $(NL_1l_2)$, $l_1$, and $l_2$ representing the recessive lethal factors of *gaudens* and *velans*, respectively, so that the constitution of *Lamarckiana* according to the customary scheme is $\frac{nl_1L_2}{NL_1l_2}$. An occasional crossover of the type $\frac{n\,|\,L_1\ l_2}{N\ l_1\ L_2}$ would then give a gamete $(n\ L_1\ l_2)$, which uniting with a normal *gaudens* gamete would give $\frac{n\ L_1\ l_2}{n\ l_1\ L_2}$, the dwarf mutant *nanella* of de Vries. It may be called, in terms of gametic composition, *gaudens-nanovelans*. *Nanella* crossed with *Lamarckiana* then gives the following results:

| | | | | |
|---|---|---|---|---|
| $P_1$ | *nanella*<br>gaudens-nanovelans | | × | *Lamarckiana*<br>gaudens-velans |
| $F_1$ | gaudens-gaudens<br>Lethal | gaudens-velans<br>*Lamarckiana* | nanovelans-gaudens<br>*nanella* | nanovelans-velans<br>lethal |

Thus although both *nanella* and *Lamarckiana* are heterozygous, only two classes are produced in $F_1$, because two only of the four possible combinations survive. Since the two which do survive are of the same constitution as the parental types, like them they breed true. Inasmuch as the *gaudens-velans* pair of chromosomes differ in numerous factors, other types may be produced by crossing-over at other levels, and thus it is possible to account for the recurrence of certain definite mutations from *Lamarckiana*. The theory that the 14-chromosome mutations of *Lamarckiana* are for the most part recombination products is confirmed by the results of crossing the mutant types with *Lamarckiana*.

The results of these crosses show clearly that *Lamarckiana* possesses the suppressed recessive factors which come to light in its derivatives. Some investigators profess to see nothing of significance in the distinction between true factor mutation and the appearance of new types by crossing-over between two chromosomes differing in numerous factors, but the demonstration that the factors concerned are already present in *Lamarckiana* in other combinations definitely sets it apart as a distinct phenomenon.

The question then arises as to the origin of the condition. Granted that these two chromosomes in *Lamarckiana* are so radically different in composition, how has the condition come about? If has been urged that *Lamarckiana*, since it has never been found growing wild in North America, the original habitat of the Oenotheras, must represent a product of hybridization between distinct species. This conception may be correct, and some investigators, notably Davis, have succeeded in synthesizing *Lamarckiana* types from crosses between other species of Oenothera, but the suggestion loses force because of the demonstration that species taken from the wild have themselves been shown to possess genetic complexities similar to those in *Lamarckiana*. While this question of the origin of the heterogametic condition remains unsolved, the mystery as to the present constitution of *Lamarckiana* and its derivatives is unquestionably on the road to solution, although a vast amount of additional work will be necessary before the difference between the *gaudens* and *velans* chromosomes may be stated in terms of factors and of positions of these factors in the chromosomes.

**Balanced Lethals.**—The most satisfactory demonstration of the validity of the above conception has come from Muller's analysis of a constant stock of beaded wing type in *Drosophila melanogaster*. Normally beaded × beaded in certain stocks gives approximately 2 beaded + 1 wild type as with other dominant mutants which are lethal when homozygous; but by selection a stock was eventually established which bred true for beaded, except for the very occasional appearance of a wild-type fly.

Muller found, however, that flies from the constant beaded stock were still heterozygous for beaded, for crosses of beaded × wild type gave 1 beaded + 1 wild type in $F_1$, just as the original beaded individuals from inconstant stock. The analysis showed that the constancy of the new beaded was due to the origin of a recessive lethal in the chromosome-III paired with the one carrying the factor for beaded; so that one chromosome carried the factor *Bd* for beaded, and the other a recessive lethal factor *l* which produced no visibile effect. Beaded, homozygous, dies; and the zygotes which formerly represented the wild-type segregants also die because of the lethal action of *l*. As a consequence only the heterozygous beaded flies live, and the stock is constant although heterozygous. Crossing-over between the chromosomes was also reduced by the presence of a crossover suppressor, a factor which practically eliminated crossing-over in that region of the chromosome so that the condition was relatively stable. The term balanced lethal is applied to conditions such as have been described in the constant beaded stock.

By the use of balanced lethals, particularly in association with crossover suppressors, it is possible to set up genetic conditions comparable to that of *Lamarckiana*. Thus in the second group of Drosophila there is a dominant, curly factor *Cy*, which is associated with a dominant suppressor of crossing-over in chromosome-II. Homozygous curly occasionally appears, but the flies are weak, and hence it may be neglected. It is readily possible to make up a stock having *Cy* in one second chromosome and *S* (star) and *b* (black) in the other, *viz.*,

$$\frac{\quad Cy \qquad Cr\text{-II} \qquad cn^2}{S \qquad\qquad\quad b \qquad\qquad}$$

Such a stock will be star curly in appearance and, aside from the production of a few homozygous curly flies, it will breed true for those characters, according to the following scheme:

| | *Sb* | *CyCr*-II*cn*² |
|---|---|---|
| *Sb* | *Sb* *Sb*<br>dies | *SbCyCr*-II*cn*²<br>star curly |
| *CyCr*-II*cn*² | *SbCy* *Cr*-II*cn*²<br>star curly | *CyCr*-II *cn*²*CyCr*-II*cn*²<br>usually dies |

Although these factors occupy loci scattered between *S* at 0.0 and $cn^2$ at 55.5, the presence of the cross-over suppressor represented by *Cr*-II practically eliminates crossing-over in this region; consequently the stock breeds true. The factors *b* and $cn^2$ present in the stock would remain undetected save as a rare crossover would permit the appearance of viable flies homozygous for them. As a matter of fact the presence of $cn^2$ for cinnabar eye color in the curly chromosome-II was detected from a cross of curly × cinnibar, which gave curly cinnabar and wild-type $F_1$ offspring; and when homozygous curly flies were discovered later, they were found to have an eye color somewhat darker than cinnabar, but readily distinguishable from wild-type eye color.

If a crossover should occur in the region immediately to the left of *b*, the two chromosomes would be reconstituted as below:

$$\frac{\quad Cy \qquad Cr\text{-II} \mid b \qquad}{S \qquad\qquad\quad \mid \qquad cn^2}$$

(*CyCr*-II*b*) with (*Sb*) would give rise to a constant star-curly black derivative, comparable to those of Oenothera. This star curly black crossed with the original star curly would give the following results:

| | *Sb* | *CyCr*-II*b* |
|---|---|---|
| *Sb* | *SbSb* dies | *SbCyCr*-II*b* star curly black |
| *CyCr*-II*cn*² | *SbCyCr*-II*cn*² star curly | *CyCr*-II*bCyCr*-II*cn*² usually dies |

Two $F_1$ types, twin hybrids, would be produced and both would breed true. Obviously these results would constitute a close parallel to the behavior of *nanella*, described above.

**Gametic Lethals in Oenothera.**—In crosses between *Oe. biennis* and *Oe. muricata* (also called *syrticola*), two species which bear a comparatively close resemblance but which differ in some distinctive features, $F_1$ is of the male-parental type according to the following scheme:

*muricata* ♀ × *biennis* ♂ = $F_1$ *biennis*-like
*biennis* ♀ × *muricata* ♂ = $F_1$ *muricata*-like

Renner has shown that these results are due to the fact that both of these species are heterozygous in the same sense as *Lamarckiana*. In terms of genetic composition *biennis* = *albicans-rubens;* but the *albicans* pollen grains and the *rubens* ovules fail to function; consequently only *albicans-rubens* = *biennis* combinations are produced. Similar *muricata* = *curvans-rigens,* with *curvans* ovules and *rigens* pollen grains functionless. Consequently crosses between these two species may be pictured as follows:

*biennis* ♀ × *muricata* ♂ = $F_1$ *albicans-curvans* = *muricata*-like
*muricata* ♀ × *biennis* ♂ = $F_1$ *rigens-rubens* = *biennis*-like

The *albicans-curvans* combination is similar to *muricata* but not completely identical with it, and the *rigens-rubens* is similar to *biennis.* Both forms apparently breed true, as is to be expected from this analysis, although pollen sterility is high.

It has been shown in these instances that a high percentage, at least 50, of pollen sterility occurs, and evidence has been secured of a difference in form and content of the pollen grains. These facts point to the existence of a pollen lethal which destroys half the pollen grains, as the breeding evidence indicates. Similarly the suppression of one of the classes of ovules has been taken to indicate the presence of an egg lethal, which destroys the contrary class of ovules. Renner, however, has been unable to demonstrate the presence of 50 per cent of functionless egg cells, as called for on this theory, but he has presented evidence for the suppression of one class of ovules by differential functioning of members of the four cells produced by reduction, which amounts to the same thing genetically. In addition to zygotic lethals which destroy zygotes homozygous for them, there are, therefore, gametic lethals in some Oenotheras which destroy or inhibit the functioning of certain gametic classes. According to Shull's experiments, these lethals also are borne by members of the chromosome-I and it is possible to secure various results by obtaining them in different combinations with factors for recognizable morphological characters.

Thus let $S$ = yellow, $s$ = sulphur flower color; and let *le* stand for an egg lethal which destroys or inhibits the functioning of gametes bearing it, and *ls* a corresponding pollen lethal, both pairs of factors being borne in chromosome-I. Then the results secured will depend upon the various combinations of these factors with the lethals as follows:

Patroclinous $P_1$
  yellow = $(l_eS)(l_ss)$; sulfur = $(l_cs)(l_ss)$
  sulfur ♀ × yellow ♂ = $F_1$ yellow; $F_2$ yellow
  yellow ♀ × sulfur ♂ = $F_1$ sulfur; $F_2$ sulfur
Matroclinous
  yellow = $(l_sS)(l_es)$; sulfur = $(l_es)(l_ss)$
  sulfur ♀ + yellow ♂ = $F_1$ sulfur, $F_2$ sulfur
  yellow ♀ × sulfur ♂ = $F_1$ yellow; $F_2$ yellow
Normal  yellow = $SS$  sulfur = $ss$
  sulfur ♀ × yellow ♂ = $F_1$, yellow; $F_2$, 3 yellow: 1 sulfur
  yellow ♀ × sulfur ♂ = $F_1$, yellow; $F_2$, 3 yellow: 1 sulfur

Shüll has actually obtained confirmation of behavior according to all of the above schemes from his Oenothera investigations.

**15-chromosome Mutants.**—The trisomic mutants of *Lamarckiana* exhibit peculiarities of genetic behavior similar to those of Datura. They are transmitted exclusively through ovules and to a rather variable extent. Apparently no transmission at all occurs through pollen; and many of the forms exhibit a high degree of pollen sterility.

One of the outstanding features of the 15-chromosome Oenothera mutants is the large number of them. Since $n = 7$, there should be only seven types due to presence of an extra chromosome of normal constitution. Possibly in view of the difference existing between the *gaudens* and *velans* chromosomes, a *gaudens-gaudens-velans* trisomic might differ sufficiently from a *gaudens-velans-velans* to be recognized as distinct.

De Vries, however, has found it possible to show that there are seven groups of these mutants just as there are twelve in Datura. He has also found it possible to identify the primaries from the type of progeny produced by the triploid, *semigigas*, crossed with *Lamarckiana*. Seven types in approximately equal numbers were found among the 15-chromosome offspring of this cross, namely, *lata*, *scintillans*, *cana*, *pallescens*, *liquida*, *spathulata*, and *pulla*. The last named was unrecognized before this cross was made, and because it resembles *Lamarckiana* most closely, it is presumed to be triplo-I. Final demonstration, however, must rest on genetic studies.

The remaining 15-chromosome mutants occurred only infrequently as the progeny of triploids. De Vries has shown that they exhibit certain evidences of relationship to one or another of the primary types; so that they are probably secondaries analogous to those in Datura. There is some evidence that *semilata* has a very small extra chromosome, so that it may have an extra half-chromosome of *lata*. It also seems to produce a rather high proportion of *lata* offspring as well as *Lamarckiana*

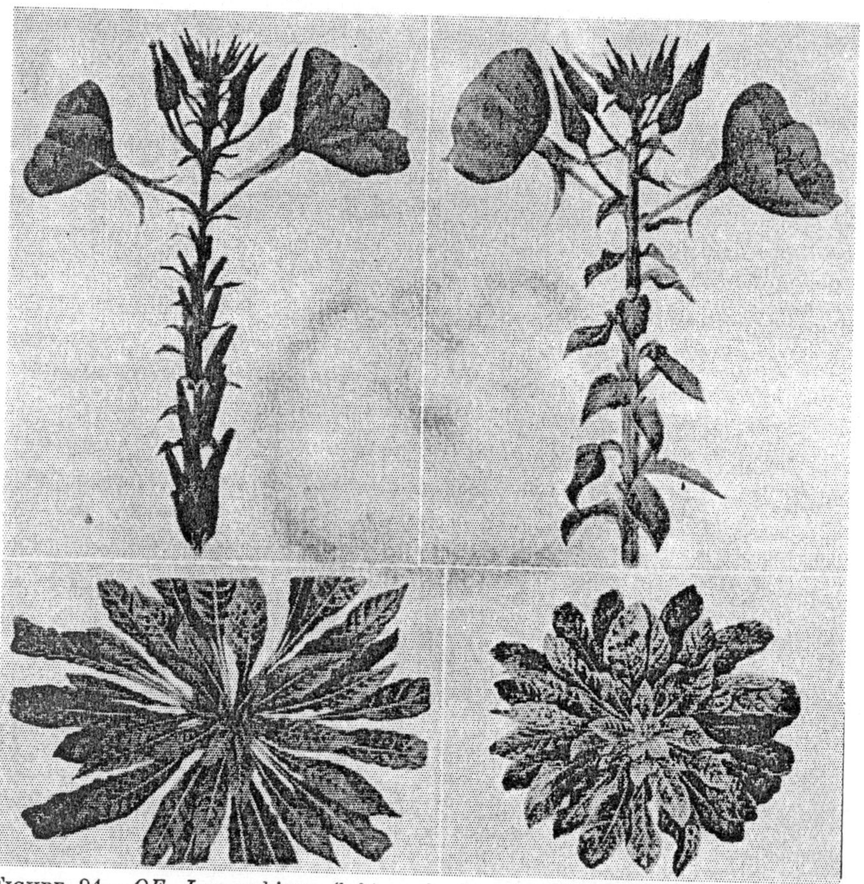

FIGURE 94.—*OE. Lamarckiana* (left) and *gigas* (right), flower stalks and rosettes. (*From de Vries, Mutationstheorie, 1st. ed., Von Veit and Co.,* 1901; *Gruppenweise Artbildung, Gebrüder Borntraeger,* 1913. *Reproduced by permission.*)

and *semilata*. *Oblonga,* which is one of the most frequent direct variants of Lamarckiana, is produced in a high percentage from *scintillans*. As in Datura, however, it has not yet been found possible to offer a final demonstration of the genetic constitution of the secondary trisomics.

**Polyploid Oenotheras.**—Two polyploid types of *Lamarckiana,* the triploid *semigigas* and the tetraploid *gigas* (figure 94), have been extensively investigated. Both have arisen spontaneously, *gigas* on a few occasions and *semigigas* somewhat more frequently. Presumably *semigigas* may arise from occasional production of diploid gametes either by non-reduction or previous production of tetraploid somatic areas which then gave rise to diploid gametes. Such gametes meeting normal haploid gametes would produce triploids. Other methods have been suggested, and also various methods for the production of tetraploid mutants; but none of them seems to have more than a speculative value. It is entirely possible that several methods of origin, rather than one, have been in operation. The origin of tetraploids by doubling after fertilization and in somatic mitoses has at least been demonstrated in other forms as shown in Chapter XXIV. *Semigigas* may also be produced from crosses of *Lamarckiana* with *gigas*. Phenotypically *gigas* is distinguishable as a giant form of comparatively high fertility; *semigigas*, as intermediate between *Lamarckiana* and *gigas* and of very low fertility.

*Semigigas* is evidently highly sterile on account of the unbalanced chromosomal combinations arising as a result of reduction in a triploid form. Assuming a $2 <> 1$ distribution in each chromosome set, a range of chromosome numbers in gametes from 7 to 14 is to be expected according to expansion of $(a + bc)^7$, coefficients of terms representing comparative frequencies and the sum of the exponents of *abc*, the number of chromosomes in the gamete. According to recent work of van Overeem, who made chromosome determinations in offspring of crosses of the types, *Lamarckiana* × *semigigas*, *semigigas* × *Lamarckiana*, *gigas* × *semigigas*, and *semigigas* × *gigas*, the functional pollen grains are restricted to those having 7 or 14 chromosomes; but the functional ovules may represent the entire range of values from 7 to 14. In *semigigas* × *Lamarckiana* a higher proportion of 7- and 8-chromosome gametes was indicated than that stipulated from chance distribution; but in *semigigas* × *gigas* progenies, variation in chromosome number from 21 to 28 was found in fairly close agreement with expectation. As in Datura, a great variety of types is produced by triploids; but the effects of unbalanced chromosomal constitution do not appear to be so serious.

*Gigas* with 28 chromosomes breeds true for the most part, but as in the tetraploid Datura, a fringe of modified tetraploids with $28 \pm$ chromosomes is produced. These appear to bear some relation to the trisomic types, as might be expected, but sufficient study of them has not been made. Tetraploid mutants equivalent to the 14-chromosome mutants have also been secured, in particular a *gigas nanella*, but the relation of this type to *gigas* has not been determined sufficiently accurately to make it possible to conclude whether or not ratios conform to those expected under tetrasomic conditions.

**Resumé.**—From the foregoing account it is evident that the Oenothera phenomena present instances of every category of mutation which has been discussed in the preceding chapters. The early investigations on these forms suffered from the fact that methods of dealing with such complex situations had not been devised; consequently, for a long time the phenomena appeared to be inexplicable on the same basis as those studied in other species. With the advent of cytological studies, a large number of the forms were found to depend upon variation in chromosome number; and these in the main have been found to give results of the same type as those now known in other species, such as Datura, which may be studied more advantageously from this point of view. A vast amount of additional study will be necessary before these forms shall have been satisfactorily related to each other, and, in particular, studies of the distribution of such characters as *brevistylis*, *nanella*, etc., in relation to chromosomal composition may be expected to yield significant results.

The 14-chromosome mutants present numerous complications which clearly remove most of them from the category of simple factor mutation. The recognition of the existence of gametic and zygotic lethals and the apparent restriction of the phenomena to a single pair of chromosomes exhibiting a balanced lethal condition clear the way for an interpretation of the puzzling genetic phenomena disclosed by crosses among these forms. The mutations are for the most part revealed as recombination products arising by crossing-over at different levels between the two chromosomes. A beginning has even been made in determining the constitution of these two chromosomes; but much additional work will be necessary before this study can be completed. The difference between these two chromosomes may be a consequence of factor mutation, but evidently for the most part these mutations have already occurred in *Lamarckiana* and merely await the necessary type of crossing-over for their disclosure.

These latter phenomena are in truth very complex and do not now appear to be very widespread in species other than Oenothera, but there are numerous puzzling cases of inheritance, particularly among garden ornamentals, which may upon analysis disclose conditions no less involved. Fortunately, however, most of the commercially important species of animals and plants do not appear to present such highly complicated situations. The highly diverse phenomena exhibited by *Oenothera Lamarckiana* and its derivatives may, however, well be regarded as an indication of the complexity which may arise in any species which is carefully and extensively investigated.

**References**

Gates: The Mutation Factor in Evolution.
Lehmann: Die Theorien der Oenotheraforschung.
de Vries: The Mutation Theory (translation by Farmer and Darbishire).

## CHAPTER XXVII

## INTERSPECIFIC HYBRIDIZATION

The discussion of variation and heredity in the preceding chapters has dealt with the phenomena exhibited within species. It is also often possible to cross different species together and thus to secure interspecific hybrids. Although these for the most part exhibit phenomena which do not conform, without marked modifications, to the laws which govern variation and heredity within a species, nevertheless, interspecific hybridization has been such an important source of improvement in domesticated animals and cultivated plants that, even though present knowledge of the laws governing the phenomena in such cases is as yet rather slight, there is every justification for an attempt to present a general survey of the present status of the problem.

It might seem desirable to open the discussion by a consideration of the criteria which distinguish species; but this problem is so involved and debatable that it is impossible to arrive at any acceptable conclusion, even from a genetic point of view. The diverse varieties of maize, dent, flint, sugar, popcorn, etc., are all reckoned as members of a single species, *Zea mays*, and this is also true of the breeds of European cattle, all of which are included in *Bos taurus;* so, in general, a species, whether wild or domesticated, consists of an assemblage of forms which interbreed freely and produce fertile hybrids conforming to Mendelian laws. In the majority of instances there is no difficulty in grouping individuals into assemblages of this character; but when the problem is considered from the evolutionary point of view, it is not surprising to find that in some instances divergence of forms has not gone far enough to make the distinctions clear and undebatable.

In most cases the taxonomist's classification may be accepted as a working basis, although disagreement among taxonomists indicates that no generally accepted criteria have as yet been discovered, if indeed such an ideal is thinkable. The geneticist may well take the point of view that one of the objects of his inquiry into this field is to discover, if possible, what the distinctions between species are in terms of genetic constitution.

In this chapter an attempt will be made to discuss in a general way some of the prominent features of interspecific hybridization, such as crossability, uniformity, form, vigor, and fertility of interspecific hybrids, and the cytological and genetic phenomena exhibited in a few selected cases. A really adequate account of the phenomena would lead far beyond the scope of this book. The student may pursue the subject further, if he so desires, by consulting the selected list of references at the end of the chapter and original sources.

**Crossability.**—An interesting question is raised as to the limits within which species crosses may be secured and the criteria which may be employed as a guide to the possibility of obtaining crosses between species. In general, relationship as indicated by taxonomic classification is the most useful guide. Usually species standing close together in the taxonomic system cross readily and the difficulty increases as the species stand farther apart; but as yet no valid criterion aside from trial has been discovered.

In the snapdragons, Antirrhinum, the several species in the section Antirrhinastrum cross readily for the most part and yield fertile hybrids; but these species apparently cannot be crossed with members of other sections of the genus. In most instances, however, correspondence between taxonomic classification and results of hybridization are not so simple. The common cabbage, *Brassica oleracea*, for example, will cross with the radish, *Raphanus sativus*, but not with the more nearly related turnip, *B. rapa*, or the rape, *B. napus*. While crosses between species of different genera, heterogenic hybrids, are comparatively rare, a great number of them have been secured both among animals and plants. Among domesticated birds, numerous heterogenic hybrids have been secured between pheasants, *e.g.*, the ring pheasant, *Phasianus torquatas*, with the silver pheasant, *Gennaeus nycthemerus;* the gold pheasant, *Chrysolophus pictus*, with the hunting pheasant, *Phasianus colchicus*, and with the silver pheasant. The hunting pheasant has also been crossed with the common fowl, *Gallus gallus;* and the guinea fowl and the peacock have been crossed.

Whether two species will or will not cross is not solely a matter of genetic constitution of their gametes. Among animals, mating between species is often attended with difficulties due to differences in the sex organs or in the stimuli which lead to mating. Artificial impregnation has sometimes been successful when normal mating does not occur; by this means hybrids have been obtained between the rat and mouse. In plants the well-established fact that crosses may in some cases be made in one direction but not in the other indicates the operation of little-understood accessory factors. The hybridization of *Raphanus sativus* ♀ × *Brassica oleracea* ♂ is rather easily accomplished, but numerous efforts to obtain the reciprocal cross have been unsuccessful. Wheat may be crossed with rye rather readily in one direction, *Triticum sativum* ♀ × *Secale cereale* ♂; but the reciprocal cross is rarely successful. Moreover, in many instances interspecific hybridization gives only a very small percentage of the normal complement of seed. *Nicotiana tabacum* ♀ × *N. glutinosa* ♂, common tobacco crossed with a wild species, usually gives not more than fifteen viable seeds per capsule, whereas a full capsule of *N. tabacum* may contain as many as 1,500 viable seeds. A particular variety of *N. tabacum* called "Cuba," however, gives full capsules of viable seed when crossed with *N. glutinosa*, although this same variety gives no more than the usual poor production of seed when used as a male parent in the reciprocal cross. Evidently varieties of a given species may differ in their ability to cross with other species.

Chromosome number is certainly not an accurate criterion as to crossability. *Crepis biennis* with $n = 20$ crosses readily with *C. setosa*, $n = 4$; but *C. capillaris*, $n = 3$, crosses with *C. tectorum*, $n = 4$, only with difficulty and usually gives non-viable hybrids. The moth, *Lycia hirtaria*, with $n = 14$ large chromosomes, has been crossed with another moth, *Ithyria zonaria*, with $n = 56$ very small chromosomes. In the tobaccos there are species with $n = 9$, 10, 12, 16, and 24 chromosomes, and almost every conceivable combination of chromosome numbers has been obtained in hybrids, but curiously enough 12 × 12 combinations are very rare, but many 12 × 24 combinations are readily secured. Cytological conditions no doubt must be considered in a study of interspecific hybridization; but the chief value of these studies will probably lie in an elucidation of the genetic phenomena exhibited by the hybrids.

On the whole, crossability must be determined by trial, and the results frequently have little significance as indicative of relationship between species.

**Uniformity of $F_1$ Interspecific Hybrids.**—In general it has been found that interspecific hybrids obtained by crossing constant races of two different species are uniform. Naturally this is to be expected for, if the races representative of the parental species are homozygous, the $F_1$ individuals should be uniform as a consequence of their identity in genetic constitution. When, however, the individuals representative of the species are not themselves homozygous, a certain degree of variability may be expected in the $F_1$ population, even though all members are true species hybrids. Such is usually the case among plants propagated asexually, for the varieties employed in hybridization are rarely even approximately homozygous, and as a consequence a considerable diversity is usually exhibited in $F_1$. In these cases it is necessary to reduce the parental races to constancy before acceptable evidence on this problem can be obtained. In some cases, as in the Oenothera investigations described in Chapter XXVI, the appearance of diverse types in $F_1$ has been shown to depend upon a heterozygous condition of such a nature as to be preserved constant from generation to generation. Such instances are, however, extremely rare; as a rule, interspecific hybrids from constant races are uniform.

For the same reasons, reciprocal hybrids in the case of species crosses are usually identical; but a few authentic instances of failure to conform to this rule have been observed. It has frequently been stated that the mule, obtained by mating a mare to a jack, differs from the hinny, the

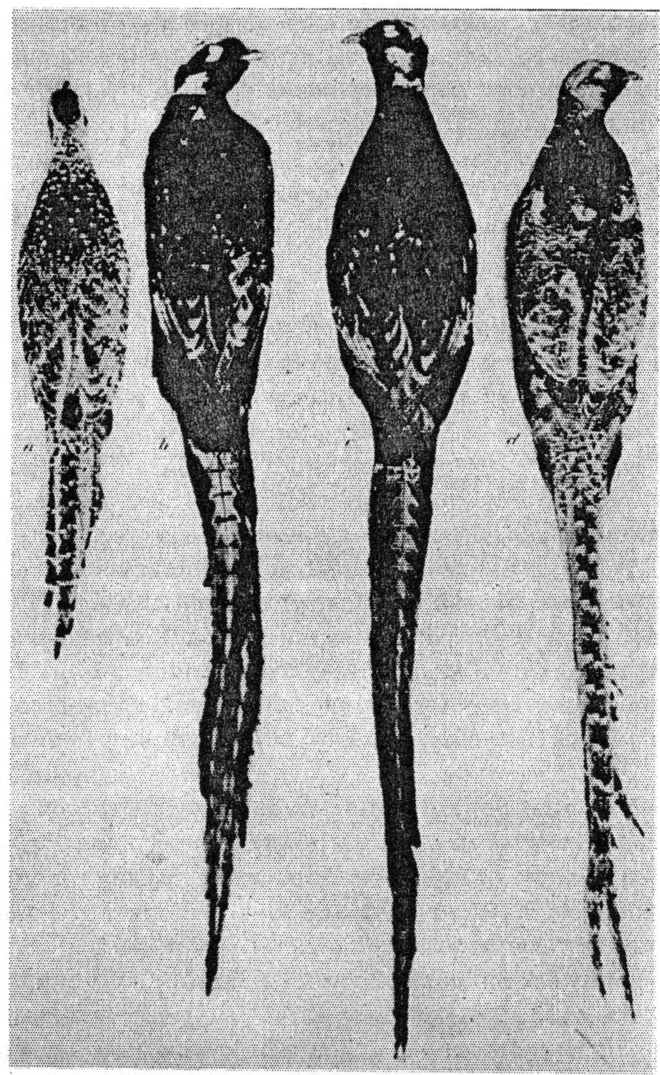

FIGURE 95.—Results of reciprocal crosses of Reeves with Prince of Wales pheasant: *a* female and *b*, male $F_1$ from Prince of Wales ♀ × Reeves ♂; *c*, male and *d*, female $F_1$ from the reciprocal cross. (*From Phillips.*)

hybrid obtained by mating a jennet to a stallion. No authentic account, however, appears available; and the differences which have been observed are probably such as fall within the normal range of variation in the two cases. In certain species crosses in pheasants, however, it has been shown that the males are identical in appearance but the females are distinctly different. Such, for example, is the case in crosses of the Reeves pheasant, *Syrmaticus reevesi*, with the Prince of Wales pheasant, *Phasianus principalis*, reported by Phillips and illustrated in figure 95. This result is perfectly intelligible, if the mode of sex determination is considered.

Nothing is known about cytological details in these species, but sex determination is doubtless of the WZ type as in other birds. The situation may, therefore, be represented as follows:

*Syrmaticus reevesi:* ♂ = $A_sA_sZ_sZ_s$ ♀ = $A_sA_sZ_sW_s$

*Phasianus principalis:* ♂ = $A_pA_pZ_pZ_p$ ♀ = $A_pA_pZ_pW_p$

$F_1$ *S. reevesi* ♀ × *P. principalis* ♂ = $A_sA_pZ_sZ_p$ ♀ = $A_sA_pZ_pW_s$

$F_1$ *P. principalis* ♀ × *S. reevesi* ♂ = $A_sA_pZ_sZ_p$ ♀ = $A_sA_pZ_sW_p$

In these formulae, A represents a haploid set of autosomes; $A_s$ of Reeves pheasant and $A_p$ of Prince of Wales pheasant; and similarly $W_s$ and $Z_s$ represent the sex chromosomes with subscripts as before. As may be seen from the formulae, the $F_1$ male hybrids of the reciprocal crosses are identical in constitution; but the female hybrids differ in their sex chromosomes. If these differ in constitution, as might reasonably be expected, then the female hybrids should differ. When, as in mammals, the male is the heterozygous sex, reciprocal crosses should give identical female hybrids but different male hybrids for the same reason.

Some investigators have sought to establish a cytoplasmic basis for a difference between reciprocal crosses based on the fact that very little, if any, cytoplasm is introduced with the male gamete and on the assumption that, although cytoplasmic differences are for the most part of no moment in varietal crosses, in interspecific crosses the cytoplasm of the two species may differ so strikingly as to necessitate its consideration. The almost universal validity of the rule that interspecific hybrids obtained from reciprocal crosses are identical contradicts this assumption. The relatively rare cases in which they differ will probably require an explanation of a different kind.

**Morphological Features of Interspecific Hybrids.**—As a general rule, interspecific hybrids are intermediate between their parents in the sum total of their characters. One parent may contribute some dominant characters; the other, others; but usually most of the characters are intermediate. In this respect interspecific hybrids are comparable to hybrids between varieties differing in numerous characters.

While most interspecific hybrids are so obviously intermediate in expression that an investigator acquainted with the genus would be able to determine the parentage of a hybrid by examination, there are some notable exceptions to the rule. The older plant hybridists, such as Gärtner and Focke, recognized this fact, and stated that any condition may be found from that of obvious intermediacy to a condition so closely approaching one of the parents as to be indistinguishable from it save as respects vigor and fertility. Striking examples are afforded by the hybrids of *Nicotiana sylvestris* with varieties of commercial tobacco, *N. tabacum* (figure 96). The varieties of *N. tabacum* differ strikingly in leaf form, flower color and form, etc., but any variety of *N. tabacum* crossed with *N. sylvestris* gives a hybrid so closely approximating the particular variety of *N. tabacum* employed as to be indistinguishable from it, aside from the enhanced vigor and almost complete sterility of the hybrid and a general proportional increase in size of all parts (figure 97). An explanation of this phenomenon is as yet impossible.

FIGURE 96.—Typical plant of *Nicotiana sylvestris.*

**Vigor of Interspecific Hybrids.**—One of the most characteristic features of interspecific hybrids is the surprising vigor which they exhibit. which is true of animal as well as plant hybrids. Enhanced vigor is not. however, of universal occurrence; but among plants an entire series of conditions, ranging from production of imperfect, non-viable seed to production of hybrids exceeding either parent in rate and extent of growth has been recorded. The entire range of conditions may occur within a single genus, as has been shown by many investigators of interspecific hybridization in Nicotiana, a favorite subject for such experiments. In some instances good seed is produced from crossing, but it fails to germinate; in other cases the seed germinates weakly; the seedlings, however, are yellowish and deformed, often apparently incapable of forming roots; in still other instances the seed germinates well, the seedlings appear to be strong and well developed, but they soon fail, often

FIGURE 97.—*Nicotiana sylvestris* (left), *N. tabacum angustifolia* (right), and the $F_1$ hybrid (center), illustrating vigor of interspecific hybrids and close correspondence of morphological characters of the $F_1$ hybrid with its *tabacum* parent. (*From Godspeed and Clausen.*)

never going beyond the cotyledon stage. If seedlings pass this stage, they usually, but not always, develop to maturity. Among hybrids which develop to maturity a complete range exists from hybrids which are conspicuously weak, although exhibiting no other imperfections of growth, to hybrids which greatly exceed either parent in vigor.

When enhanced vigor is exhibited by an interspecific hybrid, it may be expressed in a variety of ways. The total amount of growth may be greater and all parts enlarged; the stems may be thicker, the internodes longer, the leaves larger, the flowers larger and more numerous. The hybrids may be hardier than the parental forms; some interspecific hybrids may survive winters of such severity as to kill the parental forms. The hybrid may reproduce asexually with greater ease; in certain cases interspecific hybrids are so easily multiplied by this means that a twig dropped on the ground will strike root and grow into a new plant. In some instances interspecific hybrids exhibit increased resistance to disease as compared with the parents. Sometimes the growth reactions in the hybrid appear to be rather delicately balanced; for, if grown under favorable conditions, they develop into strong vigorous plants, but if they meet unsuitable conditions, particularly in early life, they may grow rather weakly and never recover from the original stunting, although the difference in conditions responsible for this effect may not be sufficient to have any noticeable influence on the growth of the parental species. The vigor of interspecific hybrids is inherent in them, not a matter of stimulation, for under asexual methods of propagation it is retained indefinitely.

Among plants some experiments show very definitely that different varieties of a species may give different results as respects vigor of the hybrids with some other species. A striking instance is that of a series of varieties of *N. tabacum*, each of which was crossed with two varieties of *N. glauca*, one derived from a California source, the other from the Argentine. With the California variety of *N. glauca* the hybrids were uniformly weak, with yellowish leaves, short internodes, and a total growth not much greater than 2 feet. With the Argentine variety they were exceptionally vigorous, with large dark leaves, long internodes, and a height ranging from 6 to 12 feet according to the variety of *N. tabacum* used. Some varieties of *Crepis capillaris* crossed with *C. tectorum* give a weak hybrid which dies soon after germination; others give a comparatively vigorous hybrid which develops to maturity. These results indicate the necessity for more precision in statements as to the results of interspecific crosses. It is necessary not only to record the species which were crossed but also the particular varieties representative of the species. The whole range of expressions may possibly be secured in a given species hybrid depending upon the varieties employed, although such a result is certainly exceptional.

**Sterility of Interspecific Hybrids.**—Interspecific hybrids exhibit a range of expressions from full fertility to complete sterility; but in the majority of instances there is a certain degree of sterility, often very marked. The degree of sterility is not obviously correlated with other features. Enhanced vigor, for example, may be shown on the one hand by completely fertile hybrids, such as the crosses between different species of Antirrhinum belonging to the section Antirrhinastrum, and on the other hand, by completely sterile hybrids like the radish-cabbage hybrid. Weak interspecific hybrids apparently are usually sterile.

Among animal hybrids one sex is sometimes sterile and the other fertile or at least partially so. Thus the domestic cow, *Bos taurus*, may be crossed with the bison, *Bison americanus;* the yak, *Poëphagus grunniens;* the gayal, *Bibos frontalis;* and the gaur, *Bibos gaurus.* In these crosses the female hybrids are fertile, or at least partially so. Hybrids of domestic cattle with the Indian zebu, *Bos indicus*, are fertile in both sexes. Hybrids of the horse, *Equus caballus*, with the ass, *E. asinus*, and zebras of the species, *E. grevyi*, *E. burchelli*, etc., and other equus crosses are sterile in both sexes.

In animal crosses an attempt has been made to draw the generalization that when the sexes differ in fertility, the most pronounced sterility will be exhibited by the heterozygous sex. In that case it would be expected that the male sex shows the greater sterility in mammalian crosses, as in fact appears to be the case when there are any differences. In birds and lepidopteran insects, on the contrary, the females should exhibit more marked sterility. There are so many exceptions to the rule, however, that it is impossible to ascribe general validity to it. Nevertheless, there are so many instances which conform to it that it would seem to have some significance. Perhaps the reason lies in the fact that sex determination in interspecific hybrids is sometimes disturbed, in which case the heterozygous sex may be expected to show the most marked effects.

In plants, the sterility of interspecific hybrids appears to be for the most part gametic, *i.e.*, due to failure of the gametes formed by the hybrid to function normally. This is shown very clearly by cytological studies on the reduction phenomena in such hybrids. The mitotic divisions occurring during development are perfectly normal and the figures show the combined haploid chromosome complexes of both parents, as is to be expected. In the meiotic divisions, however, irregularities of distribution occur, which are followed eventually by abortion of a large percentage or even all of the resulting gametes. Details of the phenomena are so varied that they cannot be presented adequately in the space here available. A brief account of a few different types of meiotic phenomena is given in a later portion of this chapter. For more complete information, the student should consult some of the references listed at the end of the chapter.

On the whole, sterility is the most characteristic feature of interspecific hybridization, and constitutes its outstanding problem. If any one feature may be used as an index of relationship, the degree of sterility of the hybrid is perhaps most useful, but even it cannot be employed without due regard for details of individual cases. The elucidation of the problem of sterility is obviously one which will require a combination of cytological and genetic attack.

**Genetic Phenomena in Interspecific Hybridization.**—In very few cases have studies of interspecific hybridization been continued long enough, and perhaps in none have they been carried out in such a way as to disclose the fundamental difference between species hybridization and ordinary varietal hybridization. Some of the studies have, however, led to interesting disclosures of the nature and variety of segregation products obtained in such cases. Illustrations will be drawn for the most part from plant hybrids, because the phenomena have been studied most completely in them.

There appears to be a fairly close relationship between the genetic phenomena and the degree of fertility exhibited by $F_1$. With respect to fertility, four arbitrary classes may be distinguished for the purposes of this discussion, *viz.:*

1. Complete fertility, a comparatively rare condition in true interspecific hybridization.
2. Incomplete fertility, but with fertility sufficient to produce viable seeds on self-fertilization.
3. Incomplete sterility; sterility so nearly complete that a few viable seeds are produced only by backcrossing to the parent species.
4. Complete sterility; no functional ovules or pollen grains produced.

The last class of hybrids obviously cannot be investigated further, since they produce no progeny. It may remain, however, an open question whether any interspecific hybrids which produce flowers are really completely sterile, as will be seen later. Probably even the most extreme cases produce an occasional offspring, if sufficiently extensive attempts are made to secure progeny from them.

**Completely Fertile Interspecific Hybrids.**—While complete fertility appears to be an unusual condition in interspecific hybrids, a few acceptable cases have been studied rather extensively. The most notable of these are the hybrids between species of Antirrhinum belonging to the section Antirrhinastrum. The section contains eight recognized species, and according to Baur, a number of others as yet undescribed. Of these *A. majus* is the garden snapdragon, of which literally hundreds of varieties are grown. The others are not cultivated: but in nature they exhibit marked diversity in lesser characters. Cultivated together they intercross freely with the exception of *A. siculum*, which crosses with the others with difficulty and produces partially sterile hybrids.

Baur has made numerous crosses between wild species and between *A. majus* and the wild species and has grown $F_2$ progenies in each case consisting of over a thousand individuals. The results from crosses of *A. majus* and *A. glutinosum* (*molle*) have been reported in greatest detail.

*A. majus,* the common garden snapdragon, is familiar to every one. It is a self-fertile species, existing in numerous cultivated varieties and wild forms. The cultivated varieties are characterized by differences in comparatively distinct Mendelian characters, mostly flower colors, but also flower shapes, habits of growth, leaf characters, etc. *A. glutinosum* is distinctly different. It is a wild, self-sterile species, of prostrate habit, profusely branched, with smaller, rounder leaves than *A. majus* and flowers only one-third as large (figure 98).

The $F_1$ is fully fertile and self-fertile like *A. majus.* In general characters it is intermediate between the two parental species, but recessive characters introduced through *A. majus* are recessive in the interspecific hybrids. $F_1$ is fairly uniform and reciprocal crosses are

FIGURE 98.—Beginning at the left, a peloric *majus* of the type used in crosses with *glutinosum;* a plant of *glutinosum;* a plant much resembling *glutinosum* obtained in $F_2$; and on the extreme right the $F_3$ progeny of such a plant. (*From Lotsy.*)

identical. A slight lack of uniformity is to be ascribed to the prevailing heterozygous condition found in *A. glutinosum* as a consequence of its self-sterility.

$F_2$ progenies are extraordinarily variable in habit of growth, leaf form, pubescence, flower size and shape, and in physiological characters, such as self-fertility, brittleness of stems, etc. These differences are found in every conceivable combination. The variability transcends the limits set by the parental species. This is particularly well shown in flower shape, for various teratological conditions are found in $F_2$ individuals and some resemble in form flower shapes found only in other genera of Scrophulariaceae (figure 99). The diversity is so great that in relatively large $F_2$ progenies no two identical individuals are found. Parental forms are recovered only rarely, if at all, but occasional individuals are secured which so closely resemble the parental types as to be distinguished from them only by the minutest examination.

In $F_3$ and subsequent generations, less complex segregation is shown than in $F_2$, and a few generations of self-fertilization give rise to constant races which are fully fertile. Numerous such constant derivatives may easily be established.

Baur concludes from these studies that the parental species differ in a large number of factors which are distributed according to Mendelian laws. He considers it unnecessary to assume any subsidiary hypotheses to account for the behavior exhibited by these hybrids. The number

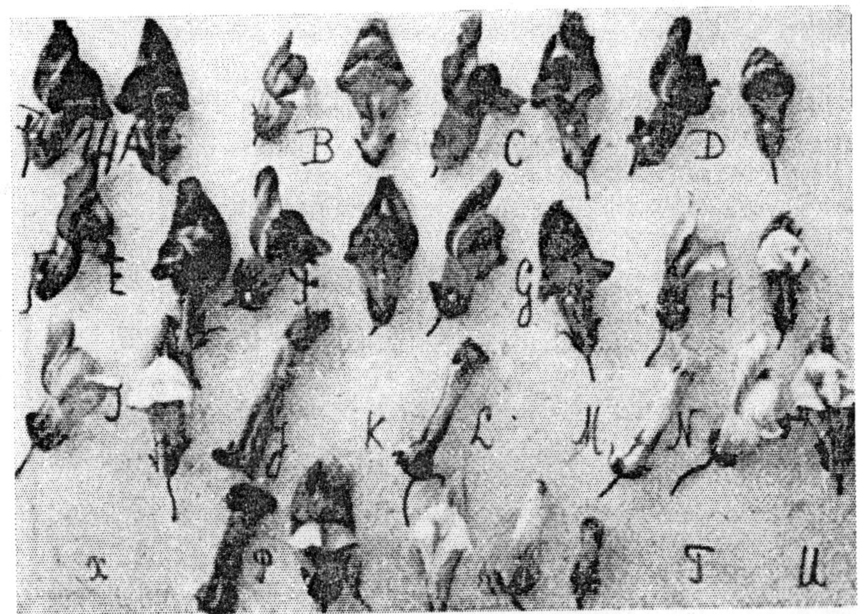


FIGURE 99.—Flower types obtained in $F_2$ of a cross between *Antirrhinum majus* (peloric) and *A. glutinosum*. (*From Lotsy.*)

of factors in which the parental species differ has not, of course, been accurately determined; but Baur estimates that in excess of a hundred such differences must exist and states that they defy accurate analysis.

Baur also investigated the possibility of occurrence of specific Mendelian factors known to exist in *A. majus* in the wild species by introducing them into the hybrid through *A. majus* and investigating their subsequent behavior. He thus determined that all these wild species contain thirty-nine factors known to be dominant in *A. majus*, one known to be recessive, the dominant being a teratological form, and only in four factors were some found to be dominant and some recessive. The studies of crosses between wild species yielded similar results; in all cases segregation was so complex as to defy accurate analysis and to necessitate the assumption of a great number of factor differences between the parental species; but each of these factors in general has a relatively slight effect.

On the basis of these facts two divergent conclusions have been reached as to the significance of hybridization in evolution. Impressed by the diversity of the $F_2$ combinations, Lotsy concluded that hybridization alone could account for all evolution; hybridization, according to his idea, gave rise to numerous combinations, and natural selection then sorted out those combinations best suited to particular conditions. Baur, however, concludes that these species have arisen largely by the accumulation of minute mutations, and that hybridization has played a secondary role in presenting for trial by natural selection various combinations of mutant factors within the species. Baur's view appears to be more in accord with the actual evidence. Lotsy's speculations are subject to the restriction that hybridization is effective in producing new forms only when the parental species differ genetically; it is hardly conceivable that hybridization actually produces new factors. Moreover, hybridization of the type described by Lotsy is probably not of frequent occurrence in nature, since these species are not generally found so situated as to give opportunity for natural crossing. Lotsy practically denies the occurrence of mutation, a phenomenon which has been certainly demonstrated under adequately controlled conditions. If mutation occurs, a proper recognition of it must be made in arriving at a true picture of the method of evolution.

The question also arises as to whether interspecific hybridization has played a part in the origin of cultivated varieties of *Antirrhinum majus*. Baur reaches the decision that it has not. This conclusion is based upon studies of the types of differences which characterize cultivated varieties and upon the types of combinations which are secured from crosses of the cultivated varieties with wild species. Baur finds that cultivated varieties are largely characterized by differences in rather large, recessive mutant characters and that the wild species largely possess the corresponding dominants. It is impossible, therefore, to obtain segregation products from wild species which duplicate the characters of cultivated varieties. He believes that cultivated forms have arisen basically through appearance and preservation of comparatively large mutations, and that hybridization has been restricted to these, so that it has been the means of trying out various combinations of mutant characters. The picture that he draws of the origin of cultivated forms is essentially the same as that of wild races and species, save that large mutations instead of small ones have been preserved in cultivated forms and man instead of nature has been the selecting agent. Man, he says, works with a coarser sieve than nature.

The idea derived from these studies is that the members of this closely related group of species included in the section Antirrhinastrum are identical in germinal organization and that the differences are confined to particulár elements in the system. Since the haploid chromosome number in all of them is 8, this conclusion is not in conflict with obvious cytological conditions. It is doubtful, however, that sufficient evidence has yet been presented to substantiate the conclusion. It may well be that the observed phenomena depend entirely upon distribution of whole chromosomes. If this were the case, the following picture might be drawn of the phenomena in the *majus-glutinosum* case:

$$\text{The } F_1 = 8\text{m} + 8\text{g chromosomes.}$$
$$F_1 \text{ gametes} = (\text{m}+\text{g})^8 = 256 \text{ combinations.}$$
$$F_2 \text{ zygotes} = (\text{m}+\text{g})^8 \times (\text{m} + \text{g})^8 = 65{,}524 \text{ combinations.}$$

This feature could be established by linkage studies in species hybridization. If no crossing-over occurs between *majus* and *glutinosum* homologues, then the phenomena must be restricted to distribution of whole chromosomes. But linkage studies of this type have not been made, although it would be highly desirable to have this information.

The hybrid of maize with annual teosinte gives similar results, although in this instance production of a certain percentage of empty pollen grains, 16.8 per cent in some counts made by Longley, may indicate that some combinations are not viable. However, cytological studies show that each parent contributes 10 chromosomes to the hybrid and that these chromosomes pair normally and are distributed without irregularities in the meiotic divisions of the hybrid.

In Nicotiana there is a group of species, consisting of *N. alata*, *N. Langsdorffii*, and *N. Forgetiana*, which differ strikingly in their characters, but all of which have nine pairs of chromosomes. As in the Antirrhinum crosses, these species give fully fertile hybrids which exhibit normal meiotic behavior and, so far as investigations go, complex Mendelian segregation. Despite their marked differences, there is some question as to the propriety of regarding these as distinct species. In Nicotiana, alleged species hybrids which are fully fertile all come from closely related forms having the same chromosome numbers. There is not a complete range of conditions from full fertility to complete sterility, for, aside from these fully fertile hybrids, all others are completely sterile, save the two described in detail in the succeeding pages, and they both exhibit a very high degree of sterility. If the groups of forms in Nicotiana which give completely fertile hybrids are considered as species rather than the individual forms within them, then an objective criterion for species separation based on failure to cross or production of a sterile hybrid might be employed. Even so, it appears unjustifiable to include maize, *Zea mays*, and annual teosinte, *Euchlaena mexicana*, in the same species. Study of linkage phenomena in such instances may disclose another basis of distinction.

**Incomplete Fertility in $F_1$.**—When $F_1$ interspecific hybrids are fertile at all, they are usually only partially so. In plants partial fertility is readily demonstrable by examination of pollen grains. In fully fertile species and varieties, the pollen grains are usually almost entirely normal in form and full of protoplasmic contents; in partially fertile interspecific hybrids a large proportion of the grains are shriveled and devoid of contents. Very frequently not all those which appear to be normal are actually functional, for germination tests will frequently show a large proportion incapable of growth.

As an example of phenomena exhibited by partially fertile interspecific hybrids, reference may be made to the extensive investigations of East on *Nicotiana paniculata* × *N. rustica*. This hybrid was one of the first interspecific crosses made for scientific purposes. It was made by Kölreuter in 1760 and reported by him in 1761–66, and it has since been studied by numerous investigators. Despite its long history, much remains yet to be done with it.

The two parental species differ widely in their characters in the way in which many species differ, not in a few striking features, but rather in many intangible respects. *N. paniculata* is a wild Peruvian species discovered in 1752 and since widely distributed in botanical gardens throughout the world. It is a low plant, freely branching from the base and bearing comparatively small cordate petioled leaves; each branch is terminated by a loose panicle of green flowers. It appears to be a remarkably constant species, for no varieties of it are known. *N. rustica*, on the contrary, is a highly polymorphic species which was widely distributed among American Indians as a cultivated plant, particularly along the Atlantic seaboard. Numerous well-characterized varieties of it exist, some of which are still cultivated, particularly by the peasants of central Europe, as a home source of tobacco. Brief descriptions of some of the recognized varieties may be presented as an evidence of the range of forms exhibited within the species. *Texana* is a tall, moderately branched variety, with flowers in a loose panicle. It is a generalized type of the species, medium robust in growth. *Brasilia* is a tall, stout plant, with thick stems and branches and large bullated leaves. The flowers are borne in a close panicle, almost like a head. *Humilis* is a dwarf form, otherwise very much like *texana*. The leaves are flat and the flowers are borne in an open panicle. *Scabra* is a tall slender form with numerous slender, ascending branches, and a loose rather delicate inflorescence. These running comments merely indicate the range of forms found within the species. The varieties differ chiefly in quantitative features rather than in definite, easily recognizable Mendelian characters.

East studied in particular a hybrid between *humilis* and *paniculata*. The $F_1$ closely resembled the *rustica* parent, but it also showed minor influences of *paniculata* in its intermediate habit of growth, loose panicles, and intermediate flower size. Reciprocal hybrids were identical, and the $F_1$ hybrid was notable for its uniformity and vigor. In the latter respect it exceeded the taller parent. Its fertility was estimated to be between 3 and 4 per cent for ovules and perhaps less than one-tenth of 1 per cent for pollen grains. By careful hand pollination of hybrid plants *inter se*, shriveled capsules were obtained which contained from three to twenty-two viable seeds.

An $F_2$ population of 246 plants exhibited such great diversity that no duplicates were found among them. For leaf size and shape and size of plant, the range extended beyond that set by the parents; for size and shape of flowers the range extended between the parents. A general survey of $F_2$, however, indicated that two groups of plants were secured; one consisting of four plants which closely resembled *paniculata* and the other consisting of the remainder of the plants which exhibited *rustica* characters. Apparently the midportion of the expected range was omitted. In the *rustica* group of plants, combination products were obtained which closely resembled other varieties of *rustica*, such as *texana*, *scabra*, *brasilia*, *pumila*, etc. Parental forms were recovered about once among 100 to 200 plants. The range of fertility extended from about that of $F_1$ to a highly fertile condition, equivalent to that of the parental species, with a marked tendency to an average fertility considerably higher than $F_1$.

In subsequent generations the *paniculata* derivatives became identical with the original *paniculata* and fully fertile, and no other *paniculata* derivatives were produced. From the *rustica* derivatives, however, a number of different types were established duplicating those found in existing varieties of *rustica*. These strains were fully fertile and constant. They were also fertile *inter se*. In their establishment a number of interesting observations were made. In the first place, the average fertility of the progeny of a partially sterile plant was usually greater than that of the parent plant. The range of segregation in the progeny of a plant was generally less than that of the population of which it was a member. In the third place, it was observed that segregation still occurred after full fertility had been acquired.

Obviously the phenomena observed in this instance cannot be accounted for in the same way as those connected with the Antirrhinum species cross. In the first place, there is sterility to be accounted for; in the second place, the $F_2$ progenies and derivatives established from this cross are apparently less diverse than those of the Antirrhinum case. They are rather obviously divided into two groups: one consisting of a few plants closely approximating *paniculata* and the other larger group of *rustica* derivatives duplicating the known varieties of this species.

Cytological conditions must be taken into account in arriving at an explanation of these phenomena; but unfortunately even then the explanation is highly speculative, because the necessary connection between cytological and genetic phenomena has not been fully established. However, *paniculata* has twelve pairs of chromosomes and *rustica*, twenty-four. The formula for the hybrid, therefore, is (12P + 24R), *i.e.*, 12 *paniculata* chromosomes and 24 *rustica*. In meiosis there are $12_{II} + 12_{I}$ chromosomes; *i.e.*, 24 chromosomes conjugate to form 12 pairs, and the remaining 12 chromosomes remain unpaired. This is interpreted to mean that the 12 *paniculata* pair with 12 *rustica* chromosomes, leaving 12 *rustica* chromosomes unpaired, a situation which may conveniently be represented by the formula $F_1$ = 12PR + 12R. In the meiotic divisions the 12 pairs behave normally; they separate into their constituent members in the first division and divide equationally in the second. The univalents are distributed irregularly, apparently at random, in the first division, so that 6< >6, 7< >5, 8< >4, etc. types of distribution occur. In the second division they divide equationally. In many instances, however, they do not follow this scheme throughout. Sometimes the univalents lag on the spindle and fail to be included in the major nuclei, and rarely apparently only a single division may occur instead of two.

Assuming that gamete formation follows the scheme outlined, it is obvious that the gametes will have $(12 + i)$ chromosomes, $i$ having values from 0 to 12, depending on the distribution of the univalents. The twelve chromosomes, however, may consist of various combinations of *paniculata* and *rustica* chromosomes. Assuming random assortment, these combinations may be determined by expansion of the binomial $(P + R)^{12}$, and similarly the distribution of univalents by $(R + r)^{12}$, $r$ representing the absent mates of the univalent chromosomes. The entire gametic series may then be worked out by multiplying these two expanded binomials together, $(P + R)^{12} \times (R + r)^{12}$. This is a good exercise in the application of principles of probability to a specific situation, but unfortunately our present methods fall far short of making an accurate comparison of theory and actual result.

In figure 100 an attempt is made to portray the final results of these experiments in terms of the cytological situation. The twenty-four chromosomes of *rustica* are represented in ordinary type, and the twelve *paniculata* chromosomes are represented in bold-faced type. In $F_1$ pairing is presumed to occur between homologues as shown. The final constant *paniculata* derivatives are represented as containing only *paniculata* chromosomes in view of their identity with the original species. In the group of *rustica* derivatives, however, replacement of some *rustica* by *paniculata* homologues apparently is responsible for production of new *rustica* forms. Theoretically these segregants may range from pure *rustica* to forms in which twelve *rustica* chromosomes have been replaced by *paniculata* homologues, 4,096 different combinations being possible. In addition to failure to secure *paniculata* derivatives containing *rustica* replacements, it is assumed that all forms which have

| N. rustica | N. paniculata | $F_1$ | Paniculata derivative | Rustica derivatives | | | |
|---|---|---|---|---|---|---|---|
| AA | **AA** | A**A** | **AA** | AA | AA | AA | **AA** |
| BB | **BB** | B**B** | **BB** | BB | BB | BB | **BB** |
| CC | **CC** | C**C** | **CC** | CC | CC | CC | **CC** |
| DD | **DD** | D**D** | **DD** | DD | **DD** | DD | **DD** |
| EE | **EE** | E**E** | **EE** | EE | EE | EE | **EE** |
| FF | **FF** | F**F** | **FF** | FF | **FF** | **FF** | **FF** |
| GG | **GG** | G**G** | **GG** | GG | **GG** | **GG** | **GG** |
| HH | **HH** | H**H** | **HH** | HH | HH | HH | **HH** |
| II | **II** | I**I** | **II** | II | II | II | **II** |
| JJ | **JJ** | J**J** | **JJ** | JJ | **JJ** | JJ | **JJ** |
| KK | **KK** | K**K** | **KK** | KK | KK | **KK** | **KK** |
| LL | **LL** | L**L** | **LL** | LL | LL | LL | **LL** |
| MM | .... | M | .... | MM | MM | MM | MM |
| NN | .... | N | .... | NN | NN | NN | NN |
| OO | .... | O | .... | OO | OO | OO | OO |
| PP | .... | P | .... | PP | PP | PP | PP |
| QQ | .... | Q | .... | QQ | QQ | QQ | QQ |
| RR | .... | R | .... | RR | RR | RR | RR |
| SS | .... | S | .... | SS | SS | SS | SS |
| TT | .... | T | .... | TT | TT | TT | TT |
| UU | .... | U | .... | UU | UU | UU | UU |
| VV | .... | V | .... | VV | VV | VV | VV |
| WW | .... | W | .... | WW | WW | WW | WW |
| XX | .... | X | .... | XX | XX | XX | XX |

FIGURE 100.—Results of the hybridization of *Nicotiana paniculata* with *N. rustica* presented in terms of chromosomes. Chromosomes of *paniculata* are represented in bold-faced type. The only *paniculata* derivative is represented as containing nothing but *paniculata* chromosomes since it was identical with the original parent. *Rustica* derivatives theoretically may range from pure *rustica* to types with twelve *rustica* chromosomes replaced with *paniculata* homologues.

more than twelve and less than twenty-four pairs of chromosomes are not secured. Under this interpretation, of 16,777,216 possible stable combinations, not more than 4,097 are capable of realization. While the available evidence indicates the validity of the interpretation, it will obviously require additional supporting evidence before it may be considered as proved, particularly as to the extent to which *rustica* chromosomes may be replaced by *paniculata* homologues.

**Triticum Species Hybrids.**—More complete studies of cytological conditions have been made in wheat hybrids. There are three groups of species of wheat characterized by different chromosome numbers: einkorn, $n = 7$; emmer group, $n = 14$; spelt, $n = 21$. The hybrids of emmer with spelt types, the so-called pentaploid hybrids, since they have $5 \times 7 = 35$ chromosomes, have been investigated most extensively, particularly by Kihara. These hybrids are rather highly fertile, but numerous sterile forms appear among the progeny, and the cytological phenomena in the succeeding generations are highly illuminating.

The $F_1$ exhibits a condition represented by the formula $14_{II} + 7_{I}$. The bivalents behave regularly in meiosis; the univalents split equationally in the first division and are distributed at random in the second. Some irregularities were observed, such as lagging, formation of microcytes, etc., but in the main gamete formation may be assumed to follow the scheme indicated. The gametic chromosome numbers may be represented by $(14 + i)$, $i$ having values ranging from 0 to 7 with frequencies corresponding to the coefficients of $(a + b)^7$. Since this series of gametes occurs, in both pollen grains and ovules, two such series must be multiplied together in order to give the complete $F_2$ zygotic series. But Kihara found that the $F_2$ plants actually obtained belonged to a much more limited category, with very few exceptions, as shown in table 44. The viable combinations, on the basis of their subsequent behavior, are grouped into two classes, one ranging in chromosome number from 28 to 34, called the diminishing group, because in subsequent generations the chromosome number gradually reverted to 28; the other, ranging from 36 to 42, called the increasing group, because their progeny eventually returned to the 42-chromosome condition of the other parent. The 35-chromosome condition, equivalent to $F_1$, may be included in both categories. The chromosome formula of the diminishing group may be represented by $(14_{II} + i_{I})$, with $i$ having values ranging from 0 to 7. Since in the progeny of any such plant any zygote in which any of these univalent chromosomes becomes doubled up, giving combinations such as $15_{II} + i$, $16_{II} + i$, $17_{II} + i$, etc., the sum of the number of pairs plus the number of univalents being less than 21, is a non-viable zygote, it is perfectly evident that this group of plants will ultimately revert to the $14_{II}$ condition automatically. In the increasing group, on the contrary, the number of bivalents plus the number of univalents uniformly equals 21. Among their progeny any combination of bivalents and univalents, the sum of which is less than 21, represents a non-viable or poorly viable zygote; consequently, this group automatically reverts to the $21_{II}$ condition eventually and rapidly.

Many interesting speculations are here raised as to the genetic features involved, but these have unfortunately not been fully investigated. Kihara found that the 42-chromosome segregants, while all of the spelt type, differed in individual features from the original parental variety, and that the 28-chromosome segregants, likewise, while all of the emmer type, differed from one another and from the original emmer parent. This must mean that exchanges of homologous chromosomes (or parts of chromosomes?) have taken place between the two species. How far this exchange may be carried is an interesting question. Kihara assumes that an emmer form may be built up entirely from the 14 chromosomes in the spelt complex which are homologous with the 14 chromosomes of emmer, and that conversely a spelt type may be constructed consisting of $14_{II}$ emmer chromosomes $+7_{II}$ spelt chromosomes, *i.e.*, by complete replacement of the 14 spelt chromosomes with the homologous emmer chromosomes. This problem requires further genetic study.

TABLE 44.—INTERPRETATION OF CHROMOSOME BEHAVIOR IN $F_2$ OF A CROSS BETWEEN EMMER AND SPELT WHEATS (*After Kihara*)

| Type | Chromosome number | Viable combinations | Non-viable combinations | | |
|---|---|---|---|---|---|
| | 28 | $14_{II}$ | | | |
| | 29 | $14_{II} + 1_{I}$ | | | |
| | 30 | $14_{II} + 2_{I}$ | $15_{II}$ | | |
| Diminishing group.. | 31 | $14_{II} + 3_{I}$ | $15_{II} + 1_{I}$ | | |
| | 32 | $14_{II} + 4_{I}$ | $15_{II} + 2_{I}$ | $16_{II}$ | |
| | 33 | $14_{II} + 5_{I}$ | $15_{II} + 3_{I}$ | $16_{II} + 1_{I}$ | |
| | 34 | $14_{II} + 6_{I}$ | $15_{II} + 4_{I}$ | $16_{II} + 2_{I}$ | $17_{II}$ |
| $F_1$ condition........ | 35 | $14_{II} + 7_{I}$ | $15_{II} + 5_{I}$ | $16_{II} + 3_{I}$ | $17_{II} + 1_{I}$ |
| | 36 | $15_{II} + 6_{I}$ | $16_{II} + 4_{I}$ | $17_{II} + 2_{I}$ | $18_{II}$ |
| | 37 | $16_{II} + 5_{I}$ | $17_{II} + 3_{I}$ | $18_{II} + 1_{I}$ | |
| | 38 | $17_{II} + 4_{I}$ | $18_{II} + 2_{I}$ | $19_{II}$ | |
| Increasing group.... | 39 | $18_{II} + 3_{I}$ | $19_{II} + 1_{I}$ | | |
| | 40 | $19_{II} + 2_{I}$ | $20_{II}$ | | |
| | 41 | $20_{II} + 1_{I}$ | | | |
| | 42 | $21_{II}$ | | | |

In both of the hybrids discussed above, there is an illustration of an observation made repeatedly in species hybridization, namely, that the descendants eventually revert to the parental condition. In neither case is the reversion complete in the sense that the ultimate segregants are absolutely identical with the original parental forms, aside from the *paniculata* descendants of the *N. paniculata-rustica* hybrid; but the chromosome numbers eventually attained are those of the parental species, and only exchanges of homologous chromosomes (or parts of chromosomes?) are affected by hybridization.

**Incomplete Sterility.**—An incompletely sterile plant is one which forms only an occasional viable gamete. Thus the hybrids of rye with wheat, according to recent investigations, produce no seeds with their own pollen, but an occasional seed is set by using pollen either of rye or of wheat. Similarly the hybrid of *N. sylvestris* with varieties of *N. tabacum* produces no seed by selfing; but a few seeds are secured when pollen of either *N. sylvestris* or *N. tabacum* is applied to the hybrid. The subsequent generations have been followed in some detail in this latter case, so that it will be treated more fully. The student should compare the results with those secured from *N. paniculata* with varieties of *N. rustica.* The two crosses are closely parallel but differ in the degree of gamete abortion, which is greater in the *sylvestris-tabacum* cross.

When $F_1$ is backcrossed to *sylvestris,* the progeny consists of a few plants which approach *sylvestris* in the sum total of characters, and the rest of the plants make up a heterogeneous assemblage of abnormal and otherwise diverse forms incapable of any logical classification. A few plants may be very close to $F_1$ in appearance. Those plants which resemble *sylvestris* are as a rule partially fertile, some comparatively highly so. Occasionally, in fact, a plant is found which is fully fertile and absolutely identical with *sylvestris.* The partially fertile individuals on self-fertilization for a few generations revert completely to the *sylvestris* type and become fully fertile. No constant derivative types are produced which show any characters whatever of the *tabacum* parent.

When $F_1$ is backcrossed to the *tabacum* parent, the progeny consists again of a varied assemblage of forms, many of which are highly abnormal. Among them are a few plants which are partially fertile, and these approach more or less closely to the *tabacum* parent in the sum total of their characters. When these partially fertile plants are self-fertilized for a few generations, they automatically return completely to the original parental condition, and they become completely fertile. The derivative types thus established are again completely identical with the original *tabacum* parent; they exhibit no characters whatever of *sylvestris.*

This phenomenon of complete return to the parental condition is all the more striking when different varieties of *tabacum* are employed in the original hybridization. The species *N. tabacum* includes all the numerous cultivated varieties of tobacco, a highly diverse assemblage of forms. If the variety *macrophylla* is employed, the *tabacum* derivatives secured are pure *macrophylla;* if *purpurea,* they are *purpurea;* if *calycina,* pure *calycina;* and so on for other varieties of *tabacum.* There is absolutely no suggestion of establishment of diverse types of *tabacum* as a result of this cross, corresponding to the *rustica* derivatives obtained by crossing that species with *paniculata.*

The chromosome situation in this cross is identical with that in the *paniculata-rustica* cross. *Sylvestris* has twelve pairs of chromosomes; *tabacum,* twenty-four. The $F_1$ hybrid, therefore, has twelve *sylvestris* and twenty-four *tabacum* chromosomes, and cytological studies show that in the hybrid meiosis the twelve *sylvestris* chromosomes pair with twelve *tabacum,* and the remaining twelve *tabacum* chromosomes remain unpaired. Cytological studies of the *sylvestris* derivatives show that they have twelve pairs of chromosomes and the *tabacum* derivatives have twenty-four pairs. The return to the parental condition, therefore, involves a return in chromosome number as well as in morphological characters. Obviously exchange of chromosomes between these species would be limited to those *sylvestris* and *tabacum* chromosomes which pair in $F_1$; but the complete return indicates that no such exchanges of chromosomes give rise to stable recombinations, unless the chromosomes which pair are identical, which is highly improbable. Evidently in this instance exchange of material between the two species is incompatible with formation of fertile offspring; consequently, no recombination products whatever are secured. There are, therefore, limits to the extent to which new forms may be secured from interspecific hybridization.

To what extent return to the parental condition follows upon interspecific hybridization, and what principles govern it, cannot yet be stated. It has often been observed that the progenies of partially fertile hybrids run back to the parental condition, especially when the parents differ in chromosome number, but the return is apparently rarely as complete as that shown by the *sylvestris-tabacum* hybrid. It may be that the return bears some relation to the degree of sterility of $F_1$; that is, that it is more complete the higher the degree of sterility exhibited by $F_1$; but further experimentation will be necessary before the truth of this generalization will have been determined.

**Didiploid Interspecific Hybrids.**—In certain instances it has been observed that interspecific hybrids have double the number of chromosomes to be expected from the known cytological conditions of the parents. Apparently the first instance of the kind was a cross between two Primula species, which was produced accidentally at Kew. From *Primula verticillata* $\times$ *P. floribunda*, a sterile hybrid was secured which exhibited obvious characters of both species and exceeded its parents in vigor. The parent species each had nine pairs of chromosomes, and the hybrid had eighteen, 9V + 9F. The hybrid was propagated vegetatively for some time. Finally, it produced a branch which bore seeds, and the seeds gave rise to a fertile progeny identical with the hybrid except for a somewhat enhanced vigor. It has since remained comparatively constant and has become widely distributed as a garden ornamental under the name of *P. Kewensis*. Cytological investigations showed that it has twice as many chromosomes as the parental species and the original hybrid, and its meiotic divisions were regular for the most part, with $18_{II}$ chromosomes. Assuming that the original eighteen chromosomes of the sterile hybrid must have split longitudinally to give the thirty-six in *P. Kewensis*, the formula of the latter must be 9VV + 9FF instead of 9V + 9F. Its chromosome complement has become balanced, its gametes are all of the same constitution, and its fertility and constancy follow as a matter of course.

Another similar case has been discovered by Blackburn and Harrison in Rosa. The situation may be diagrammed in the following way:

| | *Rosa pimpinellifolia* × *R. tomentosa* | |
|---|---|---|
| zygotes | $14_{II}$ | $7_{II} + 21_{I}$ |
| gametes | 14P | 7T |
| $F_1$ | 14P + 7T = $7_{II} + 7_{I}$ | |
| *Rosa Wilsoni* | 14PP + 7TT = $21_{II}$ | |

*Rosa Wilsoni* also exhibits regular meiotic divisions, forming twenty-one gemini and gametes with twenty-one chromosomes. Its gametes are viable; hence, it is fertile, but its constancy has not been determined.

A third case has been reported by Clausen and Goodspeed, and it may be diagrammed as follows:

| | *Nicotiana glutinosa* × *N. tabacum* | |
|---|---|---|
| zygotes | 12GG | 24TT |
| gametes | 12G | 24T |
| $F_1$ | 12G + 24T = 12GT + 12T | |
| *Nicotiana digluta* | 12GG + 24TT = $36_{II}$ | |

The normal hybrid is completely sterile; the tetraploid hybrid has the same morphological features, aside from a slightly increased vigor; but it is fertile and constant.

From these examples the general hypothesis may be advanced that any true interspecific hybrid may give rise to a constant, fully fertile derivative, a didiploid hybrid, by doubling its chromosome number; or, what amounts to the same thing, tetraploid representatives of two species crossed together should give at once a constant, fertile hybrid. While enough instances have not yet been studied to establish the general validity of this assumption, it is of both theoretical and practical importance. The practical importance lies in the fact that by this means it may be possible to obtain fertile, constant forms exhibiting all the favorable features of interspecific hybrids, such as have been noted previously.

**Evolution in Chromosome Number.**—Attention has been called to the variation which occurs in chromosome number within the confines of a single genus. The existence of these differences in chromosome number raises a distinct problem in evolution which cannot be accounted for on the basis of factor mutation alone. In some cases the evidence seems to indicate that cross-division and fusion of chromosomes may be responsible for the changes in number, as in the various species of Drosophila, where to a certain extent both cytological and genetic investigations support this view. In plants, however, as in the roses, wheats, and numerous other groups, the numbers are in arithmetical series, representing multiples of some common base number, seven in Rosa, Triticum, and Avena, nine in Chrysanthemum and Erigeron, six in Viola, and eight in Hyacinthus.

It has been suggested that variations in chromosome number, such as have occurred in Datura, may be responsible for permanent changes in number; but these investigations carefully examined lend but little support to this view. Aside from tetraploidy, these forms are not constant for any new chromosome numbers, and the investigations do not give any clue as to secondary changes which might lead to their constancy. In the case of tetraploidy, the new form, while relatively constant, exhibits other conditions which set it aside from the behavior exhibited by true species. In the first place seed production is low; but more important the chromosomes unite as quadrivalents in meiosis, whereas the so-called tetraploid species in nature show bivalents in meiotic divisions. Moreover, the series of chromosome numbers found in nature are in arithmetical progression, whereas tetraploidy, if continued, should lead to geometrical series. This objection should not, however, be too strongly urged, for crossing between polyploid species might fill up the series to make an arithmetical progression.

Purely on theoretical grounds, Winge offered a suggestion which overcomes these difficulties and for which experimental evidence has since been obtained. As an illustration he considers Chrysanthemum, in which species with $9_{II}$, $18_{II}$, $27_{II}$, $36_{II}$, and $45_{II}$ chromosomes occur, and pictures the process as follows: Assuming a stem form with $9_{II}$ chromosomes, he assumes that by evolution a number of distinct species were produced without changes in chromosome number, which may be designated *A*, *B*, *C*. The evolution of chromosome number then proceeds according to the following scheme.

| | *A* | | *B* | *C* | |
|---|---|---|---|---|---|
| Zygotes.......... | $9a_{II}$ | | $9b_{II}$ | $9c_{II}$ | |
| Gametes.......... | 9a | | 9b | | |
| A × B............ | | 9a + 9b | | | |
| Doubling.......... | | $9a_{II} + 9b_{II}$ | | | *n. sp.* D, $n = 18$ |
| Gametes.......... | | 9a + 9b | | 9c | |
| D × C............ | | | 9a + 9b + 9c | | |
| Doubling.......... | | | $9a_{II} + 9b_{II} + 9c_{II}$ | | *n. sp.* E, $n = 27$ |
| Gametes.......... | | | 9a + 9b + 9c | | |

Essentially his suggestion is that first the species diverged to such an extent that crosses between them gave hybrids in which the chromosomes failed to pair, and that as a consequence each chromosome split longitudinally, by which it obtained a homologue, and thenceforth normal pairing occurred along with full fertility and constancy as consequences, an hypothesis which has since received experimental verification in the didiploid hybrids discussed above.

To a certain extent this hypothesis is supported by cytological studies of interspecific hybrids. It was first shown by Rosenberg that the thirty chromosomes of *Drosera obovata*, a hybrid of *D. rotundifolia* with *D. longifolia*, conjugate in meiosis to form ten bivalents, the ten remaining chromosomes being left unpaired, a phenomenon represented by the formula $10_{II} + 10_{I}$. Inasmuch as *D. rotundifolia* has ten and *D. longifolia* twenty pairs of chromosomes, this behavior was interpreted as a conjugation of the ten *rotundifolia* chromosomes in the hybrid with ten of those contributed by *longifolia*, and the ten unpaired chromosomes were assumed to represent the rest of the *longifolia* chromosomes. If, in the above hypothetical instance, a cross is made between species A, $n = 9a$, and species D, $n = 9a + 9b$, the hybrid would be expected to exhibit the Drosera scheme of conjugation, *viz.*, $9a_{II} + 9b_{I}$. Since the Drosera type of conjugation is so frequently exhibited in crosses between species of this category, it may indicate evolution according to Winge's hypothesis.

**Resumé.**—From the foregoing account it is evident that the scientific study of interspecific hybridization is in a preliminary stage. As yet none of the investigations answer all of the questions which may be raised in connection with the problem; but the way has been cleared for notable advances in the near future. A combination of cytological study and genetic experimentation is obviously necessary in most cases.

Evidently in some instances new stable recombination products may be established from interspecific hybridization, but as yet the limitations here involved have not been definitely determined and they probably vary from case to case. What relations such products bear to the parental species is another problem worthy of scientific inquiry. Obviously interspecific hybridization may be an abundant source of new forms in practical breeding operations.

Another problem of utilization is raised by the observation that doubling of chromosome number after hybridization may lead to production of fertile, constant forms. While it is as yet premature to venture an opinion as to the role which this phenomenon may play in evolution and in creation of new forms, it obviously opens up an interesting field for investigation. It may be possible that this method will provide a solution of the problem of utilizing interspecific hybrids in practical operations.

## References

Clausen and Goodspeed: Interspecific hybridization in Nicotiana. II. A tetraploid *glutinosa-tabacum* hybrid, an experimental verification of Winge's hypothesis, *Genetics*, **10**: 278–384.

Darwin: Origin of Species, Chapters I and IX.

DARWIN: Variation in Animals and Plants under Domestication, **2**: Chapters XV–XIX.

DENDY: Outlines of Evolutionary Biology, Chapter XVI.

DIGBY: The cytology of *Primula kewensis* and of other related Primula hybrids, *Annals of Botany*, **26**: 357–388.

EAST: A study of partial sterility in certain hybrids, *Genetics*, **6**: 311–365.

HALL and CLEMENTS: The Phylogenetic Method in Taxonomy, *Carnegie Inst. Pub.* 326 (introduction).

HURST: Chromosomes and characters in Rosa and their significance in the origin of species (bibliography) *in* Experiments in Genetics, Chapter XXXVIII.

LOTSY: Evolution.

MORGAN: Theory of the Gene, Chapter XVIII.

POULTON: Essays on Evolution, Chapter II, What is a species?

SHARP: Introduction to Cytology, 2nd ed., Chapter XVII.

THOMPSON AND HOLLINGSHEAD: Preponderance of *dicoccum*-like characters and chromosome numbers in hybrids between *Triticum diococcum* and *T. vulgare*, *Jour. Genetics*, **17**: 283–307.

DE VRIES: Species and Varieties, Chapters I, IV, and IX.

WILSON: The Cell in Development and Heredity, pp. 841–853.

WINGE: The chromosomes. Their numbers and general importance, *Compt. Rend. Trav. Lab. Carlsberg*, **13**: 131–275.

# PART II—PLANT BREEDING

## CHAPTER XXVIII

## INTRODUCTION

In Part II will be discussed the application of the principles of genetics to problems of plant breeding. In this treatment general problems mainly will be considered and their solution illustrated. The many details involved in the breeding of all the various crop and ornamental plants would be out of place in this text and, moreover, published information on these subjects is becoming so voluminous that a very extensive treatise would be required to cover it adequately. The most complete treatise of the sort is Fruwirth's "Handbuch der landwirtschaftlichen Pflanzenzüchtung" in five volumes, but this gives no attention whatever to ornamentals or to the pome fruits, stone fruits, or other important tree fruits and nuts of the temperate zones. Even the history of plant breeding, if considered in detail, would require considerable space. A brief historical review, however, is desirable as a background for an appreciation of the contribution of genetics to scientific plant breeding.

**Beginnings of Plant Breeding.**—Long before any one thought of making a philosophical study of plant improvement, the hybridization of flowers and the preservation of choice strains of favorite varieties was common among gardeners. The oldest records of intentional preservation of plants are found in ancient Chinese encyclopedias. The best plants and fruit trees were used for propagation and an imperial edict recommended the choice of large seed. Nearly as ancient are the fig drawings on the Great Pyramid of Gizeh. According to Fruwirth hybridization was practiced in ancient times in China with various flowers, in Italy during the Roman Empire with roses, and in the seventeenth century in Holland with tulips and primulas. From Darwin's account it seems that the earliest methodical work on the production of new varieties, of which there are authentic records, was done by the Dutch flower fanciers.

**Pioneers in Plant Breeding.**—The systematic breeding of crop plants was begun in Europe during the latter part of the eighteenth century. Van Mons in Belgium, Knight in England, and Cooper in America each made important contributions to horticulture by his efforts at plant breeding. Others to be mentioned among the early breeders of crop plants are Le Couteur, who seems to have been the first to learn the importance of selecting individual plants in the improvement of cereals, and Shirreff and Hallet in England, who also improved their local varieties of wheat by means of isolating pure lines.

The work of the Vilmorin seed establishment in France with wheat and sugar beets yielded results of the greatest importance to the entire world. After 10 years of selection of carrots, in 1843 the improvement of wheat and other grains was systematically undertaken. First, a large collection of varieties from many countries was accumulated. Then a method of line selection plus progeny test was developed which came to be known as the "Vilmorin Method" and which later played an important role in cereal breeding in other countries. Sugar beets, previous to 1850, had been selected according to form only. Selection on the basis of specific gravity was practiced from 1850 to 1862, when the method of determining sugar content by means of polarized light was introduced. By means of the progeny test the increase in sugar content was so marked that beet-sugar production became economically important.

**Progress in Plant Breeding.**—The great world movements of the nineteenth century following the improvement of transportation facilities, the migration of peoples, industrial development, and the growth of international trade, together with the improvement of farm machinery, resulted in the extension of agricultural industries and gave a greater impetus to plant breeding. This activity was manifested first in Europe and later particularly in the U. S. Department of Agriculture and the state experiment stations in America. Efforts at improvement were concentrated mainly on the crop plants producing the raw materials of importance in the world's markets. The methods of crop improvement employed were those which had been used in the past, but they were systematized and combined for more effective utilization. These methods may be designated as (1) mass selection; (2) line or genotypic selection and progeny test; (3) hybridization followed by direct utilization and fixation of new varieties; and (4) clonal selection. All but the first of these methods will be considered in following chapters.

**Mass Selection in Plant Breeding.**—Mass selection was defined in Chapter XXI as the selection in each generation of a number of individuals most nearly approaching the desired type for progenitors of succeeding generations. The limitation of improvement by means of mass selection was explained on the basis of modifying genes, the numerous genetic factors which condition the degree of expression of the various characters. The method of mass selection in plant breeding consists simply in picking out choice plants from the main crop and sowing the mixed seed from them for the next crop. It has long been used with many crops, especially the cereals. Usually, however, it has been found necessary continually to repeat the selection of best plants in order to maintain the improvement already gained, but despite this limitation, mass selection has been of very great value to agriculture.

One of the earliest breeders to use this method was A. L. de Vilmorin, who began selecting carrots about 1833. Soon thereafter selection of sugar beets for seed production was begun in France and Germany, first according to form of root alone, but later according to specific gravity and actual analyses of sugar content. Mass selection became the principal method of improving small grains in Germany, and it has been known as the German method of "broad breeding." The earliest prominent breeder of small grains was W. Rimpau, who began his work with rye in 1867 and developed the famous Schlanstedt variety. Later he worked with wheat extensively, first by mass selection and later by hybridization

FIGURE 101.—Delta Farm White Dent, a superior strain of maize, the result of thirty years of continuous mass selection.

of varieties and species. Although there have been scores of successful breeders of each of the important small grains in Germany, Rimpau was the first to engage in this work on a large scale.

Mass selection has also been of great importance in the development of both maize and cotton as crops of importance in world economics. In maize such improvement was begun as early as 1825, when J. L. Leaming, of Ohio, began the selection of the best ears from his field for seed corn. By repeating this process he gradually developed a superior strain that came to be known as the Leaming variety. The same simple method was used in originating Reid Yellow Dent (1847), Morley Prolific (1876), and Boone County White (1885). An interesting illustration of what can be accomplished with maize by means of mass selection when a definite ideal is maintained and seed is selected in the field before harvesting is found in the Delta Farm White Dent variety shown in figure

101. The improvement of cotton by mass selection has doubtless been practiced for centuries. Authentic records are scarce, but the characteristic variability in length of fiber, combined with the very practical value of increasing the average length, must have appealed to growers, at least in the more progressive cotton-growing regions of the world. In the South Carolina islands for example, the sea-island types of cotton have been developed by consistent mass selection for early maturity, increased length of lint, and greater productiveness from a West Indian perennial type which was originally unsuited to conditions under which its derivatives are grown so successfully.

The importance of mass selection in the early development of our principal crop plants is indicated by the evidence of its utilization among primitive peoples. The prominence given to such plants as maize and the date palm in ancient architectural decoration shows the esteem in which these plants were held. According to Darwin, the lake dwellers of Switzerland, during the ages of stone and bronze, possessed an apple considerably larger than the one now growing in the Jura Mountains. Virgil and Celsus refer to the importance of selecting largest seeds or best ears of grain for the maintenance of the crop. It is reported by Mitra that the illiterate peasants of Assam regularly practice mass selection of the best rice plants for the next year's sowing. These facts indicate the important role that mass selection must have played in agricultural development. Despite its limitations, it will doubtless continue to be used not only among primitive peoples but also in programs of scientific breeding where it may serve a number of useful purposes, such as the purification of mixed varieties and hybrid populations; as a first step in the amelioration of wild species; and in combating some types of plant diseases.

**Empirical Origin of Plant Breeding.**—The influence of scientific discovery on the early history of plant breeding is not marked. The pioneer plant breeders undertook the production of new and improved varieties while the Linnaean theory of the catastrophic origin of all living things was still accepted by most scientists. Even Hovey, Sutton, and Vilmorin completed most of their work before the publication of Darwin's "Origin of Species." Thus the beginnings of plant breeding were made by florists, horticulturists, and agronomists, who observed the defects of commonly grown varieties and sought to improve them or to find better ones. Each attacked the problem in the light of his own knowledge or theories, the later ones sometimes profiting by the experience of their predecessors.

**The Early Hybridists.**—While the early plant breeders were working along empirical lines, however, the first efforts to obtain scientific knowledge of plant hybrids were being made. The conception of sexuality in higher plants began to be formulated during the last quarter of the seventeenth century. The knowledge of sexuality in flowering plants was actually established by Rudolph Jacob Camerer, Professor of Natural Philosophy in the University of Tübingen, who was better known under his Latinized name, Camerarius. In 1694 he wrote his fifty-page "Epistola de sexu plantarum," in which he described his experiments with spinach, hemp, hops, and maize. According to Werkenthin, Camerer was the first botanist to discover, 200 years after maize had been introduced into Europe from America, that on removing the pollen-bearing flowers from the tassel of an isolated corn plant, the seeds on the ears remain unfertilized.

For nearly 70 years after Camerer wrote his letter on sexuality in plants, there was little or no real progress in the scientific knowledge which underlies plant breeding. Meanwhile Joseph Gottlieb Kölreuter was laying the foundation for the modern study of hybridization in plants. After many fruitless experiments he produced in 1760 the first plant hybrid ever obtained in a scientific experiment by crossing two species of tobacco, *Nicotiana rustica* ♀ × *N. paniculata* ♂. The hybrid plants bloomed in 1761 and the same year appeared his first contribution on sexuality in plants. He called attention to the role of insects in cross-pollination, estimated the number of pollen grains, and accumulated data of the greatest significance on the characters of interspecific hybrids. Besides Nicotiana he worked with several other genera of plants. But little interest was taken in his work by others, as he was decades ahead of his time.

Toward the close of the eighteenth century and during the first half of the nineteenth century, the work of Kölreuter was extended by other scientists, notably by Sprengel, Wiegmann, Sagaret, and Gärtner. Sprengel studied the relations between flowers and insects in great detail.

His conclusion that nature in most cases intended that flowers should not be fertilized by their own pollen and that the peculiarities of flower structure can be understood only when studied in relation to the insect world was revolutionary for his time.

Wiegmann repeated Kölreuter's experiments and gave special attention to hybrids between genera such as Pisum, Ervum, and Vicia. He noted and discussed such phenomena as dominance in hybrids, chance crossing in nature, hybrid origin of varieties, and the significance of these facts for botany and agriculture. Sagaret, from his experiments on plants of the melon family, made discoveries which anticipated those of Goss, Naudin, and Mendel. For the first time in the history of plant hybridization, he studied pairs of contrasted characters. He even used for the first time in literature of this subject the expression "dominate" in referring to the behavior of characters in crossing. His studies led to the conclusion . . . "that, in general, the resemblance of the hybrid to its two parents consisted, not in any intimate fusion of the diverse characters peculiar to each one of them in particular, but rather in a distribution, equal or unequal, of the same characters." Gärtner's experimental work with plant hybrids extended over a quarter of a century, during which period he carried out nearly 10,000 separate experiments among about 700 different species representing 80 different genera of plants, obtaining in all some 250 hybrid plants. His work seems to have been characterized by scrupulous care. Apparently he did not realize, in spite of Sagaret's contemporaneous experiments, that some of the characters from a certain parent might dominate in the hybrid and others not. Yet he observed an inequality of one parent over another in a cross. Gärtner recognized the occurrence of unusual vigor in hybrids but without distinguishing as to the generation.

The last half of the nineteenth century brought contributions and discoveries of the utmost importance not only to biology and plant breeding but to human welfare. Especially significant for biology, but of the greatest general interest also, was the work of Charles Darwin. In 1859 appeared the first edition of that epoch-making book, "The Origin of Species by Means of Natural Selection." Part of the great mass of evidence presented therein in support of this principle of evolution is a discussion of hybridization and hybrids in relation to the species question. In Darwin's final summary of the evidence he states:

> First crosses between forms, sufficiently distinct to be marked as species, and their hybrids are generally but not universally sterile. The sterility is of all degrees and is often so slight that the most careful experimentalists have arrived at diametrically opposite conclusions in ranking forms by this test.

After showing that first crosses between varieties and their mongrel offspring are very generally, although not invariably, fertile, Darwin remarks that he finds nothing in his knowledge of hybridism which is opposed to the belief that species existed previously as varieties. This would indicate that Darwin did not attach much significance to hybridization as an evolutionary process but he was fully aware of its great importance to agriculture and horticulture.

During this same period two Frenchmen made original investigations of significance in plant breeding. Godron, by hybridization of wheat and spelt, demonstrated the hybrid origin of *Aegilops triticoides* and so destroyed the old tradition that spelt had been transformed into wheat. Naudin opposed the conception, still maintained by Cuvier, that species are hard and fast entities and, based on his experiments in hybridizing different species of the nightshade family, he almost discovered the essential principle of Mendelism but in the principle which he enunciated: "*La disjonction des deux essences specifiques a lieu dans le pollen et dans les ovules de l'hybride,*" the *essences specifiques* were not separate entities but rather the male and female contributions considered as wholes.

It remained for Mendel to contribute one of the most important discoveries ever made to biological science and one which was destined to revolutionize the study of heredity and greatly to modify breeding methods. In his monastery garden at Brünn, Austria (now Brno, Czechoslovakia), this monk made many experiments upon the inheritance of pairs of contrasted characters, particularly in peas; and through these experiments he demonstrated the laws of heredity which later were recognized as one of the greatest discoveries of the nineteenth century. Mendel's attitude in planning his experiments contrasted sharply with that of his predecessors. He realized the necessity not only of concentrating upon particular characters but also of tracing them through a sequence of generations and testing the individual offspring. By his critical experiments and his keen interpretations of the results Mendel laid the cornerstone of the foundation for the future science of genetics.

With the announcement of the discoveries of Mendel, de Vries and Johannsen, during the first three years of the present century, there was a great awakening of interest among biologists in the problems of variation, heredity, and evolution. In the enthusiasm of those first years it was thought by some that the application of these laws of heredity and mutation in practical breeding would be comparatively simple and rapid. Many plant breeders went zealously to work only to obtain further conflicting and disconcerting results. At the same time research students the world over began new investigations on variation and heredity. The occurrence of mutations and the existence of pure lines in species that reproduce by self-fertilization have been verified. But to explain the heredity of most plants requires numerous modifications or extensions of the original Mendelian laws. These neo-Mendelian principles have been presented in Part I. The application of these principles to plant breeding will be further discussed in succeeding chapters.

**Summary.**—Since the early, sporadic efforts to find better varieties of fruits and grains, there has been a gradual broadening of the great movement to increase agricultural output and raise the quality of raw materials by means of plant improvement. This has come about largely as a result of economic conditions and without much definite knowledge concerning the science underlying plant breeding. The usual method of plant improvement during this prescientific era has been mass selection, but certain pioneers discovered empirically the effectiveness of pure-line selection and hybridization. Throughout the later stages of this development, however, scientific knowledge has become increasingly important until now the science of genetics has developed to a stage where it is capable not only of furnishing a rational explanation for the phenomena of variation and heredity which in the past seemed obscure and contradictory, but also of guiding the breeder who will familiarize himself with the established principles of the science, so that he will reach his goal with greater speed and economy.

### References

BAILEY: Plant Breeding, 4th ed., 1908 (including bibliography).
Survival of the Unlike, pp. 138–162 (Van Mons and Knight).

CASTLE: Genetics and Eugenics, 3d ed., Chapter XIII, 1924.

DARWIN: "Variation in Animals and Plants under Domestication," **1**: 381-385; **2**: 260–265.

FAIRCHILD: *Yearbook U. S. Dept. Agr.*, **1911**: 411–422.

HAYES AND GARBER: Breeding Crop Plants, pp. 1–15 (historical).

*Proc. First International Conference on Plant Breeding*, 1902 (interesting early contributions).

ROBERTS: Founders of the art of breeding, *Jour. Heredity*, **10**: 99–106; 147–152: 229–239; 257–270.

DE VRIES: Plant Breeding, pp. 1–106 (historical, especially the work of Nilsson).

WEBBER and BESSEY: Progress in plant breeding in the United States, *Yearbook U. S. Dept. Agr.*, **1899**: 465-490.

WERKENTHIN: The founders of the art of breeding, *Proc. Iowa Acad. Sci.*, **29**: 290–310, 1922.

## CHAPTER XXIX

## VARIETIES

The cultivated plants of importance in the world's commerce comprise relatively few species but many varieties. The cereals, fiber plants, legumes, root crops, and tree fruits which are of importance in agriculture include some fifty species. Within this small group of species several thousand varieties are known. Among ornamental plants, there is the same diversity. The common and widely cultivated garden flowers represent not more than 200 species, but of roses alone as many as 1,000 named varieties are listed, while old varieties are continually being discarded and new ones introduced. In general, the longer and more widely cultivated species contain larger groups of varieties because of the greater opportunity afforded for their origination.

The term variety is here used in its common horticultural and agricultural signification, *viz.*, a minor form within a species which has come to be recognized as worthy of a name of its own, the "Kulturvarietät" of Fruwirth. For taxonomic purposes species, like genera, have been subdivided according to various systems of classification. The following order of rank is now widely accepted as both logical and convenient: species, subspecies, variety, form, individual. Horticultural varieties stand in about the same relation to species or subspecies and forms as do taxonomic varieties in the above system; and just as taxonomic varieties often contain a number of minor forms, which may or may not be considered as worthy of receiving distinctive names, so most horticultural varieties contain numerous minor forms which are designated by such terms as sort, strain, biotype, race, or line. It is these subvarietal forms which serve in plant improvement by selection as starting points in the development of new cultivated varieties.

**Varieties and Agriculture.**—It is in comparatively recent times that the great importance of the variety in agriculture has come to be generally realized, and not yet is this principle universally recognized. The degree of appreciation of this matter, as reflected by common practices of people on the land, is an index of their biological enlightenment. This is well illustrated in China where, in spite of its ancient civilization and the early recognition of varieties, the rural people of the western and northern provinces frequently plant impure or mediocre varieties of their most important crops. The Russian government has recently undertaken an extensive reorganization of that nation's agricultural program in which the introduction of new varieties, as well as new crops, from other countries is an important feature. Even in countries which have been more fortunately situated in recent years, it is not difficult to find certain farmers who fail to appreciate the importance of good seed. Yet nothing is more fundamental to increase or improvement in crop production than the choice of varieties. Moreover, the maintenance of a permanent agriculture involves not only the testing and selection of varieties for a given district by some well-qualified agency, but also consideration of the different sets of environmental conditions within the area, including pests and diseases. A thorough investigation into all the requirements for the production and marketing of a given crop in a given region very often brings forward the need of improved varieties.

On the other hand, critical tests of commercial varieties have revealed some very interesting facts. In oats and barley, for example, Wiggans has found that the same name has been applied, not only to varieties which are distinct morphologically, but also to strains which are morpho-

TABLE 45.—AVERAGE YIELD, 1918–1922, OF TWELVE REPRESENTATIVES OF THE SWEDISH SELECT VARIETY OF OATS (LEFT) AND OF THE SILVERMINE VARIETY (RIGHT) IN PERCENTAGE OF THE AVERAGE OF THE CHECK VARIETIES. A DIFFERENCE OF 9.18 PER CENT IS NECESSARY FOR ODDS OF 31:1 IN THE FIRST; IN THE SECOND, 9.54 PER CENT (*Data from Wiggans*)

| | | | |
|---|---|---|---|
| White Russian | 110.5 ± 2.89 | Great Dakota | 98.3 ± 2.06 |
| Garton 435 | 98.9 ± 2.46 | Great American | 96.3 ± 1.98 |
| Banner | 98.2 ± 2.16 | Big Four | 95.9 ± 2.04 |
| Czar of Russia | 97.8 ± 2.01 | Illinois German | 95.8 ± 2.04 |
| Regenerated Swedish Select | 97.6 ± 2.16 | Garton 364 | 95.4 ± 2.08 |
| President | 95.1 ± 2.56 | Canadian | 95.1 ± 2.08 |
| Siberian | 91.9 ± 1.89 | Lincoln | 95.0 ± 1.63 |
| Mortgage Lifter | 91.5 ± 1.69 | Curel 6 | 94.8 ± 2.09 |
| Minnesota 270 | 90.9 ± 1.75 | Funk | 94.3 ± 2.58 |
| Tartar King | 84.7 ± 1.74 | New Zealand | 93.2 ± 2.02 |
| White Waverly | 82.7 ± 1.50 | Funks Great Dane | 92.6 ± 1.97 |
| Lincoln | 81.0 ± 1.56 | American Banner | 83.2 ± 1.88 |

logically alike but significantly different in yield. Thus in the case of the Swedish Select variety of oats, samples were obtained from widely separated localities under the twelve names shown on the left in table 45, which gives the average yield in percentages for a period of years. The amount of difference, 9.18 per cent, which is necessary for odds of 31:1 was calculated as follows: In order to obtain such odds it is necessary to have an actual difference of 3.2 times the probable error of the difference. The approximate difference necessary for these odds was obtained by using the average probable error of all the strains of a variety as the approximate probable error for each strain. The probable error of the difference was then determined by the formula $E_d = \sqrt{2(\text{A. E.})^2}$, where A. E. = average probable error, and multiplied by 3.2. It will be noted that there is not only a great difference from the lowest to highest yields but also a fairly continuous gradation in yield with the exception of White Russian which is distinctly better than the others. This wide variation in yield in a morphologically uniform group of strains is considered indicative of (1) a wide range of adaptability within the Swedish Select variety; (2) distinct possibilities of improvement; (3) a lack of purity in some strains; and (4) various physiological strains.

Of the ten varieties of oats tested by Wiggans, eight displayed physiological diversity similar to that of Swedish Select, while two were very uniform. One of these is Silvermine, of which the average yields of twelve different strains, all obtained under varietal names, are shown on the right in table 45. Of the eleven varieties of barley tested, five contained strains showing even greater differences in yielding capacity than those found in the oats, while six were fairly uniform. From these results it is evident that little value can be attached to the name on a sample of grain unless its exact source and past performance are known. Obviously some system of control, including registration and certification of seed, is needed; but on account of the technical methods which are necessary in order to distinguish between the strains within a variety the actual work of testing varieties and strains must be carried on at experiment stations which are already doing most of the breeding of new varieties.

**Synonyms.**—This demonstration of the existence of physiologically distinct strains within morphologically uniform varieties raises an important question as to just what shall constitute a synonym among horticultural varieties. In botanical nomenclature a synonym is a name which has been applied to a given taxonomic unit subsequently to the date of the original description. According to customary practice among taxonomists the identification of the taxonomic unit referred to is necessarily accomplished by comparing external morphological characters. In agriculture, however, differences in yield are so important that two morphologically identical strains which differ consistently in productivity should be recognized as distinct and so designated. It may be found practicable, however, to retain for such strains the same varietal name with some sort of subsidiary designation such as a letter or a Roman numeral. This would tend to prevent the unnecessary multiplication of varietal names.

**Origin of Domestic Varieties.**—The production of varieties is dependent upon germinal diversity. Broadly speaking, germinal diversity may arise in either of two ways, *viz.*, by hybridization, or as a result of some sort of germinal change such as factor mutation or chromosomal variation. Although the artificial induction of germinal variations is an almost unexplored field, it may develop into a most important method of plant improvement. Aside from this, there are just two general methods for the production of new varieties: (1) the employment of hybridization as a means of securing new combinations of characters found among existing varieties or species; and (2) selection, including the systematic search for new mutations of possible value in breeding. The utilization of these two methods in the improvement of various types of crop plants will be discussed in succeeding chapters. The role of mutation and hybridization in the origin of cultivated varieties may be illustrated by considering in some detail the horticultural history of certain plants. As the ancestry of most crop plants is obscure, it will be well to turn for this purpose to a more recently domesticated species, the history of which is fairly well known.

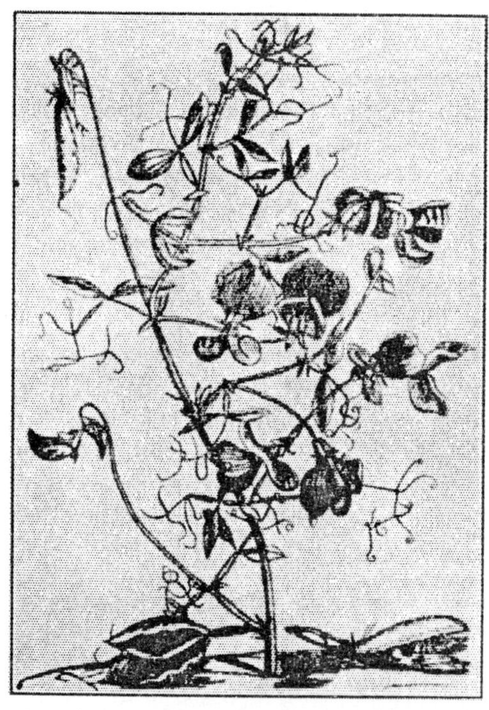

FIGURE 102.—Commelin's drawing of the sweet pea in Horti medici amstelaedamensis. (*From Beal.*)

**Origin of Varieties in the Sweet Pea.**—The sweat pea provides an excellent illustration of the origin from a single species, *Lathyrus odoratus*, of varieties, first by the utilization of mutations, and later by means of hybridization. This species was introduced into Holland and England from Sicily via Italy in 1699, and was first illustrated in a description published in 1700 (figure 102). In habit it was similar to the tall sweet peas now cultivated and the height to which it would climb was 6 to 7 feet, but the flower stems were short and bore only two flowers while the flowers were relatively small, with erect or reflexed standard and conspicuous, depressed wings. In color the standard was reddish purple and the wings light bluish purple. From this modest beginning there have been developed several distinct types of plant and flower forms and a list of named varieties, even within the most highly developed type of flower (the Spencer or waved form), which includes over 500 different colors, tints, shades, and combinations. In fact there are probably 700 distinct varieties of the Spencer peas of which about 350 have been introduced during the last 10 years. By far the greatest amount of all this improvement in the sweet pea has been accomplished within the past 60 years, during which period hybridization has been used extensively in producing better varieties. But before hybridization was resorted to, there were a dozen distinct color varieties which had arisen by mutation. In addition to color mutations, there have occurred spontaneous changes in flower form, flower size, and number of flowers on the stem, in stature and habit of the plant, and in season of bloom, some of which are described below.

**Flower Color in Sweet Peas.**—The chronology and probable ancestry of the color varieties of the sweet pea, which appeared during the first 180 years of its horticultural history, are shown in table 46. Apparently the course of events was about as follows: From the original wild type there first appeared mutants with white flowers. If the simple flower-color factor complex is called $CRB$, in which $C$ and $R$ are complementary factors producing red, and $B$ a factor which modifies the color to purple, as was first determined by Bateson, then the mutations apparently depended upon a change in either $C$ or $R$ to the recessive condition for white. The Painted Lady variety, red instead of purple, appeared very soon after this, apparently as an independent mutation in the factor $B$. By the close of the eighteenth century two other color types, black and scarlet, had been added to the list. The wild form and Painted Lady are bicolor types; that is, the wings are lighter in color than the standard. The new color, scarlet, apparently resulted from mutation in a factor for bicolor flowers $E$, giving a recessive factor for development of full color in the wings along with a certain intensification of color in the standard. Probably black was also due to a factor mutation for more intense pigmentation which arose from the wild type. Early in the eighteenth century a "blue" form was listed in catalogues, but its genetic relationships have not been determined. Further additions followed shortly in the form of a "striped" variety, and of a "yellow" variety. The latter unquestionably came as a mutation from white; the former may have arisen as a mutation in purple. Plants with primrose-yellow flowers have since been observed in white cultures, but never in red ones. This practically closes the account of the origin of color mutations up to the year 1880, after which time hybridization was employed extensively in the creation of new varieties. However, mutations affecting flower color undoubtedly continue to occur from time to time.

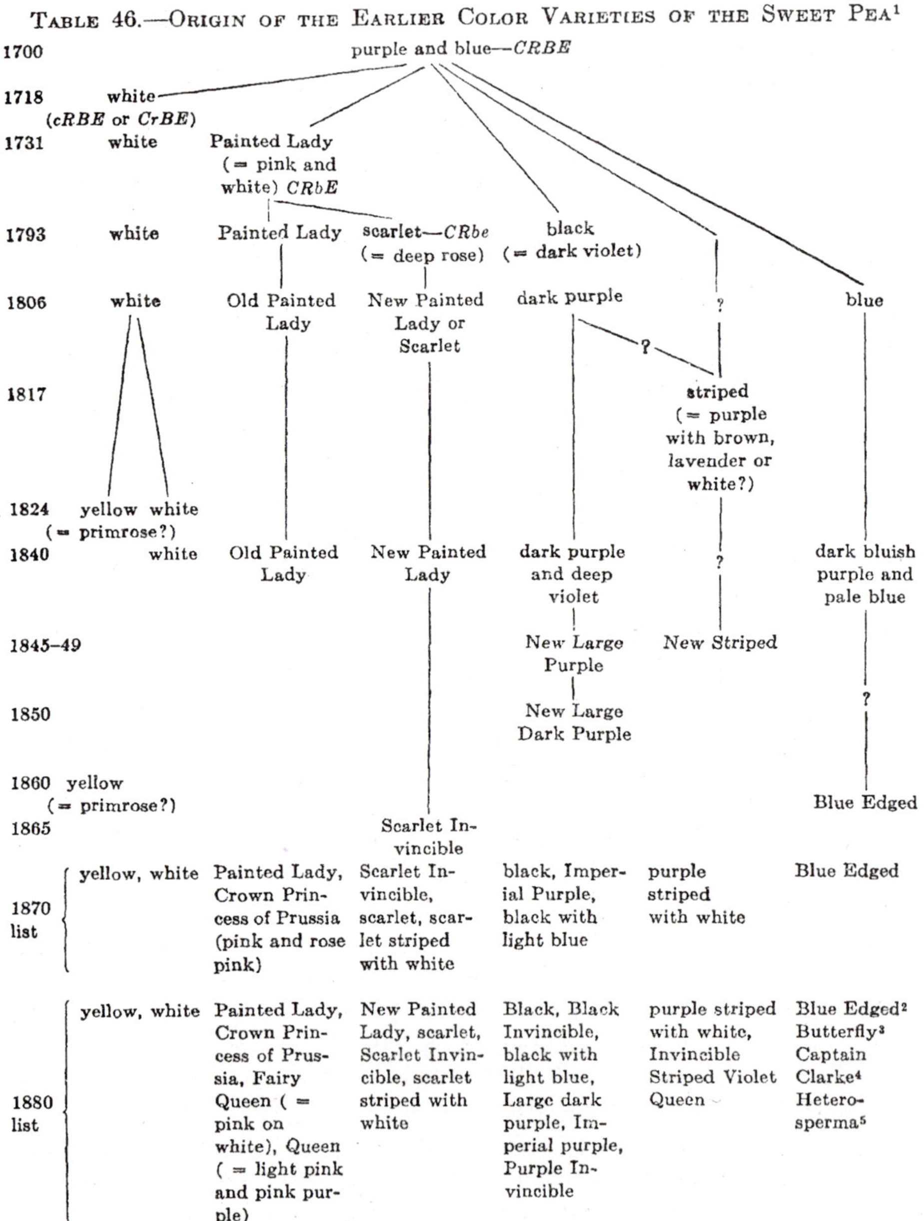

TABLE 46.—ORIGIN OF THE EARLIER COLOR VARIETIES OF THE SWEET PEA[1]

| Year | | | | | | |
|---|---|---|---|---|---|---|
| 1700 | | | | purple and blue—*CRBE* | | |
| 1718 | white (*cRBE* or *CrBE*) | | | | | |
| 1731 | white | Painted Lady (= pink and white) *CRbE* | | | | |
| 1793 | white | Painted Lady | scarlet—*CRbe* (= deep rose) | black (= dark violet) | | |
| 1806 | white | Old Painted Lady | New Painted Lady or Scarlet | dark purple | ? | blue |
| 1817 | | | | ? | striped (= purple with brown, lavender or white?) | |
| 1824 | yellow (= primrose?) white | | | | | |
| 1840 | white | Old Painted Lady | New Painted Lady | dark purple and deep violet | ? | dark bluish purple and pale blue |
| 1845–49 | | | | New Large Purple | New Striped | |
| 1850 | | | | New Large Dark Purple | | ? |
| 1860 | yellow (= primrose?) | | | | | Blue Edged |
| 1865 | | | Scarlet Invincible | | | |
| 1870 list | yellow, white | Painted Lady, Crown Princess of Prussia (pink and rose pink) | Scarlet Invincible, scarlet, scarlet striped with white | black, Imperial Purple, black with light blue | purple striped with white | Blue Edged |
| 1880 list | yellow, white | Painted Lady, Crown Princess of Prussia, Fairy Queen (= pink on white), Queen (= light pink and pink purple) | New Painted Lady, scarlet, Scarlet Invincible, scarlet striped with white | Black, Black Invincible, black with light blue, Large dark purple, Imperial purple, Purple Invincible | purple striped with white, Invincible Striped Violet Queen | Blue Edged[2] Butterfly[3] Captain Clarke[4] Heterosperma[5] |

[1] In each case the color of the standard or banner is given first and of the wings second; the descriptive terms and variety names are identical with those in the original descriptions.

[2] Described by Bailey and Wyman as purple-lilac in color (= purple picotee).

[3] Quite similar to Blue Edged according to Beal (= purple picotee).

[4] = "white merging into pink and purple, wings white with purplish cast, wings edged with blue (= purple picotee).

[5] No description available; mottled seeds?

**Form and Size of Sweet Peas.**—The changes in form and size of flower in the sweet pea have been no less striking than those in color, and they have made possible much of the popularity of this flower. In the original form the standard was erect, narrow at the base, notched at the top, and reflexed or slightly rolled at the sides. From it have been derived three distinct flower types, the grandiflora, the hooded, and the popular waved Spencer forms. The origin of the first two types is in some doubt. The hooded character was found in some of the earliest varieties. It

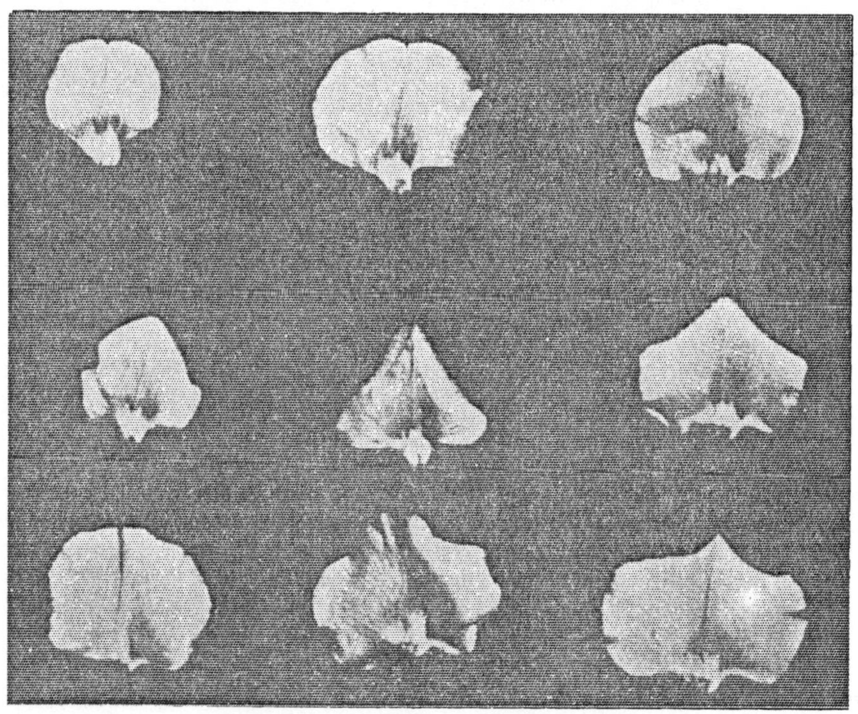

FIGURE 103.—Forms of sweet-pea flowers—the standard or banner. Open or grandiflora form (upper row left to right) Alba Magnifica, Shasta, Golden Rose. Hooded form (middle row)—Butterfly, Admiration, Dorthy Eckford. Waved form (lower row)—Elsie Herbert, Apple Blossom Spencer, White Spencer. (*From Beal.*)

was sometimes associated with notches in the sides as in Butterfly (figure 103), and this character is found in some varieties of the present day. Bateson reports that hooded is recessive to grandiflora or erect type of standard. Some of the earliest varieties of improved grandiflora form were Queen of England (1888), Blanche Ferry (1889), and Alba Magnifica (1891). The waved or Spencer form is of more recent origin, and authorities are agreed that it arose as a "sport" from the beautiful, pink, hooded variety, Prima Donna. The pronounced waviness of standard and wings which characterizes this type had not been reported before in sweet peas.

The development of flower size in the sweet pea is also indicated in figure 103. Alba Magnifica and Butterfly were great acquisitions in their day and were doubtless considerably larger than the oldest varieties. The first definite reference to size is found in New Large Purple, listed in 1845. As this occurs in the darkest color group and 15 years before the hybrid origin of a variety called Blue Edged was even suggested, it probably represents a factor mutation. That such mutations actually occurred is indicated by the fact that Countess Spencer and Gladys Unwin were both decidedly larger than Prima Donna from the first. This is true also as regards number of flowers in the cluster. Prima Donna bore two or three, usually three flowers on a stalk, while Countess Spencer has three to four flowers in a cluster. Many of the recent

FIGURE 104.—On the left, snapdragon sweet peas. On the right, double sweet pea, White Wonder. (*From Beal.*)

Spencer varieties bear almost uniformly four-flowered clusters, and a few of the unusually vigorous ones sometimes bear five, six, or seven flowers in a spray. The original form and oldest varieties had two flowers in the cluster. The oldest varieties definitely known to bear more than two flowers on a stalk are Invincible Scarlet (1865) and Crown Princess of Prussia (1868). As these antedate the era of hybridization, it is probable that the increased number arose by mutation.

Novelty forms have also appeared from time to time. In double sweet peas there are two standards instead of one. In some varieties this character has been fixed by selection so that most of the flowers come double. This gives the effect of increased size (figure 104). In the snapdragon type of flower, the standard is folded around the wings.

It is recessive to erect standard and gives a ratio of 3 erect: 1 snapdragon in $F_2$.

**Habit in Sweet Peas.**—There are several distinct types of plant, the origin of which may be definitely ascribed to mutation. The prostrate dwarf or cupid type (figure 105 *a*) is one of these. The first cupid plant appeared among plants of the tall, white-flowered variety, Emily Henderson, in 1893, and 2 years later the growers had 7 acres of the new variety in which every plant was true to type as would be expected in the case of a recessive mutant factor. The bush type (figure 105*b*) also originated from the tall form. The factor relations of bush and cupid sweet peas have been described in a preceding chapter. Semi-

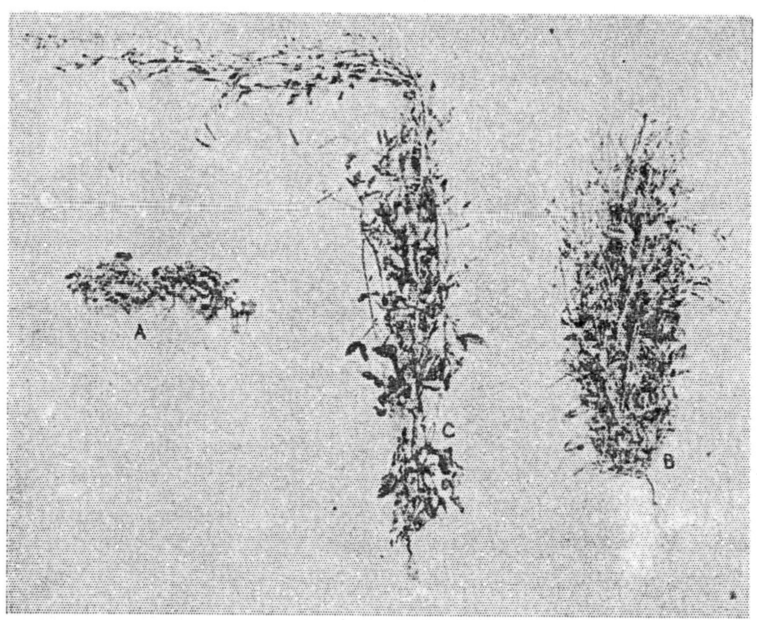


FIGURE 105.—*A*, Cupid or prostrate, dwarf sweet pea; *B*, bush or erect, tall form; *C*, cupid × bush $F_1$, the ordinary tall form (bent in order to photograph). (*From Bateson, Mendel's Principles of Heredity*, 1st *ed.*, 1909. *Cambridge University Press. Reprinted by permission.*)

dwarf, early-flowering sports have appeared even more frequently than those of the cupid type. They have furnished the basis for the winter-flowering races of sweet peas. In ordinary sweet peas, after the first few inches of growth, the rate of growth slackens until several laterals have appeared. The winter-flowering sorts, on the contrary, promptly send up a stem which begins flowering as soon as it attains a height of from 2 to 4 feet. The Blanche Ferry group of varieties apparently had their inception in a mutation of this sort among plants of the variety, Old Painted Lady. About 25 years after the discovery of the new type, the stock passed into the hands of a seedsman. From this stock a series of early-flowering varieties have arisen, probably by mutation, in the order shown below. Black-seeded varieties are indicated by (*b*) and white-seeded ones by (*w*).

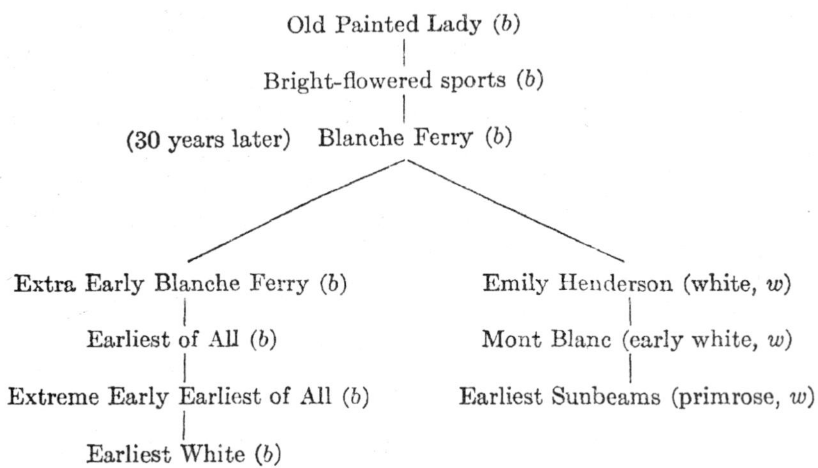


**Hybridization and Selection in Sweet Peas.**—The era of extensive hybridization in sweet peas dates from about 1880; consequently but little can be said with respect to the origin of new factors in this species after that time save in a few particularly favorable cases. Laxton's Invincible Carmine was the earliest recorded variety produced by crossing, and its parents are said to have been Invincible Scarlet and Invincible Black. Apparently it was an improved form of Invincible Scarlet resulting from the addition of the factor for intense pigmentation from Invincible Black. Similarly by hybridization it has been possible to establish groups of varieties, such as the Spencer, the hooded, the grandiflora, and the winter-flowering races. Thus hybridization has been merely a means of fully utilizing germinal differences which have arisen by mutation. It is true that in most cases it cannot be said just when the various new features of form, color, and habit have arisen, but it is known that there was only one original form, and fragments of the history are sufficiently definite to give assurance in advancing this explanation of the role of hybridization in the creation of varieties of sweet peas. There is no authentic instance of a named variety having originated through hybridization of the sweet pea with any other species of Lathyrus; consequently the possibility of such a cause of germinal diversity is precluded. Similarly in the selection of such obscure characters as number of flowers in the cluster, size of flower, and vigor of growth, improvement has been made possible by mutations and new combinations of mutant characters.

**Germinal Variations in Crop Plants.**—Both of the general categories of germinal differences—factor mutations and chromosomal variations—may be looked upon as exceptional or aberrant life processes. Genetic factors are characteristically stable; yet they are known to mutate with varying degrees of frequency. Also the processes of chromosome distribution proceed with a high degree of precision and regularity, but aberrations in these processes occur and apparently with greater frequency than has heretofore been suspected.

*Factor mutations* are of general occurrence in both animals and plants, as was set forth in Chapter XXII, where it was also shown that the chief value of factor mutations in breeding is to be found in their utilization through hybridization in the production of new series of combinations and hence new varieties of possible value. Such utilization of mutations is well illustrated in the history of the sweet pea and will be referred to again. At the same time, the fact that mutations are of general occurrence has an important bearing upon such practical considerations as the constancy of varieties, the maintenance of varieties true to type, and the effects of selection in self-fertilized plants. While the identification of factor mutation as such is often complicated by the possibility of antecedent hybridization or mixture of seed, the following two cases of sporadic appearance of new types of crop plants probably resulted from factor mutations.

The Florida Velvet Bean, *Stizolobium deeringianum,* has produced spontaneous variations of great commercial value. The original variety was limited to Florida and the Gulf coast on account of lateness of maturity of the seed, and the area planted in 1915 was about one million acres. After the discovery of early varieties which originated in at least three different places during a period of 10 years, the crop became adapted to the entire cotton belt with a resulting rapid increase in acreage planted. Although there is no proof that these early varieties resulted from factor mutation, such manner of origin is indicated by the evidence in the case.

In Coast Barley the numerous pure-line selections made by Mackie have revealed the existence of many minor differences in color, stature, diameter of culm, time of maturity, and disease resistance. This variety has been grown in California since 1876 and until recently was the only variety commonly grown. No evidence has been found during extensive observations to indicate that natural crossing has occurred between this and other varieties of barley, whereas in wheat several cases of hybridization have been discovered in the field by means of segregating populations. On the other hand the many heritable differences revealed by pure-line selection are such as would result from small mutations. Furthermore, mutations have been known to occur in pure lines of barley. Similarly, as a result of extensive pure-line selection in oats, there is considerable evidence for the occurrence of mutations affecting productivity. Mutations have also been reported in wheat and rice.

Johannsen discovered two mutations in his pure lines of Princess beans, one of which occurred as a bud sport. It produced large and relatively narrow seeds, and as it was constant from the first, it must have originated in homozygous condition. The other mutant bore relatively broad seeds and could be traced back to a heterozygous family.

Nilsson-Ehle discovered that in pure lines of oats occasional grains appear which are aberrant either in color or in morphological characters. The variations tested by him either bred true at once, or after one or two generations practically all the progeny would be uniform for the new variation.

Hayes describes a number of mutations in tobacco, the first of which was found in a homozygous strain of the Connecticut Cuban shade variety of *Nicotiana tabacum.* The new variety, called the Stewart Cuban, is characterized by indeterminate habit, extreme leafiness, and delayed blooming under field conditions (figure 106). Similar mutants have been found in the Connecticut Havana variety which has been known as a uniform variety for over half a century.

These illustrative cases will serve to indicate the important bearing of factor mutations upon the genotypic composition of varieties of both self- and cross-fertilized plants. From the evidence at hand the conclusion is warranted that even a pure-line variety cannot be expected to remain constant or homogeneous, and that the degree of heterogeneity to be expected in a variety which was originally a pure line is proportional to the opportunity there has been for mutations to occur, *i.e.*, to the time and extent of its cultivation.

*Chromosomal variations* also play an important role in the origin of new varieties and even new species (see Chapters XXIV–XXVII). The general principle, that multiplication of the haploid number of chromosomes results in increased size, is of great significance in agriculture. Owing to the limited time during which cytogenetic methods have been used in the study of crop plants, knowledge of the subject is only fragmentary, but it is becoming evident that polyploidy, or the occurrence of series of balanced chromosome groups, is the characteristic condition in some of our most valuable agronomic species, such as wheat, oats, and sugar cane; also in genera of importance to horticulture, such as Rosa, Rubus, and Fragaria. In the wheats, for example, associated with an increase in chromosome number are increased variability, adaptability, and economic value, as well as increased susceptibility to disease. According to Percival, *Triticum monococcum* ($n = 7$) is known only in a single form, while the emmer group ($n = 14$) has 151, and the *vulgare* group ($n = 21$), 536 cultivated varieties. If *T. vulgare* originated through hybridization between two primitive species, as seems highly probable, this would account for the greatly increased variability. On the other hand, as Sax points out, the great age of the three wheat groups may also explain why they differ so markedly even if they were originally derived from a single primary species by duplication of chromosome groups. Phylogenetically the situation in Triticum is not unlike that in Fragaria (strawberry), in which genus, according to Longley, the primitive species has seven as the haploid number, while the other twenty-three

FIGURE 106.—The Stewart Cuban variety of tobacco. Plants from seed sown under glass in December and transplanted to the open in May were 12 to 14 feet tall in September and had produced 80 leaves per plant. *(From the Journal of Heredity by permission.)*

forms examined by him were all either hexaploid (*F. elatior*) or octoploid, the latter group including the two American species, *F. chiloensis* and *F. virginiana* and many of the leading horticultural varieties. Longley concludes that the generic tree of Fragaria is now represented by the trunk (*F. vesca*, $n = 7$) and a few end branches, the polyploid species.

That polyploidy in the wheats, strawberries, and other genera with series of balanced chromosome groups may have originated either by hybridization or as a result of irregularities in chromosome distribution is clearly established by observations on the occurrence of polyploid forms.

FIGURE 107.—The tetraploid Campanula, Telham Beauty (left), the diploid parent species, *C, persicifolia* (upper right) and their somatic chromosomes in metaphase. (*From Miss Gairdner by permission Cambridge University Press.*)

*Tetraploidy* is known to have originated through hybridization in Primula and Nicotiana; and somatic variations in chromosome number, such as would give rise to diploid instead of haploid gametes, have been discovered in several plant species. Miss Gairdner reports a tetraploid Campanula (figure 107) which originated as a seedling of *C. persicifolia*

($n = 8$). It has the usual characteristics of a tetraploid, being a much more robust plant than the parent type while the leaves are larger and thicker and the flowers much larger. Furthermore, normal, diploid plants of *C. persicifolia* are strictly self-sterile but the Telham Beauty is self-fertile. Origin by hybridization seems to be highly improbable in this case, and the discovery that diploid gametes occur not infrequently in *C. persicifolia* indicates that the meeting of two of these may have been the origin of the tetraploid form. Tetraploidy in seedlings of apogamic origin must be the result of somatic variations in chromosome number. Frost has described a thick-leaved tetraploid type found among apogamic seedlings of orange and grapefruit (pomelo). His evidence indicates that they originate from doubling of the chromosome number in a certain cell or cells of the nucellus from which an embryo is about to develop or else in the very young embryo.

That polyploidy may result from the direct effect of the environment upon the germ cells during meiosis is also to be considered probable, although the evidence is far from conclusive. De Mol has found chromosomal variations to be rather frequent among hyacinths and narcissus, and polyploid forms are gradually supplanting the diploid varieties. He has not yet been able, however, to induce such variations by artificial methods. Belling has described irregularities in chromosome distribution in two species of Bellwort, *Uvularia grandiflora* and *U. perfoliata*, which he attributes to the effect of low temperature. The aberrations observed were such as might lead to the production of plants with chromosome groups of $2n + 1$, $2n + 2$, $3n$, $3n + 1$, $3n - 1$, $4n$, $4n + 1$, $4n - 1$, $4n + 2$, $4n - 2$, etc., besides plants showing duplication of a segment of a chromosome. Navashin has reported a triploid and a pentaploid variant among several hundred diploid seedlings from a self-fertilized plant of *Crepis capillaris* ($n = 3$) and suggests that they originated through irregularities in meisois which were induced by high temperature.

The foregoing illustrations are sufficient to indicate the wide possibilities which exist for the origin of new forms through chromosome variations. The significance of chromosomal variations for agriculture is not confined to their possible utility in plant breeding. Some of the difficulties met with in maintaining pure-line varieties probably have their origin in certain irregularities of chromosome distribution. In cotton, for example, there are undesirable and partially sterile forms which appear from time to time in pure-line varieties and the latter especially may be due to chromosomal variations. The application of cytological methods in the solution of such agricultural problems is still in its infancy.

**The Breeding of New Varieties.**—From the foregoing discussion of the processes by which new varieties originate, the student will hardly infer that the breeding of superior new varieties is a simple matter or one to be lightly undertaken. On the contrary the production of really superior varieties of our important agricultural and horticultural crop plants is continually becoming more complicated and difficult. Success in such an undertaking calls for not only a knowledge of genetics and all of the cultural requirements, including problems of diseases and insect pests, but also a familiarity with the range of cultivated varieties and with the requirements of the market. Because of the complexity of the problem of meeting all these requirements, most of the breeding work with important crop plants is now being done by employees of national and state experiment stations, industrial organizations, associations of growers, or large private establishments. Even with such popular ornamentals as the sweet pea, rose, dahlia or delphinium, while these still offer an attractive field for the amateur, experience usually leads the amateur to become a specialist in some one plant or at most two or three. Naturally the next step is the organization of these specialists into societies which formulate rules for the testing, naming, and introduction of new varieties. Even some of our experiment stations now arrange to have proposed new varieties of fruits certified or registered by some organization of growers such as the state horticultural society. Thus there is a growing appreciation of the heavy responsibility which rests upon the introducers of new varieties. The methods by which new varieties are produced will be discussed in succeeding chapters.

**The Maintenance of Varieties.**—Because of the increasing value to agriculture of really superior varieties, it is becoming correspondingly important that due attention be given to the maintenance of such varieties true to type. The degeneration of varieties has been frequently observed and is often discussed in plant-breeding literature, but it is only in recent years that the causes of these phenomena have been determined. In general, these causes are either genetical or pathological in nature. When genetical, the degeneration is due to germinal diversity arising from either natural crossing or mutation in the broad sense. Obviously the best cure for degeneration is prevention when that is possible. Stock material of a superior variety should be isolated if necessary to prevent natural crossing and, if possible, the invasion of disease. But isolation cannot prevent the occurrence of new germinal variations. As these may be expected to occur and as the great majority of them will be degenerative in their effects, the only recourse is continual attention to the maintenance of stock material true to type. In this connection, it is well to note the practical value of maintaining collections of varieties of the important field crops and orchard fruits at our state and federal experiment stations. These collections should serve as dependable sources of established varieties and as the best places for testing new varieties of possible value.

**Summary.**—Agriculture is dependent upon comparatively few species of crop plants but calls for the utilization of numerous varieties. There remain vast possibilities for increase and improvement in crop production through the improvement of old varieties and the breeding of new and better varieties. Actual realization of these possibilities, however, must be gradual; awaiting the progress of research. Real progress in plant breeding will be conditioned by recognition of the laws governing the origin of varieties. In general, the production of new varieties depends upon germinal diversity which in turn is caused either by hybridization or by some type of germinal variation. The methods of scientific plant breeding must be based upon these principles.

### References

BAILEY: Survival of the Unlike, pp. 364–376 (on the loss of good varieties); pp. 400–473 (origin of varieties of strawberries, plums, grapes, carnations, petunias, and tomatoes).

BELLING: The origin of chromosomal mutations, *Jour. Genetics*, **15**: 245–266.

DARWIN: Variation in Animals and Plants Under Domestication, **1**: 368–447.

FRUWIRTH: Handb. d. landw. Pflanzenzüchtung V. Allgemeine Züchtungslehre, pp. 7–23 (the many forms of cultivated plants).

GAIRDNER: *Campanula persicifolia* and its Tetraploid Form, "Telham Beauty," *Jour. Genetics*, **1**: 341–352.

GATES: Polyploidy, *Brit. Jour. Exp. Biol.*, **1**: 153–182.

LONGLEY: Chromosomes and their significance in strawberry classification, *Jour. Agr. Research*, **32**: 559–568.

NAVASHIN: Polyploid mutations in Crepis, *Genetics*, **10**: 583–592.

PUNNETT: Mendelism 6th ed., Chapter VIII (Reversion) and Chapter XIII (Wild and Domestic Varieties).

SAX: The relation between chromosome number, morphological characters, and rust resistance in segregates of partially sterile wheat hybrids, *Genetics*, **8**: 301–321.

THOMPSON and HOLLINGSHEAD: Preponderance of *dicoccum*-like characters and chromosome numbers in hybrids between *Triticum dicoccum* and *T. vulgare*, *Jour. Genetics*, **17**: 283–307.

DE VRIES: Species and Varieties, pp. 32–120 (elementary species).

WIGGANS: Variations within and between morphological varieties of oats and barley, *Cornell Mem.*, **94**, 1925.

## CHAPTER XXX

# CHIMERAS

Since the middle of the seventeenth century, there have been published records of unique forms or bizarre varieties of plants which seem to combine the characteristics of two distinct varieties or species and yet which do not appear to have arisen through natural crossing. Such varieties were first reported in citrus fruits by Ferrarius in 1646, and this group of plants is outstanding at the present time because of the frequency with which such variations occur. The more commonly occurring forms contain only one or two variant sectors. Less frequent in citrus are the "Bizarria" (figure 108) and the variegated varieties (figure 109),

FIGURE 108.—Chimera in orange. (*From Ferrarius, Hesperides sive de Malorum Aureorum Cultura et Usu, Romae,* 1646.)

FIGURE 109.—Variegated variety of Citrus. (*After Nolckamer,* 1695, *from Coit, Citrus Fruits, copyright* 1915 *by The Macmillan Company. Reprinted by permission.*)

but they are found occasionally. In general, among seed plants variegation is the most common of these abnormal forms to which Winkler first applied the term chimera. But it should be noted at once that in certain cases variegated foliage is caused by disease, and such cases are not to be considered as chimeras. In figure 110*b* and *c* are shown two Abutilon specimens in which the variegation is typical of a disease known as infectious chlorosis, while in *a*, *d*, and *e* are shown typical chimeras. Variegation may also be caused by Mendelian factors. It is necessary to distinguish as to the cause of variegation and to classify as chimeras only those cases in which it is due to genetic dissimilarity within the individual plant. It was in this sense that the term was first proposed; so chimera may be defined as a mixture of genetically diverse tissues in the same shoot. It matters not whether the difference be small or large so long as it is a genetic difference of some sort, such as a difference in a single pair of genes affecting chlorophyll production, in some essential cytoplasmic inclusion, or in the entire chromosome complex. Given some such genetic diversity among the tissues of a shoot, it is only necessary that the physiological relations between the genetically unlike elements be such as to permit growth and differentiation within the shoot to proceed, and a chimera is the result. If the genetic differences are such as to cause visible morphological or qualitative variations among the differentiated parts of the shoot, or between the chimeral shoot and normal shoots, the chimera is likely to be observed or even selected as a new variety. A clear idea of the nature of chimeras may be gained by learning how they have been produced experimentally.

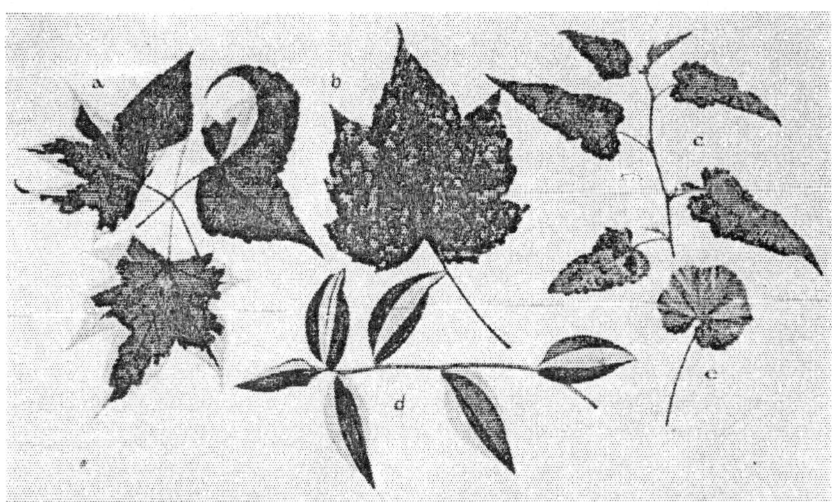


FIGURE 110.—Types of variegation. *a*, Abutilon, a periclinal chimera having green leaves with white edges. *b*, *c*, other Abutilon leaves showing variegation typical of the disease, "infectious chlorosis." *d*, sectorial chimera in Wandering Jew. *e*, a chimeral leaf of Mallow in which the light areas are pale green. (*After Chapin from the Journal of Heredity by permission.*)

**Synthetic Tomato-nightshade Chimeras.**—Some members of the nightshade family are easily grafted, even though they belong in different genera. By grafting tomato on nightshade, *Solanum nigrum*, and *vice versa*, Winkler obtained new morphological types which have been shown to be chimeras. The method is simply to graft on a scion by one of the ordinary methods and, soon after it has united with the stock, to cut it off, making the cut pass through the united tissues of stock and scion. Most of the adventitious buds pushed out are composed entirely of either stock or scion tissue, but rarely a bud will develop on or near the line of union. In such cases either one of two arrangements of the graft components may result, depending on the orientation of the bud with reference to the two types of regenerated tissue. If the two masses of tissue meet as at *a* in figure 111, the shoot will consist of sectors of nightshade and tomato, but should one of the cell masses grow over the other, producing the arrangement shown at *b*, the shoot in this instance will be composed entirely of tomato tissue, except for an epidermis of nightshade cells. The first type is called sectorial and the second, periclinal. A third type, termed by Winkler hyperchimera, consists of a complex

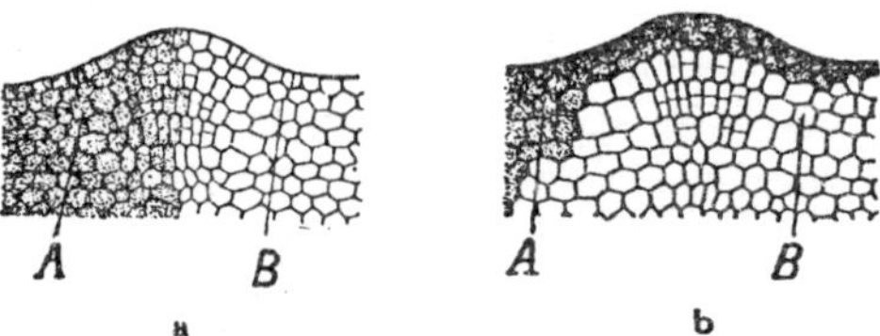


FIGURE 111.—Diagram to show an early stage in development of an adventitious bud, arising at the point of union of the two graft components, *A*, representing nightshade, and *B*, representing tomato. *a*, sectorial arrangement; *b*, periclinal arrangement. (*From Buder.*)

mixture of the two dissimilar kinds of tissue such as would result if the growing point of the shoot were a mosaic of groups of unlike cells. Certain ornamental plants are chimeras of this type of which the variegated citrus shown in figure 109 is an example.

FIGURE 112.—Leaves from *a*, tomato; *c*, nightshade; and *b*, a chimera, *Solanum Kölreuterianum*, consisting of tomato body with an outer epidermal layer of nightshade. (*From Baur, Vererbungslehre, 5 u. 6 Auflage, copyright 1922, by Gebrüder Borntraeger. Reprinted by permission.*)

Most of the chimeras obtained by Winkler were sectorial, as would be expected; but among over 3,000 shoots, produced from 268 grafts. five new morphological types appeared which represent various periclinal arrangements of tomato and nightshade tissue. The degree of resemblance to the species whose tissue comprises the inner portion of the chimeral shoot depends upon the number of layers of cells present from the other species. For example, *Solanum Kölreuterianum* (figures 112*b*; 113*b*) has only one cell-layer of nightshade epidermis on a tomato shoot and the leaf is compound, like tomato; while *S. Gärtnerianum* (figure 113*a*) is a tomato shoot with two cell layers of nightshade and the leaves are simple, like nightshade. On the other hand, *S. proteus* (figure 113*c*), whose leaves resemble those of the tomato, has a double layer of tomato cells overlaying a nightshade body; and *S. tubigense* (figure 113*d*), which closely resembles the nightshade, has all the inner portion of *S. nigrum* with only the single epidermal layer consisting of tomato.

FIGURE 113.—Four tomato-nighshade chimeras *A*, *Solanum Gärtnerianum*, tomato body covered with two epidermal layers of nightshade cells. *B*, *S. Kölreuterianum*, tomato with a single outer layer of nightshade. *C*, *S. proteus*, nightshade body with two epidermal layers of tomato. *D*, *S. tubigense*, nightshade with outer epidermal layer of tomato cells. (*From the Journal of Heredity by permission.*)

Such new forms, arising from grafted plants and morphologically intermediate between stock and scion, have been called graft-hybrids. A graft-hybrid, however, is nothing more than a special form of graft symbiosis. That the two kinds of tissue remain quite distinct, despite their intimate relations, is evidenced by the chromosome counts obtained by Winkler. The diploid numbers are 24 for the tomato and 72 for the nightshade. If a fusion of nuclei involving diploid numbers had occurred, the cells of the supposed hybrids should contain 96 chromosomes, but the only counts obtained by Winkler in these four chimeras were 24 and 72. Thus it appears that in each graft-symbiont the two kinds of tissue maintain their individuality. Yet there is a combined effect upon the morphological characters, and the physiological interactions are such as to cause reduced vigor. This effect is least noticeable in *Kölreuterianum.* This form is sterile, but from the other three chimeras seeds have been obtained which produced only nightshade or tomato plants according to the specific nature of the subepidermal cells in each case. The fifth new form, which appeared among the shoots from a decapitated graft, was named *Darwinianum* and was thought by Winkler to be a true graft-hybrid, because he counted twenty-four chromosomes in the pollen mother cells. It is more probable, however as was pointed out by Baur, that this is also a periclinal chimera with nightshade epidermis, a subepidermal layer of tomato and the inner portion nightshade.

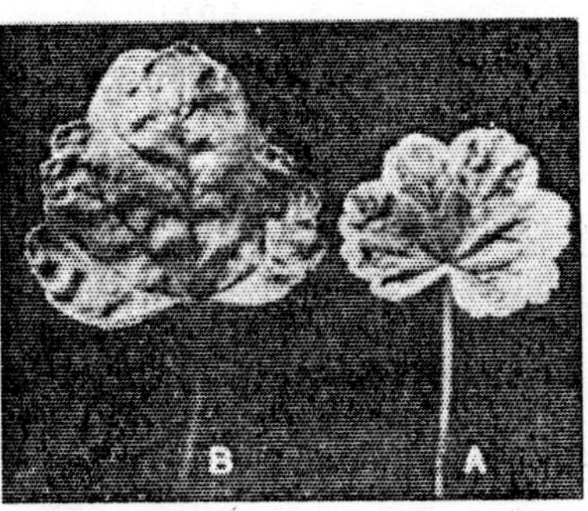


FIGURE 114.—White-edged geranium leaves. *A*, from a plant with two white peripheral cell layers; *B*, from a plant with one white peripheral layer. (*From Baur, Vererbungslehre, 2nd ed., copyright by Gebrüder Borntraeger,* 1914. *Reproduced by permission.*

**The Structure of Chimeras.**—The key to the explanation of Winkler's synthetic chimeras was furnished by Baur's discovery of the difference between the white-edged and solid-green varieties of geranium. *Pelargonium zonale.* From his study of seedlings of the white-edged variety, he learned that the color of the leaves on a seedling depends upon the nature of the cells composing the growing point. Examining the

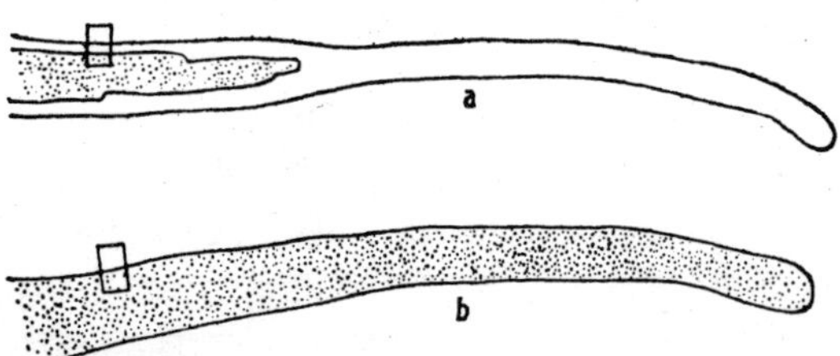


FIGURE 115.—Sections through edges of the two leaves shown in figure 114. Green tissue indicated by stippling. For enlarged views of the portions enclosed by the small rectangles see figure 116. (*From Baur, Vererbungslehre, 2nd ed., copy ight by Gebrüder Borntraeger,* 1914. *Reprinted by permission.*)

cells in white-edged and green leaves, he found that in the former there is an extra layer of colorless cells below the true epidermis (see figures 114–116) and concluded that such a plant must be considered a periclinal chimera. Baur had observed sectorial chimeras among his geranium seedlings and found that occasionally a plant having some of its leaves entirely green and some entirely white would produce a shoot bearing white-edged leaves. Finding that such shoots arose near the boundary line between the green and white tissue of the stem, he concluded that, in order to have a periclinal chimera arise from a sectorial chimera, the arrangement of the two kinds of tissue must be as shown at *b* in figure 117*B*. A bud pushing out at such a point would have an

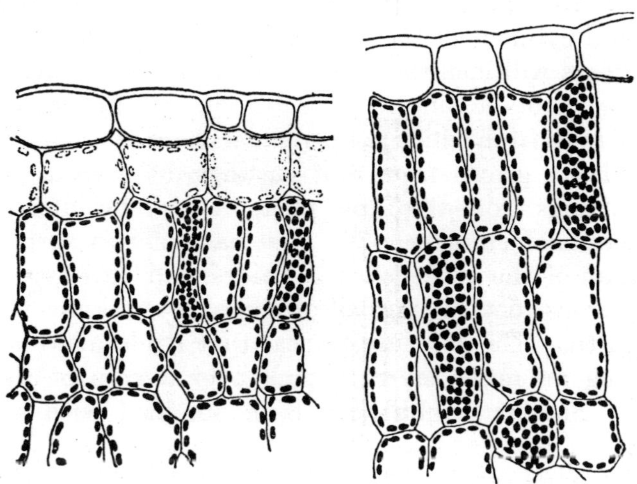

FIGURE 116.—Microscopical views of those portions of the cross-sections in figure 115 represented by the small rectangles. Colorless chromatophores are indicated in outline, green in black. (*From Baur, Vererbungslehre, 2nd. ed., copyright by Gebrüder Borntraeger,* 1914. *Reprinted by permission.*)

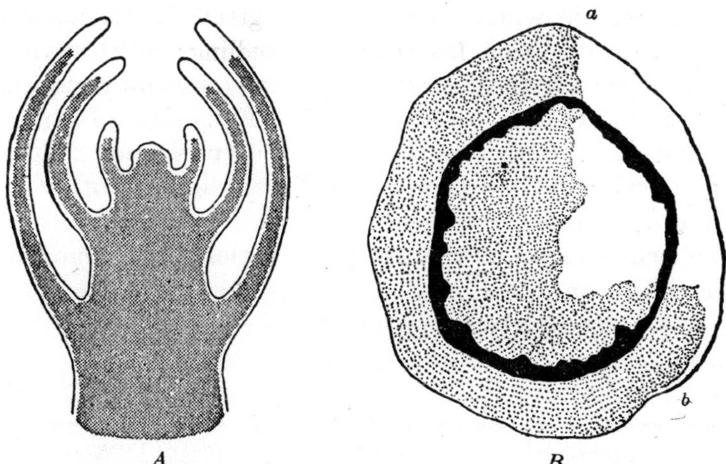


FIGURE 117.—*A*, diagram of longitudinal section through a young shoot of a plant bearing white-edged leaves—a periclinal chimera. *B*, cross-section of stem of a sectorial chimera. A bud pushing out at *a* would produce a sectorial chimera, while one arising at *b* would form a periclinal chimera. (*From Baur, Vererbungslehre, 2nd ed., copyright by Gebrüder Borntraeger,* 1914. *Reprinted by permission.*)

envelope of colorless cells in addition to the colorless epidermis. A much rarer form of variegation is the converse arrangement, a subepidermal layer of green cells covering a white core. This type has been reported in Pelargonium, Holly, and Coprosma.

Further evidence concerning the structure of periclinal chimeras has accumulated from Bateson's extensive experiments with Bouvardia and with the Fancy, Show, or Lady Washington Pelargoniums, *P. domesticum*. His attention was called to the fact that frequently these plants fail to come true from root cuttings. Now branch roots arise from the plerome, stele, or central cylinder of the shoot so that, in the case of a periclinal chimera, its roots will lack the diverse outer layers characteristic of the shoot and new shoots arising from such roots will appear as reversions to the original type from which the periclinal chimera originated. Therefore, when plants grown from root cuttings differ from plants grown from stem cuttings from the same plant, it may be inferred that this plant is a periclinal chimera. The first case of this sort reported by Bateson is that of the pink-flowered Bridesmaid variety of Bouvardia of which numerous root cuttings all produced a well-known red-flowered variety, Hogarth. The evidence is clear that Bridesmaid is a periclinal chimera having an outer layer of pink over a core of tissue normal for Hogarth. Similar results have been obtained with a number of Pelargoniums.

Similar but more conclusive evidence has been furnished by Clausen and Goodspeed from the analysis of a case of bud variation in tobacco. Fortunately the variation affected a character which had been investigated genetically so that it was possible to make a thorough investigation of this case. Two varieties of *Nicotiana tabacum* had been crossed. One, *purpurea*, bears carmine flowers and the other, Cuba, white flowers. $F_1$ is carmine-flowering and $F_2$ conforms to the dihybrid ratio 9 carmine: 3 pink: 4 white. In this instance the 200 $F_1$ plants all bore carmine flowers except one, which was "accurately divided bilaterally into two parts," one of which produced carmine flowers like the other $F_1$ plants, and the other light pink flowers like the pink segregants of $F_2$. Genetic tests of the two portions of this sectorial chimera were made by selfing and backcrossing on Cuba and on *N. sylvestris*. The white-flowered character of the latter always behaves as a recessive when it is crossed with *tabacum*. The results are summarized below:

| Parents | Carmine | Pink | White | Total |
|---|---|---|---|---|
| Carmine flowers selfed | 48 | 13 | 39 | 100 |
| Pink flowers selfed | 56 | 13 | 31 | 100 |
| Calculated 9: 3: 4 | 56.25 | 18.75 | 25 | 100 |
| Backcrosses of carmine flowers | 54 | 29 | 64 | 147 |
| Backcrosses of pink flowers | 40 | 37 | 72 | 149 |
| Calculated 1: 1: 2 | 37 | 37 | 74 | 148 |

This genetic evidence indicates that the change involved in the production of the pink-flowered variation did not affect the hypodermal layer from which the ovules and pollen grains are derived. Histological examination showed that the coloring matter in the flowers is confined to the cells of the epidermis. Stem cuttings were made from both portions of the plant, and the two clones were uniform for carmine and pink flowers, respectively. Finally root cuttings were made from plants of both clones, and the plants obtained from *all* of these root cuttings produced only carmine flowers. Thus complete evidence was obtained to prove that the pink-flowered portion of the original $F_1$ plant was a periclinal chimera. On the strength of this evidence, it would seem highly probable that the Bridesmaid Bouvardia is a periclinal chimera which originated as a bud sport from Hogarth and that this holds true for many similar cases.

**Origin of Autogenous Chimeras.**—Autogenous chimeras originate in somatic cells as a result of some natural process by which diverse cell lineages arise from certain individual cells. The processes of genetic variation by which chimeras may originate are of three categories, *viz.*, *factor mutations*, *chromosomal variations*, and *cytoplasmic variations*. The first two categories have been presented in considerable detail in Chapters XXII and XXIV, but their relation to the origin of chimeras was not discussed. The natural occurrence of genetically diverse tissues in the soma of an individual is often referred to as somatic segregation, a convenient term, the use of which in this general and superficial sense is perhaps warranted. The student should be on his guard, however, not to read into the term a meaning which was not intended. Its use would better be restricted to cases involving actual segregation of chromosomes among somatic cells.

While it is highly probable that chimeras occasionally originate as a result of a factor mutation in a somatic cell, there is very little satisfactory evidence on this point. As a matter of fact, no cases of somatic segregation in plants have actually been proved beyond all question to be due to factor mutations. There is always the possibility of explaining the phenomenon as due to a chromosomal variation or to an antecedent chimeral condition. Among animals, however, Drosophila has produced a number of cases of undoubted factor mutations in somatic cells, resulting in mosaic individuals other than gynandromorphs. Such evidence is most nearly approximated in plants by the variegated maize discussed in Chapter XXII, where it was shown that in order to harmonize some of the facts it is necessary to assume the occurrence of factor mutations so late in ontogeny that only the epidermal layer is affected. One of the earliest cases to be subjected to genetic analysis was reported by Correns in a variety of Mirabilis having green and white leaves. Occasionally a solid-green shoot appeared on such a plant, and selfed seed therefrom produced green and variegated progeny in monohybrid ratio, green being dominant. It is probable that a factor mutation occurred in one chromosome, giving rise to a group of cells heterozygous for green, but an alternative hypothesis must be recognized. Adventitious buds arising in the internodes of a stem are extensions of the plerome, just as are buds arising from roots. If the solid-green branches which were tested by Correns arose from adventitious buds, such branches would indicate that the plant itself was a periclinal chimera rather than that a factor mutation had occurred. Again, in Clausen and Goodspeed's tobacco chimera, while it is possible that a factor mutation changed the gene for carmine into the condition necessary for pink, it is also possible that the whole chromosome containing the carmine factor was lost from a cell at an early stage in the development of the plant. The trend of contributory evidence seems to favor the latter view.

FIGURE 118.—Bud sport from a white-flowered gladiolus bearing red flowers on one side of the shoot and showing one flower half red and half white; a sectorial chimera.

When such critically investigated cases are still open to question, it is obvious that enormous difficulties attend the interpretation of the nature and cause of somatic segregation in plants of unknown genetic constitution or doubtful derivation. Furthermore, there is abundant evidence that chimeras tend to produce new variations due merely to rearrangements of the component tissues occurring in the course of differentiation. These considerations emphasize the care which must be used in studying the genetic nature of bud variants and chimeras. It is altogether likely that many cases which appear to be due to mutation of a recessive gene back to the dominant condition are merely manifestations of a chimeral condition. The classical instance of reversion of the nectarine back to the peach can be explained, as East has suggested, on the assumption that nectarines which display this phenomenon are periclinal chimeras in which the endogenous tissue is peach and the epidermal layers, nectarine. Similarly, the case of a white-flowered gladiolus, The Bride, which produced a shoot sectorially divided between white and red and later a complete red-flowered shoot (figure 118) may be interpreted as evidence that The Bride is a chimeral variety. It is a horticultural variety of *Gladiolus Colvillei*, which is a red-flowered. sterile hybrid between *G. cardinalis*, red-flowering, and *G. tristis*, white or yellowish. As *Colvillei* is sterile, very likely The Bride originated from it as a white bud sport and, if only the dermatogen was changed in its production, the subsequent reversion to red might result merely from a rearrangement of the component elements.

*Chromosomal variations* sometimes originate in somatic cells of plants, although but few cases have actually been observed. The classical instance is that of the experimental induction of polyploid cells in the root tips of vetch, pea, and onion by Němec. By properly narcotizing the root tips with chloral hydrate mitosis is inhibited just sufficiently so that telophase reconstruction of the nuclei takes place but without the completion of cell division. In this way binucleate cells are formed in which, after recovery from narcotization, the two nuclei may fuse and proceed to mitosis, thus producing $4n$ daughter cells. Repeated narcotization results in giant nuclei with $8n$ or probably even $16n$ chromosomes, but after recovery from the treatment there seems to be a tendency for these higher numbers to disappear.

A tetraploid nightshade and a tetraploid tomato were obtained by Winkler during his experiments on chimeras. The tetraploid nightshade was derived from periclinal chimeras having tomato epidermis and nightshade core. After removing tho axial buds from such plants, adventitious buds developed which were composed entirely of tetraploid nightshade cells with 144 chromosomes. This of course indicates that the nightshade core of the periclinal chimera was tetraploid and that the doubling of the chromosome number occurred at an earlier stage in the process. Similarly, the tetraploid tomato was obtained from a periclinal chimera having nightshade epidermis and tomato core. Examination showed that the cells of the core had forty-eight instead of twenty-four chromosomes, the diploid number of tomato. By cutting off a young shoot and removing the axial buds below the cut new adventitious buds were obtained on the cut surface, most of which were composed entirely of tetraploid tomato tissue. How the original tetraploid cells originated is not known. Winkler first suggested that they arose by fusion of two callus cells. It seems more probable, however, that tetraploids are caused by the inhibition of cytoplasmic division during mitosis as in Němec's experiment. That tetraploid chimeras sometimes occur in diploid plants was first reported by Blakeslee and Belling. Recently Mrs. Lesley has found groups of tetraploid cells in the root tips of normal diploid tomato plants; M. Navashin has reported a sector of tetraploid cells in a root tip of a diploid plant of *Crepis Dioscoridis* L.; and Miss Hollingshead has found a similar root in a 24-chromosome derivative of *C. setosa* × *C. biennis*.

That irregularities in the distribution of individual chromosomes sometimes occur during somatic mitoses is strongly indicated by two pieces of genetic evidence. Frost has reported the occurrence of a bud sport in Matthiola in which presumably linked factors were segregated out simultaneously. It seems probable that this was due to some abnormal shifting of a chromosome or portion of a chromosome, possibly to non-disjunction. Emerson's studies of mosaic endosperm in maize furnish stronger genetic evidence of aberrant chromosome behavior in somatic mitoses, because the linkage relations of the factors involved are definitely known.

In crosses in which recessive aleurone and endosperm characters are contributed by the female parent and their dominant allelomorphs by the male parent, spots of the recessive (maternal) aleurone layer are underlaid by the recessive (maternal) type of endosperm when the genes for these aleurone and endosperm characters are genetically linked, that is, when they are carried in the same chromosome. On the contrary, similar recessive (maternal) aleurone color spots are always underlaid by the dominant (paternal) type of endosperm when the genes are not linked, that is, when they are carried in non-homologous chromosomes. The fact that linked genes separate out simultaneously while non-linked ones do not do so supports the view that mosaic seeds are the result of some chromosome aberration such as elimination or non-disjunction.

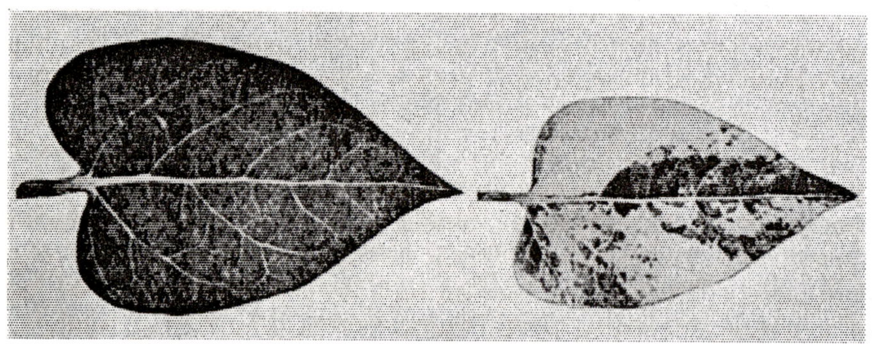

FIGURE 119.—Normal green leaf (left) of *Mirabilis jalapa* and a leaf of the variety *albomaculata*. (*From Correns.*)

This genetic evidence is supported by the discovery of unbalanced chromosomal chimeras in Datura and by actual observations of mitotic irregularities such as those reported by Rosenberg in Crepis.

*Cytoplasmic Variations.*—All of the plants thus far investigated, in which the chimeral condition was attributed to cytoplasmic variations of some sort, are characterized by a mosaic arrangement of green and white tissue in the leaves and stems. These have generally been referred to as *albomaculata* forms. Frequently, however, these forms produce wholly white or green branches. The evidence in these cases is mainly genetic, being derived from the progenies of selfed and crossed seeds from mosaic, white, and green shoots. For example, in Mirabilis (figure 119) Correns found that selfed seed from green branches produce only green plants and by selfing flowers on the white branches. only white seedlings are obtained; while from the mosaic branches the offspring will be mosaic, green, and white. But in crosses between these different types the inheritance is wholly maternal. Thus if flowers on a white branch are castrated and fertilized with pollen from a green branch, the resulting seedlings are all white and the reciprocal cross gives only pure-green offspring. These results presumably indicate that this *albomaculata* variety of four o'clock has two kinds of chloroplasts, *viz.*, normal green ones and abnormal ones which are wholly devoid of chlorophyll. In a similar variety of *Primula sinensis*, Gregory demonstrated the existence of the two kinds of chloroplasts. Similar cases have been discovered in Antirrhinum by Baur, in Lychnis by Shull, in maize by Anderson, and in several genera by Correns including the extensively analyzed case of *Stellaria media*. The phenomenon of maternal inheritance characteristic of these cases is just what would be expected if the cytoplasm of the zygote comes exclusively from the egg

FIGURE 120.—A mosaic seedling of *Pelargonium zonale* which has produced a pure-white shoot. (*From Baur, Vererbungslehre,* 5 *u.* 6 *Auflage, copyright* 1922, *by Gebrüder Borntraeger. Reprinted by permission.*)

cell, that is, if no cytoplasm is introduced with the male generative nucleus.

Different hereditary behavior was found by Baur in a mosaic form of *Pelargonium zonale* (figure 120) in which mosaic, white, and green branches occur. When flowers on green, white, or mosaic branches are selfed, the resulting seedlings are green, white, or mosaic, respectively, as in the previous instance. But when flowers on green branches are crossed with flowers on white branches, regardless of which is the maternal parent, the seedlings are always mosaic, although in the same crosses flower-color characters are inherited strictly according to Mendelian principles. Baur suggests that this mosaic is another case of plastid inheritance which differs from the foregoing only in that the cytoplasm, including chromatophores, may be transmitted through the pollen as well as through the egg. Thus far no more satisfactory hypothesis has been proposed.

An entirely different situation obtained in two other *albomaculata* forms, one in *Humulus japonicus* (figure 121), which was studied by Winge, and the other in *Capsicum annuum*, reported by Ikeno. Selfed seeds of these mosaics never produce pure-green or pure-white seedlings. Winge made reciprocal crosses of mosaic plants with normal green plants of this species of hop, and although he grew hundreds of seedlings and carried the test to the fourth generation, he found the inheritance of green *vs.* mosaic to be exclusively maternal. In Capsicum, on the other hand, Ikeno found the mosaic character to be inherited through both parents. It has been suggested by Winge that the mosaic character in these plants is due to something in the cytoplasm itself, not to plastids or other inclusions, and that in Humulus no cytoplasm is contributed to

FIGURE 121.—Leaf of *Humulus japonicus* var. *albomaculatus*. (*From Winge.*)

the zygote from the pollen, while in Capsicum the generative nucleus is accompanied by cytoplasm which passes over into the egg cell.

**Chimeral Varieties.**—Sectorial arrangements of different-colored tissues are the most common natural chimeras. They occur very frequently in citrus fruits, especially in the orange and the lemon. Color chimeras are also fairly common in apples and pears and have been found in bananas, figs, grapes, olives, and tomatoes, as well as in dahlias. gladiolus, poppies, sunflowers, and many other flowers. Many valuable ornamental shrubs are mixtures of sectorial and periclinal chimeras together with normal green shoots. Some striking examples are the Variegated Black Elderberry, the Variegated Deeringia, and the variegated forms of Japanese Spindle Tree or Strawberry Bush, *Euony-*

*mus japonicus.* Chimeras, especially of the sectorial type, are occasionally found in wild plants.

The instability of chimeral varieties is especially worthy of note, as this peculiarity reduces their horticultural value. Variegated chimeras are characteristically weaker than the normal type, so that care must be exercised to prune out the solid-green branches which would soon predominate.

Varieties which are periclinal chimeras are always intermediate between the component types and are less vigorous. They frequently display buckling of the leaves or laciniation of the petals due to lack of correlation in growth of the diverse tissues. The flowers also may be abortive and sterility is common. In the tomato-nightshade chimeras the germ cells of the two components are not equally susceptible to the influence of adjacent foreign cells. In both *tubigense* and *Gärtnerianum* the fruits contain fertile seeds which produce only pure nightshade plants. But in *proteus* only part of the seeds are viable, producing tomato seedlings, while *Kölreuterianum* is entirely sterile. In Pelargonium chimeras, Chittenden has found similar variations among the component elements in nature and degree of sterility.

**Other "Graft-hybrids."**—The discovery of the true nature of Winkler's chimeras led to investigation of a number of historical cases of graft-hybrids. As most of these cases have been fully discussed elsewhere, the results of this study are summarized in table 47. The demonstration that all these supposed graft-hybrids are really periclinal chimeras is a distinct contribution to horticulture.

**Resumé.**—From a review of present knowledge of chimeras, certain considerations stand out with special emphasis. There is, for one thing, the fact of general or superficial similarity among chimeral forms, but behind this a fundamental complexity in the diversity of causes which lead to such somatic expressions. Because of the variety of contributing factors involved in chimeras, considered as a category of phenomena, the investigation of each new case must be undertaken with due regard for the various possibilities involved. These possibilities have been by no means exhausted in the present brief discussion. The voluminous literature of this subject contains certain extensions and modifications of the general principles herein presented which may eventually lead to the solution of many obscure problems.

Of general biological interest is the evidence of great power of regeneration in higher plants and the accompanying imperfect individuality of the organism as a whole. On the other hand, it is clear that the individual cell does tend definitely to maintain its identity; there is, so far as we know, no fusion of unlike somatic cells.

The value of genetically analyzed plant material in the investigation of such complex phenomena must be obvious; also the importance of

Table 47.—List of the Most Important Periclinal Chimeras Produced by Grafting (*Adapted from Buder*)

| Name and origin | Used in grafting | | Composition |
|---|---|---|---|
| | As stock | As scion | |
| *Laburnum* (*Cytisus*) *Adami*, arose spontaneously in 1826 from an unsuccessful graft. | *Laburnum vulgare* (Shower of Gold) | *Cytisus purpureus* (purple broom) | According to *Buder* only one outer layer of *C. purpureus*, all within being *L. vulgare*. |
| The Cratægomespili of Bronvaux, originated spontaneously many decades ago at Bronvaux in Metz from places where stock and scion had overgrown on grafts nearly a century old. | *Cratægus monogyna* (whitethorn) | *Mespilus germanica* (medlar) | According to Baur and H. Mayer both forms have a Cratægus body which is covered by a Mespilus mantel: |
| *a. Cr. Asnieresi* (resembling whitethorn) | ........ | ........ | In (*a*) of one layer of cells. |
| *b. Cr. Dardari* (resembling medlar) | ........ | ........ | In (*b*) of two layers of cells. |
| The Cratægomespili of Lagrange, apparently complete analogues of the two forms from Bronvaux. | | | |
| The pear-quince "hybrid" of Frère Henri, originated about 1903 in Rennes. | Cydonia (quince) | Pyrus (pear) | Probably consists of pear tissue within a layer of quince cells. |
| The peach-almond graft hybrid of Daniel and Delphon, arose spontaneously in 1908 at Mas-Grenier (Tarn and Garonne). | *Amygdalus communis* (almond) | *Amygdalus persica* (peach) | Evidently a mixture of sectorial and periclinal chimeras. |
| The tomato-nightshade "hybrids" of Winkler, produced experimentally in 1907–1909. | *Solanum lycopersicum* (tomato) | *Solanum nigrum* (black nightshade) | According to Winkler: Outer layers / Inner tissue |
| *a. S. tubigense* | ........ | ........ | 1 of tomato / nightshade |
| *b. S. proteus* | ........ | ........ | 1 and 2 tomato / nightshade |
| *c. S. Kœlreuterianum* | ........ | ........ | 1 of nightshade / tomato |
| *d. S. Gærtnerianum* | ........ | ........ | 1 and 2 nightshade / tomato. |
| *e. S. Darwinianum* | ........ | ........ | "An actual hybrid" (but see text). |
| The tomato-eggplant and tomato-bittersweet "hybrids" of Heuer, produced experimentally in 1910. | | | |
| Form I | *Solanum lycopersicum* (tomato) | *Solanum melongena* (eggplant) | Probably only the epidermis is eggplant, tomato within. |
| Form II | *Solanum lycopersicum* (tomato) | *S. dulcamara* (bittersweet) | Epidermis of tomato, inner portion bittersweet. |
| The Populus "hybrid" of Baur, produced experimentally in 1911. | *Populus canadensis* | *P. trichocarpa* | Only the epidermis of *P. trichocarpa*, within *P. canadensis*. |

cytological examination, not only of the chromosomes but of the self-perpetuating cytoplasmic inclusions. Finally should be noted the service which genetic investigation has rendered to biology and agriculture through the removal of such misconceptions as the possibility of hybridization through the union of somatic cells, and the new fields which have been opened up for the extension of present knowledge of heredity.

### References

BAUR: Vererbungslehre, 5th and 6th ed., Chapters XII and XIV.
COIT: Citrus Fruits, Chapter VII.
FROST: Polyembryony, heterozygosis and chimeras in citrus, *Hilgardia,* **1**: (16), 365–402.
MORGAN: Physical Basis of Heredity, Chapter XVII.
MORGAN and BRIDGES: The Origin of gynandromorphs, *Carnegie Inst. Pub.,* **278.**
SHARP: Introduction to Cytology, 2nd ed., Chapters XVII and XVIII.
SINNOTT and DUNN: Principles of Genetics, Chapter XI.

## CHAPTER XXXI

## CLONAL SELECTION

The place of clonal selection in agricultural practice has been a mooted question for decades. One of the chief causes for the long controversy is the failure to distinguish clearly between the issues involved. In connection with this subject there are just two questions of fundamental importance to agriculture: (1) Is selection within a clone sufficiently effective to be of practical value in increasing yield? (2) Is the selection of new clones a practical method of producing improved varieties? Here are two entirely different questions, to the first of which the answer is negative, while the answer to the second is provisionally or conditionally affirmative. It is the purpose of the present chapter to set forth the principal reasons for each answer briefly, leaving it for those who are especially interested to follow up the details through the references cited.

**Clonal Variation.**—As was briefly set forth in Chapter XX, the variation which normally occurs within a clone is non-genetic and hence cannot serve as a basis for improvement of the clone. This principle had long been recognized by conservative horticulturists, but until Sax and Gowen completed a statistical analysis of the clonal variation in 882 Ben Davis apple trees, it had not been convincingly demonstrated. By collecting the data on yield of the individual trees for a period of 6 years and subjecting it to biometrical analysis, it was found that variability in yield was dependent upon several causes which were approximately evaluated by appropriate statistical methods. Thus a determination of interannual correlation showed that most of the variation was due to seasonal differences, while a less amount was due to such causes of variation as soil heterogeneity, stock heterogeneity, and genetic bud variations. Again, determination of the effect of grouping trees in blocks and determination of the correlation between yields of adjacent trees showed that soil heterogeneity was responsible for a large part of the so-called permanent (non-seasonal) variation in yield. A smaller part of this portion of the variation in yield remained to be attributed to stock heterogeneity and to bud variations. It was impossible to apply biometrical methods to distinguish between stock heterogeneity and bud variations, but the experimental evidence on variability of seedling stocks indicates that stock heterogeneity is responsible for most if not all of the remaining variation. Further, by applying the method of partial correlation, which in effect holds all factors constant except the one under investigation, it was possible to evaluate approximately the proportion of the total variation due to specific causes with the following results:

| | |
|---|---|
| Seasonal variations | 62 per cent |
| Soil heterogeneity | 18 per cent |
| Stock heterogeneity, genetic bud variations, and unknown causes | 15 per cent |

The results of this investigation show that variability in yield of Ben Davis apple trees is due in the main to environmental effects and seldom to bud variations depending upon genetic changes. Considerable statistical data from other sources were examined by Sax and Gowen and were found to be in general agreement with the above deductions, so that the general conclusion may be drawn that selection of scions from high-producing fruit trees would seldom if ever result in the establishment of consistently higher-producing orchards.

Experimental evidence is found to be in complete agreement with the above conclusion. The probable effect of stock heterogeneity on variability in stocks used in the propagation of fruit trees is shown by the investigations of Hedrick, working with deciduous species, and Webber, working with citrus. Seedling stocks are used almost exclusively and are highly variable in size and growth rate. The permanent effects of this variability in stocks on the scions is shown by the fact that size attained in the nursery is correlated with size attained after several years in the orchard. Furthermore, experiments on selection of scions have yielded consistently negative results.

**Selection within Clones.**—Experiments to determine the effect of selection of buds from high- and low-producing trees (Whitten, Cummings, Rawes, and others) have demonstrated that there are no significant differences in the performance of the resulting trees, if the buds are taken from healthy trees. The same conclusions are reached from experiments in selecting high-yielding hills or tuber units of potatoes (East, Whipple, and others) when diseased strains are eliminated. Working with grapes throughout a period of 15 years Bioletti concludes that mass selection of vine cuttings on the basis of the yields of the parent vines is of no value in improving or maintaining the productivity of vineyards; also that the attempt to increase the bearing of a variety of grape by the selection of buds from a parent vine which has been distinguished by continuous and heavy bearing superior to that of the average or of any of the vines of the same variety but which shows no other distinguishing characters is fruitless. He also states that the attempts of nurserymen and others to preserve or to improve the productivity of clonal varieties of fruit trees by bud selection based exclusively on yield records of the parent plants would be wasted effort if applied to vines; and that whether it is of any use for this purpose for other fruits is doubtful. This question will be referred to again.

**The Frequency of Genetic Bud Variation.**—Our first question, as to whether selection within a clone is effective in increasing yield, was found to be definitely negatived by the two most cogent lines of evidence, the statistical study of clonal variation and the experimental tests of bud selection within clones. This does not mean that genetic bud variations affecting yield and quality fail to occur, but rather that they are either so infrequent or so minute and fluctuating that they do not operate as a major cause of variation. This seems to be the situation generally in both deciduous and citrus fruits as well as in grapes, potatoes, and strawberries, and presumably it holds good in all vegetatively propagated crops. The answer to the second question, as to whether the selection of new clones is a practical method of producing improved varieties, also depends primarily upon the frequency with which genetic bud variations occur. The actual or even approximate rate of occurrence of somatic variations which are genetic in nature has not been determined for any species. Such determination, as Emerson has pointed out, is beset with grave difficulties.

> One may attempt with some assurance an estimation of the frequency of recurrence of a particular mutation, for instance, whether it appears in vegetative parts of individuals or in sexually produced progenies, but it is a hazardous undertaking to estimate the frequency of variations in general.

On the other hand, it is possible to compare crop plants in regard to the frequency with which new morphological types appear as bud variations, as shown by recorded instances and by consensus of opinion based on observation and also in regard to the number of new varieties which have originated as bud variations. Before discussing this further, it may be well to consider briefly the origin of genetic bud variations.

**Genetic Bud Variation.**—There are at least four genetic processes which may lead to the production of bud variations, *viz.*, factor or point mutation, chromosomal variation including sectional variation, cytoplasmic variation, and chimeral phenomena resulting from the rearrangements of diverse somatic tissues concomitant with growth which are so characteristic of chimeral varieties. Doubtless each of these processes varies in its rate of occurrence in different species and within the same species as regards the various elements concerned. Other genetic processes also may be found to play a role in bud variation, such for example as Eyster's hypothetical qualitative mitotic division of certain genes. It is unnecessary, however, to seek further evidence of the complicated character of bud variation in order to emphasize the danger of making assumptions as to the nature and cause of particular bud variants. In what follows, therefore, first consideration will be given to the frequency with which bud variations have been observed to occur in certain plants rather than to the genetic processes which may have caused their origin.

**Ornamental Plants.**—An outstanding case of multiplication of varieties by bud selection is that of the Boston fern and its derivatives. According to Benedict, the original Boston fern is believed to have originated as a bud sport from the tropical species, *Nephrolepis exaltata*, from which it was first recognized as distinct in 1896. In 1921 it was stated that over 100 distinct varieties of Boston fern had appeared as bud variants within 20 years. New variants continue to appear, a most interesting recent one being a fertile variety. This is the first in some 300 accessions acquired by Benedict to bear fertile spores. It produces

FIGURE 122.—A series of fronds illustrating progressive variation in ruffling and dwarfing. 1, *N. exaltata;* 2, *var. bostoniensis;* 3. *Harrisi* (or *Roosevelti*)*;* 4, Wm. K. Harris (or new sport of *Roosevelti*)*;* 5, Teddy, Jr.; 6-8, dwarf sports of Teddy, Jr.; 7, *Randolphi.* (*Courtesy Brooklyn Botanic Garden.*)

sporelings in great diversity, many of them being replicas of older bud variants. These bud variants are noteworthy on account of the progressive variation exhibited along three main lines, *viz.*, increase in ruffling or crisping, and dwarfing (figure 122), and increase in dissection of the pinna (figure 123). Whatever may be the true genetic or cytological explanation of these remarkable variations, Benedict believes it is proper to compare or even to homologize their origin with the origin of new varieties and species in the wild. At any rate the Boston fern derivatives provide an outstanding illustration of the production of improved varieties by means of clonal selection.

Stout's experiments on bud selection in Coleus furnish another important instance of the efficacy of clonal selection in the production of new color-pattern varieties. The two plants with which the experiment was begun had a definite pattern of leaf coloration consisting of a green midregion and yellow border with blotches of red in the epidermis. The green and yellow pigments exist in the subepidermal layers. The vegetative offspring from the two original plants were kept separate, and the simple habit of branching in this plant made it possible to indicate the particular branch as well as the individual plant from which a cutting was taken. In this way it was possible to trace the pedigree of any plant to its original source. During the course of the investigation sixteen new color patterns were obtained. There also appeared the

FIGURE 123.—A series of pinnae illustrating progressive variation in dissection. 1, var. *bostoniensis;* 2, *Piersoni;* 3, *Whitmani;* 4, *Goodi* (or *gracillima*); 5, *magnifica;* 6, *Craigi;* 7, *Amerpohli.* (*Courtesy Brooklyn Botanic Garden.*)

laciniate form of leaf which is seen in the younger leaves of the plant on the left in figure 124. Of the sixteen new color patterns, fifteen arose by somatic variations which produced bud sports either directly or, in some cases, indirectly from chimeras. The other new pattern arose solely as a fluctuating variation. In consisted of absence of yellow and decrease of red in the younger leaves of a few plants. Six of the fifteen patterns that arose as bud variations also appeared more or less frequently as fluctuating variations on certain plants. The derivations of the various color patterns are shown by the diagram in figure 125. It will be noted that the original pattern 2 is described as green-yellow-red blotched which means green center, yellow marginal border, and red blotches on the epidermis. In the same way the description of pattern 4 is interpreted as green center, yellow-spotted marginal border, and red blotches on epidermis.

The frequency with which these bud variations occurred is shown in table 48. Here are indicated under Plants the number of plants in which

TABLE 48.—FREQUENCY OF BUD VARIATIONS PRODUCING NEW COLOR PATTERNS AND LEAF SHAPES IN COLEUS (*From Stout*)

| | Plants | Frequency | Ratio |
|---|---|---|---|
| Increase of yellow and decrease of green.. | 827 | 27 | 1: 6,130 |
| Decrease of yellow and increase of green.. | 740 | 50 | 1: 2,960 |
| Reversal of positions of green and yellow. | 450 | 8 | 1:11,250 |
| Increase of epidermal red to solid red..... | 770 | 8 | 1:19,250 |
| Decrease of epidermal red, complete loss.. | 815 | 19 | 1: 8,580 |
| Decrease of epidermal red, all cases...... | 815 | 21 | 1: 7,760 |
| Appearance of the laciniate character..... | 765 | 13 | 1:11,770 |
| Entire leaf from laciniate leaf............ | 68 | 1 | 1:13,600 |

each type of change might have occurred, under Frequency, the actual number of bud variations that appeared, and, finally, the ratio of bud variations to constant buds. This ratio is obtained by assuming that on the average each plant produced 200 buds.

FIGURE 124.—Two Coleus plants which descended from the same branch, which at the time it was propagated was uniform as to leaf shape and bore leaves having the same general type of color pattern, *viz.*, green midregion and yellow border with red bloches on the epidermis. The right-hand plant resembles the original plant although it represents the fourth vegetative "generation." The left-hand plant shows the laciniate type of leaf which appeared several times as a bud variation. (*From Stout.*)

These data indicate the tendencies of the bud variations and give a clue to the behavior of the characters in question. Thus, decrease of yellow occurred twice as often as increase of yellow, and loss of red 2.2

times as often as increase of red. Although these data indicate a tendency toward loss rather than gain of the two colors, the fact that the number of variations involving gain is about half as large as the number involving loss has considerable interest. It has been generally considered that variations involving addition of a character are exceedingly rare. While this may be the case in many pure species, it would appear from the above evidence that among the progeny of species hybrids such variations may be relatively frequent.

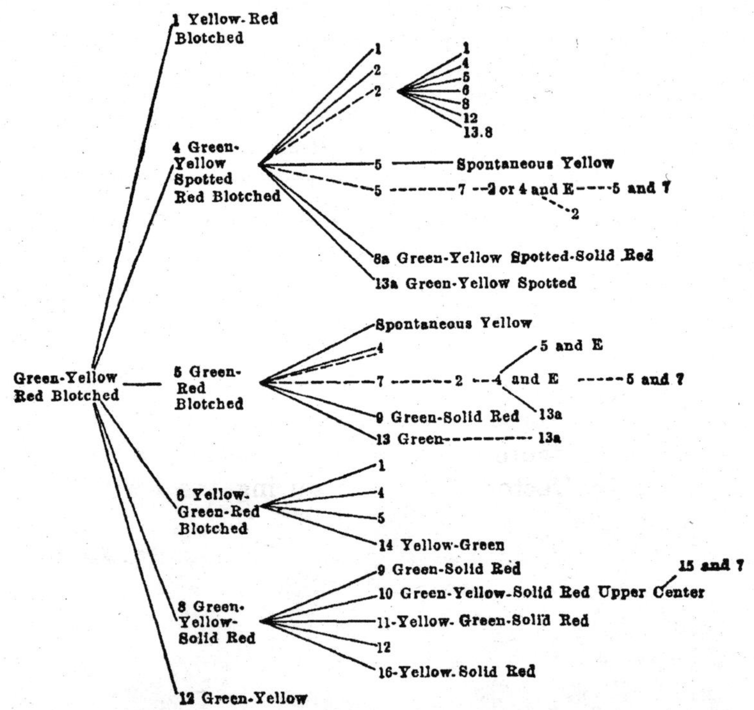


FIGURE 125.—Diagram showing derivations of color patterns in Coleus. The descriptive name of a color pattern is given only where it first appears in a line of descent. A continuous line indicates origin by bud variation and a dotted line indicates fluctuating variation. (No. 2 = original form; No. 7 = lacinate-leaved form; E = entire leaves.) (*From Stout.*)

The manner of appearance of these bud variations was such as might result from either factor mutations or chromosomal variations. For example, it is stated that "the loss of yellow, loss of green, and gain and loss of red all occurred in single branches and in sections of branches. Frequently two quite different changes occurred on the same plant." Stout tested the seed progeny of two of his plants, obtaining in all forty-five plants from selfed seed. As would be expected in such a case of composite ancestry, there was a wide range of variation in size as well as shape of leaves, and in leaf coloration every gradation between pure-yellow plants that died within a few weeks and pure-green plants.

The history of Coleus as reviewed by Stout also contains some interesting facts about the origin of the characters under discussion. The four original species that furnished the foundation stock, from which the modern Coleus has been developed, contained no yellow coloration whatsoever. They were characterized by green leaves overlaid with different shades of red, purple, or chocolate. The first appearance of yellow occurred in a leaf sport; *i.e.*, in one-half of a single leaf the green was exchanged for a decided yellow tint. "The bud at its base was propagated and gave the new variety." It does not appear, however, that this sport was used in hybridization work. "Yellow coloration also appeared as a new or spontaneous development" among the second lot of hybrids raised at the gardens of the Royal Horticultural Society about 1869. Again in regard to the laciniate-leaf character, as early as 1856 a variety of *Coleus blumei* (which was the first species introduced into Europe and was originally described in 1826) was described as being "somewhat more richly colored but differing chiefly in having the leaves deeply and doubly lobed." While there is no record of the origin of this variety, it is certain that it appeared before *C. blumei* had been hybridized with any other species. It is very probable, therefore, that it arose as a bud variation, perhaps as the result of a factor mutation. Thus it appears that two of the three characters whose presence, absence, or partial development figure conspicuously in these bud variations probably originated by factor mutations during the early horticultural history of this plant. On the other hand, the fact that the modern Coleus is a complex interspecific hybrid is sufficient justification for assuming that the majority of bud variations in Coleus are the product of chromosomal variations.

Regarding the efficacy of selection in maintaining the new forms that arose by bud variation, these results show clearly that even in such a highly variable plant as a horticultural variety of Coleus, bud selection is very effective in the production of new types and that it is necessary in maintaining strains true to type.

These two cases are sufficient to illustrate the importance of clonal selection in the production and maintenance of varieties of certain ornamentals, but not of ornamentals as a class. It is not difficult to find species or entire families in which bud variations are exceedingly rare, for example, the Begonia, Cactus, Buttercup, and Pink families. In these and similar cases it would be highly impractical to attempt to produce new or improved varieties by systematic clonal selection.

**Field Crops.**—The general inadequacy of clonal selection in potato breeding has already been mentioned. Of course, new varieties of potatoes do occasionally originate as bud mutations and, as Dorst has shown, these sometimes possess new characters which fail to appear among the seedlings of the parent variety. But these new characters are seldom, if ever, of much agronomic value. While there is some evidence, debatable in nature, that clonal selection may establish strains of higher and lower yielding ability than the average for the variety, this problem is complicated by the great susceptibility of tuber production to environmental influences. Krantz and Tolaas have suggested three possible causes which might account for variations in yield between seed stocks of a given variety of potato: (1) physiological differences in the seed stock brought about by variable growing conditions; (2) soil heterogeneity in the test plot; (3) hereditary differences in seed stocks. Chemical determinations of the nitrogen and dry-matter content of tubers from different stocks showed no relation to yield. A three years' study of relation of tuber shape to yield gave no indication of a correlation between tuber shape and productivity. Soil variation was found to be an important factor in causing yield variations. The variations in yield due to soil heterogeneity left little if any variation to be accounted for by hereditary differences in seed stocks. In the six varieties studied, no stocks were found to be consistently high or low yielders, and it was concluded that if hereditary differences do exist between seed stocks, they are too small to be of commercial importance. Most of our present commercial varieties of potato have been developed from sexually produced seed and, as Krantz points out, no notable improvement has been accomplished in our standard commercial varieties of potato by means of clonal selection. The differences in yield found when seed stocks obtained from various growers are grown for comparison are due to the following causes: (1) varying amounts of disease present; (2) lack of uniformity in storage conditions; (3) physiological differences in the tubers due to variations in the growth conditions; (4) soil heterogeneity in the test plot; (5) hereditary differences. Genetic bud variations may be regarded as too infrequent to be of practical importance in potato breeding. This means that the production of superior new varieties of potatoes must be accomplished through sexual breeding.

In marked contrast with the situation as regards clonal selection in potatoes is the success reported by Shamel in standardizing the output of Hawaiian pineapples. This industry was based upon the culture of a single variety, the Smooth Cayenne. Through insufficient care in the selection of parent plants for propagation many undesirable strains were accumulated in the commercial plantings. By the simple method of selecting plants before the fruits are harvested and marking these plants for propagation, most of the undesirable strains were eliminated and the crop correspondingly increased in uniformity, quality, and yield. As the pineapple is propagated vegetatively, the accumulation of undesirable strains in the absence of disease must have come about as a result of genetic bud variations.

In sugar cane, however, the situation is very different. Although variations in color of stalk are known to occur rather frequently, very rarely are the new types thus originating of superior commercial quality. Agronomic value in sugar cane depends upon acre-production of sugar, which in turn depends upon number of stalks per row, size of stalks and sugar content. All three of these factors are probably conditioned by multiple factors which act cumulatively. It is, therefore, very doubtful whether a mutation in any of these numerous factors would induce a somatic change which could be readily detected. Hence, although the possibility exists that desirable bud variations will be detected by means of extensive and critical tests of single eye selections, it seems unlikely that varieties of sugar cane will be much improved by this method. At any rate, the possibilities of improvement by similar tests of hybrid seedlings are incomparably greater. The same statement probably holds good for other crop plants, such as asparagus, rhubarb, hops, alfalfa, timothy, and other perennial grasses. At the same time it must be remembered that in any of these species new forms, strains, or varieties may be expected to originate from time to time by means of bud variation.

**Deciduous Fruits.**—The conclusions regarding clonal variation in the apple probably hold good for other pome fruits and the stone fruits as well. Although chimeras and bud variations are known to occur in apples, pears, plums, and apricots, they are rarely observed, perhaps because they have not been sought for diligently, but more probably because they actually occur very seldom. Hence, although a few commercial varieties of these fruits are known to have originated as bud variations, their occurrence must be looked upon as fortuitous, and the breeding of superior varieties must be attempted through hybridization.

**Citrus Fruits.**—Of the fact that distinct bud variants have been observed more frequently in citrus than in deciduous fruits, especially in the orange and lemon, there seems to be no question. Shamel and his associates have made detailed studies of variation in the principal citrus fruits, especially in the Washington Navel orange which comprises about one-half of the annual orange crop in California. As a result they have demonstrated the existence of fourteen distinct strains in the Washington Navel. These are recognizable by morphological features (figure 126). Of these fourteen types the original is best; four others were described as worthy of trial; the rest were considered undesirable.

Shamel concluded that in the Washington Navel orange the propagation of undesirable strains was the chief cause of low and poor production in many groves. That the quality of the output was lowered by the propagation of inferior strains, no one questions; but that the annual production of oranges is materially affected by this condition has been seriously questioned. The various strains were described as "heavy bearers," "less productive," or "unproductive," as compared with the original or standard strain of Washington Navel, but thus far no evidence has been furnished to prove that differences in productivity are due to bud variations. In a detailed study of individual tree performance in selected plots, 743 trees were under observation. For 151 of these trees over a period of 6 years the range was 55.8 to 394.0 pounds per tree per year; for 481 trees, including the previous 151, over a 4-year period the range was 48.3 to 412.5 pounds per tree. These results indicate wide variation in tree performance, but no statistical study of the data was made.

FIGURE 126.—Fruits of the Washington Navel orange (1), and of four strains that have originated from it by bud variation; 2, Thomson Navel; 3, Yellow Navel; 4, Corrugated; 5, Ribbed. (*Courtesy U. S. Department of Agriculture.*)

On the basis of observation alone, the possibility of improving undesirable trees by top-working has been generally accepted by the growers of citrus fruits. In California the keeping of individual tree performance records has been widely adopted and many unproductive or otherwise undesirable trees have been top-worked with selected buds. The following statement was published:

More than 40,000 undesirable citrus trees in California have been top-worked with buds selected from the most desirable trees in the investigational performance-record plots. The oldest tops grown from these buds are now more than 5 years of age. So far not a single failure has been observed in transmitting the characteristics of the parent trees by means of selected buds.

The frequency of variable fruits, attributed to bud variations, as recorded by Shamel and his co-workers is 0.66 per cent for the Washington Navel strain, 3.4 per cent for the Valencia strain, 34.35 per cent for the Lisbon strain, and 7.68 per cent for the Eureka strain. These are the strains from which buds are selected for top-working unproductive and undesirable trees. Thus bud variations even in the most desirable strains are comparatively frequent. It is, therefore, very remarkable that in more than 40,000 top-worked trees, involving over 160,000 buds, not a single undesirable or unproductive bud variant has been observed.

As Sax and Gowen point out, unwarranted conclusions concerning the efficacy of bud selection in increasing yield are mainly the result of inadequate analysis of available data. The statement has been made that "usually the most productive trees show the fewest marked variations in fruit from the type of the strain to which they belong." A similar correlation is said to exist in all the citrus fruits studied. From this it might be inferred that bud variations are less frequent in the high-yielding trees from which buds are selected. But determination of the correlation between yield and percentage of variable fruits for the 151 trees which were recorded for 6 years shows that $r = 0.01 \pm 0.05$. Evidently there is absolutely no significant correlation between yield and variability of fruit in the Washington Navel orange trees recorded. The data from Lisbon lemons are even more damaging to the conclusion that "usually the most productive trees show the fewest marked variations in fruits." Until data on the performance of the top-worked trees are presented, the conclusion that differences in productivity are transmitted by selected buds is not justified. As Sax and Gowen insist, in order to test transmission of productive ability through buds, it is not enough to top-work low-producing trees with buds from high producers.

> High-producing trees should be top-worked with buds from low-producing trees and *vice versa*. The numbers of trees involved should be large enough to eliminate chance differences in performance, and records of the top-worked trees should be kept. *Under such conditions if buds or scions from low-producing trees remain low producers when top-worked on high-producing trees, and if buds from high producers remain high producers when top-worked on low-producing trees, we would be justified in concluding that differences in productivity are inherent and due to bud variation.* At present only conflicting or unsatisfactory evidence has been presented to support such a conclusion.

In summarizing the foregoing discussion it should be emphasized that what Shamel and his co-workers have really accomplished for citrus production is strictly comparable to the service rendered Hawaiian pineapple production, *viz.*, the detection of undesirable strains and practical methods for their elimination. As a result the average quality of citrus fruit produced in America has doubtless been raised with resulting profit to the producers. But from these premises it does not necessarily follow that the general practice of keeping individual tree performance records merely for the purpose of detecting low-producing trees in order to top-work them with buds from selected high producers is warranted. In fact, it now appears that all of the high-producing trees which were selected by Shamel as sources of buds for top-working low-producing citrus trees are located on favorable spots. On the other hand, some of the oldest top-worked trees are now in the same condition as before being top-worked, which shows that poor production was due to environment, not to the nature of the scion.

The keeping of individual tree performance records is a needlessly expensive practice which fails to take account of all the factors controlling production. As Hodgson has shown, an adequate analysis of the citrus orchard can be accomplished by means of a rather simple system of records, involving the following steps: (1) establishing tree identities; (2) keeping simple estimate production records; (3) segregating the trees into classes according to yields; (4) determining the efficiency of the orchard; (5) determining the distribution of the trees in the orchard according to yields; (6) ascertaining the causes of consistent yield variation as to whether environmental or inherent in character; (7) analyzing the causes determined and applying proper remedial measures; (8) keeping a simple individual tree history record as an aid to increased tree efficiency. Such a system emphasizes the tree as the ultimate production unit and at the same time recognizes the importance of the orchard environment in relation to the inherent qualities of the trees and the cultural practices used. Some such system as this may ultimately prove to be not only warranted but essential to success in all the more highly specialized industries based on the production of tree fruits and nuts.

**Conclusion.**—From the foregoing considerations of the evidence bearing on clonal selection, the following generalizations may safely be drawn:

1. Selection within a clone is without effect because all the individuals composing a clone have the same genetic constitution.

2. The question as to whether within any vegetatively propagated variety, which from its morphological characters might be considered a clone, genetic bud variations conditioning increased yield are sufficiently significant to warrant systematic selection for the production of higher yielding clones must be answered in the negative.

3. Even in varieties like the Washington Navel orange, in which bud variations are known to occur with greater frequency than in most varieties of fruit, it is very doubtful whether bud variations, within the standard variety, which affect yield are sufficiently frequent or significant to warrant the keeping of individual tree records and the top-working of low-producing trees with scions from selected high producers. At any

rate, before such top-working is resorted to, the question of orchard environment and cultural practices should be carefully considered.

4. The keeping of tree performance records as a basis for the selection of scions does not guarantee that the new trees will be any freer from variation than the original trees because of soil heterogeneity and other environmental factors. Of far greater importance is the elimination of chimeral and off-type trees through bud selection.

5. Maintenance of valuable varieties should be safeguarded by selection of scions from normal healthy individuals of standard type.

6. Selection of stocks and eventual establishment of superior rootstock varieties are desirable procedures in the production of all important tree fruits and nuts.

7. The practicability of producing new varieties of vegetatively propagated crop plants by clonal selection depends simply upon the frequency of occurrence of desirable bud variants. Extreme variations have been noted in the frequency with which such bud variants have been observed in different species and varieties.

**References**

BAILEY: Plant Breeding, 1st and 2d eds. (Carrière on rose sports).

BENEDICT: Variations among the sporelings of a fertile sport of the Boston fern, *Jour. Heredity,* **15**: 379–391; 421–431.

BIOLETTI: Selection of planting stock for vineyards, *Hilgardia,* **2**: (1) 1–23.

COIT: Citrus Fruits.

CRAMER: Kritische Übersicht der bekannten Fälle von Knospenvariationen.

CUMMINGS: Apple scion selection from high and low yielding trees, *Vt. Sta. Bull.* 221.

DARWIN: Variation in Animals and Plants under Domestication, **2**, Chapter IX.

EAST: Transmission of variations in the potato in asexual reproduction, *Conn. Sta. Rept.,* **1909–1910**: 119–160.

EMERSON: Nature of bud variations, *Amer. Nat.* **56**: 64–79.

HEDRICK: Dwarf apples, *N. Y. Sta. Bull.* 406.

HODGSON: Analyzing the citrus orchard by means of simple tree records, *Calif. Sta. Cir.* 266.

KRANTZ: On the relative value of certain methods of potato breeding, *Proc. Eleventh Ann. Meet. Potato Assoc. Amer.,* **1924**: 40–44.

RAWES: Comparative cropping of apple trees propagated from trees of good history and bad history, *Jour. Roy. Hort. Soc.,* **47**: 163–168.

SAX and GOWEN: Permanence of performance in a clonal variety, *Genetics,* **8**: 179–211.

SHAMEL: Bud selection as applied to the Smooth Cayenne pineapple, *Hawaii Univ. Ext. Serv. Dept.,* **4**: 123–133.

SHAMEL, *et al.:* Citrus-fruit improvement, *U. S. Dept. Agr. Bulls.* 623, 624, 697, 813, 815.

STOUT: Establishment of varieties in Coleus by selection of somatic variations, *Carnegie Inst. Pub.* 218.

WEBBER: Improvement of rootstocks, *Jour. Heredity,* **11**: 291–299.

WHIPPLE: Line selection work with potatoes, *Jour. Agric. Research,* **19**: 543–573.

WHITTEN: Bud selection for increasing yields, *Mo. Sta. Bull.* 131.

# CHAPTER XXXII

## PURE-LINE SELECTION

Pure-line selection, or the isolation of pure lines from a population of sexually reproduced, self-fertilized plants, is distinguished from selection in cross-fertilized plants and from clonal selection in that the individual plant is normally homozygous and, therefore, serves immediately as the starting point of a pure line. The practice of this method of plant breeding antedates the scientific demonstration of the existence of pure lines. It was Louis de Vilmorin who first gave special attention to the value of the progeny test (1856) and, contemporaneously with Hallet, practiced the selection of single plants in wheat, oats, and barley, followed by separate tests of their progeny. The method was first used in America by Willet M. Hays, who began the improvement of small grains at the Minnesota Experiment Station in 1888. Convinced by the results of extensive variety tests that systematic breeding would be required in order to secure a marked increase in yield of first-class wheat, Hays devised the "centgener method" of grain breeding, which begins by planting about 100 seeds from each selected plant in trial plots; the more promising centgeners are then selected for testing on a larger scale. Hays' work resulted in the isolation in 1892 of two plants whose progeny within a decade were grown on thousands of acres. Although many new strains were secured, the rigid tests of several consecutive years, in which the most promising strains were compared with each other and with the best commercial varieties, resulted in securing but few really superior varieties. These made possible an increased production of wheat, however, throughout the northern states and in Canada.

The Swedish Seed Association was organized in 1886 and established an experiment station at Svalöf. During the first 5 or 6 years mass selection only was practiced, but soon after Hjalmar Nilsson became director in 1891 the "Vilmorin method" was introduced. At Svalöf it came to be known as the "system of pedigree" or separate culture. Nilsson was led to adopt this system as the method of originating new varieties by the accidental discovery that the only wheat plots that were entirely uniform were grown from single plant selections. The new varieties produced at Svalöf are now grown throughout the agricultural portion of Sweden. Since the discovery of pure lines by Johannsen (Chapter XX), very extensive use has been made of this method of crop improvement, resulting in the replacement of many of the older varieties of self-fertilized crop plants by new ones which are superior in yield, quality, or disease resistance.

**Self-fertilized Crop Plants.**—Practically all the important agronomic and horticultural species may be roughly classified either as largely self-fertilized or largely cross-fertilized. Those in the first group may again be divided as follows:

*a.* Flowers hermaphrodite, but the floral mechanism such as largely to prevent cross-pollination. Examples: wheat, oats, barley, rice, beans, peas, and many other legumes.

*b.* Flowers hermaphrodite and the floral mechanism permitting cross-fertilization which usually occurs only in a small proportion of the seeds produced. Examples: alfalfa, cotton, tobacco, tomato, flax, timothy, sorghum, and other plants having similar floral structure or mechanism.

The plants listed under (*a*) are so generally self-fertilized that it is usually not necessary to protect them to insure self-fertilization, but there are some species and varieties among them which sometimes exhibit a significant amount of cross-fertilization. The cultivated varieties of wheat are only rarely cross-fertilized, but the wild wheat of Palestine has a floral mechanism especially designed for cross-fertilization. Some varieties of rice, also, are cross-fertilized often enough in mixed plantings to make it impossible to assume self-fertilization in a given selection. In peas and beans the proportion of crossing is probably greater than in the cereals mentioned above, and in some cases it is absolutely necessary to protect them from insects. On the other hand, in extensive investigations with oats few and sometimes no cases of spontaneous hybridization have occurred. Furthermore, in most of the commonly cultivated varieties of wheat, barley, and rice, natural crossing is so rare a phenomenon in temperate and subtropical regions as to be worthy of special note in any observed case, and Johannsen's pure-line investigations with Princess beans would have been impossible had natural crossing occurred among them in any significant amount.

Among plants having hermaphrodite flowers which are usually self-fertilized, there is also vast difference in the relative proportions of self- and cross-fertilization. In cotton Kearney has found that the amount of natural crossing seldom exceeds 20 per cent and is often much lower. In tobacco self-fertilization is the rule, but it is not sufficiently assured to obviate the necessity for protection in gathering pure seed. This applies to the commercial varieties of tobacco but does not indicate the true state of affairs in all species of Nicotiana, for a few species are completely self-sterile. Thus in *N. alata grandiflora* some individuals are entirely self-sterile and others exhibit no bar whatever to self-fertilization. It is especially important, therefore, in dealing with plants in this class to determine these data for the particular species and varieties and the special conditions attending the breeding work. Working with alfalfa and the wild *Medicago falcata*, Waldron found over 10 per cent of natural crossing in the latter and 7.5 per cent in the former. Hayes and Garber suggest that the low habit of *M. falcata* may explain the difference observed and that twice the mean of the two values, or 50 per cent, may indicate the probable amount. If this is correct, these species of alfalfa occupy an intermediate position with reference to the two general categories of self- and cross-fertilized plants. As more data are accumulated, it is probable that species or varieties will be found to form an intergrading series from those exclusively self-fertilized to the obligatory allogamous species, those in which the individual is self-sterile.

**Selection within Pure Lines.**—The ineffectiveness of selection within a pure line has been demonstrated by several investigators working with various self-fertilized crop plants (Johannsen with beans; Fruwirth with lentils, vetches, beans, peas, and mustard; Pearl, Love, and others with oats). This has revolutionized the methods of many breeders and seed growers. It is now generally understood that, even in the absence of natural crossing, the initial selection of individual plants will occasionally result in the isolation of genetically superior lines but that nothing is likely to be gained by continuing such selection within the isolated lines. This simply means that mutations occur generally among plants and that they are rare. In the case of commercial varieties of self-fertilized plants, however, there is often a small amount of intervarietal crossing even in temperate climates; the amount is likely to be larger in subtropical and tropical regions. In case this happens not long before the selection of individual plants from such a population, there will, of course, be evidence of segregation in the first progenies, and further selection of individual plants will be necessary to isolate the desirable strains. The important principle, however, is the value of individual plant selections in the improvement of varieties of self-fertilized plants.

**Self-fertilized Populations.**—Repeated self-fertilization in successive generations normally results in the rapid elimination of heterozygous individuals from the population. The operation of this principle was fully explained in Chapter XX so that here it is necessary only to refer to the consequences and their significance in the improvement of self-fertilized varieties. It has been shown that, starting with a population in which every individual is heterozygous for as many independent pairs of factors as there are chromosome pairs, only a few generations of self-fertilization are required to make the proportion of homozygous individuals in the population approach 100 per cent. At the same time such a population is heterogeneous; *i.e.*, it tends rapidly to become a mixture of different homozygous genotypes, the number of such genotypes depending primarily upon the number of pairs of factors involved and secondarily upon the size of the available population and the certain elimination of some genotypes from various causes. As each homozygous genotype in the available population is a potential pure line, it is only necessary to grow the progeny of individual plants in separate cultures to establish as many pure lines as desired. Subjection of these pure lines to critical comparison and competitive tests is the most important and time-consuming part of pure-line selection.

**Selection within Varieties.**—Numerous cases are on record of the successful improvement of old, established varieties by the method of

FIGURE 127.—Spreading and erect pure lines of Gypsy wheat, 1907. (*From Williams.*)

pure-line selection, and frequently large agricultural areas have thus been provided with new and superior sorts which outyield the older commonly grown varieties. Pure-line selection has been found effective in dealing with morphological and physiological characters including yield. Some commercial varieties of wheat, for example, are mixtures of pure lines which differ strikingly. In figure 127 are shown two pure lines of Gypsy wheat as they appeared in 1907, and in figures 128 and 129 one of these appears as grown in 1909 and 1915, respectively. Differences in yield are sometimes accompanied by definite morphological differences as in the pure lines of Defiance wheat represented in figure 130. Lines 1 and 2 did not have an average of one grain per spikelet while lines 6 and 7 have from four to seven grains per spikelet. But differences in yielding capacity often exist among pure lines which are indistinguishable morphologically. This was successfully demonstrated by Surface and Zinn at the Maine Experiment Station in experiments on oat breeding. Individual oat plants were selected with reference to increase in yield and strength of straw. In 1910 they selected 460 plants from 18 different commercial varieties. Of these only 188 were selected for planting in 1911 and, on the basis of the first year's results, 80 were retained for testing in 1912. Of these pure lines 34 were sufficiently promising to be continued into field tests in 1913; 31 of these were again tested in 1914. These were all discarded except 12 which were tested in quadruplicate plots in 1915. In 1914–1915 the pure-line plots alternated in the

FIGURE 128.—The erect pure line of Gypsy wheat in 1909. (*From Williams.*)

field with plots of the best commercial varieties. After correcting the yield of individual plots for differences in soil fertility (Chapter XXXVII), it was found that the 12 pure lines averaged to yield 80.8 bushels per acre as against 75.2 bushels for the 11 commercial varieties. It was found that these pure lines closely resemble their respective parents in morphological characters. It was concluded, therefore, that mutations in the physiological characters which result in higher yield are not necessarily associated with changes in the morphological characters of the plant or grain.

Correlation between yield and certain morphological characters in small grains has been thoroughly investigated and in no case has it been possible to find any one character which would serve as a constant, reliable index of high yield, unless it be the presence of awns in wheat and barley. This does not mean there is no genetic basis for yield but rather that it is dependent upon so many genetic factors that it is impossible to discover high correlation between yield and one morphological character. This is also true of another highly desirable characteristic of small grains, freedom from lodging. Certain correlations, however, may be of value in selection, such as date of heading and time of maturity. But in breeding for increased yield, it is necessary to make comparative tests for yield of individual plant selections, and as yield is of paramount importance, it is necessary to give special consideration to this method.

FIGURE 129.—The erect pure line of Gypsy wheat (on right) now known as the Gladden variety, as grown in 1915. (*From Williams.*)

**General Features of the Pure-line Method.**—Although the principles involved are simple, the amount of work required in producing actually superior varieties is large. From the time of making the original selections throughout the years needed to establish a new variety, unremitting care is necessary in planning and executing the details of the comparative tests. The experience of the past quarter-century has proved the value of the *plant-to-row method*, the details of which are discussed in several of the references listed at the end of this chapter. In the utilization of this method for breeding the following points are of general importance:

1. At the outset a large number of individual plant selections must be made. This is especially important when selecting for increased

yield and for disease or drouth resistance. Nearly 20 years of experiments on selecting oats, by the Cornell Station and U. S. Department of Agriculture cooperating, show the importance of individual plant selections in increasing yield. By selecting a very large number of individual plants from a variety or a field and testing the progeny of each plant separately for several years, some high-yielding strains may be found. But only rarely is such a strain obtained. Several hundred selections may be made and only a single good strain obtained.

2. Field observations should precede the final selection of the individual plants to be tested. If the variety or field from which the selections

FIGURE 130.—Typical heads from seven pure lines of Defiance wheat.

are to be made is very uniform and the purpose of selection is increased yield, the field observations may be limited to records of stand, growth, and yield per acre. But if the end in view is production of a superior variety in respect to two or more characters, especially if natural crossing is suspected of having occurred, notes taken on the original selections before harvesting will make it possible to check the first generation against their parents and thereby to eliminate those which fail to give sufficient promise at the end of the first year.

3. If severe selection is exercised during the first 2 or 3 years, the future work will be greatly reduced. For the first year's test the amount

of seed will be small and little importance can be attached to differences in yield, but if morphological characters are under observation, or resistance to disease or drouth, it may be possible to eliminate many selections at the end of the first season. In the second and third years of the test, by using a system of replication and comparison with a standard variety, a sufficiently accurate indication of yielding ability is obtained to eliminate many more selections.

4. A very thorough testing of the most promising selections in comparison with the best commercial varieties is essential, and finally there must be cooperative trials on farms of selections that seem to be worthy of introduction as commercial varieties. Long before this stage is reached, however, the assurance of marketability of the proposed new variety must be determined as definitely as possible. But the final and critical test of a new variety is its acceptance by the growers.

**Wide Usefulness of the Pure-line Method.**—From the foregoing illustrations it is evident that selection of pure lines is an effective method of improvement in our most important small grains. The experiments of recent years are continually adding to the list of improved varieties secured in this way. In wheat, for example, such valuable varieties as Kanred, Blackhull, Mindum, Nodak, Fulhis, Ilbred and Progress have been produced in America by this method; in oats Kanota, Albion, Richland, Iowar, Iogren, Markton, Forward, Cornellian, and Washington; in barley, Manchuria, Trebi and Atlas. Seven new varieties of rice have been developed recently at the Louisiana Station by pure-line selection, all of them possessing certain commercial value. Similar results with grains have been obtained in other countries, especially those of the temperate zones. This method has also proved useful in improving various legumes as well as our most important fiber plants. In cotton, for example, the U.S. Department of Agriculture, beginning with the pioneer work of Webber and continued by Cook, Kearney, and others, has developed a succession of valuable new varieties by means of line selection. Similar improvements have been made in cotton by various state experiment stations and in foreign countries. This method of improvement is now being used by the best growers of the New York variety of lettuce, the standard commercial variety in California. Plants of the nightshade family are also adapted to the use of this method. In the tomato especially numerous improved varieties have been secured in this way. In these plants, as in cotton, considerable natural crossing may occur (sometimes as much as 10 per cent in eggplant), and precautions should be taken to isolate stock material. This is often accomplished in commercial seed growing by planting larger plots than are actually needed and taking seed only from the central portion.

**Limitations of Pure-line Selection.**—This method of plant improvement is, of course, limited to the naturally occurring genetic variations within the populations which are available for selection. Interchange of seed between stations nationally and internationally has brought to light many new lines of special value for particular regions. Doubtless much remains to be accomplished in this field. But many of the problems of modern agriculture cannot be solved by line selection alone; it is only by resorting to hybridization that the necessary combinations of qualities can be secured. Another limitation of pure-line selection, or rather a possible danger which may serve as a warning against its too extensive use, has been pointed out by Cook. Some inbred lines of cotton have been found to be adapted to restricted areas, and their transfer to radically different environments has resulted in marked deterioration. Such a situation can be met to some extent by the breeding of numerous locally adapted varieties, but this tends away from uniformity in output of raw material. Another solution will be suggested in the next chapter.

**Summary.**—Improved varieties are produced by the selection of pure lines in self-fertilized crop plants, and in others having a low percentage of natural crossing. The practice of this method antedates scientific discovery. Selection within pure lines is usually ineffective, but self-fertilized commercial varieties which have been long under cultivation are found to be composed of a mixture of pure lines. These differ in yielding ability as well as in morphological features. It is important to make a large number of individual plant selections at the outset and to conduct systematic testing, especially when working for higher yield.

**References**


CONNER and KARPER: Inbreeding grain sorghum, *Jour. Heredity*, **15**: 299–302.

HAYES and BARKER: Self-fertilization in timothy, *Jour. Amer. Soc. Agron.*, **14**: 289–293.

HAYES and GARBER: Breeding Crop Plants, Chapters VIII and IX.

JONES: Genetics, pp. 412–422.

LOVE: Oats for New York, *Cornell Sta. Bull.* 343.

LOVE and CRAIG: Methods used and results obtained in cereal investigations at the Cornell Station, *Jour. Amer. Soc. Agron.*, **10**: 145–157.

LOVE and CRAIG: Results of experiments with oats in New York, *Cornell Sta. Bull.* 436.

SURFACE and BARBER: Studies on oat breeding I, *Maine Sta. Bull.* 229.

SURFACE and PEARL: Studies on oat breeding II, *Maine Sta. Bull.* 235.

SURFACE and ZINN: Studies on oat breeding IV, *Maine Sta. Bull.* 250.

DE VRIES: Plant Breeding, pp. 67–102.

## CHAPTER XXXIII

## SELECTION IN CROSS-FERTILIZED PLANTS

The results of selection in allogamous species depend upon the method of selection and the genotypic nature of the population. It has been shown in Chapter XXI that allogamous species are normally heterozygous genotypes and that repeated mass selection of superior or inferior biotypes tends gradually to raise or lower the population with respect to the character for which selection is practiced. If the character is influenced by a large number of genes, which is usually the case with quantitative characters and such physiological characters as yield and chemical composition, then even though mass selection is continued through many generations, there still remains the possibility of further modification by selection. On the other hand, long-continued mass selection will so alter the genetic composition of the population that its effects will be fairly permanent so long as the selected strain is not permitted to outcross. In naturally cross-pollinated species, however, this condition is seldom met except in carefully controlled operations. Mass selection, as it is usually practiced by the growers of maize and other cross-fertilized plants, must be followed up continuously in order to maintain the improvement gained. Such methods may be utilized to increase the yield and uniformity of inferior or unadapted sorts until they begin to approach the better varieties. After that more efficient methods must be used.

**Ear-to-row Selection in Maize.**—This method of selection, as its name implies, is merely a special form of mass selection. While there is partial control of heredity in that the female parents are selected, yet the fact that the ear is made the unit of selection instead of the individual plant and that these selected ears have been open-pollinated introduces such important elements of uncertainty regarding the genotypic nature of the progeny as to render the method but little better than ordinary mass selection. It has this advantage though, that records may be kept of individual selections from year to year, and from these records something may be learned as to the processes in operation under such a system of selection (figure 131).

Thus in the Illinois corn-selection experiments the original twenty-four ears with which the high oil and protein strains were started, and the original twelve ears on which the two low strains were founded, were given numbers which were retained throughout the course of the experiments, so that the performance and ultimate fate of their progeny is known. This makes it possible to state that the gradual raising and lowering of the percentage of protein or oil in the grains (table 34) was due (1) to the inclusion of some genotypes for higher or lower protein or oil among those chosen at the beginning of the experiment and (2) to the gradual elimination of all but a few lowest or highest genotypes, respectively, so that after ten generations each strain consisted of the progeny of only one, two, or three of the original ears with which the experiment started.

Nevertheless after 25 years of ear-to-row selection the Illinois experiments had not reached a point where there seemed to be no hope of

FIGURE 131.—Result in the fifth generation of ear-to-row selection in maize for high and low position of the first ear on the stalk. The white tape marks the height of the ears in the first row of both plots. (*From L. H. Smith.*)

further differentiation in the strains. These results are just what would be expected in any allogamous species under continuous selection for a given character, provided there is no control over the fertilization of the selected mothers and the character in question is dependent upon several or many genetic factors. The first of these conditions is met by basing selection upon open-pollinated ears in the Illinois experiments; for the second there is abundant evidence. As East and Jones point out, the number of factors affecting protein elaboration by which varieties of maize may differ must be large. There are several kinds of proteins which differ in their chemical composition and they are distributed throughout the various tissues which go to make up the seed. Also it has been shown that numerous physical differences accompany change in protein content. These include different size relations within the seed and relations between such characters as size, number, and shape of seeds. These facts, as East and Jones state, point to a large number of multiple factors affecting protein in maize. Yet the uncontrolled method of these experiments would make it possible to go on for many generations without much progress even though only four or five genetic factors were involved. The fact that with ten or eleven generations all of the original lines were eliminated but one, two, or three and that East and Jones secured an equally high protein strain in 4 years of self-

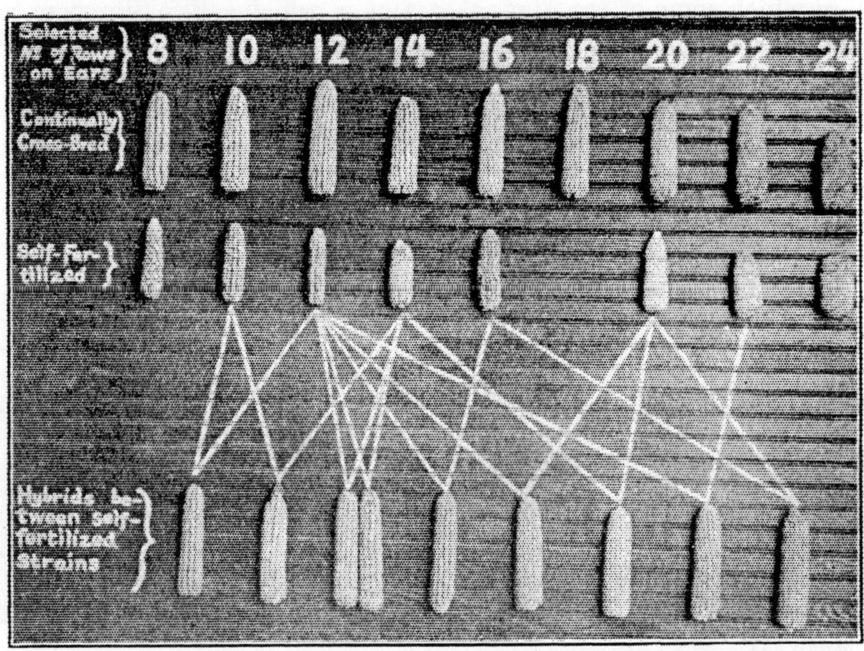


FIGURE 132.—*Upper row*, cross-fertilized ears representing the original strains, selected for number of rows per ear. *Middle row*, ears representing inbred strains after self-fertilization during five generations. In these families the variation was slight, the ears chosen being typical. *Lower row*, $F_1$ pogeny from crossing self-fertilized lines. (*From Geo. H. Shull.*)

fertilization would indicate a relatively small number of major factors. But the number of minor or subsidiary factors is probably very large.

This case is complicated, however, not merely because of multiple factors. There is a difficulty which is inherent in the breeding of all cereals, due to the occurrence of the protein partly in the embryo and partly in the endosperm. Hence, the phenotype of the seed may be influenced by the mother no matter what genetic factors are introduced by the male gametes. This "exaggerated lack of correlation between individuality and performance, between phenotype and genotype" is an important cause of many failures in cereal breeding. Even these complicated difficulties, however, can be practically overcome by the production of artificial pure lines by controlled self-fertilization of individual plants.

**Self-fertilization in Maize.**—The first application of the pure-line idea to a naturally cross-fertilized species was made by Shull and East working independently with maize. By guarding and self-pollinating individual plants and repeating this process among the self-fertilized progeny for successive generations, a number of morphologically distinct strains were isolated, thus proving that the original population was a mixture of biotypes. One of these experiments is illustrated in figure 132, in which some of the original strains in which self-fertilization was prevented are represented in the upper row of ears and the inbred strains derived therefrom in the middle row. The results of crossing inbred strains will be discussed in the next chapter. For the present, attention may be confined to the primary phase of these experiments and similar ones, namely the results of inbreeding in such a species.

FIGURE 133.—Seven successive generations of inbred maize growing at the same time. For seven generations this strain had been self-fertilized and then seed from each generation and from the parent strain was grown as shown. After five generations of inbreeding the strain became uniform. (*From Jones, Courtesy Journal of Heredity.*)

**Effects of Inbreeding.**—In a naturally cross-fertilized species like maize, in which each plant is highly heterozygous, inbreeding the progeny of individual plants produces two effects of great significance in breeding and in crop production, *viz., loss of vigor* and *isolation of biotypes.* There is soon observed a weakening in power of development in plants of such inbred strains, but this continues only to a certain point and is in no sense a genetic degeneration. Along with loss of vigor, there is an isolation of biotypes from the cross-fertilized population. The inbred strains, in the course of a few generations, are found to be distinct from each other either in morphological or physiological characters or in both, and are usually very distinct from the mean of the population from which they came. As these biotypes become more constant in the fifth to the seventh generation when they approximate homozygosity, the loss of vigor ceases to be noticeable (figure 133). That some inbred strains are homozygous for most characters in the

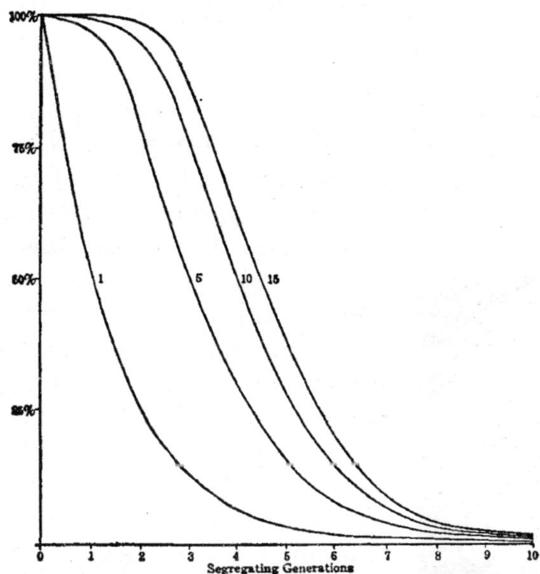


FIGURE 134.—Diagram representing the rate of reduction in percentage of heterozygous individuals in successive generations under self-fertilization when the originally selected individual is heterozygous for one, five, ten, or fifteen independent pairs of factors. (*From Jones.*)

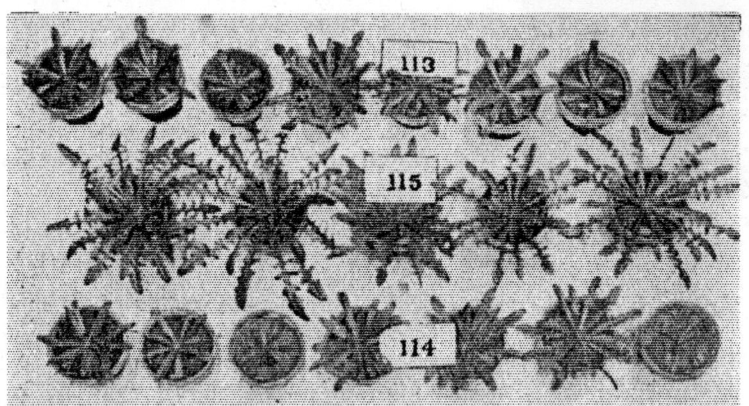


FIGURE 135.—Effects of inbreeding and crossing in the Smooth Hawksbeard (*Crepis capillaris*). Cultures 113 and 114 are two strains in the fourth generation of inbreeding, culture 115 contains $F_1$ hybrid plants of the same age. (*From Collins.*)

fifteenth generation has been demonstrated by critical tests. Jones has shown that the approach to homozygosity would be fairly close in the tenth generation even when the individual selected at the beginning of self-fertilization is heterozygous for fifteen independent pairs of factors.

His diagram (figure 134) represents the rate of reduction in heterozygous individuals under continuous selfing when one, five, ten, or fifteen independent pairs of factors are involved. While the rate of reduction to uniformity is modified somewhat by the occurrence of linkage, the population becomes homozygous just the same with accompanying isolation of lines which differ in morphological characters and in degree of vigor.

The explanation of the loss of vigor in inbred strains of normally cross-fertilized plants was presented in Chapter XX, and it was there suggested that the degree of weakening under inbreeding may be roughly proportional to the degree of heterozygosity of the originally selected individual. Reference was made to the results obtained by J. L. Collins with the Smooth Hawksbeard, a wild species (figure 135). Similar corroborative results were secured by Babcock, working with the hayfield tarweed, another wild species, in which the individual plant is self-sterile so that cross-fertilization is obligatory, a condition resulting in maximum heterozygosity in the individual plant. Abundant evidence was secured that the wild plants of this species are heterozygous for numerous lethal, semilethal, or deleterious factors, for they rapidly came to light in the inbred strains, so rapidly in fact that some strains were completely lost after two generations of brother-sister mating. These detrimental factors affected all parts of the plant—seeds germinated lacking roots, albinic and abnormal seedlings were frequent, there were numerous dwarf segregants, and even the flowers exhibited serious abnormalities which interfered with or prevented reproduction (figure 136). These results are in striking contrast to those reported on hemp by Fruwirth. In this case, although close inbreeding was practiced for 6 years, only two conspicuously abnormal plants appeared. Hemp is a dioecious species but it has been long under cultivation, during which time it has been subjected to artificial selection for desirable characters, which must have brought about a near approach to homozygosity. In general, the amount of natural cross-fertilization within a species determines the degree of loss of vigor under enforced self-fertilization or close inbreeding.

FIGURE 136.—Proliferation of inflorescence in the Hayfield Tarweed, *Hemizonia congesta*, early stage. Only the ray-florets, which never produce pollen, were functional; the inner florets were transformed into vegetative shoots. One of the detrimental characters which appeared in an inbred strain.

because the amount of natural crossing is a primary condition in determining the degree of heterozygosity in the population.

**Cross-fertilized Plants.**—Species of plants which are normally cross-fertilized may be classified roughly as indicated below.

*a.* Flowers hermaphrodite, self-fertile, but with floral devices favorable to cross-fertilization. Examples: cabbage and its relatives, rye, sugar beet, sugar cane (but some varieties of sugar cane are self-sterile).

*b.* Flowers hemaphrodite, but self-fertilization precluded on account of self-sterility of the plants. Examples: radishes, and some other cruciferous species, red clover, sunflower (but self-fertile strains of red clover and common sunflower exist).

*c.* Flowers hermaphrodite (or both hermaphrodite and unisexual on the same plant as in the olive and lemon), adapted to cross-fertilization and variable as to degree of self-fertility. Examples: pome fruits, stone fruits, citrus fruits, olive, avocado, grape (*i.e.*, most varieties of *Vitis vinifera* have perfect flowers but in wild species of Vitis many plants bear only staminate flowers).

*d.* Monoecious plants, self-fertile, but the floral mechanism such as to favor cross-fertilization. Examples: maize, watermelon, squash, pumpkin, cucumber, cantaloupe.

*e.* Dioecious plants. Flowers of different sexes on different plants, thus insuring cross-fertilization. Examples: asparagus, spinach, hemp, hops, and date palm.

A separate class might be recognized for plants having hermaphrodite and unisexual flowers on the same plant and termed polygamous. Other examples besides the olive and lemon are *Lychnis dioica* L., and *Silene nutans* L. The sunflower might be classed here also, because its ray flowers are pistillate only. Certain species of Compositae have the marginal flowers pistillate through suppression of the anthers, while the inner flowers are hermaphrodite, but in these the pistil aborts. Hence, in effect, these are really monoecious plants, but in some cases as in the hayfield tarweed the individual plant is self-sterile, so that in these cross-fertilization is necessary for seed production.

Under (*a*) have been included cabbage and its relatives, although strains occur which are highly self-incompatible, and rye and sugar cane in spite of their variable behavior as reported by different investigators. This is of interest in connection with the next following class which includes plants that are self-sterile. Certain complications arise from contradictory reports as to self-sterility in some species belonging in these two groups. Thus there are reports that flowers on a given plant of rye are sterile with their own pollen but exhibit a certain degree of fertility when pollinated from some other flowers on the same plan In effect such relations give results which are equivalent to self-fertility but in some breeding operations it is important to know the exact relations, in order to take advantage of them. Probably, in general, any difference which may be found in the fertilizing power of pollen taken from different flowers on a given plant is non-essential and dependent upon some such factor as relative maturity of the pollen with respect to the receptive period of the stigma or cyclic sterility, such as that discovered by Stout in certain forms of Brassica. In sugar cane, fertility is very susceptible to environmental conditions, such as accompany differences in latitude and altitude. The occurrence of sterile anthers is known in South Africa, and in northern India sterility of either anthers or pistil or both is common, but when the sexual organs are normal, the plant is self-fertile.

**Improving Varieties by Self-fertilization.**—Some of the crop plants mentioned above, notably maize, have been extensively investigated with reference to the possibility of improvement by the selection of artificial pure lines. Theoretically this should be possible if selection can be conducted on a sufficiently large scale. Even in a highly heterozygous population such as an ordinary field of maize, there must be some genotypes containing all or nearly all of the cumulative multiple factors conditioning rate and vigor of growth. There is a possibility, therefore, that among many self-fertilized lines, one might occasionally be found which would exhibit little or no loss of vigor under continued inbreeding and which would yield more than the average of the parent variety. Thus a very exceptional inbred strain of Sacaton June corn was isolated by G. N. Collins in 1921 (figure 137). As determined from the average length of ears for each progeny row, this strain in 1924 produced the longest thirteen out of the twenty longest ears in the inbred strains. It has been conclusively shown by Richey, however, that the practical value of self-fertilization in maize breeding is to isolate a number of different lines among which selection can be practiced, the basis of selection being *performance of the crossed progeny* of these lines. It is only necessary that the selected lines be sufficiently productive to render their maintenance practicable. This phase of corn breeding will be considered in the following chapter.

On the other hand, plants in which the amount of natural crossing is relatively low display less reduction in vigor as an effect of inbreeding in selected lines. In the tomato, for example, the experiments of Hayes and Jones show that self-fertilization simply isolates genotypic lines which may be better or poorer than the average of the original variety in size of fruit and yield. Similarly, with grain sorghums which normally show about 6 per cent of natural crossing, Conner and Karper have found that inbreeding has not caused either reduction or increase in size of head or productivity except in so far as it has isolated distinct strains which differ from each other and from the parent variety in these characters.

In timothy (*Phleum pratense*) Hayes and Clarke have found that self-fertilization does not lead to as great reduction in vigor as has been observed in maize. The reason for this is rather obscure. Timothy is supposed to be normally cross-fertilized because it is open-pollinated. Extensive breeding programs, involving mass and clonal selection, have been based upon this supposition; but Fruwirth states that while fertiliza-

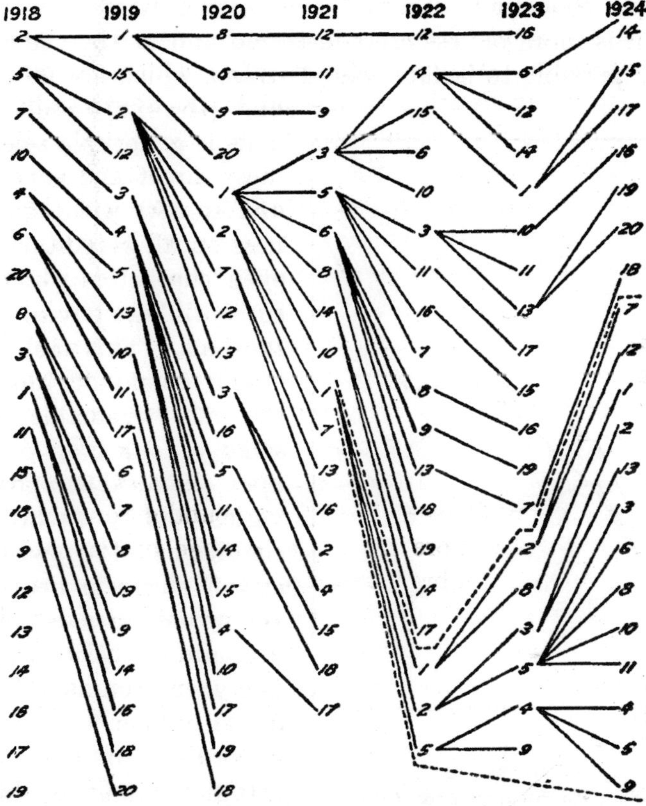


FIGURE 137 —Descent of progenies in self-fertilized strains of Sacaton June maize. Each year the progenies are numbered according to their standing as determined by the average length of ear. The dotted lines include the descendants of progeny No. 1, 1921. *From G. N. Collins.*)

tion from foreign pollen is greatly favored and self-fertilization (of the individual floret) is excluded by the floral mechanism, yet crossing between neighboring plants is rare. From this it appears that the timothy plant is largely self-fertilized. Fruwirth's statement that most timothy plants produce seed scantily is consistent with Frandsen's report that controlled self-fertilization in timothy results in low seed production, which would seem to indicate high self-sterility. But Hayes and Clarke experienced no difficulty in establishing self-fertilized lines from individual plants, although there is marked difference in seed production among the different lines and high correlation between amounts of seed produced in successive years for the several lines. Thus it appears that timothy seed is naturally produced mostly by self-fertilization of the individual plants and that there are important differences among the plants of a field of common timothy in the genes affecting self-fertility, so that improvement in yield of seed (also of hay) can be accomplished by selection and self-fertilization.

**Genotypic Selection.**—From the foregoing it is clear that in any plant which is more or less cross-fertilized and at the same time is self-fertile, it is possible to isolate inbred strains which are morphologically and physiologically distinct. This is genotypic selection of the most direct and expeditious sort, and it acts with equal effectiveness on all hereditary characters, even characters which are highly susceptible to environmental fluctuations. This fact is of considerable economic importance, as the most valuable qualities of some of our crop plants depend upon chemical composition which varies widely according to the conditions of growth. Percentage of sugar in beets and of oil and protein in maize are such qualities and doubtless are affected by every environmental factor having an influence on development. According to East and Jones, external conditions have such a marked effect on the protein content of maize seed that it may be raised or lowered as much as 40 per cent above or below the total per cent produced under average growing conditions. Yet by the simple method of self-fertilizing individual plants and planting seed from those ears which, upon chemical analysis, were found to have the highest protein content, they were able in 4 years to produce a strain yielding over 40 per cent more protein than the average for the variety. Despite seasonal variations this selected strain made a fairly consistent gain in protein content, as is shown graphically in figure 138. Aside from any question as to the agronomic importance of breeding for high protein content in maize, the result secured in this experiment serves as a further demonstration of the great effectiveness of

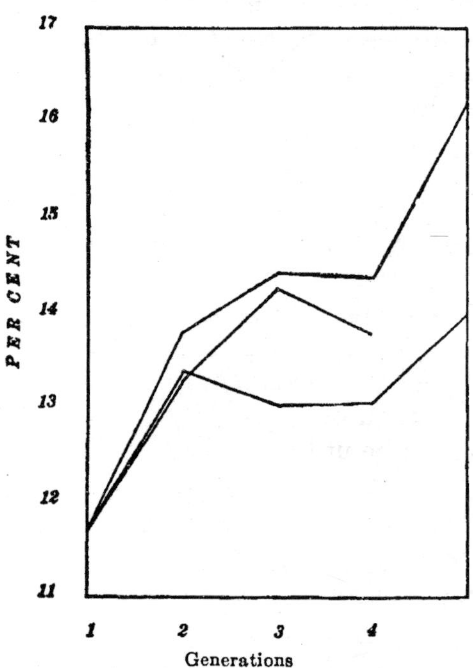


FIGURE 138.—Graphical representation of the results of selecting Burwell's Yellow Flint for high protein in three self-fertilized lines. (*From East and Jones.*)

genotypic selection by means of self-fertilization. There are two reasons for the outstanding value of this method. First, it is the most rapid way of isolating genotypes. Second, it affords the greatest opportunity of finding the best genotypes and of maintaining them once they are found. For these reasons it is the most efficient method for improvement by selection within self-fertile, cross-fertilized species, and similarly the nearest approach to this method which is possible in dioecious plants or in species in which the individual plant is self-sterile is brother-sister mating (*cf.* figure 195.).

**Summary.**—Various modifications of mass selection have been widely used in the improvement of cross-fertilized plants. The ear-to-row method of selecting maize has been proved to be effective but time consuming. Certain improvements of the old ear-to-row method, *e.g.*, Montgomery's method and the remnant system of Williams (see references at end of chapter), are still more efficacious, but they are too complicated for the average farmer who can do little better than practice mass selection in the field, giving first consideration to vigor of the plant and time of maturity. The breeders of maize, rye, cabbage,. and other cross-fertilized plants, on the other hand, must utilize the knowledge which has been gained from genetic investigations on these crop plants.

Genotypic selection as effected by means of pure lines isolated artificially by self-fertilization or, in the case of self-sterile plants, by brother-sister mating, is the most efficient method of improving cross-fertilized plants, whether the isolated lines are to be utilized directly or indirectly by means of intercrossing.

**References**

EAST: The genotype hypothesis and hybridization, *Amer. Nat.*, **45**: 160–174.

EAST and JONES: Genetic studies on the protein content of maize, *Genetics* **5**: 543–610.

——— Inbreeding and Outbreeding, Chapter VI.

HAYES and ALEXANDER: Methods of corn-breeding, *Minn. Sta. Bull.* **210.**

JONES: Genetics, Chapter XII.

JONES and MANGLESDORF: The improvement of naturally cross-pollinated plants by selection in self-fertilized lines: I, The production of inbred strains of corn, *Conn. Sta. Bull.* **266.**

MONTGOMERY: Experiments with corn, *Neb. Sta. Bull.* **112.**

RICHEY: Effects of selection on the yield of a cross between varieties of corn, *U. S. Dept. Agr. Dept. Bull.* **1209.**

——— The productiveness of successive generations of self-fertilized lines of corn and crosses between them, *U. S. Dept. Agr. Dept. Bull.* **1354.**

SHULL: The genotypes of maize, *Amer. Nat.*, **45**: 234–252.

WILLIAMS: Corn breeding and registration, *Ohio Sta. Cir.* **66.**

## CHAPTER XXXIV

## THE UTILIZATION OF HYBRID VIGOR

Hybrid vigor is a phenomenon of common occurrence among sexually reproduced organisms. It was one of the first experimental results to be observed by the early plant hybridists. Nevertheless the possibility of its wide utilization in agriculture has not been appreciated until comparatively recent years. Of course it has played a definite role in the improvement of fruits and flowers which can be propagated vegetatively; but the idea of taking advantage of this principle, that hybrids, especially $F_1$ hybrids, are frequently more vigorous and productive than either parent, by applying it in crop production is only now beginning to influence agricultural practice. Its utilization in horticulture has been largely a matter of chance, and frequently it has been used to advantage in spite of ignorance or misconception as to the cause of increased size and vigor in hybrids. In addition to size, vigor, and productiveness, other qualities which may be enhanced in $F_1$ hybrids are rate of growth, winter hardiness, drouth resistance and disease resistance. It is the purpose of this chapter to call attention to the generally unappreciated possibilities in the application of this principle in agriculture and forestry.

The theoretical explanation of increased vigor in first-generation hybrids within a single species is based upon the fundamental principle of multiple factors, as was shown in Chapter XIX. It has been clearly demonstrated that in various sexually reproduced species there are many different genes affecting both quantitative and qualitative characters. If diverse varieties or inbred strains of a species contain different dominant factors which condition size and vigor, and if these factors react cumulatively, then $F_1$ hybrids between such strains or varieties should be larger, more vigorous, or more productive than either parent. Inasmuch as these qualities are frequently displayed by such $F_1$ hybrids, it may be logically inferred that the parental strains or varieties do actually possess different dominant cumulative factors for growth. Furthermore, the generally increased size and vigor of polyploid forms is wholly consistent with the view that hybrid vigor within a species is caused by increasing the number of cumulative, dominant factors conditioning growth.

The fact that $F_1$ hybrids between species and genera may also display this phenomenon is not necessarily additional evidence in support of the foregoing explanation of intervarietal hybrid vigor. In hybrids between species or genera, development is conditioned by the interrelations of diverse chromosome groups, and such hybrids are characteristically more or less self-sterile. It is probable that, when such hybrids display greater vigor than the parents, this is due to a fortuitous relation between the two reaction systems rather than the cumulative effects of certain factors. This question is complicated, however, by the present diversity of practice among taxonomists in the recognition of species. From the cytogenetic viewpoint, many so-called species are merely subspecies or varieties of a single species because they possess a common chromosome reaction system and differ only in respect to certain genes. Unusual vigor in $F_1$ hybrids between such forms must depend upon the cumulative effect of multiple-growth factors reacting in a single reaction system. The whole problem of factorial homology between species is invol ed in this question.

The principle of multiple factors, combined with the principle of linkage, also accounts for the fact that the $F_2$ generation from an intraspecific cross which produced hybrid vigor in $F_1$ is less vigorous and productive than the $F_1$ (figures 132, 139, and 140). The multiple factors involved may be assumed to be distributed among all the chromosomes, and as a result of segregation, crossing-over, and assortment, all possible recombinations of these factors must take place with random mating.

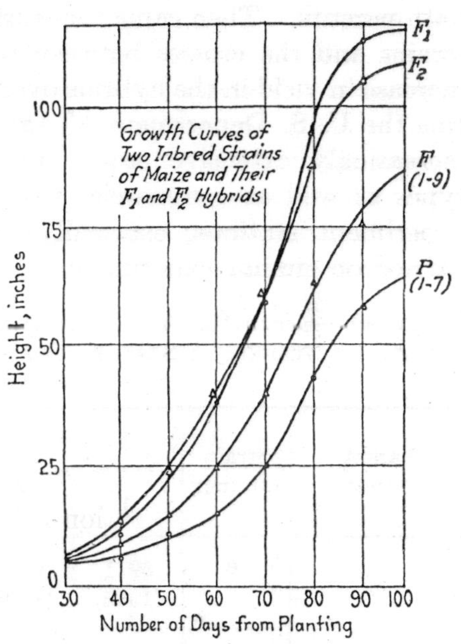


FIGURE 139.—Graphs showing height and rate of growth of two inbred strains of maize and their first- and second-generation hybrids. (*From East and Jones, Inbreeding and Outbreeding, copyright 1919 by J. B. Lippincott Company. Reprinted by permission.*)

Consequently, only a small fraction of the $F_2$ progeny possesses the same full complement of favorable growth factors that were present in the $F_1$. The average growth of the $F_2$ population exhibits considerably less increase over the parental averages than occurred in the $F_1$ population, and later generations generally display still less of the beneficial effect of the original cross. This fact is of the utmost importance in its bearing upon the utilization of hybrid vigor in agriculture. The procedure required for utilization of the increased productivity resulting from hybrid vigor in crop plants depends upon the method of propagation. If the plants are sexually reproduced, obviously maximum utilization is possible only by the use of $F_1$ hybrid seed for each crop. With vegetatively propagated plants, however, a single superior individual is potentially a new clonal variety, which is exceedingly fortunate, as most vegetatively propagated varieties are themselves heterozygous for many factors, so that relatively few desirable recombinations can be expected to occur among their hybrid progeny.

**Hybrid Vigor and Maize Breeding.**—The earliest experiments on increasing production in maize by crossing were made by Beal (Michigan, 1878–1882), Ingersoll (Indiana, 1881), Sanborn (Maine, 1889), and Morrow and Gardner (Illinois, 1892). All of these crosses were made between commercial varieties, and in each case the hybrids outyielded one or both parents. Then came the work of Shull and East (1908) with inbred strains and the crosses between them, both investigators obtaining an increase in yield in the hybrids over that of the original stock. Following this the U. S. Department of Agriculture conducted experiments on an increasingly extensive scale, including work with the most distinct types as well as commercial varieties and inbred strains; and various experiment stations, especially in Connecticut and Minnesota, have carried on similar experiments.

TABLE 49.—EFFECT OF SELF-FERTILIZATION IN STRAINS OF LEAMING DENT MAIZE. YIELD IN BUSHELS OF SHELLED CORN PER ACRE AND YEARS IN WHICH GROWN
(*After East and Hayes*)

| Parent variety | Strain number | Generations | | | | | |
|---|---|---|---|---|---|---|---|
| | | 1 | 2 | 3 | 4 | 5 | 6 |
| 88.0 (1905) | 6 | 59.1 (1906) | 95.2 (1908) | 57.9 (1909) | 80.0 (1910) | 27.7 (1911) | |
| | 7 | 60.9 (1906) | 59.3 (1907) | 46.0 (1908) | 63.2 (1910) | 25.4 (1911) | |
| | | | | 59.7 (1909) | 68.1 (1910) | 41.3 (1911) | |
| | 9 | 42.3 (1906) | 51.7 (1908) | 35.4 (1909) | 47.7 (1910) | 26.0 (1911) | |
| | 12 | 38.1 (1906) | 32.8 (1907) | 46.2 (1908) | 23.3 (1909) | 16.5 (1910) | 2.0 (1911) |
| | | | | | 28.7 (1909) | 9.5 (1910) | 2.0 (1911) |

*Crossing self-fertilized strains* produces the most striking results because the degree of increase in vigor in $F_1$ over the parental strains is enormous (as much as 250 per cent over the average of the parents). Of course it is much greater in some cases than in others, because some inbred strains possess more of the favorable growth factors than do others. East and Hayes, working with inbred strains of four different varieties, secured an average increase of 73 per cent in all $F_1$ hybrids. The data on breeding the Leaming dent variety are summarized in table 49. This table is instructive, as it shows that the effects of self-fertilization may be temporarily masked by the effect of extreme environmental variations. It will be noted that in two of the strains the second generation was not grown until 1908, in which year the general environmental conditions were much above normal, and that strains 6 and 9 yielded more than in the first generation, while strain 6 outyielded the parent variety. It also happened that in 1909 and 1911 the combination of poor soil and badly distributed rainfall made the yields too low. Even so, the strains became more and more differentiated in yielding ability as inbreeding progressed; for example, compare strains 6 and 12 in 1910. $F_1$ hybrids between these strains were earlier and taller and contained a greater total amount of dry matter per plant. Thus in 1911 the average height of all the strains of inbred Leaming dent was 84 inches, while the average height of the sixteen hybrid combinations was 111 inches. It seems that the combinations into which strain 7 was introduced were the best, while those in which the poorest strain were used were poorest. The $F_2$ generations from a number of the crosses were grown, and in every case there was a decided falling off in production. Figure 140 shows size of ears and comparative yields in strain 9 after four generations of self-fertilization, strain 12 in the fifth inbred generation, and $F_1$ and $F_2$ hybrids, all grown in 1910.

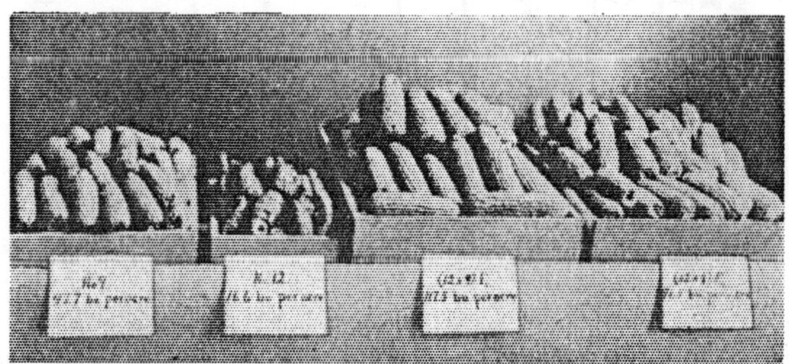

FIGURE 140.—Self-fertilized strains of Leaming dent maize compared with $F_1$ and $F_2$ hybrids. (*From East and Hayes.*)

These features of self-fertilized strains and of their $F_1$ hybrids have been in the main repeatedly demonstrated by various workers with different varieties of maize. Richey, for example, starting with foundation stock consisting of an $F_2$ population from a cross between two varieties Whatley × St. Charles White) carried out comparative tests of carefully selected inbred strains and the parent varieties over a period of 6 years and at the end of 4 years made seventy different crosses between the strains and included the $F_1$ hybrids in the yield tests. The yield of the seventy hybrid lots ranged from considerably less to considerably more than the yield of the original stock, and the average yield of three of the crosses was 30 per cent more than the foundation stock. The average yield of these three best crosses was also significantly higher than that of $F_1$ hybrids of Whatley × St. Charles White grown the same year in the test plots. At the same time it should be noted that some at least of the inbred strains used in these crosses were not at all promising in themselves as to yield, and one at least (10–3) had been marked for elimination before the results of crossing were known. The relative

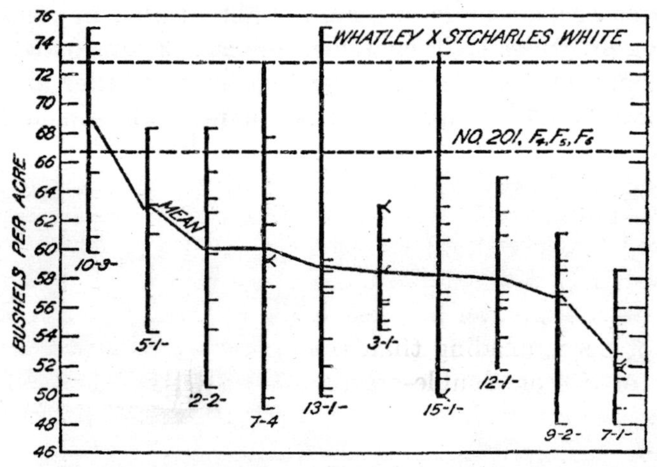


FIGURE 141.—Diagram showing the relative yields of Whatley × St. Charles White corn, of the average of generations $F_4$–$F_6$ from the same cross, and of forty-one crosses between self-fertilized lines grouped to show the productiveness by families. (*From Richey.*)

yields of forty-one crosses between self-fertilized lines are shown in comparison with the relative yields of the $F_1$ and some later generations in figure 141. It will be noted that six of the crosses gave higher relative yields than the $F_1$ varietal cross and that these six crosses were in the three families of which one parent was 10–3, 13–1, and 15–1. From Richey's data it is evident "that there is little or no relation between the productiveness of the self-fertilized lines and that of their crosses;" also "that the final value of the lines for crossing must be determined by comparison of the productiveness of their crosses." The same criterion is insisted upon by Hayes and others. This increases the number of detailed operations necessary for the final selection of the best inbred strains from a variety of maize for use in the production of $F_1$ crossed corn over what would be required if the performance of the strains themselves could be used as an index of their values.

*Double-crossed corn*, produced by the crossing of two $F_1$ hybrids between different inbred strains has been demonstrated by Jones to have certain advantages over simple $F_1$ crosses (figure 142). Hybrid vigor is kept at the maximum and productivity may even be increased. There is, according to Jones, less uniformity than in single crosses, but the greater variability may be an advantage in favoring adaptiveness to different seasons and soils. On the other hand, Kiesselbach has shown that from the standpoint of water requirement ratios there is no advantage in having more than two selfed lines combined in an $F_1$ hybrid; and that, with a number of selfed lines for producing hybrids, greater variability may be expected among the single crosses produced than among crosses involving four or more lines. His investigations included one double-double cross involving eight selfed lines; its high efficiency in the use of water indicates that efficient synthetic varieties may be produced. This is perhaps of greater promise as a method of corn breeding than the utilization of $F_1$ or double-crossed seed. The amount of detailed work essential for utilization of double-crossed corn is rather large, even after the preliminary work of discovering and proving the best strains for crossing has been accomplished. It is necessary to maintain the four inbred strains, which calls for at least two isolated plots each year, and the production of the double-crossed seed calls for another isolated plot; otherwise hand pollinating must be done. Inasmuch as the breeders of commercial seed corn have been very reluctant to depart from the familiar methods of mass selection and adopt the radically different system necessary for the production of $F_1$ seed from crosses of varieties or strains, it is very doubtful whether double-crossed maize will soon be widely used. Certainly it can hardly be expected until publicly supported agencies have accomplished the time-consuming and critical selection which is necessary to obtain the strains which will produce the highest-yielding crossed seed.

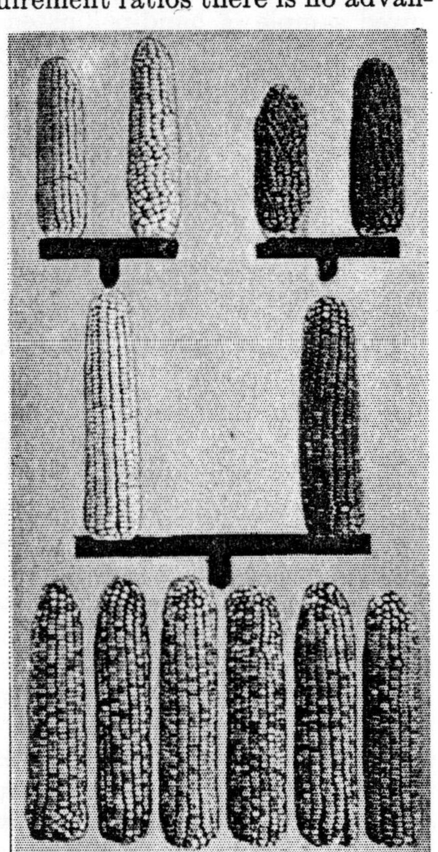

FIGURE 142.—Representation of actual field results of bringing together four inbred strains by three crossings to produce double-crossed corn. (*From Jones.*)

*Improvement of a variety through elimination of inferior strains* by means of self-fertilization followed after 3 years by intercrossing of the surviving strains was attempted by Collins, working with Sacaton June corn. This method was compared throughout with the ear-to-row method of selecting open-pollinated corn. Both methods gave substantial increases over the original variety, but so far as this one experiment is concerned, the first method appears to be of doubtful value. A comparison of the results of the two methods was made by planting alternate rows to a mixed sample of hand-pollinated seed from the ear-to-row experiments and a mixture of $F_2$ progenies from crosses between selfed progenies. The former outyielded the latter by about 25 per cent. A more comprehensive test was made involving the continuously crossbred seed from the ear-to-row experiment, crosses between the selfed lines representing the first, second, and fourth generations, and the original variety. The highest yield was from the first-generation cross of selfed lines. The superiority of the first-generation cross over the fourth and the original variety amounted to 40 to 50 per cent.

Although the highest yields were obtained from the selfed experiment, a single, high-performing selfed line (*cf.* figure 137) played so important a part in the final yields that it seems unsafe to consider the results as sufficiently typical of the method to warrant recommendations.

The failure to obtain significant increases by the elimination through selfing of the poorest-yielding strains, together with the numerous instances in which individual crosses of selfed lines have shown exceptional vigor, suggests the great importance of dominant variations. Against this view is the failure of high-performing selfed lines to show themselves uniformly superior as the parents of crosses. This failure may be due to the fact that in most of the reported experiments the parents of the crosses were closely related. There would be the liability for a favorable dominant variation to carry with it deleterious recessive characters. Or, to put it another way, the deleterious recessive characters would survive when offset by some accompanying favorable character.

If the parents of a cross are unrelated, most of the deleterious recessive characters will be kept out of expression. We may gain the advantage of the favorable factors and suppress the accompanying recessive characters by crossing unrelated high-yielding selfed strains. But there is great need for more information regarding the value as parents of high- and low-performing selfed lines in intervarietal and intravarietal crosses.

*Crossing varieties and subspecies of maize* without previous selection is a practicable way of utilizing hybrid vigor for increased production. Many experiments have demonstrated this fact, the practical utilization of which depends largely upon local efforts to find the most satisfactory crosses. Obviously, there is a great advantage in using varieties which are themselves well adapted to the region. Thus, in Connecticut. Hayes found that the $F_1$ between two early varieties, Hall's Tyler dent and Brewer's flint, proved the most productive of all varieties and crosses in one test (figure 143), while the $F_1$ of an early flint, Longfellow, and a late dent, Minnesota 13, appears to be of promise for northern Minnesota.

It has been shown by Collins that $F_1$ hybrids between subspecies of maize from widely separated localities display numerous valuable qualities. The classification indicated by Collins' descriptions are as follows: *Zea mays indentata* (starch or dent varieties)—Maryland, Kansas dent, Brownsville, Chihuahua, Mexican, Tuscarora; *Zea mays everta* (pop)—Cinquantino, Algerian, Tom Thumb; *Zea mays indurata* (flint) —Guatemala red, Salvador Black; *Zea hirta* Bonafous—Hairy Mexican, Huamamantla, Arribeño; unclassified—Hopi, Chinese (waxy endosperm), Quezaltenango Black, Quarentano. The yields of the sixteen crosses and of their parents are given in table 50.

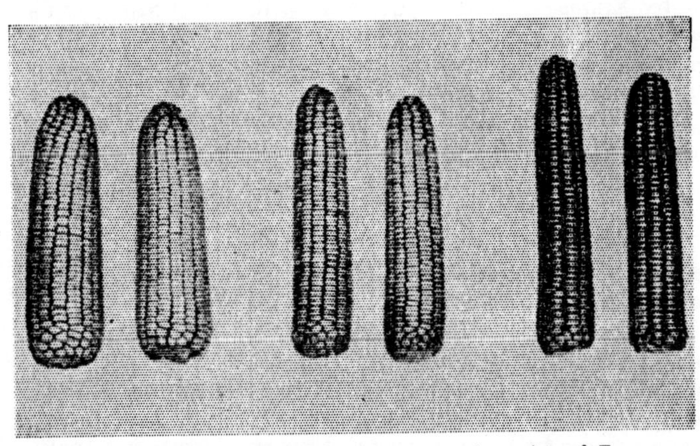

FIGURE 143.—Parents and $F_1$ hybrid between two subspecies of *Zea mays:* Hall's Tyler dent (left), Brewer's flint (right) and hybrid (center). The hybrid yielded nine per cent more shelled corn than the flint and proved the most productive of all varieties and crosses in the 1913 test. (*From Hayes.*)

The superior qualities of first-generation hybrids in maize as set forth by Collins may be summarized as follows:

1. Increased yield.
2. Uniformity equal to that of the parents.
3. Quality intermediate between parents (but Hayes' data indicate complete dominance of low protein over high protein).
4. Increased immunity from disease.
5. Extension of the industry into new territory. Especially strong evidence for this is found in several of the crosses between diverse types. "Amost without regard to the nature of the parents the hybrids remained dark green and vigorous when nearly all the pure strains were giving evidence of the lack of moisture by their curved leaves and yellow color."
6. Less localization of highly bred strains. The importance of local adjustment in highly bred strains is the chief reason for the disappoint-

Table 50.—Yields of Sixteen Maize Crosses Compared with Parental Yields
(After Collins)

| Cross | Yield of female parent, pounds | Yield of male parent, pound | Average yield of parents, pound | Yield of hybrid, pounds | Percentage of increase of hybrid over average of parents, per cent |
|---|---|---|---|---|---|
| $Ah_3$, Maryland dent by Hopi | 1.19 | 0.74 | 0.965 | 1.25 | 29 |
| $Ah_4$, Tuscarora by Cinquantino | 0.53 | 0.24 | 0.385 | 0.75 | 95 |
| $Dh_1$, Kansas dent by Chinese | 0.99 | 0.39 | 0.690 | 1.09 | 58 |
| $Dh_2$, Chinese by Chihuahua | 0.39 | 0.69 | 0.540 | 0.95 | 76 |
| $Dh_3$, Hopi by Chinese | 0.74 | 0.39 | 0.565 | 1.28 | 126 |
| $Dh_4$, Chinese by Xupha | 0.39 | 0.63 | 0.510 | 0.54 | 6 |
| $Dh_6$, Brownville by Chinese | 0.77 | 0.39 | 0.580 | 1.16 | 100 |
| $Eh_1$, Hopi by Algerian pop | 0.74 | 0.34 | 0.540 | 0.91 | 69 |
| $Gh_2$, Tom Thumb by Quezaltenango black | 0.10 | 0.10 | 0.100 | 0.42 | (*a*) |
| $Kh_{31}$, Brownsville by Guatemala red | 0.77 | 0.31 | 0.540 | 0.49 | −9 |
| $Kh_{62}$, Guatemala red by Salvador black | 0.31 | 0.27 | 0.290 | 0.33 | 14 |
| $Mh_{13}$, Guarentano by Brownsville | 0.27 | 0.77 | 0.520 | 0.48 | −8 |
| $Mh_{15}$, Huamamantla by Hairy Mexican | 0.40 | 0.18 | 0.290 | 0.31 | 7 |
| $Mh_{16}$, Arribeño by Hairy Mexican | 0.39 | 0.18 | 0.285 | 0.47 | 65 |
| $Mh_{17}$, Hairy Mexican by Chinese | 0.18 | 0.39 | 0.285 | 0.61 | 114 |
| $Mh_{25}$, Mexican dent by Tom Thumb | 0.52 | 0.10 | 0.310 | 0.54 | (*a*) |
| Average percentage of increase of hybrids over average parents | .... | .... | ..... | .... | 53 |

(*a*) Where the yield of either parent fell as low as 0.10 pound per plant, the percentage of increase of the hybrid is omitted. In dealing with these small quantities, it is believed that percentages would be misleading.

ment which sweet-corn growers experience when they purchase carefully selected strains from other localities. "First-generation hybrids are to a great extent independent of this delicate adjustment to local conditions."

7. Increased utilization of the work of experienced breeders.

8. Stimulus to the work of improvement through the possibility of protecting new productions.

**The Use of F₁ Hybrids in Other Annual Crops.**—The utilization of $F_1$ hybrids between inbred strains in annual crops is limited to maize and other cross-fertilized species such as the cucurbits, especially melons and cucumbers. It has been suggested by Krantz, however, that varieties of potatoes superior to the present commercial sorts may be produced by making $F_1$ crosses of inbred strains. It has been found in Minnesota that continued self-fertilization is possible and that both intra- and intervarietal crosses display hybrid vigor to a marked degree. In tomatoes, crosses between distinct varieties or strains frequently display hybrid vigor and intermediacy in size and season of ripening. Tomatoes are largely self-fertilized and, once having found the varieties or strains which produce an $F_1$ best adapted for a given market, they can easily be maintained and the crosses repeated from time to time. Tomato seed is sufficiently long lived so that the crosses need not be repeated oftener than once in 2 or 3 years. In Germany an $F_1$ tomato hybrid which yields 30 per cent more than the parental average has been recommended for commercial growing. It is estimated that two workers can produce 3.5 kilos of crossed seed in 30 days. While this practice may always be limited to the production of special-purpose crops, it is possible that it may eventually prove useful in meeting the requirements of uniformity of product and disease resistance in the extensive production of tomatoes and other crops of sufficient economic importance to justify the special methods of seed production which are necessary.

**The Utilization of Hybrid Vigor in Forestry.**—Mankind is slowly awakening to the importance of forest conservation. Extensive programs of investigation of the fundamental problems underlying the maintenance of forests are being considered by research foundations. Meanwhile, very little is being done in the way of utilizing such clearly established principles as hybrid vigor. That this principle operates in trees just as in other plants is shown by numerous instances. Henry mentions the following valuable trees which, on account of their vigor, botanical characters, and non-occurrence in the wild state, are presumably first-generation hybrids; Black Italian poplar, London plane, Huntingdon elm, Cricket-bat willow, and the common lime (*Tilia vulgaris*). According to Henry, the pioneer work on hybridization of trees was done by Klotzsch at Berlin in 1845. He crossed two species each of pine, oak, elm, and alder. He "claimed that by hybridization, both the rapidity of growth and the durability of timber of forest trees could be augmented considerably; but no further experiments were made, and his pioneer work fell into oblivion." The art of breeding trees was renewed by the reports of Burbank's work with the walnuts about 1890. Henry reports successful results with $F_1$ hybrids in Populus, Fraxinus, Alnus, Ulmus, and Larix. He points out that one of his most vigorous hybrids, *Populus generosa*, was "derived from two parents so little related thay they are placed in two distinct sections of the genus." At the same time, "a cross between two races of the common alder shows considerable vigor, though the parents are so closely allied that they can be distinguished by only the most trivial characters." Thus it appears that prediction as to the outcome of species crosses in trees is quite as impossible as in other classes of plants. There is great need for further experimentation. In planting wind-pollinated species, provision can easily be made for natural hybridization by planting isolated groups of different species, which renders practicable the production of $F_1$ seed on a large scale. As $F_1$ hybrids usually display greater vigor at an early stage, the hybrid seedlings can be selected directly from the seed beds. It has been found that the quality of the timber in rapid-growing $F_1$ hybrids is equal or superior to that of the parents.

Increased resistance of $F_1$ hybrid plants to insect pests and diseases is doubtless often merely another manifestation of their increased vigor. It is possible that the most serious diseases of wind-pollinated timber trees can be successfully combated by the use of $F_1$ hybrids between species one of which is immune to the disease. But in this connection it is to be remembered that disease resistance is generally a heritable character, so that in a particular instance its appearance in $F_1$ will depend on the factorial composition of the parents and the relation of the factors in inheritance.

**Summary.**—Increased vigor, growth, and productivity in $F_1$ hybrids within a species result from the combination of cumulative dominant-growth factors. In later generations relatively few individuals possess the same maximum number of such factors as were present in $F_1$. Therefore, the fullest utilization of this principle in the production of sexually reproduced crops and forest trees requires that each crop be grown from $F_1$ seed. This is practicable on a large scale only in the case of wind-pollinated species.

The application of this principle in corn growing awaits the discovery of those particular inbred strains or stabilized varieties which, when crossed, will produce $F_1$ progeny definitely superior in yield to standard varieties and equal to them in other respects. The discovery of such strains or varieties, on account of the labor and time required, is being left to the experiment stations, and many promising strains are already available. It is not improbable that future economic conditions will sometime compel commercial seed growers to produce $F_1$ hybrid seed for general use. This may await the production of synthetic "superlines" which combine disease resistance with other superior qualities. Eventually, however, it may be found practicable to breed synthetic super-varieties which can be used in crop husbandry without the necessity of growing $F_1$ hybrid seed (see next chapter).

Meanwhile, some of the experiment stations are determining those varietal crosses of maize which can be used to advantage by individual growers in certain regions. The utilization of this principle of hybrid vigor in forestry deserves immediate attention. Hybrid vigor in asexually propagated plants is of importance in the breeding of new varieties.

## References


BRUNSON: Relation of inheritance studies to corn improvement, *Jour. Amer. Soc. Agron.*, **18**: 308–313.

COLLINS, G. N.: The value of first-generation hybrids in corn, *U. S. Dept. Agr., Bur. Plant Ind. Bull.* 191.

——— A more accurate method of comparing first-generation maize hybrids with their parents, *Jour. Agric. Research*, **3**: 85–91.

——— A comparison of maize-breeding methods, *U. S. Dept. Agr. Dept. Bull.* 1396.

EAST and HAYES: Heterozygosis in evolution and in plant breeding, *U. S. Dept. Agr., Bur. Plant Ind. Bull.* 243.

EAST and JONES: Inbreeding and Outbreeding.

HAYES: Present-day problems of corn breeding, *Jour. Amer. Soc. Agron.*, **18**: 344–363.

——— Methods of corn breeding, *Minn. Sta. Bull.* 210.

HAYES and JONES: First-generation crosses in cucumbers; effects of cross- and self-fertilization in tomatoes, *Conn. Sta. Rept.*, 1916.

HENRY: The artificial production of vigorous trees, *Jour. Dept. Agr. Tech. Instr. Ireland*, **15**: 1.

JONES and MANGLESDORF: Crossed corn, *Conn. Sta. Bull.* 273.

KIESSELBACH: Comparative water economy of selfed lines of corn and their hybrids, *Jour. Amer. Soc. Agron.*, **18**: 335–343.

RICHEY: Effects of selection on the yield of a cross between varieties of corn, *U. S. Dept. Agr. Dept. Bull.* 1209.

——— The productiveness of successive generations of self-fertilized lines of corn and of crosses between them, *U. S. Dept. Agr. Dept. Bull.* 1354.

WELLINGTON: Influence of crossing in increasing yields of the tomato, *N. Y. Geneva Sta. Bull.* 346.

## CHAPTER XXXV

## BREEDING NEW VARIETIES BY HYBRIDIZATION

The scientific experiments of the early plant hybridists were briefly reviewed in Chapter XXVIII. Notwithstanding their discoveries of the possibilities in plant hybridization and the early demonstration by Knight and others of the value of varietal crosses in horticulture, hybridization did not reach a prominent place in plant breeding until the middle of the nineteenth century. Even then it was most popular among flower fanciers and fruit specialists. Later it was used to advantage by a number of practical cereal breeders, some of whom learned by experience that crosses must be carried out to the second generation, the variable generation of Farrer, in order to find desired combinations.

On the other hand, biologists who were familiar with the contributions of Verlot, Focke, and especially of Naudin understood that hybrid progeny exhibit segregation in the second generation, giving new combinations of characters; and as Webber points out, the suggestion was even then present in the minds of scientific breeders that segregation takes place in the reduction division. At that time breeders planned experiments in hybridizing different varieties and species to obtain recombinations of desired characters. For example, the experiments in citrus hybridization conducted by Webber and Swingle were planned in 1893 entirely on this basis. Yet, as stated by Webber and Bessey in 1899, the practical importance of this discovery came to be appreciated by American plant breeders only in the closing years of the century. But it was not until the announcement of Mendel's great contribution that the real possibilities of breeding by hybridization began to be appreciated. Webber writes:

> With a full understanding of the knowledge and practices of the breeders of two decades ago (the 90's), it must be admitted that the conception of unit characters and Mendelian segregation was necessary to clarify this knowledge and bring out the latent possibilities of the material presented by nature for the use of the breeder, and it is doubtful whether we even yet adequately comprehend the almost infinite possibilities open to us.

Among the first to call attention to these possibilities were Bateson, Hurst, Biffen and Punnett in England, and Castle, Emerson, East, and Shull in America. By the end of the first decade, many plant-breeding experiments had been planned along the new lines laid down by Mendelism, and the epoch-making investigations on Drosophila were under way. Since those early years of the present century, there have been numerous important achievements in plant breeding, some of which will be briefly described by way of illustrating certain general principles. Before proceeding to this, however, it is desirable to review the general method of plant hybridization.

**General Method.**—Hybridization, in order to be generally successful, must be intelligently performed, and in the long run the breeder who is most thoroughly familiar with his plants will stand the best chance of securing the improved forms desired. Each species has its own morphological and physiological peculiarities, and the method will need to be modified more or less in every case. In general, however, systematic procedure is essential. Such procedure includes six steps.

1. *Choice of Parents.*—This involves two important matters: first, decision regarding the object to be attained, which implies thorough familiarity with existing conditions affecting crop production; second, comparative study of existing varieties or of species that may yield the desired result.

2. *Culture of Parent Plants.*—Hybridization is painstaking work, and when carried on extensively, it is time consuming and, therefore, expensive. While it is sometimes necessary to use certain plants, especially shrubs and trees, wherever they happen to be growing, yet it is always advisable to concentrate materials so far as possible and to grow them under protection in the breeding garden or greenhouse. Arrangement of the details of culture should include consideration of the optimum conditions for normal fruitfulness of the intended mother plants. These plants in some cases must be kept under observation and prepared for crossing by reducing vegetative growth and restricting blooming and the setting of fruit.

3. *Protection of Pollen.*—Flowers on intended male parents should be guarded in order to prevent contamination with pollen of other plants. A convenient way to insure the use of uncontaminated pollen is to collect nearly mature anthers in a clean, dry vial or place them on a glass plate. In a short time they will burst, and the pollen can be placed in small vials or celluloid capsules for future use. If the atmosphere is very humid or it is desired to preserve the pollen for some time, the open vial or glass plate should be placed in a desiccator.

4. *Castration of Hermaphrodite Flowers.*—This must be accomplished before the anthers dehisce and is usually done shortly before the flower opens, in order to avoid needless mutilation. But in some close-pollinated species, it is necessary to emasculate very young buds. The operation consists of removal of the stamens and can generally be accomplished easily by using a pair of fine-pointed forceps or scissors. The castrated flower is then protected with some sort of covering until ready for pollination. In monoecious plants, it is necessary to guard the young pistillate flowers which are to be pollinated.

5. *Pollination.*—The transfer of pollen from guarded flowers of the male parent to the prepared flowers of the mother plant should be accomplished before or just at the time the stigma becomes receptive. In many species this receptive condition of the stigma is evidenced by the secretion of a viscid fluid on the stigmatic surface. It has been thought that premature pollination wrought disastrous effects on the setting of fruit, but evidence is conflicting on this point, and no arbitrary mode of procedure can be given. Certain it is that in some species, for example, wheat, no untoward results appear from pollination at the time of castration. Plants with small, entomophilous flowers such as clover and alfalfa may be hybridized by enclosing the insects in a cage surrounding the intended mother plant or plants. The same method is applicable to shrubs and trees.

6. *Protection of Pollinated Flowers and Developing Seed.*—The most commonly used device is the paper bag tied with a string or fastened securely with a copper wire and label on which the necessary data are written. In many cases ordinary manila bags of suitable size are entirely satisfactory. Where wasps give trouble by cutting holes, the use of bags made of ramie fiber will be found more satisfactory, since these bags are made with a glossy surface, but even these will give way, in course of time, under the attack of wasps. Bags made of thin paper which has been treated with oil or paraffine are best for withstanding insect attacks and for use on delicate plants. In some cases it is necessary to puncture such bags with numerous small holes to provide for necessary circulation of air. Many special devices, such as cylindrical muslin bags or glass or celluloid cylinders plugged with cotton and firmly supported, are used upon occasion.

**Some Difficulties Attending Hybridization.** *a. Different Seasons of Maturity.*—This is a common obstacle to the crossing of different forms. When it involves merely growing periods of unequal length, the difficulty can be overcome usually by planting at such times that the various forms will flower simultaneously. When this is not feasible, it becomes necessary to resort to some method of preserving the pollen. It has been found that pollen of certain species will retain vitality for weeks or even months if it is kept very dry. Miss Kellerman reports that the most effective method tried by the Bureau of Plant Industry was as follows: anthers of citrus were placed in dried vacuum glass tubes; *e.g.*, tube filled with anthers 1 to 2 inches, cotton ½ inch, exhausted to about 0.5-millimeter pressure in the presence of sulfuric acid, and the tube then sealed. As far as practicable, the pollen was kept at a temperature of 10°C. A simpler and very useful method is to make a double container by fitting a small vial inside a larger one and partially filling the space between the two with anhydrous calcium chloride, filling in the upper portion with absorbent cotton and tightly corking the larger vial. The anthers or pollen grains are placed in the inner vial after it has been thoroughly sterilized and allowed to dry, and it is then loosely plugged with cotton.

Various other ways of getting around difference in time of sexual maturity may be devised. For example, one of the intended parents may be forced or retarded. In connection with forcing the stimulating effects of various substances may be utilized. The recent discovery by Veret of the stimulating effect of sulfurous acid on sugar cane may prove useful with other plants. Again, by regulating the length of day this difficulty may be overcome as reported by Emerson for teosinte.

b. *Failure of Fertilization.*—This may be due to many causes ranging from simple morphological maladjustments to complex physiological relations amounting to antagonism. Probably a very frequent cause of unsuccessful crosses is failure of viable pollen to germinate. When appropriate tests show that part at least of the pollen grains are viable and failures indicate some obstacle to germination, it will be worth while to try the application of a film of water or weak sugar solution to the surface of the stigma before pollination. By the aid of this simple device, crosses have been secured between certain species of beans which had been repeatedly attempted without success. In this connection it may be well to give a word of warning. While it is always advisable to ascertain what one's predecessors have accomplished or failed to accomplish, the hybridizer should remember that both plants and local conditions are variable, and what may have been impossible at one place may be possible at another. Much perseverance is sometimes necessary. On the other hand, as investigations multiply, more and more detailed information is becoming available on the morphological and physiological peculiarities affecting reproduction in economic plant species, and due account of these should of course be taken.

c. *Susceptibility to Mutilation.*—There is great variability in the response of plants to mutilation. It appears that some are susceptible merely to removal of the anthers from the ends of the filaments. In such cases it is necessary to resort to special methods for protecting the stigma from self-pollination. The details will depend upon the structure of the flower and whether it is protandrous or protogynous.

d. *Non-crossability and Intersterility.*—Interspecific crosses frequently and wider crosses generally result in complete failure. Various factors are responsible for the failure of crosses, some of which have already been noted. When progeny are secured from interspecific and wider crosses, they are usually more or less sterile. From what is known of the relations between partial fertility and chromosome behavior in species hybrids, it is probable that most of these phenomena of non-crossability and intersterility are dependent upon cytological relations, particularly those between the chromosome reaction systems of unlike parents Sterility in intervarietal hybrids may be due to genetic factors such as zygotic and gametic lethals.

*Conditions favorable for hybridization* may be summarized as follows optimum conditions for flowering and fruiting; receptive stigmas; viable pollen; morphological and physiological compatibility between pollen and pistil; resistance of flowers to manipulation; and assuming that fertilization has occurred, sufficient affinity between the maternal and paternal chromosomes to permit embryogeny to proceed and differentiation to take place.

**Application of Genetic Principles.**—The Mendelian principles of segregation and independent assortment, with resulting recombinations of grandparental characters in $F_2$, have been extensively utilized in recent practical breeding work. Throughout all civilized countries, this scientific breeding has been quietly forging ahead until the wealth of significant data so produced can hardly be realized by those unfamiliar with the literature. The publications of agricultural experiment stations contain many interesting contributions on the genetics of crop plants, but the reports of real achievements in the production of new and valuable varieties by hybridization are not so numerous. This is partly because the creation of a really superior agronomic or horticultural variety usually involves many requirements, and partly because such a variety, once it is recognized by growers, is sufficiently in demand without the necessity of publication. Another reason may be the failure of the hybridist to recognize the improved type when it appears. But probably the chief reason is simply that most of the breeding work on crop plants has been by selection, and the synthetic breeding of really superior varieties has been followed thus far to a very limited extent.

Utilization of desirable new mutations by hybridization has already been referred to (Chapter XXII) and illustrated by the history of the sweet pea (Chapter XXIX). Many other garden flowers have been developed to a remarkable degree by this method. The large dahlia and chrysanthemum shows, for example, contain exhibits of several distinct types of flower and usually in a great range of colors. These distinct types are for the most part due to a few basic genetic differences which have occurred spontaneously and been preserved by the fancier. In the gladiolus, on the other hand, although the ruffled edge of the petals seems to have come as a mutation, more has been accomplished by crossing species, and this method may add still greater variety to the cultivated dahlia. Really desirable new mutations are rare in crop plants, and to attempt improvement by merely searching for them would be impracticable; yet it is important to be able to recognize their value and to proceed to their utilization by hybridization.

Earliness of maturity is a quality of great value in many crop plants. This character in oats was investigated by Caporn, who concluded that earliness is possibly a function of three genetic factors. Noll found it possible to produce earlier strains of oats than those now obtainable by selection through crossing certain early varieties. Some of the $F_2$ segregants were much earlier than the earliest of the parent varieties. These selections gave rise to very early strains, which is sufficient to illustrate the value of crossing in oat improvement. Stanton and Coffman state that henceforth less progress than formerly may be expected in improving oats by pure line breeding and that hybridization is being adopted more and more for the development of better oat varieties. They believe this method will reach its greatest development under irrigated conditions in the western United States during the next fifty years. Oat breeding now under way in California seems to indicate the correctness of the above statement although this particular breeding work is not being done with reference to irrigation farming.

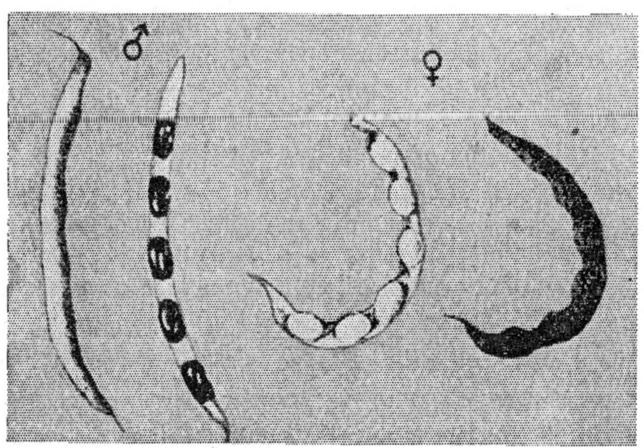


FIGURE 144.—The two varieties of common bean used as parents in producing a new type having long, kidney-shaped, white seeds and straight, "stringless" pods. The ♀ had large, strong vascular bundles and the ♂, small weak ones. (*From Stehlik and Tymich in Mendel Memorial Volume, Prague,* 1925.)

An excellent example of the application of Mendelian analysis in the creation of new varieties of agronomic value is reported by Stehlik and Tymich. It was their aim to produce a variety of common beans having long, kidney-shaped white seeds and straight pods of the "stringless" type, *i.e.*, with small vascular bundles. They had as working material two varieties, shown in figure 144, one with round, white seeds and short, curved pods, the other with long, black seeds and long, straight pods. During their breeding experiments they found evidence that shape of seed was conditioned by seven factors, while shape of pod was affected by two or three factors. Large strong vascular bundles were conditioned by another dominant factor, thus making about ten pairs of allelomorphs to be considered aside from color of seed coat.

Some idea of the amount of work required to establish this new variety may be gained from the fact that this particular genotype could be expected to occur only once in an $F_2$ population of over a million plants on the basis of independent inheritance. The situation is further complicated by linkage (the haploid number is 7) and by the fact that color of the seed coat is conditioned by five factors with complicated interrelations. On the other hand, assuming that natural self-fertilization was the rule, the work of selecting promising phenotypes was greatly simplified, as it was only necessary to test part of the seed from the most promising $F_2$ selections to approximately determine their genotypes. Then the most promising $F_2$ selections could be subjected to careful breeding tests. The desired new variety was actually obtained.

The seed characters of the parent, the $F_1$ and some of the $F_2$ plants, which by the progeny test proved to be mostly homozygous, are shown in figure 145. It will be noted that numbers 13, 14 and 17 approximate the desired seed type. The progeny test revealed that number 13 was of a genotype approximating very closely to the one desired for seed shape and that it was recessive for all the color factors. The combination of this with the desired pod type was only a matter of further crossing and selection.

A good illustration of the practical combination of a knowledge of genetic principles with agronomic requirements is found in the Sacramento barley which was created by W. W. Mackie of the University of California. There was urgent need on the part of farmers for a non-shattering barley because of the necessity of awaiting turns in the use of the combined harvester and the frequent high winds which occur in the great interior valley of California. Coast barley, the variety commonly grown, frequently suffers an annual loss in the Sacramento Valley of about 25 per cent due to shattering, which includes breaking of the stem below the head, breaking of the rachis and loss of part of the head, separation of glumes, loss of part of the grains, etc. Lodging, smut, and scald are also frequent causes of loss in the Coast variety. The first step was the testing of about 1,000 varieties of barley under Sacramento Valley conditions, and about 98 per cent of these were soon eliminated. Of the remaining varieties the best ten or twelve were crossed with Coast barley. Most of the hybrids secured were eliminated for one reason or another in $F_1$ or $F_2$, but one particular cross seemed to give great promise. *viz.*, Cape ♀ × Coast ♂. Cape is a late Australian variety introduced from South Africa. It proved to be a very valuable variety for breeding purposes. The important contrasted characters involved in this cross and the characters of the $F_1$ are indicated below.

| Coast Barley | Cape Barley | $F_1$ Hybrid |
|---|---|---|
| shattering | shattering resistant | shattering |
| lodging | lodging resistant | lodging |
| lax head | compact head | lax head |
| early | late | early |
| scald susceptible | scald resistant | scald susceptible |
| smut susceptible | smut resistant* | smut susceptible |

*Not accurately determined.

From the $F_1$ phenotype, it will be noted that all of the desirable characters except earliness were recessive, which fact greatly facilitated the selection of desirable true-breeding strains, but the original $F_2$ selections were late in maturity. Within 7 years after the original cross, the variety named Sacramento was distributed for trial by farmers.

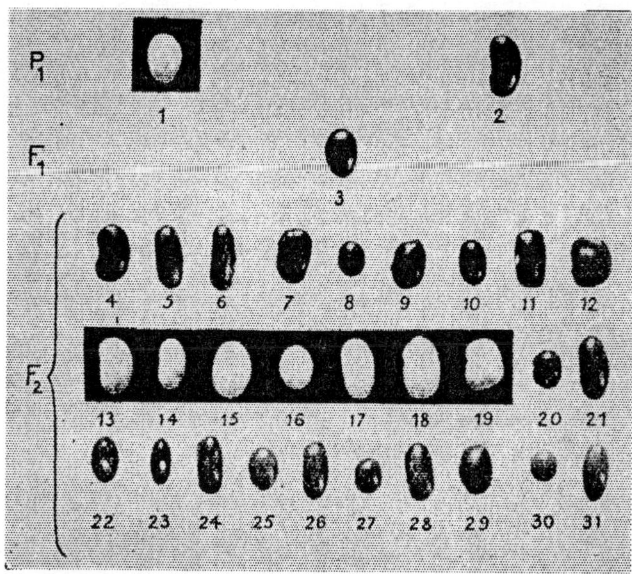


Figure 145.—Seed characters of parents, $F_1$ and some $F_2$ classes obtained from crossing the two varieties shown in figure 144. (*From Stehlik and Tymich in Mendel Memorial Volume, Prague,* 1925.)

As it had not been possible at that time to find a strain which combined earliness with the other five important characters, the Sacramento variety is not adapted to the southern interior region, but other strains are now available which combine these five characters and ripen 3 weeks earlier. However, the Sacramento barley itself is already so popular that all the grain produced thus far has been saved for seed.

Most of the serious agronomic problems which must be solved by breeding are complicated by the various demands of the ultimate consumer, the "market" or the manufacturer, and the grower. Usually complex quantitative or physiological characters are of first importance in meeting these demands, a fact which still further complicates the problem. "Making a plant to order" under such conditions is not so much a matter of simple Mendelian ratios as ability to define the ideal to be reached as nearly as possible in terms of the best varieties available for use as parents, and an understanding of the principles involved so as to be able to plan the work properly. Such, in general, was the method pursued in the creation of the Little Joss and Yeoman wheats by Biffen.

FIGURE 146.—A well-grown plant of Yeoman II, one of the successful new varieties of wheat originated by Biffen through hybridization. (*From Biffen and Engledow.*)

In wheat, as Biffen and Engledow point out, yield and quality are the two features of most concern to the grower, but others, such as time of ripening, the stiffness or standing capacity of the straw, disease resistance, and milling and baking quality, all have to be taken into account. Speaking generally, the commonly grown wheats of England were not particularly defective in any of these respects. On the ground of cropping capacity indeed they were the envy of all the great exporting countries. But in quality they compared very unfavorably with wheats imported into England. The problem then was to improve on forms which already in respect of everything except quality were fairly, if not very, satisfactory. Extensive variety tests of the wheats of the world revealed the fact that none were sufficiently adapted to English conditions to justify their introduction. It therefore became necessary to conduct extensive investigations on the production of new varieties by hybridization.

The forms used in these investigations were Red Fife, as a source of grain quality or "strength;" American Club for resistance to yellow rust (*Puccinia glumarum*); Chinese wheat, for resistance to brown rust (*P. triticina*); Persian Black, a *dicoccum* wheat, for resistance to mildew (*Erysiphe graminis*); and, for cropping capacity combined with good straw characters, such varieties as Square Head's Master, Browick, Iron, and Rivet.

Most of the investigations involving the diverse forms mentioned above are still in progress. The series of Red Fife crosses has produced a number of promising wheats, two of which, resulting from a cross with Browick, have been in cultivation for several years. These are Yeoman, introduced in 1916, and Yeoman II, in 1924. The Yeoman wheat has proved to be of sufficiently high quality that its flour does not require the addition of any imported flour to increase the gluten content, and it is consistently high producing even under adverse conditions. But some of the other forms derived from the same series of crosses have been found to possess a higher total nitrogen content, and one of these, Yeoman II, is superior in baking qualities (figure 146).

> Only the future can show whether Yeoman II will prove as good a wheat under general cultivation as it has been in the trials of the Plant Breeding Institute and elsewhere . . . From analogy with other wheats, especially perhaps the older Yeoman, it is probable, if not certain, that it will not succeed equally well in all districts. Where it is obviously outcropped by other wheats, no matter what their quality, some better form will have to be raised to substitute for it. Work in this direction is already at an advanced stage, and $F_3$ and $F_4$ generations from Yeoman II crossed with most of the best English wheats are now under trial.

In deciduous fruits and nuts, hybridization is the most efficient method of breeding. Clonal selection is of little value in general, and the opportunities of improvement by selection of promising chance seedlings are no longer worthy of consideration. Carefully planned varietal crosses, however, do present definite possibilities of producing really superior varieties. This has been demonstrated by a number of successful hybridizers. Apple breeding, for example, by crossing varieties was begun by Knight, but this method has not been used extensively in Europe. In America the cross-breeding of apple varieties probably was begun by Charles Arnold of Ontario, Canada, about the middle of the last century. Other early hybridizers of the common apple were F. P. Sharp of New Brunswick and C. G. Patten of Iowa. More recently important work on the crossing of varieties of apples has been done by Macoun in Canada, Hedrick in New York, Chandler and Evans in Missouri, and Etter in California. Such breeding work should be continued with all the important deciduous fruits and nuts. At the same time it is highly important that future breeding operations be planned so as to take advantage as far as possible of known genetic principles. It is well established that most commercial varieties of vegetatively propagated plants are heterozygous, and in this respect they are comparable to the heterozygous individuals in a cross-fertilized population like maize. When such varieties are self-fertile, therefore, it would be possible and desirable to practice selection among the self-fertilized strains derived therefrom and to recombine the best strains in new and better varieties. The application of this principle will be limited by many factors such as self-sterility, failure of seed production due to loss of vigor, and the very long time required to attain homozygosity sufficient for the practice of effective selection. The economics of tree fruit production, however, furnish abundant argument for comprehensive, well-planned breeding programs such as can be undertaken safely only under auspices which are permanently financed for such work. These programs should also include cytological studies on all the important fruits. The chromosome numbers of both cultivated and wild species should be determined, and also of important commercial varieties. This information may throw light on peculiar breeding behavior in crossed or selfed seedlings and may indicate future possibilities in securing combinations of value both for fruit bearing and for use as root stocks.

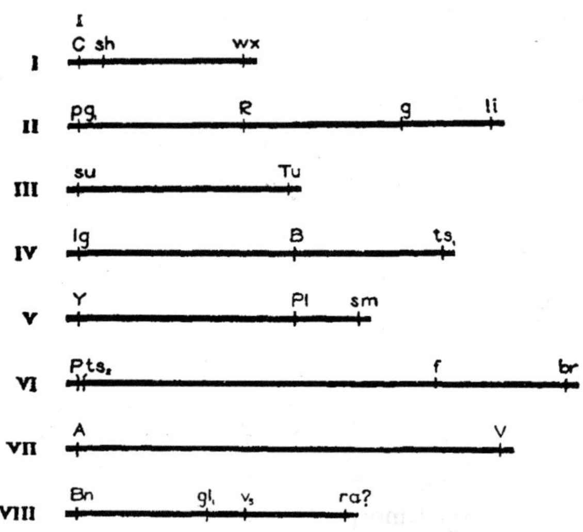


FIGURE 147.—The eight established linkage groups in maize. (*From Brunson.*)

Corn contains ten pairs of chromosomes of which eight have so far been assigned to independent linkage groups. The loci of genetic factors on these eight chromosomes are shown by symbols which refer to the following characters:

*I* dominant inhibitor of aleurone color
*C* necessary for aleurone color
*sh* shrunken endosperm
*wx* waxy endosperm
$pg_1$ pale-green seedlings
*R* necessary for aleurone color
*g* golden plants
*li* lineate leaves
*su* sugary endosperm
*Tu* tunicate condition
*lg* liguleless leaves
*B* involved in plant color
$ts_1$ tassel seed
*sm* salmon-colored silks
*Pl* involved in plant color
*Y* yellow endosperm
*f* fine striped leaves
*P* pericarp and cob color
$ts_2$ tassel seed
*br* brachytic culms
*A* anthocyan pigment
*v* virescent seedlings
*Bn* brown aleurone
$gl_1$ glossy seedlings
$v_5$ virescent seedlings
*ra* ramose inflorescence

**Genetic Analyses of Crop Plants.**—Many superior varieties of crop plants have been produced through hybridization without any very precise knowledge of the genetic factors involved in the important character expressions of the new varieties. It remains to be seen how much more can be accomplished in synthesizing supervarieties after the important crop plants have been subjected to such extensive genetic analyses as have been accomplished in *Drosophila melanogaster*. The present status of maize breeding, however, seems to indicate that still greater possibilities await the completion and application of such analyses. Maize is the most important field crop. At the same time its great variability and ease of manipulation make it unusually satisfactory for genetic investigation.

*Genetic analysis of maize* has advanced to a stage exceeded only by Drosophila. The breeding behavior of nearly a hundred characters has been determined; cytological studies have shown that there are normally ten pairs of chromosomes present in somatic cells; and linkage studies have established the existence of eight linked groups of characters which have been assigned to eight of the ten chromosomes (figure 147). As Brunson points out, further extension of the knowledge of linkage in maize offers the most promising avenue of assistance to corn improvement from corn genetics. Important scientifically as would be an experimental proof of the multiple-factor hypothesis of hybrid vigor, of much greater import for agriculture is "the possibility of locating those intangible and hitherto elusive factors which in their aggregate make for a desirable individual under a given set of conditions." Brunson states:

We have every reason to believe that the complex of characters which go together to make a desirable strain of corn are inherited just as definitely, and are represented by individual factors in the germ cell just as surely, as are the characters of aleurone color or endosperm texture, for example. In making linkage studies it is natural that the characters used thus far have been the obvious, easily distinguished morphological ones which lend themselves to easy separations and exact counts. With an outline representing the chromosomes, however, and marked at various points by factors representing easily distinguishable characters, it should be possible to learn much concerning the inheritance of the more intangible characters of so much economic importance in corn improvement. With this information at hand the production of superior strains would be transferred from an empirical to an exact basis.

At present we have no idea of the number or relative importance of the factors which may be involved. It may be inferred, however, from the great diversity of inbred lines, that the number is not small. It also seems probable that there may be a few whose importance far outweighs the others. Lindstrom has recently given a good example in the tomato of the linkage between the factor for skin color and a factor or linked group of factors which largely determine the size of fruit. If we knew the inheritance of a few important physiological factors in corn and could definitely assign them to linkage groups, it would greatly simplify our procedure and increase the probability of final achievement.

*Synthetic breeding* of the most advanced grade certainly awaits the extension of genetic analyses. In maize, for example, numerous selfed strains now exist which differ from one another in factors conditioning such physiological expressions as growth rate, water requirement, and disease resistance. Definite knowledge as to the mode of inheritance of these physiological characters would facilitate their transfer from one selfed strain to another and would aid in synthesizing supervarieties adapted to special conditions. These remarks apply generally to other important crop plants, such as alfalfa, rice, and potatoes. Wheat presents special difficulties because the cultivated species comprise a polyploid series (7, 14, 21), and the commercial varieties belong to the *vulgare* group which is hexaploid. The many excellent inheritance studies which have been made and are under way furnish a mass of genetic data which doubtless will eventually be assembled and reduced to more usable form. Meanwhile, it is fortunate that many valuable improvements can be made by the method of synthesis without waiting for the more complete genetic analyses which must eventually be made. It is desirable that more such work be undertaken, provided it is properly planned to meet definite agricultural problems.

**Chromosomal Variations.**—The rapidly increasing data on polyploidy, polyploid seriation, and other types of chromosomal variations in plants must be taken into consideration in future plant-breeding operations. In the strawberries, for example, Longley has shown that the species belong in three groups, *viz.*, the diploid group ($n = 7$), represented by *Fragaria vesca* and its varieties, the hexaploid group ($n = 21$), represented by two varieties of *F. elatior*, and the octoploid group ($n = 28$), which includes the two American species, *F. chiloensis* and *F. virginiana*, and many if not all of the outstanding horticultural varieties. Furthermore, it is found that the higher chromosome numbers are associated with the dioecious condition in the strawberries, and the genetic behavior of sex in these plants is in accord with expectation from a duplication of chromosomes such as exists in true polyploid forms.

This cytological evidence has thrown light on a number of important questions concerning the breeding of strawberries. A review of the taxonomic, genetic, and cytological findings in the genus Fragaria, according to Longley, will demonstrate that variability is associated with polyploidy, which in turn may represent a latent hybridity, or in other words. the octoploid Fragaria are crypthybrids. At the same time crossing between forms belonging to different chromosome groups may occur as in the polyploid forms of Rubus, Rosa, and Crataegus. Light is also thrown on the important question of sex determination, and the thesis presented in Chapter VII, *viz.*, that sex in plants is primarily determined by the chromosomes, is corroborated by the fact that diploid strawberries are hermaphroditic while sex differentiation occurs only in polyploid forms. This may have an important bearing upon the breeding of more highly fertile hermaphroditic varieties. Darrow states that during a period of 75 years the percentage of flowers setting fruit in hermaphroditic varieties was increased, presumably as a result of selection for high productivity, from 17 to 66 per cent. If these varieties are all octoploids, the differences in productivity would indicate important differences within the chromosomes which should make breeding for increased fertility practicable. The expected ratios from crossing true octoploids differing in allelomorphic characters will be extreme modifications of the ordinary Mendelian ratios. Actual and calculated frequencies in tetraploid Daturas are given in table 41 (Chapter XXIV). As all of the horticultural varieties of Fragaria are octoploid, the data from crossing these varieties would be expected to exhibit even greater complications than in the tetraploid Datura.

The occurrence of tetraploids in several different genera of plants has been mentioned in Chapters XXIV and XXX. Not only are such forms of possible value in themselves, for reasons already stated, but they also furnish the possibility of producing triploid forms as a result of crossing with diploids. As Belling has stated, triploid plants are already of commercial value for their flowers in the case of a few cannas. Such plants set no seeds, and the abortive ovaries drop and so make way for other trusses of flowers on the same stalk. "In the triploid mulberries, of which many clones are grown in Japan, no seeds usually appear, although ripe fruits are formed. The production of seedless fruit is sometimes a desideratum."

**Interspecific Hybrids.**—The present knowledge of interspecific hybridization (see Chapter XXVII) is sufficient to indicate important possibilities in its application to plant-breeding problems. Some of these are very recent suggestions which have not yet been tested sufficiently to indicate their future importance. Such, for example, is the possibility that doubling of chromosome number after hybridization may lead to fertile, constant forms of economic value. It is certain, however, that new, stable forms can sometimes be produced by crossing species having the same or different chromosome numbers and that in such cases these forms may represent various combinations of the characters of the two parental species. This fact has an important bearing on the breeding of annual crop plants. The production of such new types is greatly facilitated by cytological work going along with genetic experiments. In Crepis, for example, checking the chromosome numbers by the rapid (Belling) method of making smear preparations of pollen mother cells made it possible to isolate hybrid strains having even numbers of chromosomes from among the progeny derived by crossing two species with 8 and 40 chromosomes, respectively, and some of these strains have been practically constant from the beginning. As the original $F_1$ hybrids were highly sterile, it was necessary to backcross on the parent species. Even if the $F_1$ had been fairly fertile, however, backcrosses would have been made for the obvious advantages derived through simplification of the expected classes in $F_2$.

*Wheat species* have been investigated rather extensively by means of hybridization experiments combined with cytological investigations, and as a result considerable light has been shed on the genetic relationship of these species and on various breeding problems. The species of Triticum fall naturally into three groups as shown below:

*a.* Einkorn group, $n = 7$.

*T. monococcum.* Varieties interfertile.
Sterile or only slightly fertile with groups *b* and *c*.

*b.* Emmer group, $n = 14$.

*T. dicoccum, durum, turgidum,* and *polonicum.* Species and varieties interfertile.
Sterile or slightly fertile with group *a*.
Partially sterile with group *c*.

*c. Vulgare* group, $n = 21$.

*T. spelta, vulgare,* and *compactum.* Species and varieties interfertile.
Sterile or slightly fertile with group *a*.
Partially sterile with group *b*.

Sax reports that in all cases the $F_1$ plants are unusually vigorous, and sterility is not due to poor vegetative development but is caused by the formation of non-functional gametes. Sterility in $F_2$ segregates may be greater than in $F_1$ individuals due, not to greater gametic sterility *per se,* but to a combination of weak growth and gametic sterility. The differences in sterility in the various species hybrids representing crosses between the three groups are accounted for by Sax on the basis of the unbalanced relations of the chromosomes resulting from the irregular meiotic divisions. To this he added the specific interrelations of the parental chromosomes, in order to account also for the partial association of the original parental characters in the $F_2$ segregates; the absence of varieties or species with intermediate chromosome numbers; and the difficulty in obtaining homozygous segregates combining the desirable characters of the parental species in partially sterile wheat hybrids. Because of this difficulty, Sax inferred that the breeding of wheat varieties to combine disease resistance with high yield and quality of grain is much more likely to be sucessful if the parents are selected within the *vulgare* group.

Thompson has shown, however, that *vulgare* × *durum* segregates do occur which are chiefly *vulgare*-like but possess *durum* characters. He states that

> . . . even the most characteristic of the *durum* qualities may be found in plants all of whose other characters are *vulgare.* In agreement with this we find 14-chromosome plants with odd *vulgare* characters and 21-chromosome plants with odd *durum* characters. It therefore appears to be premature to conclude that it will be impossible to get any desired *durum* character in combination with *vulgare* characters in a wheat of *economic* importance.

Furthermore, rust-resistant *vulgare*-like segregates have actually been secured both by Hayes and by Thompson and have maintained their resistance at many different stations. Whether they are sufficiently resistant for practical purposes remains to be seen. Thompson concludes that the correlation of characters and partial sterility in interspecific wheat hybrids depend, in part at least, on chromosome incompatibilities, and not entirely on the elimination, due to unbalance, of gametes with intermediate chromosome numbers.

Very different results were obtained by Thompson when he used *dicoccum* instead of *durum* in crosses with *vulgare* varieties. The fact that the taxonomic differences between *dicoccum* and *vulgare* are greater than between other 14-chromosome wheats and *vulgare* would seem to indicate greater chromosome incompatibilities; and the genetic results are certainly in agreement with this conception. The great majority of the individual $F_2$ plants were *dicoccum*-like in species-distinguishing characters, while the remainder were intermediate. Thompson concludes that the difficulty of combining, through breeding operations, desirable qualities of *dicoccum* with those of *vulgare* will be very great because of the correlation of *dicoccum* characters and particularly because of the rarity of *vulgare*-like segregates.

Many other interspecific wheat hybrids have been made. A notable one is *Triticum monodurum*, derived from crossing *T. monococcum* and *T. durum*, which, according to Blaringhem, after six generations shows remarkable vigor and good quality of grain but has weak straw.

A noteworthy investigation is that of Vavilov, who crossed his new species, *T. persicum* of the durum group, with thirteen other wheat species involving forty-nine varieties both wild and cultivated and also with Secale and Aegilops. Wheats of the 14-chromosome group were easily crossed with *T. persicum*, but the *vulgare* group gave markedly sterile $F_1$ hybrids, and crosses with *T. monococcum* gave a still higher degree of sterility, as was true also of Secale and Aegilops. The hybridization data are in agreement with the grouping of wheat species given above. Vavilov believes that hybrids between *T. persicum* and *durum* varieties may bring great practical results such as extending the northern limits of spring *durum* wheats and giving earlier maturing *durums* which are less particular in their soil requirements. The extensive cytological studies of Kihara on hybrids between 14-chromosome and 21-chromosome species have been summarized in Chapter XXVII.

These illustrations are sufficient to indicate the type of investigations on interspecific hybridization in wheat which are going on in various parts of the world. When to these are added the many intensive studies on inheritance in varietal crosses of wheat which are now under way and the various wheat-breeding projects being conducted primarily with reference to disease resistance, a slight though inadequate conception is gained of the amount of human energy which is being devoted to the improvement of this important crop through cytologic, genetic, and breeding investigations. It would seem that the developments of the next decade may be awaited with considerable anticipation.

*Composite hybridization*, or the attempt to combine several different species or varieties into one by successive crossings, was first accomplished by Kölreuter, who made a triple hybrid of Nicotiana species. Triple species hybrids have been made in numerous other genera, especially among ornamental plants, and some of them have proved valuable. Some of the earlier hybridists, notably Wichura, succeeded in uniting even four to six species in such genera as Salix, Rhododendron, Vitis, Nicotiana, Hippeastrum, Gladiolus, Begonia, Pelargonium, and Dianthus.

The cultivated roses of today furnish a striking illustration of the creation of varieties through composite interspecific hybridization. The genus Rosa is widely distributed in the northern hemisphere and contains several hundred species of which about twenty-six have been utilized in the production of our garden roses. These twenty-six species fall into fifteen distinct groups, and in habitat they represent Asia, Europe, and North America.

The most important group of modern roses are the Hybrid Teas, for they include garden and forcing varieties which combine marvelous beauty of form and color with vigor and hardiness. Four or five distinct species enter into the ancestry of the group as shown by the following pedigree.

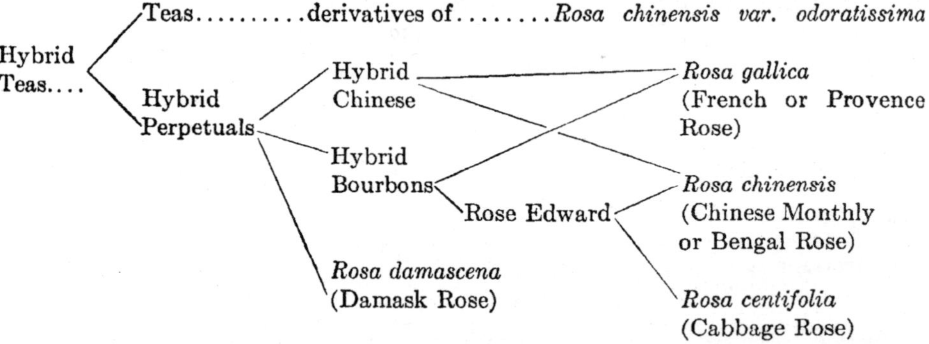


The practicability of producing new varieties by this method depends both on the comparative ease with which the different species intercross and the possibility of vegetative propagation. A few valuable rose varieties have originated as bud mutations, but these usually differ from the parent variety only in some particular character like color of flower or habit of growth. In the creation of varieties for special purposes hybridization must usually be employed. The success of breeders in combining the desirable qualities of several species in the best modern varieties of the rose suggests untold possibilities which may yet be realized when breeding is methodically planned with reference to the chromosome numbers of the species to be used.

The cytological investigations of Täckholm, Blackburn and Harrison, and Hurst have revealed many interesting facts regarding the chromosomes in Rosa. The multitudinous forms in this genus have long been the despair of taxonomists, but at last a dependable basis for classification seems to have been found. The situation is much too complicated for a detailed presentation here. Briefly, it has been stated by Täckholm that the fundamental haploid number is 7, and that with reference to chromosome number the genus may be divided into two unlike groups of species, *viz.*, (1) the *Caninae* section (in the widest sense), and (2) all other sections. The latter group is distinguished by only paired chromosomes in diakinesis: 7, 14, or 21 pairs. The *Caninae* section, containing most of the European species, exhibits the Drosera type of heterotypic division; *i.e.*, there are both bivalent and single chromosomes, usually in the proportions 7 paired + 14 unpaired, 7 paired + 21 unpaired, or 7 paired + 28 unpaired. In somatic numbers the species form a polyploid series, 14, 21, 28, 35, and 42. Occasionally plants are found with somatic numbers which are not a multiple of 7, and these are explained as hybrid derivatives.

Composite hybridization has been a popular method with various empirical breeders, notably, Farrer, Jones, and Garton, working with wheat, and Burbank, working with plums. An illustration of the latter is shown by the following pedigree of the Alhambra plum which is found in deVries' account of Burbank's work in his book on plant breeding.

Alhambra..........
- *d*..........
  - *nigra*
  - *americana*
- *e*..........
  - *c*..........
    - *triflora*
    - *Simoni*
  - *b*..........
    - French prune
    - *a*..........
      - *Pissardi*
      - Kelsey

While valuable new varieties are sometimes produced by this method, the breeder is working wholly in the dark so far as genetic principles are concerned, and although intimate knowledge of the material may be used in the choice of parents, absolutely no predictions can be made as to the final outcome. The chief essentials are the growing of very large populations and skill in making selections, and these are always most important especially in fruit breeding.

**Intergeneric Hybrids.**—That species belonging to different genera will sometimes cross and produce viable hybrids has been known for a century. While all such hybrids are of scientific interest, the great majority of the new forms so produced are of little or no value to agriculture. A few exceptions exist, however, and these will be noted, but only briefly as further details are available in the references cited. Before mentioning these it may be well to point out that many reports of intergeneric crosses are erroneous.

The cucurbits furnish an outstanding illustration of mistaken beliefs as to the occurrence of hybrids between genera. The idea is still prevalent that melons, watermelons, squashes, and cucumbers will freely intercross if planted in close proximity. The genetic relations of these groups of plants have been studied very thoroughly by Vavilov, who reports that the genera of cultivated Cucurbitaceae differ in the character of their pollen, in chromosome number, and in their behavior towards parasites. Over 1,000 crosses between different forms of watermelons, muskmelons, canteloupes, and squashes were made, and it was found that none of these genera would intercross, whereas crosses between varieties and species within a genus were successful. It appears that melons and squashes are more closely related than squashes and watermelons or melons and watermelons. The so-called natural hybrids of melon and watermelon which were reported by Korshinsky are regarded by Vavilov as phenomena of character-overlapping of genera and species. Thus a race of watermelon with entire leaves and a variety of muskmelon with dissected leaves were found. A race of watermelon having fruits indistinguishable in external appearance and color from squashes was also discovered. There is, therefore, no factual basis for the widespread belief in the crossing of watermelons, other melons, and squashes when planted together.

Certain cases of actual hybrids between genera must be considered as no more remarkable than many wide species crosses. Such, for example, is *Amarcrinum Howardii*, which is a hybrid between *Amaryllis belladonna* ♀ and *Crinum Moorei* ♂, the two genera being very closely allied. This particular hybrid is noteworthy on account of its great vigor and, being a bulbous plant, it is easily propagated.

Two intergeneric types of crosses which are of possible agronomic value are the maize-teosinte hybrids and the wheat-rye or rye-wheat hybrids. Crosses between maize and teosinte (Euchlena), which differ widely in morphological characters, have been made several times, and the hybrids were found to be resistant to the corn aphis; but G. N. Collins was the first to produce hybrid derivatives from this cross which give promise of being valuable. Many of the $F_2$ plants combined the abundant production of suckers characteristic of the teosinte with the sturdy, upright character of maize and produced very leafy, compact plants of a type that should prove valuable for forage purposes (figure 148). As a result of his study of teosinte, Collins was also able practically to disprove the repeated claims that teosinte could be made by means of selection to metamorphose into maize. The ears supposed to represent steps in metamorphosis were shown to be identical with those of hybrid segregates from maize × teosinte. Furthermore, in Mexico, where the teosinte which was supposed to have changed into corn was obtained, no truly wild plants of *Euchlena mexicana* have been found. It occurs typically as a weed in maize fields and frequently shows contamination with maize. In all likelihood the teosinte used in the selection experiments referred to was mixed with maize. The strain used in Collins' experiments was the Florida teosinte, which is presumably pure.

Fertile wheat-rye hybrids have been reported by Love and Craig, Gaines and Stevenson, Blaringhem, and Thompson, and in the first three cases the investigators believe that their hybrids possess some characteristics which give promise of further value. A number of other experimenters have reported similar hybrids, but in all cases the $F_1$ plants were completely sterile. Fertile hybrids have been rare thus far, but it may be possible to find certain varieties or species of wheat which will produce fertile hybrids with rye more frequently. Blaringhem used "Bourdon," a poulard wheat, and a Schlandstedt rye selection. Gaines

FIGURE 148.—Maize-teosinte hybrids, $F_2$ generation, showing diversity in size and time of maturity. (*From Collins.*)

used Rosen rye and four varieties of common wheat. Thompson used a wheat of Chinese origin, which he states is typical *Triticum vulgare* var. *albidum*, and several varieties of rye including a spring type known locally as Prolific. He finds that this particular wheat crosses more easily with rye than with any other wheat. "It is not difficult to get 90 per cent of pollinated flowers setting seed while Jesenko, working with a number of *vulgare* wheats, secured only 35 seeds from 6,129 flowers pollinated with rye (and 12 of these from one head)." Thompson has a large number of lines in late generations which have a preponderance of wheat characters but also certain definite rye characters. Love and Craig crossed Dawson's Golden Chaff with common rye (figure 149). In the last case the hybrids exhibit an increase in winter hardiness.

The orchid family is preeminent in the number of intergeneric crosses which have been made. No other family of cultivated plants has yielded so many beautiful hybrids in so short a time, according to Hurst, who states

> . . . that at least 40 genera and 300 distinct species have been utilized by orchid breeders. About 2,000 of these are primary hybrids while no fewer than 300 are generic hybrids. Beside these botanical hybrids, there are on record some thousands of horticultural—secondary and multiple—hybrids, to say nothing of multitudinous varietal and individual forms which no man can number.

When it is realized that these remarkable results have been achieved in less than a generation and that the raising of such delicate exotics from

FIGURE 149.—Dawson Golden Chaff wheat (left) common rye (right) and hybrid progeny. (*From Love and Craig, Courtesy Journal of Heredity.*)

seed under artificial conditions is no easy matter, one cannot but admire with Hurst the practical genius of the orchid breeders which has brought about such a consummation. Doubtless if one could see the wonderful exhibits at the International Shows, his admiration would be correspondingly increased. Hurst points out that enormous numbers of seedlings have been reared in order to obtain the really valuable hybrids and that there is need for systematizing orchid breeding in the light of genetic principles. Hurst himself has made important contributions on the genetics of orchids (see reference), as a result of which he asserts that no useful purpose can be served by indiscriminate matings of complex hybrids which are far too common in orchid breeding. Viewed economically, it is extremely doubtful whether it is necessary to go beyond the mating of primary hybrids, as any two good qualities can be combined in this way in the second generation in a single experiment with two, three, or four species.

**Limits of Crossing.**—The fact that true intergeneric hybrids are rare in the plant kingdom and that the majority of such hybrids are completely sterile is in itself sufficient evidence that still wider crosses seldom if ever occur. It is sometimes stated that interfamily hybrids have been made, but no authentic instance can be given. An illustration of the way such unfounded reports originate is found in the oak-like varieties of black walnut, the first of which appeared in California in 1907 among seedings of a wild tree of the southern species, *Juglans californica* (figure 150). Later a parallel variant was derived from the northern species, *J. Hindsii.* Because the original wild walnut tree happened to stand very close to a live oak tree, it was assumed by the nurseryman who discovered the new form that it was a natural hybrid, and his statement was widely accepted even by certain scientists. Genetic investigation, however, proved that, although the oak-like type differs in many details from the parent species, it behaves as a simple recessive when the two are crossed. On the other hand, experiments in pollinating walnut with oak resulted only in pure walnut seedlings even in the second generation when large numbers were grown. It is concluded, therefore, that the first generation seedlings derived from oak pollinations were "false hybrids," which resulted from some form of apogamic development. Such seedlings are known to occur in intergeneric crosses in the walnut family.

The limits of crossing are dependent upon three factors, *viz.*, crossability, viability, and fertility. The first of these has been discussed in Chapter XXVII and the present chapter. Experience has shown that variation in viability sometimes occurs among the progeny resulting from interspecific crosses. For example, in the first crosses made between *Crepis capillaris* ($n = 3$) and *C. tectorum* ($n = 4$), Babcock and Collins obtained seedlings which could not develop beyond the cotyledon stage. Later crosses between other strains of these same species, however, produced seedlings which developed to maturity and which were more or less fertile.

Fertility is also a variable character, as has been indicated in several discussions of intervarietal, interspecific, and intergeneric hybrids. The phenomena of sterility in hybrids may be grouped under three heads, *viz.*, true sterility, self-sterility, and intersterility. True sterility is due to absence of fecundity or the ability to produce functional gametes

While this may be caused by environmental conditions including disease, in hybrids it is most often a manifestation of incompatibility between the two chromosome reaction systems contributed by the unlike parents. Such manifestation may range from complete abortion of the sexual organs to mere inability to develop functional germ cells.

FIGURE 150.—A seedling of the oak-like walnut, *Juglans california* var. *quercina* (left) and of the California black walnut, the parent species (right).

Self-sterility may be due to incompatibility between reaction systems, as in wide crosses, or to the effect of specific genetic factors within a single reaction system, as in many varieties of tree fruits which are almost completely sterile when self-pollinated but will produce heavily when fertilized with pollen from other varieties. Gowen has shown that in self-sterile varieties it makes no difference whether the pollen comes from the same flower, from other flowers of the same plant, or from different plants of the same clone; the number of fruits set is approximately the same as long as the pollen used has the same genetic composition.

Self-sterility of interspecific hybrids does not necessarily preclude crossability of such hybrids with other species of the same genus. This has been demonstrated in Crepis by placing self-stertile $F_1$ interspecific hybrids in the open near species other than their parents, with the result that a few viable seeds were obtained. These produced progeny which are indubitably triple hybrids in some of which the third species has a different chromosome number from the first two. For example, it is rather difficult. to obtain primary hybrids beteewn species of Crepis having four and five pairs of chromosomes; but self-sterile $F_1$ hybrids between two four-paired species will cross naturally, through insect pollination, with certain five-paired species. This indicates one method by which it is possible to secure composite hybrids between species with unlike chromosome numbers.

Intersterility is likewise a manifestation of either reaction-system incompatibility or individual genes. Thus in certain wide crosses the highly self-sterile $F_1$ plants are also intersterile with each other but are fertile when backcrossed on one or both parents (Chapter XXVII). Intersterile genotypes have been found in the hayfield tarweed, *Hemizonia congesta*, by Babcock and Hall. In these tarweeds the individual plant is self-sterile so that in the original ($P_1$) population grown from seed collected in the wild, pairs of plants were chosen and crossed reciprocally. That some of these plants carried factors for intersterility is shown by the proportion of $F_3$ families which grew to those which completely failed to germinate. This is illustrated in the following pedigree in which all matings are reciprocal:

| | | | | |
|---|---|---|---|---|
| $P_1$ | | C16—5 × C18—6 | | |
| $F_1$ | 16.27—1 × 3 | 16.27—8 × 10 | 16.30—1 × 2 | |
| $F_2$ | 17.62—2 × 10 | 17.64—5 × 10 | 17.66—15 × 20 | |
| | 17.62—1 × 8 | 17.64—13 × 16 | 17.66—22 × 27 | |
| | 17.62—14 × 15 | 17.64—4 × 6 | 17.66—41 × 42 | |
| | 17.62—9 × 19 | 17.64—21 × 22 | 17.66—45 × 46 | |
| | 17.63—1 × 9 | 17.65—8 × 10 | 17.66—17 × 35 | |
| | 17.63—5 × 7 | 17.65—1 × 7 | 17.66—43 × 44 | |
| | | 17.64—14 × 23 | 17.67—5 × 7 | |
| | | 17.64—15 × 17 | | |
| | | 17.65—6 × 38 | | |
| | | 17.65—9 × 11 | | Totals |
| $F_3$ | grew 4 families | 12 families | 10 families | 26 |
| | failed 8 families | 8 families | 6 families | 22 |

The ratio of viable to non-viable $F_3$ families is 1:1, from which it appears that one of the original $P_1$ plants was heterozygous and the other homozygous for a recessive factor for sterility. Thus if $a$ represents the gene in question and $A$ its dominant allelomorph, the $P_1$ mating was $Aa \times aa$, and each of the three $F_1$ pairs mated must have resembled the parents in genotype. The selection of $F_2$ pairs for intercrossing was entirely at random as regards sterility. As the eleven $F_2$ pairs (underlined in the pedigree) which failed to reproduce when crossed reciprocally were derived from a single pair of grandparents, they comprise an intrasterile class comparable to classes known to occur in Nicotiana, Veronica, Linaria, and Bursa (see review of recent literature by Anderson). This evidence that factors which condition intersterility between individuals exist in wild plants is in agreement with the Mendelian interpretation of self-sterility advanced by East and Park. For more extensive treatment of sterility and fertility the student may consult the selected references.

**Summary.**—Success in breeding new varieties by hybridization calls for orderly procedure. The general method includes choice and culture of intended parent plants, protection of pollen, castration or the protection of female organs, pollination, and guarding the pollinated flowers and the developing seed. Various obstacles to the production of viable hybrid seed must be considered and in some cases they may prove insurmountable.

In crop plants the really superior new varieties which have been produced through hybridization are not very numerous, while in ornamental plants the reverse is true. Two good examples of such varieties are Yeoman wheat and Sacramento barley. Similar projects with deciduous fruits and nuts are highly desirable.

Genetic analyses of the most important crop plants are essential to the fullest realization of the possibilities of improvement in such plants. The synthetic breeding of supervarieties may proceed meanwhile, but such breeding can advance most rapidly and surely only when based on genetic principles.

Chromosomal variations are now known to be of fundamental importance in breeding by hybridization. It is highly desirable that one of the first steps in every important project of this sort should be a comparative study of the chromosomes in the forms intended for use. Such preliminary studies are also of the greatest value in breeding by interspecific hybridization. Many puzzling problems connected with interspecific hybridization in wheat have been cleared up by cytological studies combined with breeding experiments, and thereby new possibilities have been discovered.

Interspecific hybrids involving three or more different species have been obtained in several important genera. Cultivated roses possess unusual complexity in their ancestry, involving some twenty-six different species. Cytology has again come to the assistance of classifiers and breeders by discovering a fairly dependable basis in the chromosome groups of this complex and highly variable group of plants.

Intergeneric hybrids are much more rare and are mostly of slight value, but a few give some promise of utility. Such are the teosite-maize hybrids and the wheat-rye hybrids of which a few are fertile. The orchid family is outstanding in the large number of intergeneric hybrids which have been made.

The limits of crossing depend upon crossability, viability, and fertility. The role of these factors in each particular cross can be determined only by experiment, and they display extreme variations.

## References

ANDERSON: Studies in self-sterility, *Genetics*, **9**: 13–40.

BAILEY: Plant Breeding, 1st and 2d ed., pp. 215–251 (translation of Focke, Die Pflanzen-Mischlinge, Chapter 4).

BRUNSON: Relation of inheritance studies to corn improvement, *Jour. Amer. Soc. Agron.*, **18**: 308–313.

COLLINS and KEMPTON: A teosinte-maize hybrid, *Jour. Agric. Research*, **19**: 1–38.

EAST and PARK: Studies in self-sterility, *Genetics*, **2**: 505–609 (bibliography).

GAINES and STEVENSON: Rye-wheat and wheat-rye hybrids, *Jour. Heredity*, **13**: 81–90.

HAYES and GARBER: Breeding Crop Plants, Chapters X, XVI, XVII, and XVIII.

HURST: Experiments in Genetics, Chapters I–IV, VI, VIII, XXII and XXXII (Orchids) Chapter XXXVIII (Rosa) (bibliography).

JONES: Genetics, Chapter XIII (Sterility); pp. 423–428 (Round-tip tobacco); pp. 463–471 (Heterozygosity of vegetatively propagated plants).

LONGLEY: Chromosomes and their significance in strawberry classification, *Jour. Agric. Research*, **33**: 559–568 (bibliography).

LOVE and CRAIG: Fertile wheat-rye hybrids, *Jour. Heredity*, **10**: 195–207.

PUNNETT: Mendelism, early editions; 6th ed., Chapter XVI.

THOMPSON: Correlation of characters in hybrids of *Triticum durum* and *T. vulgare*, *Genetics*, **10**: 285–304 (bibliography).

THOMPSON and HOLLINGSHEAD: Preponderance of *dicoccum*-like characters and chromosome numbers in hybrids between *Triticum dicoccum* and *T. vulgare*, *Jour. Genetics*, **17**: 283–307.

WEBBER: Progress in plant breeding in the United States, *Yearbook U. S. Dept. Agr.*, **1899**: 465–490.

## CHAPTER XXXVI

## BREEDING DISEASE-RESISTANT PLANTS

The term plant disease has been restricted by some authors to those disorders and abnormalities caused by fungous and bacterial parasites only. Other authors have employed the term in a more general sense, including thereunder all abnormal conditions of structure and function which are caused by the different elements of the environment. The term will be used here in this more general sense, and for the purpose of this discussion it may be defined as follows: Plant diseases include all the ailments and injuries which can be traced to specific causes or agencies, as well as certain functional disorders the causes of which are obscure or difficult of analysis. In order to discuss profitably the breeding of disease-resistant plants, it is necessary to consider more fully the various categories of causes.

**The Causes of Plant Diseases.**—In general the diseases of plants are caused either by unfavorable conditions among the inanimate elements of the environment or by the invasions of other organisms. While every case of disease must be considered as the result of interrelated causal agencies, yet it is usually possible to discover specific agents that are primarily responsible for the pathological condition. It is then possible to determine the nature of disease resistance in particular instances with more or less definiteness according to the nature of the specific causes.

The most important non-living elements of the environment affecting the health and vigor of cultivated plants are the soil, the water supply, light, and the temperature and humidity of the atmosphere. These environmental factors influence plant development in so many ways that the opportunities for maladjustment between plants and their environment are practically endless. Such conditions as excess of alkali or lack of sufficient moisture in the soil or the combination of excessively high temperature and low relative humidity are typical and important illustrations of specific environmental conditions which induce disease in plants.

The living organisms of chief importance in causing plant diseases are insects, fungi, and bacteria. Pathogenic fungi and bacteria are wholly or partially parasitic and all insects which are injurious to plants may be classed as parasites. The specific relations between parasites and their hosts vary from a condition of tolerance of the parasite without serious injury to the host to one in which the destruction of the host finally ensues.

Thus there are many agencies, some non-living as well as many living things, which threaten the normal development of cultivated plants. Even among the parasitic fungi themselves there are many devices for invading the host plant and many instances of specific physiological relationship between parasite and host.

**The Nature of Disease Resistance in Plants.**—Disease resistance in a plant may be defined as the ability to develop and function normally under conditions such that other plants of the same species fail to develop or are destroyed. Resistance is always either partial or complete, the latter condition is usually termed immunity. The avoidance of disease by such means as precocious or delayed maturity is hardly to be considered as true disease resistance. Since there are so many agencies which may cause disease in plants, it is evident that the ability to resist disease may depend on any one of many characters or it may involve every function of the plant. In either case this ability is a manifestation of the physiological individuality of the plant, and hence it may be inherited. Nowhere is this more strikingly shown than in the disease resistance of certain natural species.

**Disease Resistance in Plant Species.**—The nature of disease resistance in a particular instance is indicated by the nature of the cause of the disease. In the case of non-living causes, resistance on the part of certain plants can be explained only as a manifestation of the inherent properties of the protoplasm. Thus the alkali resistance of salt grass, the Australian salt bushes, the common beet, asparagus, and species of Eucalyptus is a heritable character. If it were not so, these species could not perpetuate themselves on soils which are too strong in alkali content for most plants. Similarly with many plant troubles that are referred to adverse soil conditions, such as chlorosis and die-back, it has been found that some species are much better able to resist such conditions than other species, and within a particular species certain varieties may be more resistant than other varieties. This holds true in the case of other non-living agencies, such as excess and deficiency of moisture and heat. For every plant there is a set of optimum conditions, and these conditions are very different in different species and among varieties of the same species. For example, rice flourishes in standing water while maize requires well-aerated soil. But there are thousands of varieties of rice, each one adapted to the conditions peculiar to a certain locality, and there are many varieties of maize which make possible the culture of this species under conditions varying from the humid corn belt to the arid regions of northern Mexico, Bolivia, and central China. Similarly, in other field crops and in fruits in various parts of the world, there exist species and varieties which are adapted to certain local conditions that would be inimical to normal development of related species and varieties. Agricultural exploration cooperating with systematic seed and plant introduction has already made available for the plant breeder a large number of distinct forms of economic plants which in course of time may revolutionize many productive and manufacturing industries.

Turning now to the phenomena of resistance to the attacks of animal or plant parasites, it is found that natural species are characterized by as great diversity in this respect as was observed in the case of resistance to alkali, drouth, and other physical elements of the environment. A

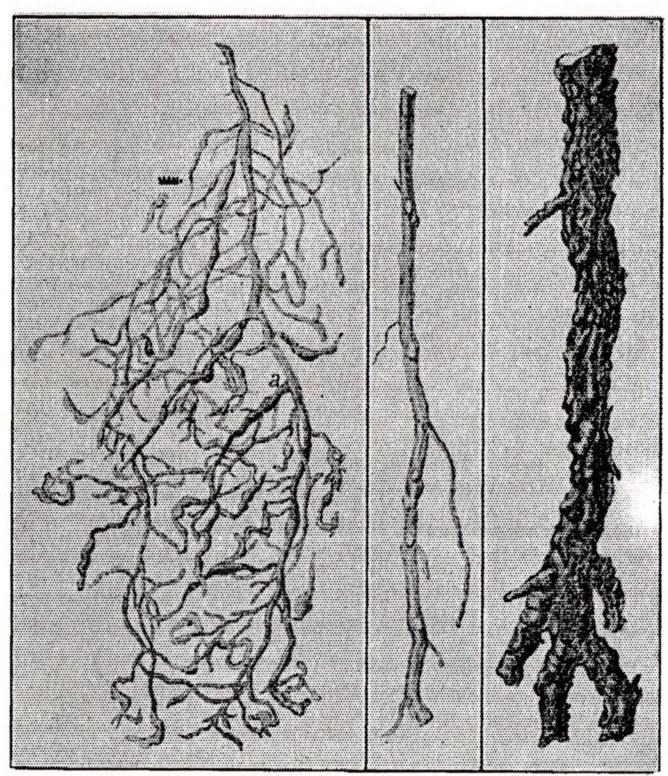


FIGURE 151.—Effects of Phylloxera on vine roots. On *left* affected root tips or nodosities; in same figure incipient tuberosities are shown at *a*. *Center*, nonpenetrating tuberosities on an American vine. *Right*, penetrating and confluent tuberosities on *V. vinifera*, the most serious form of the disease. (*From Viala and Ravaz.*)

few specific examples will serve to illustrate this general principle. The relation of different species of the grape to the phylloxera, *Peritymbia vitifoliæ* Fitch (*Phylloxera vastatrix* Planchon), is representative of a great number of reported instances of insect parasitism on vegetation. Also in their general aspects the phenomena of variation in phylloxera resistance among species of the vine are representative of the facts of disease resistance in general. Moreover, on account of the great economic importance which this particular vine disease assumed in Europe some 50 years ago, and later in California, there has been a large amount of investigation on the culture of grapes in phylloxera-infested regions. The life cycle of this insect includes both leaf-feeding and root-feeding forms. It is the root-feeding form which inflicts serious damage to susceptible vines (figure 151). The root-tip galls or nodosities are commonly found even on resistant species if phylloxera are present. The principal difference between resistance and susceptible vines as regards reaction to phylloxera attacks is found in the number, size, and penetration of the lesions on the larger roots. This phylloxeran is a native of eastern North America, and the species of Vitis which are native to this region all exhibit some resistance to its attacks. This resistance of species native to the habitat of a disease-causing parasite is a general fact of great significance to agriculture on account of its potential value in both plant and animal breeding.

The phylloxera was introduced into France through the importation of American vines, and it soon became a most serious obstacle to the culture of the choice wine, table, and raisin grapes of the Mediterranean region, all of which varieties belong to a single species, *Vitis vinifera.* In fact, every member of this large and valuable plant group has been found to be susceptible to phylloxera, thus making impossible its culture as a direct producer, *i.e.*, on its own roots, in a phylloxera-infested region. After striving in vain to exterminate the insect in all infested areas, European vineyardists gradually adopted the only other practicable method of grape growing, *viz.*, the grafting of *vinifera* varieties upon resistant roots. The problem of determining which species of Vitis were both highly resistant to phylloxera and well adapted to the soil and climatic conditions of various European localities required extensive investigations. Eighteen native American grapes have been tested as well as several Asiatic species, but the latter were all less resistant than the most susceptible American species. The American vines which have come into most prominence on account of their proved value in the reconstitution of phylloxera-devastated vineyards may be listed according to relative resistance about as follows, if the maximum or absolute immunity be taken as 20:

18–19 *V. rupestris*
18 V. *riparia* and *cordifolia*
17 V. *berlandieri*
16 V. *cinerea*
14–15 V. *aestivalis, linsecomii* and *candicans*

All of the above species belong to the subgenus or section, Euvitis. Two of these, *rupestris* and *riparia,* together with certain hybrids between these and *vinifera,* are now considered the most valuable resistant stocks. Another American species belonging to the section *Muscadinia, viz.*, *rotundifolia,* has been found to have a resistance of 19 or higher, inasmuch as the insect has never been observed on its roots. It is also free from the common fungous diseases of the vine, but the difficulty of propagating it from cuttings and its slight affinity for grafts of other species make it a valueless species for the reconstitution of vineyards. On the other hand the American species, *labrusca*, has become of great economic importance since it is the parent of the Concord, Isabella, Niagara, and many other cultivated varieties. Yet its resistance to phylloxera is ranked at 5, and when grown in California it is no more resistant than is *californica* when used as a rootstock for producing vines, and the resistance of the latter is ranked at 4. Yet the *labrusca* derivatives are extensively grown in the northeastern states and in other northern temperate regions. This is explained by the fact that the phylloxera itself does not thrive below a certain minimum temperature. Thus it is found that resistance to phylloxera in the species of Vitis varies all the way from zero in *vinifera* to practically absolute resistance in *rupestris*, *rotundifolia*, and certain hybrids and that the existence of highly resistant forms which are also suitable for vineyard culture has made possible the preservation of an important agricultural industry.

The question of the nature of the cause of resistance to phylloxera has received rather wide attention among investigators, but it has not yet been definitely answered. The probability that the phylloxera itself varies in its specific effects on the vine has not been overlooked. Variability in resistance of several varieties of grape when grown in different infested localities is accepted by Grassi as evidence of the existence of "benignant" and "malignant" races of phylloxera. But this does not explain the high resistance or immunity of some American species. Having in mind the fact that the phylloxera sucks its nourishment from the leaf or root by inserting its prolonged rostrum into the living tissue, it seems most probable that resistance is to be explained as absence of response to a specific stimulus. At any rate, the fact that the problem concerns itself here with distinct natural species makes it reasonably certain that resistance and susceptibility to phylloxera infestation are somatic expressions of genotypic diversity.

Another important example of variation in disease resistance among species of the same genus is found in the relation of various chestnuts to the very destructive bark disease caused by the fungus, *Endothia parasitica*. The parasite is a native of eastern Asia where it is parasitic upon native species of chestnut. Some of these species, particularly the Japanese chestnut, *Castanea japonica*, and the hairy Chinese chestnut, *C. mollissima*, are highly resistant to the parasite. When the fungus was introduced into America, presumably in nursery stock some 35 years ago, it found in the native species, *C. americana*, a very susceptible host (figure 152). The parasite has already completed the extermination of the American chestnut as a timber tree. Investigations have determined that the European chestnut is also susceptible to the attacks of this fungus, so that the future existence of this species is also jeopardized. The American chestnut was one of our most valuable forest trees and its destruction is a serious loss. The hairy Chinese and the Japanese chestnuts seem rather promising, as a number of these trees are known to have survived the blight. While they are scarcely timber trees as compared with the native species, yet the nuts are of good quality and it is hoped that they will be useful in producing a substitute for the vanishing American species. Possibly $F_1$ hybrids will be resistant and valuable timber trees. Even the culture of the American species for commercial nut production in western North America will be constantly threatened. Hence it is fortunate that breeding experiments with chestnuts have already begun to yield results, to which reference will be made again.

FIGURE 152.—An advanced stage of the chestnut bark disease, caused by *Endothia parasitica*, a virulent pathogenic fungus from China. (*From the Journal of Heredity by permission.*)

A bacterial organism which finds similar wide diversity in the resistance of possible hosts is the fire-blight pathogene, *Bacillus amylovorus*. Being indigenous in eastern North America, this organism must have maintained itself on the native species of apples and related genera previous to the introduction of European apples, pears, and quinces, since it cannot survive long even in the dead tissues of the host. The disease is spread naturally by insects that visit infected plants; it may also be carried on pruning tools. Fire blight is the most widely destructive of all pomaceous fruit diseases; but the pathogene manifests different degrees of virulence in different species. Its most susceptible hosts are the commercial varieties of the pear, which are all derivatives of the European species, *Pyrus communis.* In several regions naturally well adapted for pear growing, the culture of this fruit has been abandoned on account of the destructiveness of pear blight. Even the more resistant varieties of *communis* as well as certain hybrids between *communis* and other species, such as the Kieffer, a supposedly resistant pear, have all proved to be susceptible to the disease when grown in the humid climate of the southern states. Therefore, in discussing the problem of blight resistance in pears, it must be remembered that the pathogene itself is very susceptible

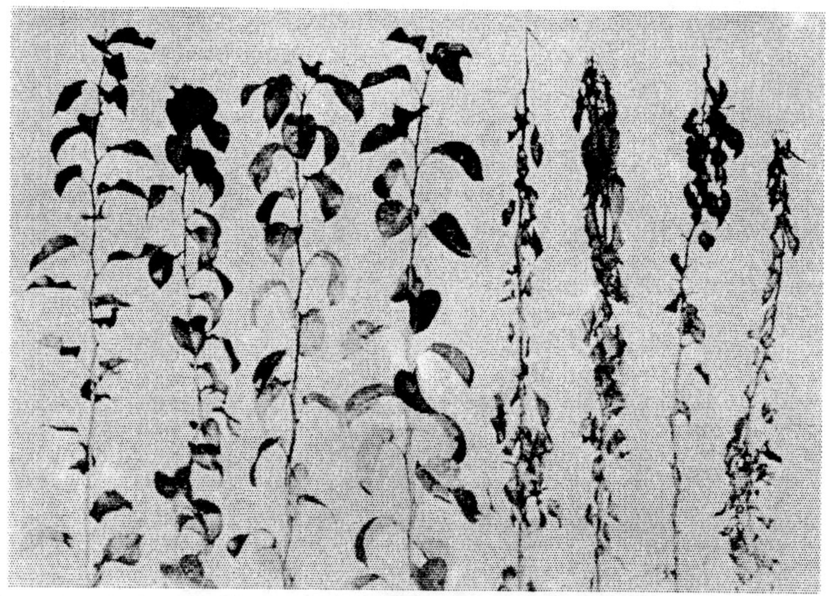

FIGURE 153.—On the left, four cultivated varieties of *Pyrus ussuriensis* all highly resistant to blight; and four blighted Bartlett trees. These four Chinese varieties were inoculated on four different dates; on each date a Bartlett was inoculated with the same culture. The Bartletts were all killed while the four Chinese varieties remained perfectly healthy. (*From Reimer.*)

to environmental conditions and that a particular host which is known to be resistant under one set of conditions will not necessarily prove to be generally resistant. Hence the breeding of blight-resistant pears should be carried on in a region ideal for pear culture in every respect except that it is ideal for the fire-blight organism also. Such conditions exist in southern Oregon where Reimer has made a complete collection of the known species of pears and has conducted scientific tests of their resistance to blight by means of inoculations with pure cultures of the bacillus. At the end of 12 years, 34 per cent of all the varieties in the collection had been killed as a result of natural infection. The two most promising species are of Chinese origin, *viz., P. calleryana* and *P. ussuriensis;* the former because it is highly resistant and otherwise promising as a rootstock; the latter because it is also very resistant and exists in numerous varieties some of which are of commercial importance in China and may prove of value in breeding resistant commercial varieties. The resistance of four of these Chinese varieties to artificial blight inoculations which destroyed Bartlett pears is illustrated in figure 153. The birch-leaved pear, *P. betulifolia,* which is used as a stock in China, exhibits varying degrees of susceptibility, and a small percentage of individuals are fully immune. These immune strains will be useful for stocks on sandy or gravelly soils.

One plant disease has now been discussed which may be considered typical of each of the three great classes of disease-causing organisms and in each the same diversity is found among natural species as regards disease resistance. It is unnecessary to multiply instances further. In all likelihood the resistance of the hairy Chinese chestnut to *Endothia parasitica* and of Chinese pears to the fire-blight bacillus is due to some specific quality of the protoplasm which inhibits the normal development of the pathogene. That this quality is inherited is shown by the results of hybridization experiments.

**Breeding Disease-resistant Varieties by Hybridization.**—Allusion was made in Chapter XXXIV to the fact that first-generation maize hybrids are often more drouth resistant than either parent. Presumably this is merely one manifestation of hybrid vigor. Hybridization is a very important means, however, for the production of improved varieties which are better adapted to specific adverse elements of the environment. Witness the important results already secured in the production of cold-resistant varieties of fruits, grains, and forage plants by Hansen, Patten, and Saunders and at the U. S. Agricultural Experiment Stations in Alaska.

At one stage in the antiphylloxera campaign in France and California, viticulturists held definitely to the ideal of securing through hybridization

> . . . a vine that, while resisting the phylloxera, the two mildews, the black rot, etc. (all of which diseases are natives, and which the American vines resist more or less well), will give *without grafting* a grape that has size and the quantity and quality of the *Vitis vinifera.*

With this object in mind many crosses were made, but they have produced no hybrids between *vinifera* and American species that can be substituted for the choice *vinifera* varieties. It, therefore, became necessary to utilize resistant species and hybrids as stocks on which to graft the producing varieties. It is still possible, however, that by growing large numbers of $F_2$ and $F_3$ seedlings from some of the most promising $F_1$ hybrids, the dream of the viticulturist might be realized. It seems that few grape breeders have carried out extensive tests of hybrids beyond the first generation from the cross. This is not strange, inasmuch as grape breeding for phylloxera resistance was at its height during the latter part of the nineteenth century and before the importance of testing for several generations after a cross was generally appreciated. That phylloxera resistance and suceptibility are conditioned by specific genotypic elements is evidenced by the results of Rasmuson who tested $F_2$ seedlings from several crosses between certain American species and between American species and *V. vinifera,* as well as crosses between different varieties of *vinifera.* The latter, he reports, yielded only susceptible offspring while the crosses between different American species gave both resistant and susceptible offspring, the latter being in the minority. Resistance appeared to be dominant and susceptibility recessive in the progeny of crosses between American species and *vinifera.* The data are not given, but he believes the observed numbers of resistant and susceptible vines favor the assumption of two factors that condition immunity when either is present alone or when both are present together.

Resistance to aphis in the stone fruits is thought to be a heritable character from the result of crosses made at the Iowa Experiment Station. The data permit no reliable conclusions regarding the genotypic relation of aphis resistance and susceptibility in these plants, but the indications are that these characters are conditioned by a single factor difference. Another interesting case of the inheritance of resistance to aphis was observed by Gernert in $F_1$ hybrids between teosinte and corn. Both corn root-aphis, *Aphis maidiradicis,* and the corn plant-aphis, *A. maidis,* were involved, and both the teosinte and the hybrids were completely resistant while the corn was badly infested. The desirability of securing aphis-resistant varieties of maize will be apparent when it is realized that most of the corn-growing regions of North America are infested with these insects and that the loss in reduction of yield caused by them is enormous.

The work of Van Fleet on hybridizing various species of chestnuts was begun 10 years before the bark disease had worked havoc with the chestnut trees near New York City, which is the oldest known center of infection. Hence many crosses were made with either the American or European chestnut as one parent, but in 1907 these were all destroyed by the Endothia. Fortunately, however, numerous controlled pollinations were made on the bush or Virginia chinquapin, *Castanea pumila,* using pollen of a Japanese species, *C. crenata,* as well as other Asiatic chestnuts. It is asserted that the Asiatic species and the chinquapin-Asiatic hybrids are highly resistant, because few have shown any appearance of infection although surrounded by diseased trees, and that even when infection takes place the injury is quite local in character. Van Fleet adds that second-generation seedlings of chinquapin-*crenata* crosses show no disease although constantly exposed to infection (figure

154). Thus a beginning has been made in what promises to be an important branch of nut breeding, and the orchard production of commercial chestnuts has been insured against future encroachments by a deadly disease through the timely efforts of a zealous and farsighted plant breeder.

The breeding of blight-resistant commercial varieties of pears is being attempted by using promising varieties of the hardy *Pyrus ussuriensis*, from northern China and southern Siberia, crossed with well-known commercial varieties of *P. communis*, the European pear. Although *P. ussuriensis* is the hardiest pear known and seedlings from wild trees have endured a temperature of −40°F. in Iowa and Dakota, some of its horticultural varieties from China are well adapted to the mild, moist winters of southern Oregon. These are accordingly being used by Reimer in crosses with the best American varieties such as Bosc, Anjou, Bartlett, and Comice, and 21 per cent of the $F_1$ hybrids have proved to be immune to blight when inoculated. None of these have borne fruit, but it is

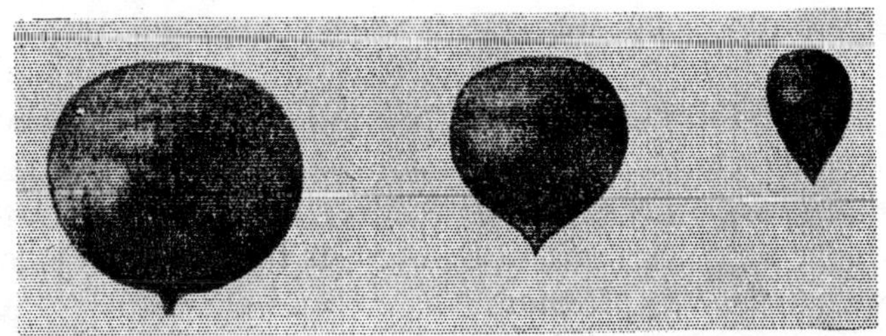

FIGURE 154.—In the center is a nut produced by a cross between the American bush chinquapin *Castanea pumila* (right), and Japanese chestnut, *C. crenata* (left). Although intermediate in size the hybrid nut is disease resistant and of good quality. (*From the Journal of Heredity by permission.*)

hoped that a few will prove to be of immediate value; the growing of as many $F_2$ progeny as possible, assuming that the $F_1$ trees will be fertile, will greatly increase the chances of finding desirable combinations. In South Dakota Hansen has made numerous crosses on hardy forms of *P. ussuriensis* using pollen of numerous horticultural varieties which was obtained by visiting the large collections of cultivated pears in Arkansas, Missouri, Iowa, and Illinois. At least three of the $F_1$ hybrids produce fruits of full commercial size and appear worthy of propagation and distribution for trial. The prospect of resistant commercial varieties and the demonstration that fully resistant rootstocks exist seem to insure a supply in the future of this excellent fruit.

**Breeding Rust-resistant Wheats.**—The grain rusts are the most important of all fungous plant diseases. The annual losses they entail for the grain crops of the world must be estimated in the hundreds of millions of dollars. For example, over 200,000,000 bushels of wheat were lost in the United States alone in 1916 because of wheat rusts.

Although prevention of wheat rust to some extent is possible by giving careful attention to the water and soil relations of the wheat plant and by early seeding or planting of early varieties which sometimes escape attacks by rust, yet these diseases are a serious menace to the maximum production of wheat. Hence, the creation of rust-resistant varieties has become a very important problem. The diversity among varieties of wheat as regards resistance and susceptibility to rust fungi was recognized by Knight in 1815 and the desirability of creating new varieties which should be resistant to rust as well as highly productive and of good milling quality was fully realized by such breeders as Pringle, Blount, and Farrer. Although they were not familiar with the Mendelian principles of segregation and recombination of characters, these breeders of wheat, a self-fertilized annual crop plant, were naturally led to persist in their efforts beyond the $F_1$ generation. The work of Farrer especially was thorough and reliable. He found that he could not secure complete resistance to the black stem rust, *Puccinia graminis* Pers., combined with good milling quality in his wheat crosses, even when rigorously selected in $F_2$. Most of the soft bread wheats are very susceptible to rust, and when crossed with resistant durums, poulards, and spelts, they give rise to strains which are either poor bread wheats or are rust susceptible. Biffen discovered in 1903 that resistance to the yellow rust, *Puccinia glumarum* Eriks. & Henn., in his cross between Rivet, a slightly susceptible wheat and Red King, a very susceptible variety, was recessive in the $F_1$ generation but appeared in approximately one-fourth of his $F_2$ population. Tests of later generations proved that this character bred true. However, it appears that resistance of the wheat plant to other species of rust fungi may be inherited as a dominant character. Vavilov reports that he crossed Persian wheat, *Triticum vulgare* var. *fuliginosum* Al., which alone out of 540 varieties was immune to mildew, *Erisiphe graminis* DC., but which was susceptible to brown rust, *Puccinia triticina* Eriks., with other varieties of common bread wheats and secured $F_1$ hybrids which were immune to both diseases. Thus it is clear that the inheritance of rust resistance is dependent upon the specific relation existing between the parasite and the host.

The practical aspect of breeding rust-resistant cereals is greatly complicated by the fact that resistance in a single variety of wheat, for example, is likely to vary geographically. While this is due in part to the responsiveness of the wheat plant to radical changes in environment, it is probably more often due to physiological variations in the rust fungi. The virulence of a given parasite appears to vary not only with the host but with the geographical location. A striking example of this was observed by Mackie in the behavior of Kubanka, a durum wheat of Russian origin. Although this wheat is remarkably rust resistant in the northern Great Plains region, yet when grown on the west coast of

Mexico, it succumbed completely to the stem rust, *Puccinia graminis* var. *tritici*, which it had resisted successfully in the Dakotas. The explanation of this failure of a supposedly resistant wheat is found in the existence of physiologic races of the species *P. graminis*. Forty or more such races of the black stem rust of wheat, *P. graminis tritici*, are known, according to Hayes, Stakman, and Aamodt, and these can be distinguished by their action in the greenhouse on seedlings of twelve varieties representing five groups of wheat. Similarly, at least four distinct forms of the oat stem rust, *P. graminis avenae*, are known and the existence of such races is indicated in various other parasitic fungi.

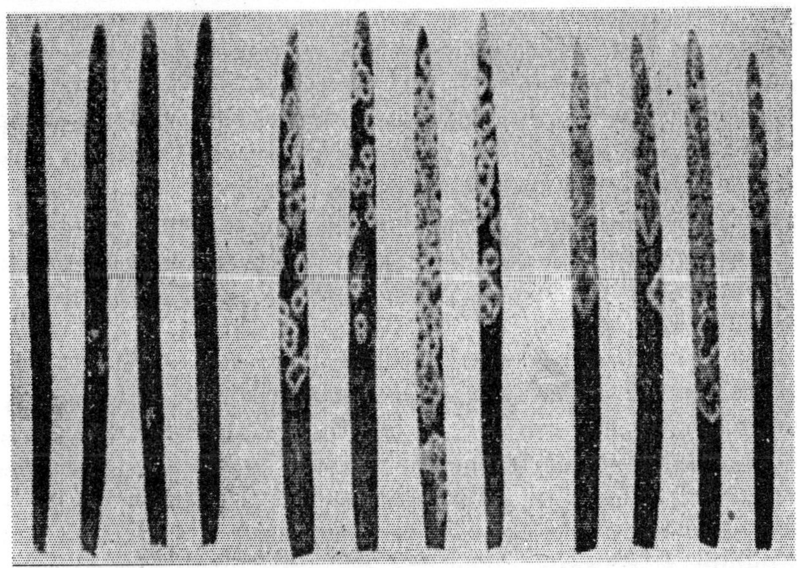

FIGURE 155.—Result of testing three $F_3$ families from the cross Marquis × Kota wheat with black stem rust race No. XXVII. Each family is represented by a leaf from each of four different plants: (left) an immune family; (center) a resistant family; (right) a susceptible family. (*From Hayes and Aamodt.*)

The breeding of superior varieties of wheat which are resistant to rusts, not to mention other important fungous diseases, might well seem a hopeless undertaking, when the most destructive species of rust exists in forty distinct races, each with its own peculiar relations to various wheat varieties. Nevertheless, definite progress has been made toward the creation of new rust-resistant wheats of high producing ability and good quality. The inheritance of resistance to *Puccinia graminis* has been studied by a number of investigators with encouraging results. In a study of Marquis and Kota wheats, for example, Hayes and Aamodt first tested the two varieties by inoculations in the greenhouse with eleven different races of stem rust. For further work they chose race XIX to which Marquis was resistant and Kota moderately susceptible, and race XXVII from which Kota was immune and to which Marquis was resistant. In figure 155 are shown the results of testing three $F_3$

families for resistance to race XXVII. The resistance of Marquis to race XXVII and the immunity of Kota to race XIX were combined in three $F_3$ families out of a total of 372 families studied. This together with certain other evidence indicates that there are several genetic factors which condition the differential reactions of Marquis and Kota to the two races of rust in question. With Mindum and Pentad wheats, on the other hand, Harrington and Aamodt, using stem rust races I and XXXIV, found that a single main factor difference explained fairly satisfactorily the results obtained in $F_3$ families inoculated with race I, while the same genetic difference in the two parental varieties was indicated by the results with race XXXIV. No relation was found between seed color and rust resistance. All combinations of susceptibility and resistance to these two races of rust appeared in the $F_3$ families. In a total of 110 $F_3$ families, six were highly resistant to both races of rust, and two of these families were from white-seeded parents. The results of these two investigations prove that it is possible to combine in a single variety resistance to two physiologic races of stem rust of wheat and support the belief, expressed by Harrington and Aamodt, that eventually a wheat can be produced which will be resistant to all the different races of this destructive parasite.

It is well known that certain durum and emmer wheats are more generally resistant to physiologic races of stem rust than are any known varieties of the *vulgare* group. To test the possibility of combining rust resistance from 14-chromosome wheat with the desirable qualities of a 21-chromosome wheat in a hybrid having 21 chromosomes, Hayes, Stakman, and Aamodt made two such crosses, *viz.*, Iumillo × Marquis and Mindum × Marquis. The first cross produced a variety (Minn. 2202; C. I. 6887) with 21 chromosomes, which is rather rust resistant and under field conditions yields well, and is not apparently greatly inferior to Marquis in quality, though further tests are needed. From the second cross one $F_3$ family of *vulgare* habit gave $F_4$ progeny resistant to two physiologic races, and this resistance was derived from the durum ancestor. By means of a double cross (Iumillo × Marquis) × (Kanred × Marquis), these investigators have produced new types which are homozygous for at least three factors for resistance to stem rust. It appears that the resistance of the Kanred-Marquis hybrid is protoplasmic; but whether the resistance of the Iumillo-Marquis variety is physiologic or morphologic in nature, or both, is not known definitely. The latter is resistant to many physiologic races of stem rust under field conditions, and the former is immune to eleven of the twenty-one physiologic races found in the hard red spring wheat region. The Kanred-Marquis type of immunity is dependent upon a single genetic factor, while at least two factors are necessary to explain the resistance of the Iumillo-Marquis variety, and these are apparently inherited independently of the factor for immunity from the other parent. These results furnish additional basis for anticipating the ultimate production of a 21-chromosome wheat combining the desirable characters of bread wheats with certain desirable durum characters such as rust resistance.

**Inheritance of Disease Resistance in Other Plants.**—The conclusions reached in respect to rust resistance hold good in a general way for other parasitic plant diseases. In addition to the typical cases already described, brief reference may be made to other notable examples of the successful creation of disease-resistant varieties by hybridization and subsequent selection. The next case, however, will be considered somewhat in detail because it serves as an excellent model in method of procedure. The ravages of a group of wilt diseases caused by closely related fungi of the genus Fusarium have been checked through the successful efforts of the U. S. Department of Agriculture. As reported by Orton these are the cotton wilt, *Fusarium vasinfectum* Atk., the cowpea wilt, *F. tracheiphilum* Erw. Sm., and the watermelon wilt, *F. niveum* Erw. Sm. It is clear that these fungi possess a high degree of adaptation to the parasitic mode of existence. Also that, while the cause of resistance in certain varieties of the host species is not fully established, yet the resistance itself is a physiological quality. No constant morphological differences have been detected between immune and susceptible plants; neither are there observable differences in time of germination, rate of development, or period of maturity. Furthermore, the resistance is specific; varieties that resist the wilt may be susceptible to bacterial blight and *vice versa.*

That wilt resistance is a heritable character was strikingly demonstrated by Orton's creation of a wilt-resistant edible watermelon, *Citrullus vulgaris.* All watermelons appear to be very susceptible to the disease. Extended tests in 1900 and 1901 failed to show any basis for selection among the 120 or more varieties tested. Recourse was had, however, to an inedible form of *Citrullus vulgaris* known as citron or stock melon which was immune to wilt. From a cross between this citron and the Eden variety of watermelon $F_1$ hybrids of "wonderful vigor and productiveness" were raised. The fruits were intermediate in character, having oval form and stripes of the watermelon and the hard flesh of the citron. The $F_2$ population was extremely variable in every respect, the various citron characters appearing to be dominant in the majority of plants. From among 3,000 or 4,000 plants ten fruits were selected on the basis of resistance and quality, and the seeds were planted the following year, 1904, in isolated, infected plots. Of these ten plots, two were found to bear melons of uniform appearance and quality one of which resembled the Eden parent. These were backcrosses from the $F_1$ pollinated by Eden. Again all the best melons were selected and planted separately the following year and further variations were found. After 5 more years of selection, a variety, the Conqueror, was obtained which has great uniformity and disease resistance, while the fruit has a thin, tough rind which enables it to endure long railway shipments. The flesh is so juicy that the melons are heavier than Eden melons of the same size; the quality and flavor are good although not equal to the finest (figure 156). These qualities have been preserved and resistance maintained at various localities in the South and Middle West, but on the Pacific coast the resistance was not maintained.

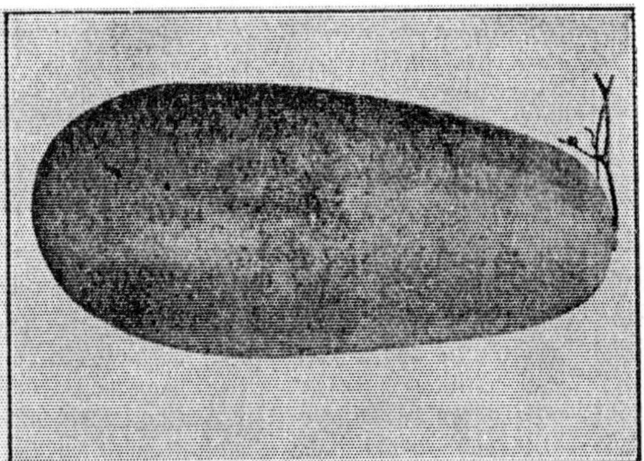

FIGURE 156.—Parents and third generation of the watermelon-citron hybrid. (*From Orton.*)

This failure of the supposedly resistant variety when grown in a far-distant locality is not strange when it is remembered that the wilt fungi are highly specialized in their adaptation to hosts. According to Orton, *Fusarium niveum* attacks no other living plant than the watermelon and certain species of Fusarium "in this respect, coupled with their close morphological resemblance and their common geographical distribution . . . seem to be analogous to the biological strains of Puccinia and Erisyphe." In combating all such diseases, locally adapted varieties are of value, but the importance of developing synthetic resistant varieties must not be overlooked.

The specific nature and heritability of disease resistance is also evidenced by the results of numerous other experiments among which may be cited the following. In the tomato, wilt resistance was found by Norton to be recessive to susceptibility and varieties of Fusarium-resistant tomatoes from Tennessee were found to be susceptible in Maryland. Stuckey found that cherry, pear, and currant tomatoes were immune to the blossom-end rot, a so-called physiological disease and, when crossed with commercial varieties, they transmitted resistance as a dominant character. Resistance to leaf blight in the cantaloupe, *Alternaria Brassicae*, was found by Blinn to be inherited as a dominant character. Jesse B. Norton when breeding for resistant varieties in combating the asparagus rust, *Puccinia asparagi* DC., found resistance dominant in all the $F_1$ offspring in his crosses between the female plants of the rust-susceptible American varieties and a rust-resistant European asparagus. The resistance was somewhat variable but was fixed by selection in succeeding generations. These few cases, taken almost at random, together with the typical illustrations already discussed, amply justify the recommendation that the breeding of disease-resistant varieties of economic plants by hybridization and subsequent selection should receive more attention from plant pathologists, horticulturists, and plant breeders in the future.

**Breeding Disease-resistant Plants by Selection.**—Selection alone is a powerful means of improving plants with respect to disease resistance when used either in variety tests or in the improvement of a single variety. The testing of varieties for disease resistance is an exceedingly important service which can be done most satisfactorily by experiment stations and commercial seedsmen in connection with their routine work. However, the geographical variability in many parasitic organisms and the importance of local adaptation of varieties in many economic plants make it imperative that each important agricultural region should have its own station for variety testing.

The diversity between varieties in respect to disease "resistance" is sometimes due to morphological or anatomical peculiarities which prevent the invasions of parasites. For example, pubescence or waxy excretions on the surface sometimes prevent disease; the number of stomata or the arrangement of cells beneath them may condition fungus infection. Also some varieties escape certain diseases by virtue of their seasonal adaptation or because of precocity. Thus certain grains are less troubled with smut than others because they germinate more quickly. A difference of 2 days in time required for germination may be the deciding factor in smut infection. Certain varieties of potatoes are able to form a corky layer in about 6 hours after being cut, while others require

3 or 4 days. Bacteria require from 12 to 24 hours to commence putrefaction through enzyme action. In addition to these and many other varietal differences, there is always the possibility of immunity which is due to some specific physiological character of the variety. A probable instance of considerable importance is the immunity of milo to smut fungus, which is infectious to all other sorghums.

Although there are numerous valuable reports on the disease relations of certain varieties of our important crop plants, much remains to be done in the way of both extensive and intensive testing. The following citations merely illustrate the kind of information that is now available. Observations at the Kansas Station on 119 varieties of winter wheat showed infection with orange rust, *Puccinia triticina,* varying from 5 to 90 per cent. Weimer and Harter report notable differences in resistance of sweet potatoes to nematodes. According to Gough, there are varieties of the potato which are resistant to potato wart disease, and others partially resistant to late blight and probably also to scab, a matter which has received considerable attention abroad but very little in this country, although there are undoubtedly great possibilities in this work. Orton's success in producing wilt-resistant varieties of cowpea was made possible by the discovery of one immune variety, the Iron, which was apparently of chance origin. In future breeding work, much time and effort might be saved if agromomists and horticulturists would insist on accuracy with respect to the specific diseases observed in all records of disease resistance or immunity.

The selection of disease-resistant strains is a simple but effective method of improving commercial varieties of many crops. In the majority of crop plants there is more or less crossing, and consequently more or less likelihood of selecting desirable combinations of disease-resistant factors. As a result of efforts to find a wilt-resistant sea-island cotton, several strains were obtained by planters and by the U. S. Department of Agriculture, all of which are resistant enough to grow on the worst infected land. Upland cottons are even more susceptible to wilt, but varieties are now grown which are very resistant to wilt and of excellent productiveness. All of these improved varieties of cotton have been secured by continuous selection, beginning with resistant individuals.

As a result of his experience in improving varieties of flax for wilt resistance, Bolley emphasized the importance of continually subjecting the select strains to conditions favorable to the disease but otherwise optimum for the plant. Stakman and others have corroborated this practice by experiments in selection of individual plants and recommend distribution of seed to flax growers from the highest-yielding selections available. The grower should plant this resistant seed on "flax-sick" soil and continue to select seed therefrom for the main crop. It will be noted that this method succeeds with flax, although it failed to develop rust-resistant wheats. It will be recalled that the most serious rust of wheats exists in numerous physiologic races which react differently with different host varieties and in some cases even with the same variety in different localities or at different times. Although, in most of the cases thus far investigated, rust resistance is conditioned by one or two main factors, these factors provide resistance against only certain races of rust. Under these conditions the production of fully rust-resistant wheats by individual plant selections would be highly impracticable. In the case of flax wilt, the parasitic fungus, *Fusarium lini* Bolley, may also exist in physiologic races; but if so, they may be comparatively few in number, and any one strain may occupy an extensive geographic area, as in the case of watermelon wilt. Furthermore, from Tisdale's study of the inheritance of wilt resistance in flax, such resistance is apparently conditioned by multiple factors, a fact which is sufficient to explain the value of long-continued selection for wilt resistance in flax.

The menace of the destructive cabbage yellows, caused by *Fusarium conglutinosum*, which rendered impossible the successful culture of the cabbage in large areas of the United States, has been successfully met by L. R. Jones and associates. Relatively resistant strains, which have proved resistant in other states as well as Wisconsin, so far as tested, have been obtained by selection from several of the most valuable commercial varieties. By following proper methods, according to Jones, any skillful cabbage grower who has fusarium-sick soil can develop a resistant strain of his own, or, having secured one of these strains, he can maintain its resistance and produce his own seed. It has been found that, through careful and repeated selection, resistance to yellows may be found combined with any of the other important qualities of standard varieties.

All our experience indicates that Tisdale's conclusions relative to the flax wilt hold true for cabbage, that resistance is probably determined by multiple factors. This explanation is consistent with our experience that after proceeding to a certain stage with our present methods of selection little or no further progress as to disease resistance is made. This is also consistent with our general experience that the best results have in each case been secured through growing a selected head in isolation and thus securing seed through self-pollination, but that when the benefits were once secured in this way with our best selections, mass culture has been followed to advantage.

A more baffling disease of still greater economic importance is the blight or curly top of the sugar beet, which is caused by a filterable virus introduced into the beet plant by the beet leaf hopper, *Eutettix tenella*. Sugar beet production became impracticable over large areas of the western United States because of this disease and satisfactory preventive measures have not been devised. The breeding of disease-resistant varieties should be the most effective way of combatting this disease in the long run. The U. S. Department of Agriculture and the Spreckels Sugar Company, working independently, have succeeded in isolating strains which appeared to be fairly resistant to blight in certain localities and one of which at least is of sufficiently high sugar content to make it of direct usefulness (figure 157).

The importance of chance in the discovery of disease-resistant crop plants may be illustrated by a few of the many practical results already achieved. L. R. Jones points out that two wheat varieties which are outstanding for rust resistance are Kanred and Webster. The former was a selection from an imported Crimean wheat, and was grown in variety

FIGURE 157.—Yield of roots of a blight resistant strain of sugar beets (right) compared with yield on equal area of common stock seed. From 100 plants of the former in 125 feet of row 220.9 pounds were harvested while 116 plants of the latter in 125 feet of row produced 48.75 pounds. *(From Carsner.)*

trials for a decade by the Kansas Experiment Station before its rust-resistant character was defined. The Webster descended from another strain of Russian wheat and was so unpromising a type that it was not even named until a seed grower chanced to observe its rust-resisting character. Stakman and associates found that Webster wheat is resistant to more physiologic forms of black stem rust than any other wheat thus far tested in the United States and that it may be correspondingly "potentially valuable as a parent of rust-resistant hybrids." With the bean, according to L. R. Jones, our practical progress to date in developing disease-resistant strains rests upon the chance discovery of two basic types, *viz.*, the anthracnose-resistant strain, the Wells Red Kidney, found by a New York bean grower, and the mosaic-resistant Robust bean plant, which was first selected by Spragg for general vigor and yield rather than for recognized disease resistance.

These observations suggest the desirability of cooperation between farmers, seed growers, and the scientists who are concerned directly or indirectly with the breeding of disease-resistant plants. L. R. Jones has emphasized this point as well as the great importance of the fullest possible cooperation between the several groups of scientists concerned with improvement of crop production. The practical value to the farmer of such simple selection as the roguing of all diseased or weakly plants, an ancient practice, must not be overlooked. In the case of severe epidemics, the saving of seed from plants that seem likely to be actually resistant or immune is the first step toward the production of more valuable varieties; but the cooperation of an experiment station may be needed to determine the real value of such selections. Similarly, with more complicated problems of breeding for disease resistance, the production of superior synthetic varieties calls for synthetic biological attack; and no phase of modern agriculture is calling more insistently for the efforts of well-trained students or offers greater possibilities of reward for such efforts. The breeding of disease-resistant plants has become indispensable to agriculture and is, accordingly, a common field of endeavor for scientist and layman, for student and amateur.

### References

BROOKS: Relation of plant pathology to genetics, *Rept. Imperial Bot. Conf.*, **1925**: 104–110.

CARSNER: Resistance of sugar beets to curly top, *U. S. Dept. Agr. Cir.* **388**-C.

HARRINGTON and AAMODT: Mode of inheritance of resistance to *Puccinia graminis* in crosses between varieties of durum wheat, *Jour. Agric. Research*, **24**: 979–996.

HAYES and AAMODT: A study in rust resistance in a cross between Marquis and Kota wheats, *Jour. Agric. Research*, **24**: 997–1012.

HAYES, STAKMAN, and AAMODT: Inheritance in wheat of resistance to black stem rust, *Phytopath.*, **15**: 371–387.

JONES, L. R.: Fusarium resistant cabbage, *Wis. Sta. Res. Bull.* **48**.

——— Securing disease resistant plants, *Science*, **63**: 341–345.

ORTON: Development of farm crops resistant to disease, *Yearbook U. S. Dept. Agr.*, **1908**: 453–464.

RASMUSON: Über Vererbung bei Vitis, *Mitteil. K. Biol. Anstalt für Land-und Forstevirtschaft*, **15**: 29–34.

REED: Varietal resistance and susceptibility of oats to powdery mildew, crown rust, and smuts, *Mo. Sta. Res. Bull.* **37**.

——— Varietal resistance to smuts, *U. S. Dept. Agr. Bulls.* **1275** and **1284**.

REIMER: Blight resistance in pears, *Ore. Sta. Bull.* **214**.

TISDALE: Flaxwilt: a study of the nature and inheritance of wilt resistance, *Jour. Agr. Research*, **11**: 573–605.

## CHAPTER XXXVII

# PLANT-BREEDING METHODS

In no phase of agriculture is there greater need of careful planning with reference to economy of time and resources as well as efficiency of method than in plant breeding. In annual species the individual plant is of small intrinsic value as compared with a domestic animal, and generations follow each other in rapid succession. These facts tend to encourage methods which are wasteful or inefficient, or at any rate methods which fail to accomplish all that might be accomplished in a given time. In work with perennial species the need of a well-planned system of breeding is even more urgent because of the greater intrinsic value of the individual plant and the longer time required to obtain results.

In planning a plant-breeding program the following general propositions must always be considered: (1) The plants chosen for improvement must be adapted to the locality where the work is to be done or the necessary special facilities for their culture must be available. (2) The plant breeder must know how to grow plants; failing in this, all his efforts are likely to be in vain. (3) Culture and breeding methods must be adapted to particular species and varieties and to local conditions of climate and soil.

**Pedigree Culture Methods.**—The pedigree culture was first used in a systematic manner by Vilmorin in breeding wheat. Later it was adopted by Hays of Minnesota and by Nilsson in Sweden. The essential feature of the pedigree culture consists in rearing successive generations of organisms under such conditions that the ancestry of each individual is known. Its purpose is to insure absolutely accurate knowledge of ancestry. To attain to this ideal many precautions are necessary, and some sources of error, due principally to accident, cannot be entirely eliminated. The same accident, however, is not likely to happen twice in the same culture, and when an accident does occur, sometimes the culture can be repeated. Any material whose pedigree is in doubt should be eliminated at once or, if rare or of special value, relegated to the class of unknown until its own behavior in breeding indicates its genotypic nature.

Seedage methods in pedigree culture work with plants are very important. Handling seeds and seedlings requires quite as much care as does castration and pollination, especially with very small seeds, and this work should be done by the person in charge if possible or by specially trained assistants. The same holds true in the recording of data.

In collecting seed, the original label should be preserved with the seed until the permanent records are made; then it should be compared with the record before being discarded. In many cases it is well to count the seeds in each bag inasmuch as hybrid seed may prove to be partially sterile or may exhibit delayed germination or other abnormalities.

Planting pedigree cultures involves some of the most difficult problems, especially in handling small seeds. Larger seeds such as wheat, corn, peas, and beans are handled satisfactorily by planting a single seed in a paper planting pot containing thoroughly sifted soil. These individual plants should then be set out at equidistant points in the rows, and each row labeled. Also at the time of planting a memorandum should be made of the plan of the plot with the position or number and contents of each row as a safeguard in case of loss of labels. Small seeds like those of tobacco and petunia and even those like the strawberry may be sown in or on soil or sand which has been sterilized so that any seeds of the same species already in the soil will be destroyed. (Seeds of certain species are said to germinate poorly in sterilized soil. In such cases special methods must be devised to insure purity of cultures.) The most satisfactory method of sterilizing soil for seedage is to steam it under pressure. The soil is placed in the clean earthenware pots or seed pans in which the seeds are to be sown, and these go directly into the autoclave, where they should remain under pressure of 15 pounds to the square inch for at least 1 hour. If it is not convenient to prepare the sterilized soil each day as needed, the pots not intended for immediate use should be covered before being sterilized with a piece of fine-meshed fabric which is securely tied below the flange of the pot. This covering is left in place until the pot is used. The manipulator should have clean hands, clothes, and utensils and should handle only one lot of seed at a time, using due precaution between each lot to avoid mixing. Each lot should be labeled as soon as sown. The seed pots or pans are then moistened by setting them in a vessel containing water, care being taken not to let the water overflow into the seed vessels. All subsequent watering should be by means of subirrigation rather than surface watering when dealing with small seeds; but the soil or sand used must be practically free from injurious salts. When ready to prick out, the little seedlings may be transplanted into unsterilized soil, provided they are set at uniform distances or marked in some way so that any foreign seedlings that appear later can be distinguished with certainty.

Protection of cultures also involves problems which are highly important in pedigree plant breeding but difficult to discuss without entering into considerable detail. The danger may appear in the form of curious or ignorant persons who do not understand the importance of keeping hands off. Or a strong wind may succeed in forcing into the greenhouse seeds of the same species as the one undergoing investigation.

Or pests of various sorts may appear—damping off, insect pests, snails, slugs, mice, and in the breeding garden birds, gophers, moles, rabbits, etc. In this connection it may be sufficient to say that eternal vigilance is the price of success. Any one with ordinary ingenuity will usually be able to provide the necessary protection. The important thing is to realize its need in time to prevent loss or contamination of cultures. If the breeding garden is located in or near cities, the sparrows will work havoc with developing seeds, especially of cereals. This menace can be completely overcome only by enclosing the threatened cultures with something that will keep out the birds and at the same time cut off the minimum amount of light. It has been found that 1-inch-mesh poultry netting is satisfactory. Many plants used in genetic investigations can be handled most satisfactorily in the greenhouse.

A systematic method of recording and preserving data is *sine qua non* for the pedigree culture. It is absolutely unsafe to trust to memory if any degree of accuracy is to be attained. For work on a small scale, a serial number (using Arabic numerals) for each culture is satisfactory. This number then becomes the permanent designation of the given culture, each plant in the row receiving a subscript number. Thus plant 5 of culture 3 would be designated as $3P_5$. If this plant is selected for further testing with self-pollinated seed, its progeny in the next generation will be labeled $3F_1P_5P_1$, $3F_1P_5P_2$, and so on down the row of plants. This method, however, is rather cumbersome, and for work upon a large scale the "annual-notebook-page" method first described by Shull is much more satisfactory. In this system each culture of a given year is numbered chronologically receiving the number of the page in the notebook for that year on which it happens to be recorded. The label bears this number preceded by the distinctive numerals of that year. Thus the particular culture recorded on page 1 of the 1926 notebook will be labeled 261 or 26.1. The use of the decimal point is a convenience, especially if one is working at an institution where serial numbers are in use in another department. Writing the current year number in one corner of the page and the corresponding number for the preceding year in the other corner gives a convenient system for securing the complete pedigree of a given culture. To complete this method, some designation is necessary for the individual plants selected in any year. This may be a number in parentheses, a subscript, a letter, or, where plants are set at equal distances from a given base line and each plant is thus numbered automatically, the letter P with subscript is satisfactory. Whatever the individual designation may be, it becomes the name of the particular plant for the remainder of its existence, but its progeny will receive a new number when the seed is sown. An adequate system of recording data is obviously essential. Forms of general usefulness may be procured from experiment stations, and these will suggest special modifications which may be needed for particular lines of work.

**The Svalöf System.**—At the Swedish Institute for the Improvement of Field Crops, Nilsson and his associates have worked out a very complete and efficient system of plant breeding. Gradually, as increased appropriations of funds have permitted expansion, a corps of experts has been employed, each investigator concentrating on one or two species and thereby training himself to distinguish all the different forms so as to judge of the relative value of different combinations of characters. Furthermore, a definite course of procedure has been developed as a result of many years of experience, during which time marked success has been achieved in the improvement of Swedish field crops. To begin with, the work consisted mainly of variety testing and extensive effort at improvement through mass selection. These methods still find a place in the routine work, but they are of insignificant value as compared with the coordination of intensive methods which makes the Institute's system a model which institutions engaged in similar work may profitably follow. From Nilsson's description, the Svalöf system may be briefly outlined under three heads, *viz.*, *genotype selection*, *strain tests*, and *hybridization*. In all this work the methods of pedigree culture are followed so that the original source and performance record of each form grown at the station or distributed for trial can be accurately stated. In connection with the preliminary selection of promising forms the intensive study of the specialists at Svalöf has made each member skillful in detecting different forms in the species with which he is working and in judging the relative value of the characters displayed. Having separated from the population some of the biotypes of which the "variety" is composed, it next becomes necessary to subject all these strains to comparative tests in order that the few superior forms may be discovered and propagated more extensively. This requires long years of careful work and the overcoming of certain difficulties. The handling of thousands of cultures each year has required the devising of many practical arrangements to insure exactness and order. Each distinct strain that has been retained for more extensive trials is regarded as an established variety.

Hybridization of cereals had been started at Svalöf upon rather an extensive scale previous to 1900 but it had not given encouraging results. After the announcement of Mendel's discovery, this phase of the work was greatly enlarged, particularly with wheat and oats, by Nilsson-Ehle. Through the application of Mendelian principles, new varieties of these cereals have been produced which combine general good qualities with cold resistance or earliness, which makes them better adapted for more northern localities. It has been found, as would be expected, that in breeding for a number of different characters at the same time the recombinations in $F_2$ following hybridization of select strains are so complicated as to render very illusory the idea of reaching definite results by theoretical calculations. But in spite of this difficulty, hybridization offers a means of creating new forms which cannot be obtained otherwise.

Progress in plant improvement by means of hybridization will always be limited by the available supply of experts as well as by facilities and time. Any method, therefore, that enables the breeder to secure desirable new combinations of characters without the enormous amount of detail involved in a system of pedigree culture, is worthy of serious consideration. Such a method was devised by Nilsson-Ehle and has been used at Svalöf with success. The essential feature is as follows: After crossing two varieties, all the progeny of the second generation are sown together *en masse*, the object being to allow extreme climatic conditions to destroy or expose the weakness of as many of the delicate combinations as possible. Natural selection plus mass selection may thus eliminate the unfit combinations. With self-fertilized plants the continuation of this method for a few generations will automatically produce a population of near-homozygous individuals, the most promising of which can then be selected for more intensive work. As applied to wheat, in which winter hardiness is the all-important character for Sweden, this method is essentially as follows: (1) Cross a variety of very high quality, like Biffen's Yeoman, with numerous most winter-hardy sorts; (2) grow the $F_1$ plants widely spaced to insure abundant seed production; (3) plant large $F_2$ populations *in flats* and place these flats in the open with bottoms exposed throughout the winter (from this severest possible test very few plants may survive in the spring); (4) transplant the survivors to the field and interplant with wheat plants already inoculated with rust spores; (5) save seed separately from those individuals which prove to be resistant to rust; (6) make yield tests of selected individuals in rows and later in plots. This is really a short-cut method by which potential varieties may be produced in two or three generations.

**Other Time-saving Methods.**—The economical use of time is of the utmost importance in plant breeding. Many special methods have been devised for the purpose of saving time, of which a few may be mentioned by way of illustration.

Greenhouses furnish many possibilities of hastening progress in plant breeding. Simply increasing the number of generations grown each year is of obvious advantage. With some annual species, it is possible to grow several generations in one year, a fact which is of enormous advantage in genetic investigations. Similarly, in such investigations it is sometimes highly important to rear tropical species or delicate individuals and this can be done only under greenhouse conditions. Light requirements are of considerable value in differentiating between certain species and varieties, and this factor can be best controlled in the greenhouse. Richey finds the use of specially constructed greenhouses to be of practical value in maize breeding, when it has reached an advanced stage such that the expense of construction is warranted.

Growing cultures of the same material in two places having radically different climates is sometimes more economical and gives better results than the use of greenhouses. In their extensive breeding operations with cereals, Love and Craig find it satisfactory to grow duplicate cultures of much of their material at Ithaca and at Davis, Calif. The loss of pedigreed stock from winterkilling is thereby avoided, and double tests under diverse conditions are procured. It is also possible to grow two generations of cereals in one year by using two locations having different climates. Sometimes it is necessary, when working with seed which naturally requires a rest period before germination, to hasten germination by subjecting the seed to a temperature of about 0°C. for a short time. These illustrations will indicate the kinds of devices which may be resorted to in order to expedite plant-breeding projects.

**Correlation an Aid in Breeding.**—Extensive use is now made of correlation studies as a guide in selection and hybridization. The literature is voluminous and the degree of usefulness of correlation is variable; but it is of sufficient importance to require emphasis. An excellent illustration of the practical value of correlation studies is furnished by Davin and Searle's botanical-genetic survey of the flax plant as a preliminary step to breeding operations. Four years of study yielded extensive data on the more important characters of the plant considered from the point of view of fiber yield.

The linen industry has certain definite requirements as to length of fiber and proportion of fiber to other stem tissue, so that it is important that the correlation between these and other characters, such as diameter of stem and diameter of fibers, be determined. With definite information on such points and on the role played by environmental factors, it becomes possible to proceed with the breeding of improved varieties with some degree of assurance of ultimate success.

In the diagram (figure 158) correlations are indicated, together with other results of practical importance in breeding, such as indicators of vigor and possible linkages. It was found that variations in the characters represented in the upper row (Nos. 1–5) are predominately hereditary; that differences in Nos. 6–11 are largely the result of environmental effects; while variations in number and size of fibers are correlated both with length, which is definitely inherited, and with diameter, which is highly variable within selected strains.

From these studies it is considered possible to breed new varieties of flax combining qualities of tallness, high percentage of fiber, and high mean number of seeds per capsule, and to include also the quality of early maturity if found desirable. From the practical point of view, the two most important facts thus far determined seem to be the inheritance of variations in percentage of fiber and the relationship between the number and size of the ultimate fibers and length of stem.

Preliminary surveys of the type above described are much to be desired for many important crop plants. In some cases, such as wheat, much detailed study has been done, but the scattered data should be compiled and made available to breeders. In many other cases the necessary surveys remain to be made.

**Variety Tests.**—The principal purposes of variety testing are, first, establishment of varietal types of field crops and vegetable plants,

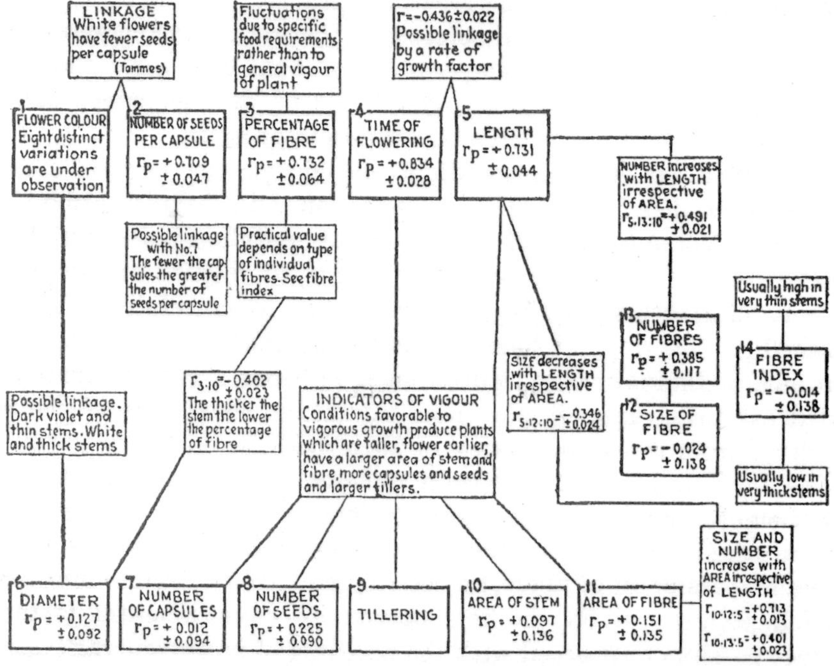


FIGURE 158.—Chart depicting correlations between various characters of the flax plant. $r$ = coefficient of correlation, 1921; $r_p$ = coefficient of parental correlation, 1921–1922; $r_{1.2:3}$ = coefficient of partial correlation, 1921. Note that (10) area of stem and (11) area of fiber show very close linkage as $r = +0.944 \pm 0.003$. (*From Davin and Searle by permission of the Linen Industry Research Association.*)

and second, determination of the best varieties for a given locality. In the establishment of varietal types, many difficulties are encountered of which the following are most important:

*a. Confusion in Nomenclature.*—Horticultural and agricultural plant nomenclature is in a state of great confusion, a condition which makes it necessary at the outset of scientific variety testing to make taxonomic studies of the known types and to adopt a system of nomenclature which shall be followed consistently throughout the investigations.

*b. Inaccurate Descriptions.*—The type descriptions of many cultivated varieties are either too meager or faulty to be serviceable in critical work or are wanting altogether. This often makes it necessary to secure seed bearing a given varietal name from various growers, to make comparative trials, and then to choose the particular lot which may be considered most nearly typical of the variety. Much of this expenditure of time and money could be saved by the general adoption of definite, mechanical methods of describing varieties. The Bureau of Plant Industry of the U. S. Department of Agriculture has advocated for years the wider use of variety description blanks.

*c. Impurity of Commercial Seed.*—This has been one of the most serious difficulties in the way of accurate work in variety testing. Time was when practically every sample of seed of any of the turnip-rooted beets would be found to contain 50 per cent or more of roots resembling more or less closely the form and other characteristics of the variety, but the remaining population would be made up of an admixture of all possible variations of the turnip-rooted beets. The condition in potatoes was also serious, according to Corbett, and before Stuart undertook by progeny-row and hill-selection methods to establish pure strains of commercial varieties any pretense at variety testing consisted merely in comparing one mixed lot with another mixed lot. Conditions are gradually improving, however, and the larger seed firms and many specialists are endeavoring to maintain fairly high standards of varietal purity. Adoption of seed certification, which tends to insure uniformity in seed stocks, is coming into wide favor.

*d. Complexity of Variation.*—The fact that fluctuating variations may be both heritable and non-heritable makes it necessary to use pedigree culture methods in any effort to compare the genetic constitution of varieties. In determining the best varieties for a given location, the chief desiderata are yield, quality or chemical composition, and uniformity. The chief requisites for success are (1) an adequate system of records for which loose-leaf printed forms are by far the most practicable; (2) proper interpretation of the results.

The experimental errors involved in variety testing may be considered under two heads: (1) accidental errors such as incorrect weighing, faulty computation, unobserved variations in field treatment, sampling, etc.; (2) residual errors such as variations caused by soil heterogeneity due to natural conditions or to non-uniform treatment, uneven distribution of soil moisture, etc. The practical questions involved in reducing accidental errors and the experimental determination of the probable error have received considerable attention particularly from English and American workers (see papers by Carleton, Farrell, Hall, Hall and Russell, Lyon, Mercer and Hall, Montgomery, Olmstead, Pritchard, Stockberger, Surface and Barber, Fisher, Student, Sax, Love, Hayes and others). The net result of all this discussion seems to be a certain measure of dissatisfaction with the established methods of applying the probable-error concept in determining the significance of differences in results of variety tests, for example. For this reason it is perhaps desirable to call attention briefly to a simpler method of determining the significance of such results.

**Student's Method of Interpreting Paired Experiments.**—In contrast with the widely used formulae of Bessel and Peter, Student's method, according to Love, takes into consideration the parallelism of paired observations and gives weight to the amount and consistency of the individual gains in estimating the significance of the mean difference. The application of Bessel's and Student's methods in the reduction of data on yield in oats may be compared by examining tables 51 and 52. Whereas the

TABLE 51.—THE YIELDS OF TWO VARIETIES OF OATS FOR A PERIOD OF 9 YEARS. PROBABLE ERRORS OF THE MEANS CALCULATED ACCORDING TO BESSEL'S METHOD

*(From Love)*

| Year | Great Northern, bushels per acre | $D$ | $D^2$ | Big Four, bushels per acre | $D$ | $D^2$ |
|---|---|---|---|---|---|---|
| 1912 | 71.0 | 12.8 | 163.84 | 54.7 | 3.9 | 15.21 |
| 1913 | 73.9 | 15.7 | 246.49 | 60.5 | 9.7 | 94.09 |
| 1914 | 48.9 | −9.3 | 86.49 | 45.1 | −5.7 | 32.49 |
| 1915 | 78.9 | 20.7 | 428.49 | 71.0 | 20.2 | 408.04 |
| 1916 | 43.5 | −14.7 | 216.09 | 40.9 | −9.9 | 98.01 |
| 1917 | 47.9 | −10.3 | 106.09 | 45.4 | −5.4 | 29.16 |
| 1918 | 63.0 | 4.8 | 23.04 | 53.4 | 2.6 | 6.76 |
| 1919 | 48.4 | −9.8 | 96.04 | 41.2 | −9.6 | 92.16 |
| 1920 | 48.1 | −10.1 | 102.01 | 44.8 | −6.0 | 36.00 |

$M = 58.2$ $\quad \Sigma D^2 = 1{,}468.58$ $\quad M = 50.8$ $\quad \Sigma D^2 = 811.92$

$$E_m = \pm 0.6745\sqrt{\frac{1{,}468.58}{9 \times 8}} = \pm 3.05 \qquad E_m = \pm 0.6745\sqrt{\frac{811.92}{9 \times 8}} = \pm 2.26$$

Great Northern, $M = 58.2 \pm 3.05$
Big Four, $M = 50.8 \pm 2.26$
Difference $7.4 \pm 3.80$

$$\frac{D}{E_D} = 1.95 = \text{odds } 4.3 \text{ to } 1$$

odds of 4.3 to 1, as derived by Bessel's method, are far too small to permit the conclusion that the difference in yielding capacities of the two varieties is statistically significant,

> . . . the use of Student's method gives very high odds against the possibility that such a difference between these varieties is due to chance, or in other words, if the same variety has been sown on the two plots each year, the odds are

908 to 1 against such a difference occurring. Unlike Bessel's formula, the results obtained from the use of Student's method are greatly influenced by the pairing of the comparisons, and this is what the investigator is interested in emphasizing.

TABLE 52.—THE YIELDS OF TWO VARIETIES OF OATS AS SHOWN IN TABLE 51. PROBABILITY OF THE DIFFERENCE CALCULATED ACCORDING TO STUDENT'S METHOD (*From Love*)

| Year | Yields per acre | | Gain of Great Northern over Big Four | $D$ | $D^2$ |
|---|---|---|---|---|---|
| | Great Northern, bushels | Big Four, bushels | | | |
| 1912 | 71.0 | 54.7 | 16.3 | −8.9 | 79.21 |
| 1913 | 73.9 | 60.5 | 13.4 | −6.0 | 36.00 |
| 1914 | 48.9 | 45.1 | 3.8 | 3.6 | 12.96 |
| 1915 | 78.9 | 71.0 | 7.9 | −0.5 | 0.25 |
| 1916 | 43.5 | 40.9 | 2.6 | 4.8 | 23.04 |
| 1917 | 47.9 | 45.4 | 2.5 | 4.9 | 24.01 |
| 1918 | 63.0 | 53.4 | 9.6 | −2.2 | 4.84 |
| 1919 | 48.4 | 41.2 | 7.2 | 0.2 | 0.04 |
| 1920 | 48.1 | 44.8 | 3.3 | 4.1 | 16.81 |

$M = 7.4$ $\Sigma D^2 = 197.16$

$Z^* = \frac{7.4}{4.68} = 1.58$ $\sigma = \sqrt{\frac{197.16}{9}}$

$P^* = 0.9989$

Odds = 908 to 1 $= 4.68$

To calculate $\sigma$ directly from the annual differences in yield between the two varieties, square values in column 4 and summate, giving 690. Divide this by number of items (9) = 76.6667. Square mean difference, 7.4, and subtract from 76.6667 = 21.9067. Extracting square root of this gives $\sigma = 4.68$ (*From Love*).

* Values of $Z$ and $P$ are found in Student's original article, in Pearson's "Tables for Statisticians and Biometricians," and in Love's article on a modification of Student's table.

Love points out, however, that

. . . although Student's method may be used in calculating the significance of as few as two pairs of observations, extreme caution should be used in drawing conclusions from any experiment in which the number of replications is very small. In common with other measures of fluctuations in sampling, a strictly numerical interpretation of constants determined from small populations may give a false sense of security as to the reliability of the sample.

**Reduction of Experimental Error.**—The requirements for the reduction of errors are steadily becoming more and more refined. Only a suggestion can be given here as to the nature of the precautions found necessary. Following are some of the conclusions reached by Montgomery regarding the reduction of experimental error in variety testing of cereals.

*a.* Systematic repetition constantly reduces error as the number of repetitions increases, but with 16-foot row plots, ten to twenty repetitions must be made, depending on the degree of accuracy required.

*b.* It is probable that the greater the number of strains to be compared the more repetitions will be necessary, because of the greater area they will cover.

*c.* Small blocks, 5.5 feet square, give results similar to those of the row plots, except that the reduction of experimental error is somewhat greater as a result of repetition. Blocks repeated eight or ten times give results apparently about as accurate as rows repeated fifteen or twenty times.

FIGURE 159.—Row-plot nursery (wheat) in which the rows are 16 feet in length with a 4-foot alley adjacent, thus making the beds 20 feet in width. (*From Montgomery.*)

*d.* The rate of planting, within certain wide limits, has little influence on yield.

*e.* There is some competition between adjacent rows, especially when varieties very different in habit of growth are planted side by side. The use of blocks [and guard rows] does away with this source of error.

*f.* Block plots and row plots at the usual rates of seeding will probably correlate more closely with results in field plots than in plots where the plants are spaced as in centgeners.

*g.* Where error is corrected by the system of repetition plots, check plots would be used for the purpose of determining the experimental error. When the variation in checks equals the variation in strains, no possible selection can be made.

In general the above conclusions apply to most plot experimentation, although certain modifications are required in working with other types of crops, such as potatoes. In regard to (*a*), however, it should be understood that, whatever the system of replication adopted, *serial repetition is to be avoided.* This has been emphasized by Fisher, as follows:

> For our test of significance to be valid the difference in fertility between plots chosen as parallels must be truly representative of the differences between plots with different treatment [or different varieties under the same treatment]; and we cannot assume that this is the case if our plots have been chosen in any way according to a prearranged system; for the systematic arrangement of our plots may have, and tests with results of uniformity trials show that it often does have, features in common with the systematic variation of fertility, and thus the test of significance is wholly vitiated.

**Strain Tests.**—The chief purposes of strain tests are (1) separation of types with a view to standardization of varieties and (2) selection of the

FIGURE 160.—Increase plots of 1⁄30 acre each. Selected strains of wheat from the nursery are tested in these plots for 3 years. (*From Montgomery.*)

most profitable strains within a variety. The mixed condition of varieties of vegetables calls for continual attention on the part of the seedsmen and experiment stations in an effort to bring existing varieties up to some definite standard. One difficulty is found in the widespread use of synonyms, according to Work, who advocates the organization of a recognized board of review which will decide, after trial and comparison, whether a submitted sample is worthy of standing as a new variety or simply as a strain or stock. Myers has conducted extensive strain tests of tomatoes and has reached the conclusion that the best way to insure success in procuring high-yielding strains of vegetables is to secure seed a year in advance of the time it will be needed and submit it to a preliminary test. The difficulties encountered in selecting the most profitable strains within a variety involve the same sources of experimental error as are met in variety testing. The necessity of distinguishing between heritable and non-heritable variations calls for individual plant selection and pedigree culture methods. Plant-to-row tests and subsequent plot tests of the progeny of individual wheat plants are shown in figures 159 and 160. In all such work the use of loose-leaf record blanks is advantageous. Plant-to-row tests will always be indispensable in the purification of varieties and the production of superior strains. The general requirements to insure dependability in the interpretation of results are the same whether the object be to test varieties or selections within a variety.

**The Future of Plant Breeding.**—An open-minded survey of the facts pertaining to world population growth in relation to rate of increase in available food supply leaves no reason for doubt as to the future status of plant breeding. As the problems of human welfare become more acute, the demands upon agriculture will stimulate a corresponding increase in efforts to apply scientific knowledge in the production of the world's supply of food, textiles, rubber, and wood. Plant breeding will play a definite role in the necessary increase of these essential commodities, how important or successful a role will depend primarily upon the advancement which is possible in genetic research.

It is, therefore, incumbent upon all agencies and institutions invested with responsibility for conserving and increasing agricultural production to make adequate provision, along with other applications of physical, chemical, and biological science, for the fullest possible development of genetic research and of scientific breeding. It may safely be assumed that the real achievements of future plant breeders will be made through the application of genetic principles to problems of plant production. The present trend of genetic research is very definitely along two lines, *viz.*, the factorial analysis of individual species, and the study of chromosomal variations within individual species and among the several species of a genus, together with genetic research on interspecific hybrids. The fundamental relation of these two phases of genetics to the breeding of synthetic varieties has been considered, all too briefly, in foregoing chapters. It remains for the future to disclose just how much improvement in crop plants can be accomplished by applications of the most advanced genetic discoveries along the lines already mentioned and in the still unexplored field of artificially induced germinal changes. Definite beginnings, however, have been made in the creation of superstrains of maize and the synthesis of disease-resistant wheats, and indications exist which suggest fruitful lines of research on the artificial induction of chromosomal variations and the causation of factor mutations. While awaiting the front-line advance in genetic research, there is much to be done in organizing plant-breeding work with reference to present and future needs of agriculture.

Comprehensive plant-breeding programs should be based on careful surveys of the present status and probable future development of the principal crops and the possible development of new industries. Such problems as the effect of the rapid development of the artificial-silk (wood pulp) industry upon the future of cotton as a world crop, or the distribution of the world's supply of rubber, as well as all the immediate and pressing problems of crop production and marketing, are involved in the administration of public funds. But even though the utilization of public funds for purely theoretical genetic research done with non-economic plants may be soundly defended, the practical difficulty remains of securing sufficient support for such investigations. Many plant-breeding investigations necessarily extend over long periods during which no practical applications may be forthcoming. There is, therefore, urgent need of great institutions of scientific and applied botany, including genetics and plant breeding, supported by permanent endowment. Such an institution should be equipped for botanical exploration; plant interchange between countries; maintenance of extensive collections of living plants for use in scientific and practical work; and experiments on the amelioration of wild-plant species of possible value in agriculture. Only by such permanently endowed institutions can adequate attention be given to the development of crops adapted to the great uncultivated areas such as the arid regions of western North and South America and northern Africa, and those regions of vast but little known possibilities, the tropics.

Meanwhile, individual workers in genetics and plant breeding may not ignore the burden of responsibility which rests upon them—responsibility for the economical use of time and energy, which implies serious consideration of the problems to be selected and methods followed; and responsibility for generous and whole-hearted cooperation with others in the noble cause of advancing research both pure and applied. In pure research, perhaps, there is no problem of more immediate importance to agriculture than the genetic analysis of yielding ability in crop plants, and no aspect of applied genetics looms with greater import than the synthetic breeding of superior, disease-resistant varieties.

### References

Belling: Production of triploid and tetraploid plants, *Jour. Heredity,* **16**: 463–464.

Collison and Harlan: Statistical study of some field plat yields, *N. Y. Sta. Tech. Bull.* **94.**

East: Mankind at the Crossroads.

Engledow: Economic possibilities of plant breeding, *Rept. Proc. Imperial Bot. Conf.* (London), **1924**: 31–40.

Fisher: Statistical Methods for Research Workers.

Haskell: Technique of plot experiments, *Rept. Proc. Imperial Bot. Conf.* (London), **1924**: 373–381.

HAYES: Control of soil heterogeneity and use of probable error concept in plant breeding studies, *Minn. Sta. Tech. Bull.* **30**.

HAYES and GARBER: Breeding Crop Plants, Chapter IV.

LOVE: A modification of Student's table for use in interpreting experimental results, *Jour. Amer. Soc. Agron.*, **16**: 68–73.

LOVE and BRUNSON: Student's method for interpreting paired experiments, *Jour. Amer. Soc. Agron.*, **16**: 60–68.

LOVE and CRAIG: Methods now in use in cereal breeding and testing at the Cornell agricultural experiment station, *Jour. Amer. Soc. Agron.*, **16**: 109–127.

PEARL: The Biology of Population Growth.

SWEENEY: The Natural Increase of Mankind.